MATHEMATICAL PROGRAMMING FOR AGRICULTURAL, ENVIRONMENTAL, AND RESOURCE ECONOMICS

MATHEMATICAL PROGRAMMING FOR AGRICULTURAL, ENVIRONMENTAL, AND RESOURCE ECONOMICS

HARRY M. KAISER
Cornell University

and

KENT D. MESSER
University of Delaware

WILEY

John Wiley & Sons, Inc.

Vice President & Publisher	George Hoffman
Acquisitions Editor	Lacey Vitetta
Project Editor	Jennifer Manias
Senior Editorial Assistant	Emily McGee
Assistant Marketing Manager	Diane Mars
Media Editor	Greg Chaput
Production Manager	Janis Soo
Assistant Production Editor	Yee Lyn Song
Cover Designer	Seng Ping Ngieng
Cover Photo Credit	(sunflowers) Travel Library Limited/SuperStock; (turbines) Robert Huberman/SuperStock; (boats) Axiom Photographic Limited/SuperStock; (bear) James Urbach/SuperStock

This book was set in 10/12 Times Roman by MPS Limited, a Macmillan Company, and printed and bound by Hamilton Printing Company. The cover was printed by Hamilton Printing Company.
This book is printed on acid free paper. ⊚

Founded in 1807, John Wiley & Sons, Inc. has been a valued source of knowledge and understanding for more than 200 years, helping people around the world meet their needs and fulfill their aspirations. Our company is built on a foundation of principles that include responsibility to the communities we serve and where we live and work. In 2008, we launched a Corporate Citizenship Initiative, a global effort to address the environmental, social, economic, and ethical challenges we face in our business. Among the issues we are addressing are carbon impact, paper specifications and procurement, ethical conduct within our business and among our vendors, and community and charitable support. For more information, please visit our website: www.wiley.com/go/citizenship.

Library of Congress Cataloging-in-Publication Data
Kaiser, Harry Mason.
  Mathematical programming for agricultural, environmental, and resource economics / Harry M. Kaiser, Kent D. Messer.
    p.   cm.
  Includes index.
   ISBN 978-0-470-59936-5   (hardback)
     1. Programming (Mathematics) 2. Nonlinear programming. 3. Evolutionary economics—Mathematical models.   I. Messer, Kent D.   II. Title.
  QA402.5.K25 2011
  658.4'033—dc22

2010039902

Printed in the United States of America

10 9 8 7 6 5 4 3 2 1

Brief Contents

Part 1   **LINEAR PROGRAMMING**   **1**

Chapter 1   INTRODUCTORY CONCEPTS AND THE GRAPHICAL APPROACH TO LINEAR PROGRAMMING   **2**

Chapter 2   THE SIMPLEX METHOD TO SOLVING LINEAR PROGRAMMING PROBLEMS   **55**

Chapter 3   SENSITIVITY ANALYSIS USING THE SIMPLEX METHOD AND DUALITY   **88**

Chapter 4   FARM-LEVEL LINEAR PROGRAMMING MODELS   **135**

Chapter 5   TRANSPORTATION AND ASSIGNMENT MODELS FOR FOOD AND AGRICULTURAL MARKETS   **173**

Chapter 6   NATURAL RESOURCE AND ENVIRONMENTAL ECONOMICS APPLICATIONS OF LINEAR PROGRAMMING   **211**

Part 2   **RELAXING THE ASSUMPTIONS OF LINEAR PROGRAMMING**   **247**

Chapter 7   INTEGER AND BINARY PROGRAMMING   **248**

Chapter 8   OPTIMIZATION OF NONLINEAR FUNCTIONS   **283**

Chapter 9   GLOBAL APPROACHES TO NONLINEAR OPTIMIZATION   **321**

Chapter 10   RISK PROGRAMMING MODELS   **347**

Chapter 11   PRICE ENDOGENOUS MATHEMATICAL PROGRAMMING MODELS   **401**

Chapter 12   GOAL PROGRAMMING   **427**

Chapter 13   DYNAMIC PROGRAMMING   **453**

Index   **485**

Contents

Preface **xiii**

Part 1   **LINEAR PROGRAMMING**   **1**

Chapter 1   INTRODUCTORY CONCEPTS AND THE GRAPHICAL APPROACH TO LINEAR PROGRAMMING   **2**

1.1   Applications of Linear Programming in Agricultural, Environmental, and Resource Economics   **3**
1.2   Components of the General Form of the Model   **5**
1.3   Standard Assumptions of Linear Programming Models   **7**
1.4   Formulating Linear Programming Problems   **7**
1.5   The Graphical Approach for Solving Linear Programming Maximization Problems   **13**
1.6   The Graphical Approach for Solving Linear Programming Minimization Problems   **26**
1.7   Sensitivity Analysis with the Graphical Approach   **33**
    Summary   **43**
    Exercises   **44**

Chapter 2   THE SIMPLEX METHOD TO SOLVING LINEAR PROGRAMMING PROBLEMS   **55**

2.1   The Simplex Method for a Simple Maximization Problem   **56**
2.2   The Simplex Method for Maximization Problems: General Case   **64**
2.3   The Simplex Method and Minimization Problems   **72**
    Summary   **79**
    Exercises   **80**

Chapter 3   SENSITIVITY ANALYSIS USING THE SIMPLEX METHOD AND DUALITY   **88**

3.1   Simplex-Based Sensitivity Analysis for Maximization Problems   **89**
3.2   Simplex-Based Sensitivity Analysis for Minimization Problems   **93**
3.3   Duality   **96**
3.4   Solving Linear Programming Problems Using Solver   **107**
    Summary   **116**
    References   **117**
    Appendix: Summation and Matrix Notation   **117**
    Exercises   **124**

Chapter 4   FARM-LEVEL LINEAR PROGRAMMING MODELS   **135**

4.1   Static Models of a Crop Farm   **136**
4.2   A Multiple-Year Model   **145**
4.3   Crop-Livestock Enterprises   **148**
4.4   Dynamic Models   **151**
4.5   Model Validation   **155**
4.6   Research Application: Crop Farm Model   **158**
4.7   Research Application: Economic Feasibility of an Energy Crop
       for a South Alabama Cotton–Peanut Farm   **164**
       Summary   **165**
       References   **166**
       Exercises   **167**

Chapter 5   TRANSPORTATION AND ASSIGNMENT MODELS FOR FOOD AND
                  AGRICULTURAL MARKETS   **173**

5.1   General Transportation Model   **174**
5.2   Extensions of the Model   **179**
5.3   The Transshipment Model   **182**
5.4   The Assignment Model   **189**
5.5   Research Application: U.S. Dairy Sector Simulator   **191**
       Summary   **196**
       References   **196**
       Exercises   **197**

Chapter 6   NATURAL RESOURCE AND ENVIRONMENTAL ECONOMICS
                  APPLICATIONS OF LINEAR PROGRAMMING   **211**

6.1   Forest Management   **211**
6.2   Land Use Planning   **215**
6.3   Optimal Stocking Problem for a Game Ranch   **222**
6.4   Efficient Irrigation and Cropping Patterns   **228**
6.5   Research Application: Optimizing Grizzly Bear
       Corridor Design   **233**
       Summary   **237**
       References   **237**
       Exercises   **237**

Part 2   **RELAXING THE ASSUMPTIONS OF LINEAR
                  PROGRAMMING**   **247**

Chapter 7   INTEGER AND BINARY PROGRAMMING   **248**

7.1   Background on Integer Programming   **249**
7.2   The Branch-and-Bound Solution Procedure   **250**
7.3   Mixed-Integer Programs   **255**
7.4   Solver's Integer and Binary Programming Options   **256**
7.5   Capital Budgeting Problems—A Case of Water Conservation   **257**
7.6   Distribution System Design   **263**
7.7   Sensitivity Analysis in Integer Programming   **266**
7.8   Research Application: Optimizing Agricultural Land Protection
       in Delaware   **268**
7.9   Research Application: Farmland Conservation with a Simultaneous
       Multiple-Knapsack Model   **271**
       Summary   **274**
       References   **274**
       Exercises   **275**

Chapter 8   OPTIMIZATION OF NONLINEAR FUNCTIONS   283

8.1   Slopes of Functions **283**
8.2   Shortcut Formulas for Derivatives **286**
8.3   Unconstrained Optimization **287**
8.4   Multivariate Functions **289**
8.5   Constrained Optimization with Equality Constraints **290**
8.6   Kuhn–Tucker Conditions and Constrained Optimization with
         Inequality Constraints **293**
8.7   Solving Constrained Optimization Problems with Solver **299**
8.8   Fishery Management Using Nonlinear Programming **305**
8.9   Research Application: Optimal Advertising **309**
8.10   Research Application: Water Pollution Abatement Policies **312**
         Summary **314**
         References **315**
         Exercises **315**

Chapter 9   GLOBAL APPROACHES TO NONLINEAR OPTIMIZATION   321

9.1   Development of Nonlinear Problems **322**
9.2   Second-order Conic Problem Barrier Solver **324**
9.3   Evolutionary Solver **328**
9.4   Interval Global Solver **336**
9.5   A Forestry Example Using Nonlinear Excel Functions **338**
9.6   Research Application: Crop Farming in Northeast Australia **341**
9.7   Research Application: An Analysis of Energy Market
         Deregulation **342**
         Summary **342**
         References **343**
         Exercises **343**

Chapter 10   RISK PROGRAMMING MODELS   347

10.1   Expected Value, Variance, and Covariance **348**
10.2   Agricultural Decision Analysis under Risk and Uncertainty **349**
10.3   Quadratic Risk Programming **353**
10.4   Linearized Version of Quadratic Risk Programming **360**
10.5   Target Minimization of Total Absolute Deviations **367**
10.6   Chance-Constrained Programming **369**
10.7   Discrete Stochastic Sequential Programming **371**
10.8   Issues in Measuring Risk in Risk Programming **376**
10.9   Research Application: Quadratic Risk Programming **377**
10.10 Research Application: Discrete Stochastic Sequential
         Programming **379**
10.11 Research Application: Agriculture and Climate Change **385**
         Summary **391**
         References **393**
         Exercises **394**

Chapter 11   PRICE ENDOGENOUS MATHEMATICAL PROGRAMMING MODELS   401

11.1   The Market under Perfect Competition **402**
11.2   The Market under Monopoly/Monopsony and Imperfect
         Competition **404**
11.3   Spatial Equilibrium Models **407**
11.4   Industry Models **409**
11.5   Research Application: A Spatial Equilibrium Model for Imperfectly
         Competitive Milk Markets **410**
11.6   Research Application: Climate Change and U.S. Agriculture **419**

Summary **421**
References **422**
Exercises **423**

Chapter 12   GOAL PROGRAMMING   **427**

12.1   Goal Programming   **428**
12.2   Nonpreemptive Goal Problem   **428**
12.3   Preemptive Goal Programming   **431**
12.4   Deriving Weights for Goal Programming   **438**
12.5   Research Application: Optimal Parasite Control Programs   **438**
12.6   Research Application: Forest Land Protection   **443**
Summary   **446**
References   **447**
Exercises   **448**

Chapter 13   DYNAMIC PROGRAMMING   **453**

13.1   A Network Problem   **454**
13.2   Characteristics of Dynamic Programming Problems   **458**
13.3   A Production Inventory Problem   **458**
13.4   A Capital Budgeting Problem   **463**
13.5   Comments on Dynamic Programming   **465**
13.6   Research Application: Animal Health in Developing Countries   **466**
13.7   Research Application: Conversion to Organic Arable Farming   **471**
Summary   **478**
References   **479**
Exercises   **479**

Index   **485**

Preface

This book provides a comprehensive overview of mathematical programming models and their applications to important problems confronting agricultural, environmental, and resource economists. Mathematical programming, which includes linear and nonlinear programming models, is one of the most powerful and widely used problem-solving approaches in quantitative methods. It is used by researchers in businesses, governments, nongovernmental organizations, and academics to address problems involving the efficient allocation of scarce resources.

Unlike most mathematical programming books, the principal focus of this book is on applications of these techniques and models to the fields of agricultural, environmental, and resource economics. While applied to these important sectors of the economy, the models described here are also useful to other areas of applied economics. The three fundamental goals of the book are to provide the reader with (1) a level of background sufficient to apply mathematical programming techniques to real-world policy and business to conduct solid research and analysis; (2) a variety of applications of mathematical programming to important problems in the areas of agricultural, environmental, and resource economics; and (3) a firm foundation for preparation to more advanced, Ph.D.-level books on linear and nonlinear programming.

This book is designed to be an introductory book in applied mathematical programming. The reader is not required to have any formal background or training in this area. All techniques covered in this book are based on this assumption. Unlike more theoretical mathematical programming books, this book is written at a more basic mathematical level, which consists primarily of algebraic and geometric concepts, but a few of the later chapters include some basic calculus. The book is geared towards upper-level undergraduate and M.S-level graduate students majoring in economics, agricultural economics, environmental and resource economics, applied economics, business, and operations research. The book will also be useful to undergraduate and graduate students majoring in agricultural and food disciplines, such as food science, animal science, agronomy, and veterinarian medicine, as well as students majoring in environmental and resource studies.

Despite its introductory nature, the book places significant emphasis on real-world applications of mathematical programming to decision problems. A wide array of examples and case studies are used to convey the various programming techniques available to decision analysts. Readers will learn (1) how to set up programming models of real-world problems; (2) how to solve them graphically, algebraically, computationally, and with computer software; (3) how to interpret the results; (4) how to validate the model; (5) how to conduct sensitivity analysis; and (6) how to judge and verify the model's performance relative to the real-world decision process it depicts. Upon completing this book, students should be able to use mathematical programming in independent applied research, including applications in academic, business, nonprofit, and governmental research.

While the major focus is on applications, this book also integrates neoclassical economic theory with applied examples. The problems will almost entirely consist of areas within microeconomic theory, primarily theory of the firm, as well as applications of consumer theory, welfare economics, and environmental and resource economics. Hence, the book is a nice supplement to many courses in applied economics.

Because the overall goal of this book is to demonstrate how to use mathematical programming in real-world problem solving, the book provides many case studies from published research. Each chapter includes up to three case studies involving the use of mathematical programming in agricultural, environmental, and resource economics. The reader will be exposed to a thorough range of interesting applications, and teachers in agricultural and applied economics will find the inclusions of these case studies quite helpful in illustrating the power behind this quantitative method.

MATHEMATICAL PROGRAMMING

Mathematical programming is a branch of quantitative methods concerned with finding optimal ways to achieve a certain objective when faced with constraints on the ways the objective is achieved. For example, a farm enterprise is interested in choosing a mix of crops to grow and/or livestock to raise that will maximize profits while satisfying resource restrictions it faces on land, labor, machinery, animal numbers, and capital. An example from environmental economics is a manufacturing firm desiring to maximize profits while meeting constraints on carbon dioxide emissions. Mathematical programming models feature several common elements, including (1) an objective to be maximized or minimized; (2) activities or decision variables, which are the ways to carry out the objective; (3) objective function coefficients, which translate an overall numeric value to the objective through interaction with the values of the activities; and (4) a set of constraints that model the restrictions that the decision maker must operate within. Mathematical programming problems are modeled as a set of equations, linear or nonlinear, that define the decision-making environment.

Mathematical programming has its roots in the 1940s, when solution procedures for linear programming (LP) were developed. Linear programming was used extensively by the U.S. military during World War II, primarily to minimize various costs associated with the war effort. Techniques for solving LP problems were invented during this period by Leonid Kantorovich (LP problem), George Danzig (simplex method), and John Von Neumann (duality). After the War, mathematical programming techniques and applications were rapidly adopted in the private sector, academia, and government as a quantitative technique to handle a huge variety of problems. Today, it is one of the most widely used quantitative approaches in decision analysis.

THE USE OF THE RISK SOLVER PLATFORM FOR EDUCATION

All of the examples in this textbook have been developed as Microsoft Excel spreadsheets and can be solved using the Risk Solver Platform for Education (generally referred to in this textbook as Solver). In cooperation with Frontline Systems Inc., the developers of Solver, this program has been made available for free to students who purchase this book.

We decided to use Solver in this textbook because we have found that our students appreciated having a program that is user-friendly and works within the context of Excel spreadsheets. Therefore, students can take advantage of a variety of Excel functions in the development of their models and spreadsheet features, such as a variety of graphing options in the presentation of their results. Furthermore, the skills that students learn in Excel through the development of models for Solver can be transferred to other data analysis activities within Excel. An additional advantage of Solver is that it can be incorporated into customized programs within Excel through the use of Visual Basic for Applications (VBA). Students interested in this topic may want to consult *VBA for Modelers: Developing Decision Support Systems with Microsoft Office Excel* by S. Christian Albright.

We have observed that students find using Solver relatively easy as it builds upon skills they have already developed with Excel, and they do not have to learn a program-specific programming language. Students also tend to appreciate Solver's interactive visual menus. We recommend that instructors allow for some time at the beginning of the course for their students to get the Risk Solver for Education program installed on their personal computers—as well as having the program installed on classroom and laboratory computers, if applicable—before assigning exercises that require the program. This can be especially important if some students decide not to purchase the book until after attending a couple of classes.

Throughout the book, we have provided tips on how to use the tools of Excel to enhance model development and how to develop models that are easily interpreted by others. Instructors reviewing problems will appreciate the well-designed models that make the identification of problems straightforward. Readers interested in further discussions of related topics may want to consult books dedicated to this topic, such as *The Art of Modeling with Spreadsheets: Management Science, Spreadsheet Engineering, and Modeling Craft* by Stephen G. Powell and Kenneth R. Baker.

We note that some instructors may have more experience, and therefore comfort, with other mathematical programs, such as LINDO/LINGO, GAMS, AMPL, AIMMS, and MPL. Each one of these programs has its own strengths and weaknesses and we appreciate the challenges that come with learning new software. Certainly, users not previously accustomed to Solver will experience some initial challenges as they learn to navigate around its interface while setting the objective function, decision variables, constraints, Solver engine, and related parameters. To help instructors with this transition, we have developed supplemental materials that include the initial problem, a solved version of the problem, and related sensitivity analysis for every problem outlined in the book. These supplemental materials are provided as instructional aids to instructors and students and are available at www.wiley.com/college/kaiser.

One of the traditional advantages of stand-alone mathematical programs has been the ability to solve large-scale problems, especially when constraints are indexed over many different dimensions. The magnitude of this problem has decreased in recent years, and as computer power continues to grow, we encourage students and instructors to review the current editions of Frontline's Solver products (www.solver.com). Some of these Solvers can handle larger and more complex problems, not only in terms of the number of variables and constraints, but also in the incorporation of other important techniques from operations research and management science. While these advanced products come with an additional cost, Frontline traditionally has offered educational discounts.

ORGANIZATION OF THE BOOK

Since mathematical programming consists of linear and nonlinear programming, this book is divided into two major sections. Part 1 consists of six chapters involving LP and its applications. Part 2 features seven chapters involving nonlinear programming (NLP) models or linear models that relax the standard assumptions of LP.

In Part 1, the first three chapters provide a thorough overview of LP concepts, including the basic elements of the LP model, standard assumptions, tips on formulating an LP problem, sensitivity analysis, duality, and solving LP models with graphs, algebra, and Solver. While entire books have been written on these topics, Chapters 1, 2, and 3 provide enough detail and sufficient background on LP concepts to give the reader an ample foundation for applying this method to real-world problem solving. Instructors wishing to de-emphasize the theoretical concepts of LP may want to select sections from these chapters to cover in order to emphasize the remaining application chapters.

Chapters 4, 5, and 6 are concerned with applications of LP in agricultural, environmental, and resource economics. Chapter 4 examines the use of LP for farm-level decision making, and includes analyses of static and dynamic models for grain and livestock farmers, order preserving sequencing constraints, and multiperiod models. This chapter also features two research applications of farm models.

Chapter 5 examines the use of network and transportation LP models in the agricultural, food, and resources sector. The chapter also illustrates how to model product transformation problems. These models are extremely useful in developing efficient networks to minimize flows of commodities from a research application of a large transshipment model with product transformation is included in this chapter.

Chapter 6 is devoted to environmental and natural resource economic LP models. Popular models applied to problems in environmental and resource economics are presented in this chapter including application to forestry, land use planning, water conservation, and game management. In addition, a research case study looks at designing migratory corridors for grizzly bears.

Part 2 features seven chapters that cover applications of nonlinear and more advanced LP models. Chapter 7 covers integer and binary programming models. Integer programming (IP) is basically the same as LP, with the exception that some or all variables are restricted to be integers. In this chapter, the basic concepts underlying IP are presented. Specifically, the most efficient IP and general solution procedure to date, known as the branch-and-bound method, is examined. This is followed by several important applications of binary programming to the conservation of agricultural and ecologically valuable lands.

Chapter 8 provides an introduction into NLP problems that can be solved using calculus. This chapter looks at unconstrained and constrained optimization and shows how some nonlinear problems can be solved using Solver. This chapter is intended to give a conceptual foundation for NLP models. The chapter concludes by providing an application to fishery management and summarizing two research examples of NLP, one from agricultural economics and the other from environmental economics.

Chapter 9 continues this examination of nonlinear optimization and discusses a variety of techniques available in Solver that can be used for these problems. Methods include the SOCP Barrier Solver, Evolutionary Solver, and Interval Global Solver. Sensitivity analysis of nonlinear optimization is discussed in the context of a forest example and two research applications are presented. The first example is related to agricultural economics, and the second comes from environmental economics.

Chapter 10, which deals with risk programming models, relaxes the assumption of parameter certainty. Considerable evidence exists that suggests that farmers adjust their farm plans according to their risk posture, and that profit-maximizing models, which ignore risk preferences by farmers, have failed to give accurate normative or positive economic results when applied to many farming situations. Thus, in order to properly study most farm-level decision-making problems, one must formulate the decision environment in such a way that risk and uncertainty is a critical component in the model. This chapter presents several risk programming models that have been extensively used in food and agricultural applications, including quadratic risk programming (i.e., mean-variance analysis), minimization of absolute deviations (MOTAD), target-MOTAD, chance-constrained programming, and discrete sequential programming. The chapter concludes with three research applications of risk programming in agricultural and environmental economics. This chapter is one of the more advanced in terms of mathematical complexities, and it is geared more toward graduate students than undergraduate students. Therefore, instructors of undergraduate courses may want to selectively use sections in this chapter.

Chapter 11 focuses on price endogenous programming models, which relax the assumption that price is a constant parameter. When one moves from the individual-firm level to the market level, the assumption of constant price is no longer valid. At the market level, price is determined by the interaction of market supply (the collection of all individual firms' supply curves in the market) and market demand (the collection of all individual consumers' demand curves in the market). Consequently, if one is interested in modeling a market or sector rather than an individual firm, then a "price endogenous" or "sector programming" model is necessary. Price endogenous models are also necessary at the firm level if the firm has some degree of market power because in such cases, the firm can influence price by altering its output. Several popular models are presented to illustrate price endogenous programming, along with two research applications in agricultural and environmental economics.

Chapter 12 examines goal programming (GP) models, which is a technique that relaxes the sole objective assumption. Under this approach, one can specify multiple goals or targets for the decision maker and minimize the deviations from not achieving each goal. Goal programming has been used extensively in environmental, natural resource, and agricultural economics as a planning tool. There have been numerous applications in forestry management, land use planning, pollution mitigation, and farm planning. Numerous examples of GP are presented, along with two research applications: one relating to parasite control and one to forest conservation.

Chapter 13 examines the technique of dynamic programming (DP). Dynamic programming is a method used to solve large and complicated problems by splitting them up into smaller subproblems that are both easier to solve and yield the same optimal solution as the original large problem. Three examples of the DP solution procedure are presented. In addition, two research applications from agricultural economics are summarized.

The book is intended for both upper-level undergraduate as well as introductory graduate courses in mathematical programming. Not all chapters or parts of chapters are intended for undergraduate students, and each instructor should use discretion in choosing which material to cover. The book is fairly comprehensive in addressing all the important mathematical programming topics typically covered in introductory courses. Indeed, there is probably more material in this book than can be covered in a single semester course. We intended this to be the case as it offers greater flexibility to the instructor to cover the topics the teacher prefers.

EXERCISES

We believe that students learn how to use mathematical programming best when they have ample opportunity to practice the techniques presented in the chapters as part of assigned problem sets. Therefore, we have developed approximately 25 to 30 exercises, for each chapter. Additionally, we have developed answers to the exercises, which are available to instructors at www.wiley.com/college/kaiser. Instructors will want to review the exercises and the answers before assigning them to their students as the difficulty of the questions varies. Generally, we have organized the exercises from easiest to hardest. However, we recognize that an exercise that might be relatively easy for us may be difficult for some students, and vice-versa, as some students are undoubtedly better at using spreadsheets and Solver than us.

ACKNOWLEDGMENTS

This book was a major undertaking, and there are numerous people to thank for their assistance. From Cornell University, we wish to thank Mike Adler, Thomas Carman, Amanda Chan, Daniel Ochs, Kate Pennington, E. Bea Smith, and Lynette Tsai for their extensive and excellent beta testing and edits of the book. We also wish to thank Thomas Drennen, Dale Rothman, and the approximately 300 undergraduate students at Cornell University who used and commented on earlier drafts of some of the chapters in the book as a basis for an upper-level undergraduate and graduate-level course in mathematical programming, as well as Anita Vogel, who provided valuable assistance in preparing the book. Thanks also to Jeffrey Apland of the University of Minnesota, who taught many useful concepts of mathematical programming. From the University of Delaware, we wish to thank Juan Castellanos, Desiree Davidson, Jacob Fooks, Yuan Li, Xing Tang, Shen Zhen, and approximately 75 graduate students their assistance with proof-reading earlier versions of the textbook, and with the preparation of figures, supplemental materials, and exercises for the book. Double thanks needs to be given to Thomas Carman at Cornell University for thoroughly proofing the galley proofs for this book and to Jacob Fooks at the University of Delaware for his significant contributions throughout the book including the supplemental materials. We appreciate the assistance of Daniel Fylstra and Marc Kellison of Frontline Systems, who helped ensure that the examples in the textbook were compatible with the most recent versions of Solver and Excel available. We also appreciate a number of reviewers who have provided valued comments that have improved the book, including Ken Baerenklau (Univesity of California, Riverside), Carl Dillon (University of Kentucky), Patricia Duffy (Auburn University), Mohamed Said Gheblawi (United Arab Emirates University), Donald Liu (University of Minnesota), Bruce McCarl (Texas A&M), Gregory Perry (Oregon State University), Gerald Shively (Purdue University), and several other anonymous reviewers. Of course, any errors, which we hope there are none, remain entirely our fault.

Part 1
LINEAR PROGRAMMING

1

Introductory Concepts and the Graphical Approach to Linear Programming

The focus of this book is on applications of optimization models for the fields of agricultural, environmental, and resource economics. Before such techniques can be applied to real-world problems, an elementary foundation needs to be established on basic concepts, definitions, and approaches of mathematical programming in order for you to fully understand the usefulness and limitations of mathematical programming. This is not a book on theory, so the basics established here are intended to supplement the applications that follow, which are the real focus of the book. Readers interested in obtaining a broader understanding of theoretical concepts of **linear programming (LP)** may wish to consult a book in LP theory. In this and the next two chapters, we concentrate on building this foundation for LP models, and in later chapters we develop a similar set of introductory concepts for nonlinear models.

There are several objectives of this chapter. The first objective is to define LP, explain how it is used, and outline the assumptions necessary to apply LP to agricultural, environmental, and resource economics problems. The second objective is to describe how to set up simple decision problems as LP problems, a task that is often more of an art than a science. The third objective of this chapter is to demonstrate how to solve two variable **maximization** and **minimization** LP problems using graph paper and a straight edge. The reader will also see how to use simple algebra to verify whether solutions obtained via the graphical method are indeed correct. Finally, the last objective is to discuss the notion of **sensitivity analysis**. Sensitivity analysis involves examining how sensitive the solution to a problem is with respect to the problem's parameters. The discussion in this chapter will focus on graphical techniques to accomplish various types of sensitivity analyses.

Linear programming is a category of mathematical programming models. One way to categorize mathematical programming is to divide it into two classes of models: linear and nonlinear programming. **Nonlinear programming (NLP)** is less restrictive than LP, in that, equations may have nonlinear, as well as linear forms. The majority of the chapters that follow will deal exclusively with LP models, since these models represent the majority of mathematical programming models.

Linear programming is a widely used problem-solving approach in quantitative methods. The **LP problem** is to determine the optimal value of a linear function (which defines the objectives of the problem) subject to a set of linear constraints (which defines the limits or decision environment of the problem). The term **optimal**, in this context, means minimizing or maximizing a given objective, for instance, maximizing profit, or minimizing costs. Linear programming models are used to help people and organizations make decisions. Decisions involve a process of formulating a set of alternatives to complete a goal, weighing each alternative based on some choice criterion, and selecting among these alternatives to accomplish this goal.

It should be emphasized that linear (and nonlinear) programming models are decision models or aids, not the means and ends for making the decision in question. Like all decision aids, the LP technique is there to assist people in their decision-making process. The management skills of the decision maker, which include qualitative as well as quantitative abilities, are the key attributes of the basis for one's decision. Nevertheless, quantitative techniques like LP have become powerful tools which are often used to improve managerial decision making.

1.1 APPLICATIONS OF LINEAR PROGRAMMING IN AGRICULTURAL, ENVIRONMENTAL, AND RESOURCE ECONOMICS

Linear programming has been used in a wide variety of applications of decision analysis. To provide a glimpse of such applications, which is by no means exhaustive, consider the following areas.

1. **The Diet Problem.** The problem is to determine the least-cost diet for a person, based on food prices, subject to the person receiving an adequate diet. The solution to this problem gives the combination of foods that a person should purchase to minimize food expenditures. Such applications are useful in developing countries, where food is scarce and starvation and malnutrition are major problems, as well as in food manufacturing and farming, where individuals are interested in minimizing the cost of producing food. This problem also applies to livestock producers wishing to minimize feed costs of livestock production.

2. **The Carbon Abatement Problem.** The problem is to determine the least-cost way to reduce carbon emissions by a firm in response to new legislation against global warming. The solution to this problem provides the combination of carbon-reducing activities for the firm to follow in a way that achieves the targeted reductions mandated by the law.

3. **The Product Mix Problem.** The problem is to determine the product mix (combination of outputs to be produced and sold), given limited resources, that maximizes profits, gross revenue, cash flow, net revenue, or utility for a firm. For example, a farmer needs to determine how to best allocate land among crops so that profits are maximized, given the level of control over all factors of production: for instance, the farmer owns and controls 600 acres of land, has two sons to supply family labor, owns one tractor, and so on. Small and large businesses often use LP to help determine product mixes.

4. **The Portfolio Problem.** The problem is to allocate a fixed amount of a resource (e.g., corn harvest) among alternative prospects so as to maximize the returns or minimize the risk from marketing the crop. For instance, corn could be sold at harvest, forward

marketed, stored and sold in future months, or hedged and sold on the futures market. The farmer's objective is to either maximize profit, minimize risk, or some combination of the two. Banks, investment institutions, private investors, universities, state and federal governments, and others also use LP to assist in their portfolio decision process.

5. **The Transportation Problem.** The problem is to determine how to move a product, such as oranges, produced on farms located in different geographic locations to different demand destinations in the most cost-efficient (least-expensive transportation costs) way. Linear programming applications for this class of problem are common.

6. **The Allocation Problem.** The problem is to determine how to allocate scarce resources among competing projects. For example, a conservation organization seeks to maximize the ecosystem services provided in an ecoregion, but has to select which conservation projects to fund. Given the different outcomes provided by the projects and the different objectives and priorities of the funding sources, **binary linear programming** can be used to determine which services should be used.

7. **Capital Budgeting Problem.** The problem is to invest capital, which is finite (scarce), to alternative projects. What is capital? Capital can mean money, or it can mean man-made resources, such as machinery. Business school types often define capital as some sort of money or financial measure, such as cash, stocks, bonds, savings, and so on. Economists generally define capital more broadly to include tools, equipments, factories, machinery, and all man-made items used to produce goods and services. Hence, the uses of capital budgeting may include monetary investments among alternative projects or the allocation of man-made aids to production to alternative projects.

All of these problems have four general properties that are inherent in any LP model. These properties are:

1. The **objective** is to be optimized by either maximization or minimization.
2. There are **constraints** restricting the activities that are required to carry out the objective.
3. All equations are **linear**.
4. The **activities** (or **decision variables**) are generally non-negative.

To illustrate these properties, consider the following example. Suppose that a grain farmer's objective is to maximize profit by producing two types of crops: wheat and sorghum. The farmer knows that the net profit of producing wheat is \$135 per acre, while the net profit of producing sorghum is \$100 per acre. The farmer's **objective function**, then, is to maximize profit from the production of the two crops, which can be expressed mathematically as:

$$\text{Max: } Z = 135 wheat + 100 sorghum,$$

where *wheat* is the number of acres of wheat produced and *sorghum* is the number of acres of sorghum produced. The variables *wheat* and *sorghum* are called activities in LP language. If there were no constraints placed on these activities, the optimal solution to this problem would be to produce only wheat because it has a higher unit profit (135 versus 100). Furthermore, the optimal solution would be to produce an infinite amount of wheat because there are no restrictions currently placed on the problem's activities. In reality, the farmer would likely face many restrictions such as constraints on the availability of land, labor, machinery, and raw materials needed to produce crops. For example, suppose that the farmer has a labor force of 10 people and that each acre of wheat requires two people to produce, while each acre of sorghum requires one person to produce.

The following constraint can be added to this problem to reflect the scarcity of labor for this situation:

$$2wheat + 1sorghum \leq 10.$$

This constraint has the following interpretation: each acre of wheat produced requires two people, each acre of sorghum produced requires one person, and the total amount of labor used in raising both crops cannot exceed 10 people. Similar constraints could be added to this problem to reflect the scarcity of other resources such as land, capital, raw materials, and other resource endowments. These constraints are referred to as **structural constraints**. Note that both the objective function and the resource constraint to this problem are linear. Finally, in most applications it is appropriate to add a non-negativity constraint, which requires all activities to be non-negative. In this example, this implies that the farmer cannot produce negative quantities of either wheat or sorghum. The LP model for this example is:

Max: Z = 135*wheat* + 100*sorghum*   Objective function,

Subject to (s.t.):

$$2wheat + 1sorghum \leq 10 \quad \text{Labor constraint,}$$
$$wheat, \quad sorghum \geq 0 \quad \text{Non-negativity.}$$

1.2   COMPONENTS OF THE GENERAL FORM OF THE MODEL

There are several ways to express an LP model. The first of these is called the **general form** of the model, which was used in the example above. The general form of a generic LP model for n activities and m structural constraints is:

Max or Min: $Z = c_1 x_1 + c_2 x_2 + \cdots + c_n x_n$ (0)

s.t.:

$$a_{11} x_1 + a_{12} x_2 + \cdots + a_{1n} x_n \ \{\leq, =, \geq\} \ b_1 \tag{1}$$
$$a_{21} x_1 + a_{22} x_2 + \cdots + a_{2n} x_n \ \{\leq, =, \geq\} \ b_2 \tag{2}$$
$$\vdots \quad \vdots \quad \vdots \quad \vdots \quad \vdots \qquad \qquad \vdots$$
$$a_{m1} x_1 + a_{m2} x_2 + \cdots + a_{mn} x_n \ \{\leq, =, \geq\} \ b_m \tag{m}$$
$$x_1, \quad x_2, \ldots \quad x_n \qquad \geq \ 0 \tag{m+1}$$

The first component of the model will always be the objective function, which is expressed in equation (0). The objective function is a mathematical formulation of the decision maker's objective. The objective is expressed as a function of the activities (x_i) that are under the control of the decision maker: that is, $Z = f(x_1, x_2, \ldots, x_n)$. The objective function value (Z) measures the alternative solutions to the problem, such as profit, costs, sales, production, and so on. The objective function will either be maximized or minimized depending upon the problem. The activities (also referred to as "decision variables" or just "variables") are the unknown endogenous (model-determined) variables of the problem. The model solution provides the decision maker with the optimal activities levels. The c_is in the objective function are called the **objective function coefficients**. These are fixed parameters (or coefficients), which give the contribution of each activity to the value of the objective function. For example, if the objective is to maximize profits from the sale of two products such as wheat and sorghum, then the objective function coefficients could be unit net profit per acre for each crop.

Equations numbered (1) through ($m+1$) represent the **constraint set** for this problem. The objective function is optimized subject to (s.t.) satisfying all of the constraints, which define the restrictions on the activities in the problem. Intuitively, the constraints model the restrictions that the decision maker must operate within. Notice that there are two types of constraints for an LP model: structural constraints and a non-negativity constraint.

Mathematically, there are three possible directions for the structural constraints in an LP model. Constraints may be (1) less-than-or-equal-to ($\leq$) restrictions, (2) greater-than-or-equal-to ($\geq$) restrictions, or (3) equal-to ($=$) restrictions. The structural constraints are the first m constraints, which define the technical relationship between resource usage ($a_{ij}x_i$) for each activity and the resource endowment (b_j). The technical coefficients (a_{ij}) define how much of resource i it takes to produce a unit of activity j. The **resource endowment** or **right-hand-side (RHS) value** (b_j) either represents the amount of resources that the decision maker controls in the decision process, or represents a minimum condition that must be met.[1] For example, one $\leq$ type of structural constraint is a land constraint for a farm problem that limits total acreage planted for all crops to not exceed total acres controlled by the farmer. An example of a $\geq$ type of constraint is a minimum amount of some nutrient needed to survive for a balanced diet problem. As was previously mentioned, the non-negativity constraint, which is included in most LP models but is not a structural constraint, requires that all activities be non-negative (i.e., zero or positive). For example, one cannot have negative seven acres of corn being produced in the optimal solution.

The collection of all fixed coefficients (not variables) in the LP model (i.e., c_i, a_{ij}, and b_j) is called the parameters of the model. Linear programming assumes that all parameters are known by the decision maker in order to completely determine the model's solution. This assumption is relaxed later on in this book when a special class of mathematical programming known as stochastic or risk programming is examined.

Notice from the general form of the model that all LP models are composed of linear functions. A **linear function** is a function whose form is the following:

$$c_1x_1 + c_2x_2 + \cdots + c_nx_n,$$

where c_1, c_2, ... , c_n are numerical constants (parameters) and x_1, x_2, ... , x_n are variables (or activities). In other words, linear functions are characterized by all variables having an exponent of 1 and no multiplicative terms, for instance, no $x_i\,x_j$ terms.

The **optimal solution** (denoted as x_i^*) to an LP model gives the values for the activities that optimize the objective function, that is, gives the best way to achieve the desired objective while satisfying all the restrictions. If the objective function is to minimize the cost of producing a certain amount of an output, y, given two inputs, x_1 and x_2, the optimal solution to this problem, x_1^* and x_2^*, provides the least-expensive way to utilize the two inputs in achieving the desired amount of output. The optimal solution for a maximization problem is a feasible solution that yields the largest value of the objective function.

---

[1] The term "resource endowment" is more appropriate for LP problems involving allocating fixed resources. On the other hand, there are also LP applications that involve RHS parameters that are not resource endowments. In these cases, such parameters are generally known as RHS values.

1.3 STANDARD ASSUMPTIONS OF LINEAR PROGRAMMING MODELS

There are four assumptions of all LP models.

1. **Proportionality**. The contribution of each objective function coefficient (c_i) to the activities is the same regardless of the level of the activity. In other words, if the value of an activity is tripled, then its contribution to the objective function will also be tripled. Proportionality also implies that the contribution of each resource requirement (a_{ij}) to activity j is the same regardless of the level of the activity. In other words, if the value of an activity is tripled, then it will require three times as much resource as it previously required.

2. **Additivity**. The contribution of each activity does not influence the contribution of all other activities. Additivity requires that for any level of activities ($x_1, \ldots, x_n$), the value of the objective function and the total resource usage are found by the summation of the individual activities times their associated parameters.

3. **Divisibility**. The optimal values of decision variables are real numbers, for instance, $x_1 = 22.34527$. That is, optimal solutions to the problem may be continuous variables.

4. **Certainty**. All parameters are constants that are known by the decision maker.

   The first two assumptions above guarantee that all equations in a linear program are in fact linear. These two assumptions suggest that if a firm does not experience constant returns to scale, or its activities are not independent, then traditional LP should not be used. In this case, NLP may be more appropriate because it can handle the relaxation of these two assumptions. Alternatively, linear approximations such as separable LP can be used to handle cases such as decreasing returns to scale.

   What about divisibility and certainty? While the divisibility assumption may be fine for many applications, it may be unrealistic for some, such as capital budgeting. **Integer** or **binary programming** is necessary when this is the case. Certainty is an assumption that is often unrealistic. For example, when lags are involved between deciding upon input allocation and realization of outputs, some of the parameters in the objective function and/or constraint set may not be known with certainty. This is clearly the case for crop farming. If this is the case, then a linear or nonlinear **risk programming** model is necessary.

   While these assumptions are usually violated in reality, LP is still the most commonly used quantitative decision aid among decision makers. In many applications, one can "live" with these assumptions, and LP provides an approximation to reality.

1.4 FORMULATING LINEAR PROGRAMMING PROBLEMS

Formulating LP problems is the most important part of LP; if the problem cannot be set up properly, a correct answer cannot be found. Unfortunately, there is no theory or scientific method for formulating LP problems. This task is more of an art than a science. The only way to learn this task is to practice, practice, and practice some more by working through example after example.

Formulating LP problems is similar to translating word problems encountered in introductory algebra into mathematical problems. The problem is written out in words, and the task of the analyst is to convert the language into a suitable mathematical translation. The first step is to read and study the verbal problem carefully until a complete understanding of what it entails is developed.

Study the verbal problem well enough to have a complete understanding of what it entails.

Read the problem once or twice carefully without worrying about how to set it up in LP "language." Concentrate on what the problem is, rather than what the LP problem will look like.

Once the problem is fully understood, the next step is to read it again, but this time pay closer attention to sorting out the relevant information required for the LP model. There are several pieces of information that need to be sorted out in this phase of constructing an LP model. The pieces of information pertinent for an LP problem are defined by various steps listed below:

1. Identify the **objective function** for the problem. For example, the objective may be to maximize profits, or maximize sales, or maximize production. Or the objective may be to minimize costs or some other goal.

2. Identify the **activities** (decision variables) for the problem. Recall that an activity is a way (or ways) of reaching the objective, that is, the objective is a function of the activities of the problem. The units of measurement for each activity should also be identified.

3. Identify the **objective function coefficients** for each activity. These coefficients give the per unit contribution of each activity to the value of the objective function.

4. Identify the resources controlled by the decision maker (**resource endowment**), their levels, and their units of measurement. In problems not involving fixed resources, identify the **RHS parameters** of the problem.

5. Identify the **technical coefficients** that give a correspondence between the activities and the resources. The technical coefficients define how much of each resource it takes to produce a unit of an activity. Also, identify the units of measurement for the technical coefficients.

6. Set up the appropriate **structural constraints** in the constraint set.

After following these procedures, the last step is to put all the pieces of information together into the general form of the model (see definition on page 5).

Example 1   The Polluter's Problem

A factory emits four types of greenhouse gases into the air, all of which cause global warming. The four greenhouse gases are (1) carbon dioxide, (2) methane, (3) chlorofluorocarbons, and (4) nitrous oxide. The federal government has just passed a new environmental bill designed to slow the growth of greenhouse gases in the atmosphere. Under the new law, this factory must reduce its annual emission of carbon dioxide by 100 million pounds, methane by 25 million pounds, chlorofluorocarbons by 50 million pounds, and nitrous oxide by 75 million pounds. Assume there are four abatement techniques, A, B, C, and D.[2] Each abatement method has the following cost and per unit reduction for each greenhouse gas:

Greenhouse Gas	Gas Abatement Technique (per unit reduction in million pounds)			
	A	B	C	D
Carbon Dioxide	10	15	25	25
Methane	2	5	7	6
Chlorofluorocarbons	8	10	9	15
Nitrous Oxide	5	12	13	16
Cost/Unit ($ million)	$0.50	$1.20	$3.30	$5.00

---

[2]Real-world examples of pollution abatement techniques include flue-gas desulfurization, electrostatic filters, selective catalytic reduction, and use of lower-sulfur coal.

Suppose that the firm's sole objective is to minimize the total cost of reducing emissions of these four greenhouse gases to the federal government's new standard by using any combination of the four abatement techniques.

To formulate this problem as an LP model, consider the steps previously outlined. Let's start out with an easy question.

What is the objective of the decision maker?

The firm's objective is to minimize total greenhouse gas abatement costs.

What are the activities or decision variables, which describe how the objective (minimum costs) is made?

In this case, there are four activities corresponding to the adoption of each abatement approach. Let:

a = number of units of technique A used to reduce greenhouse gas emissions,

b = number of units of technique B used to reduce greenhouse gas emissions,

c = number of units of technique C used to reduce greenhouse gas emissions,

d = number of units of technique D used to reduce greenhouse gas emissions.[3]

Notice that the units of all four activities are measured in terms of number of greenhouse gas abatement techniques implemented, for instance, four of technique A implemented. Given the objective function coefficients, which are the per unit costs for a, b, c, and d, we can now define the objective function for this problem. The objective function is to minimize total abatement costs (Z), which is equal to:

$$\text{Min: } Z = 0.5a + 1.2b + 3.3c + 5d,$$

where Z is measured in million dollars.

The reader is now ready to ask the next question.

What are the restrictions faced by the firm, what are their minimum requirements, and what are their units of measurement?

The restrictions have been set by the new law, that is, the firm has to reduce its annual emissions levels of the four greenhouse gases. Specifically, the firm must reduce its annual emission of carbon dioxide by 100 million pounds, methane by 25 million pounds, chlorofluorocarbons by 50 million pounds, and nitrous oxide by 75 million pounds. These minimum requirements, which are all measured in terms of million pounds of reduced greenhouse gas emissions, will be modeled as four separate constraints in the LP model. So the firm's LP problem is to minimize the total cost of abatement, while satisfying these four minimum levels of reductions in greenhouse gases.

What are the technical coefficients, which define how many million pounds each greenhouse gas is reduced, for each abatement technique?

Carbon Dioxide

The first constraint of the problem is to reduce carbon dioxide emissions by 100 million pounds. From the table above, it is given that one unit of technique A reduces carbon dioxide emissions by 10 million pounds, one unit of technique B reduces carbon dioxide

---

[3]While activities were previously denoted as x_1, x_2, etc., it is often useful to use abbreviations for names to define activities. In this book, both naming conventions for activities will be followed. For generic models, activities will generally be denoted as x or some other letter, while in specific applications, abbreviations will generally be used.

emissions by 15 million pounds, one unit of technique C reduces carbon dioxide emissions by 25 million pounds, and one unit of technique D reduces carbon dioxide emissions by 25 million pounds. Combining the required reduction (100 million pounds) with the technical coefficients for a (10 million pound reduction per device), b (15 million pound reduction per device), c (25 million pound reduction per device), and d (25 million pound reduction per device), the first constraint can be written as:

$$10a + 15b + 25c + 25d \geq 100.$$

This constraint says that the minimum amount of carbon dioxide reduction by the combination of the four abatement techniques cannot be lower than 100 million pounds. For example, if the firm wanted to only use technique A, then, at a minimum, 10 units of a would be needed (i.e., 100/10 = 10). Likewise, if only technique D were used, a minimum of four units would be necessary (i.e., 100/25 = 4).

Methane

The second constraint of the problem is to reduce methane emissions by 25 million pounds. From the table above, it is given that one unit of technique A reduces methane emissions by 2 million pounds, one unit of technique B reduces methane emissions by 5 million pounds, one unit of technique C reduces methane emissions by 7 million pounds, and one unit of technique D reduces methane emissions by 6 million pounds. Combining the required reduction (25 million pounds) with the technical coefficients for a (2 million pound reduction per device), b (5 million pound reduction per device), c (7 million pound reduction per device), and d (6 million pound reduction per device), the second constraint can be written as:

$$2a + 5b + 7c + 6d \geq 25.$$

This constraint says that the minimum amount of methane reduction by the combination of the four abatement techniques cannot be lower than 25 million pounds. For example, if the firm wanted to only use technique A, then, at a minimum, 12.5 units of a would be needed (i.e., 25/2 = 12.5). Likewise, if only technique D were used, a minimum of 4.17 units would be necessary (i.e., 25/6 = 4.17). Notice the LP assumption of perfect divisibility here may or may not be appropriate: that is, fractional amounts of each of the four abatement techniques can be used.

Chlorofluorocarbons

The third constraint of the problem is to reduce chlorofluorocarbons emissions by 50 million pounds. Similar to the two previous constraints, the third constraint can be written as:

$$8a + 10b + 9c + 15d \geq 50.$$

This constraint says that the minimum amount of chlorofluorocarbons reduction by the combination of the four abatement techniques cannot be lower than 50 million pounds. For example, if the firm wanted to only use technique A, then, at a minimum, 6.25 units of a would be needed (i.e., 50/8 = 6.25). Likewise, if only technique D were used, a minimum of 3.33 units would be necessary (i.e., 50/15 = 3.33).

Nitrous Oxide

The fourth constraint of the problem is to reduce nitrous oxide emissions by 75 million pounds. The fourth constraint can be written as:

$$5a + 12b + 13c + 16d \geq 75.$$

This constraint says that the minimum amount of nitrous oxide reduction by the combination of the four abatement techniques cannot be lower than 75 million pounds. For example, if the firm wanted to only use technique A, then, at a minimum, 15 units of A would be needed (i.e., 75/5 = 15). Likewise, if only technique D were used, a minimum of 4.69 units would be necessary (i.e., 75/16 = 4.69).

Non-negativity   Finally, every LP problem requires that the activities be non-negative: for instance, a firm cannot use −5 units of A. The non-negativity constraint is typically written as the last constraint in the LP problem as:

$$a, b, c, d \geq 0.$$

The last step in setting up an LP problem is writing out the full general model. In this case, we have written the objective function and all constraints for the problem. Hence, we just need to combine them. Students are encouraged to number the equations, or descriptive labels, or both, when presenting the model in general form. Using both equation-numbering and descriptive labels, the general form for this problem is:

[Objective Function]   Min: $Z = 0.5a + 1.2b + 3.3c + 5d$ (0)
s.t.:

[Minimum carbon dioxide reduction] $10a + 15b + 25c + 25d \geq 100$ (1)

[Minimum methane reduction] $2a + 5b + 7c + 6d \geq 25$ (2)

[Minimum chlorofluorocarbon reduction] $8a + 10b + 9c + 15d \geq 50$ (3)

[Minimum nitrous oxide reduction] $5a + 12b + 13c + 16d \geq 75$ (4)

[Non-negativity constraint] $a, \quad b, \quad c, \quad d \geq 0$ (5)

The solution to this problem (Z^*, a^*, b^*, c^*, d^*) yields the least total cost to attain the minimum reductions, given the parameters (unit cost, technical coefficients, and minimum reductions) of the problem.

Example 2   The Brewer's Decision

A local brewery manufactures two types of beer for sale. The first is a relatively inexpensive beer (*cheap*) and the second is a premium beer (*prem*). The net profit for *cheap* is $1.00 per case while the net profit for *prem* is $2.50 per case. It is assumed that the brewery can sell all the beer it produces.

There are three workers who specialize in separate production operations to brew both beers. One works 50 hours per week on pre-fermenting operations. The second devotes 40 hours per week to bottling the beer. Finally, the third spends 25 hours per week on quality control. The average time requirements per case of beer are as follows:

	(*cheap*) Inexpensive Beer (minutes)	(*prem*) Premium Beer (minutes)
Pre-Fermenting Operations	3	3
Bottling	2	4
Quality Control	1	3

The LP problem is as follows: given this set of requirements and the profit per unit, what combination of inexpensive and premium beer should be produced to maximize profit during the week?

There are two activities for this problem, *cheap* and *prem*. The objective function for this problem is to maximize profit, and the objective function coefficients are the unit profits for each type of beer. Since it is given that per unit profit of the inexpensive beer is $1.00 per case, and the unit profit for the premium beer is $2.50 per case, the objective function is:

$$\text{Max: } Z = 1cheap + 2.5prem,$$

where the activities are measured in cases of beer produced each week. There are three resource constraints for this problem: (1) labor for the pre-fermentation operations (resource endowment is 50 hours per week, or 3,000 minutes per week), (2) labor for the bottling operations (resource endowment is 40 hours per week, or 2,400 minutes per week), and (3) labor for the quality control operations (resource endowment is 25 hours per week, or 1,500 minutes per week).

Combining the resource endowments with the technical coefficients, the following constraints are derived:

$$3cheap + 3prem \leq 3,000 \quad \text{(fermentation constraint, units = minutes),}$$

$$2cheap + 4prem \leq 2,400 \quad \text{(bottling constraint, units = minutes),}$$

$$1cheap + 3prem \leq 1,500 \quad \text{(fermentation constraint, units = minutes).}$$

The general form of the model, including the non-negativity restriction, is:

[Objective Function]   Max: $Z = 1cheap + 2.5prem$ (0)

s.t.:

[Fermentation Constraint]	$3cheap + 3prem \leq 3,000$	(1)
[Bottling Constraint]	$2cheap + 4prem \leq 2,400$	(2)
[Quality Control Constraint]	$1cheap + 3prem \leq 1,500$	(3)
[Non-negativity]	$cheap, \quad prem \geq 0$	(4)

Example 3   The Dairy Farmer's Feeding Decision

A dairy farmer can purchase two kinds of feed for his cows. Each cow requires 60, 84, and 72 units of nutrients A, B, and C, respectively, per day. The nutrient contents and costs per pound of feed 1 and feed 2 are the following:

Feed	Nutrient Content/Pound			Cost (cents/pound)
	A	B	C	
Feed 1	3	7	3	10
Feed 2	2	2	6	4

The farmer's LP problem is to determine what would be the least expensive diet to feed his cows given the minimum requirements for nutrients A, B, and C, and the feed prices. It is assumed that feeding more than the minimal nutrient requirements will not be harmful to the animals. Note that the objective function will be expressed on a cost-per-cow basis. This problem is quite similar to the first example of greenhouse gas abatement.

There are two activities in this problem, how many pounds of feed 1 and how many pounds of feed 2 to purchase and feed the farmer's dairy herd. The objective function coefficients for each activity are the unit costs of each feed. Denoting feed 1 as $feed_1$ and feed 2 as $feed_2$, the objective function is to minimize feed costs:

$$\text{Min: } Z = 10feed_1 + 4feed_2$$

There are three structural constraints: (1) each cow receives a minimum of 60 units of nutrient A, (2) each cow receives a minimum of 84 units of nutrient B, and (3) each cow receives a minimum of 72 units of nutrient C. Notice that unlike the previous maximization problem which restricted resource use to not exceed the resource endowments, in this minimization problem the constraints restrict the decision maker to feed at least the minimum daily nutrient requirements for each cow. The technical coefficients for this problem translates the nutrient content (A, B, and C) for each feed. Using this information, the three constraints can be written:

$3feed_1 + 2feed_2 \geq 60$ (Minimum Daily Requirement, Nutrient A, units = nutrients),

$7feed_1 + 2feed_2 \geq 84$ (Minimum Daily Requirement, Nutrient B, units = nutrients),

$3feed_1 + 6feed_2 \geq 72$ (Minimum Daily Requirement, Nutrient C, units = nutrients).

The general form of the model is:

[Objective Function]   Min: $Z = 10feed_1 + 4feed_2$ (0)

s.t.:

[Minimum Daily Requirement A] $3feed_1 + 2feed_2 \geq 60$ (1)

[Minimum Daily Requirement B] $7feed_1 + 2feed_2 \geq 84$ (2)

[Minimum Daily Requirement C] $3feed_1 + 6feed_2 \geq 72$ (3)

[Non-negativity] $feed_1, \quad feed_2 \geq 0$ (4)

1.5   THE GRAPHICAL APPROACH FOR SOLVING LINEAR PROGRAMMING MAXIMIZATION PROBLEMS

In the previous section, we learned how to set up very simple (two-dimensional) LP problems. The next step in the process is to find a solution for the problem using the logic of LP. Generally speaking, there are three ways to solve LP models. The first is the **graphical method**, which is useful for illustrating the intuition behind LP. However, the graphical method can only be used for very small problems (typically two activities, sometimes three activities). To tackle larger problems, algebraic approaches are necessary. One of the most powerful and efficient of these approaches is called the **simplex method**. As will become clear in the next chapter, this method can be used to solve larger problems involving many activities. The last approach is to use **LP computer software**, which is based upon the simplex method or other solution algorithms. When computers are used to solve LP problems, the size of the problem can be very large. Problems with thousands of activities and thousands of constraints may be readily solved using computer software. Later chapters will describe how to use and interpret the output of LP software.

Solving Two-Activity Maximization Problems

Linear programming problems with only two activities can be solved easily using the graphical approach. The first step is to determine what the possible **feasible solutions** and, hence, **feasible region** are, given the constraints of the problem.

Any combination of values for the activities that satisfies all the constraints (including non-negativity) constitutes a **feasible solution**. An **infeasible solution** is a specification of values for the decision variables that violates one or more constraints. The **feasible region** is the set of all possible feasible solutions.

To determine the feasible region, we need to graph all of the constraints. To illustrate, consider the following maximization problem.

$$\text{Max: } Z = 40x + 45y \tag{0}$$

s.t.:

$$x + \quad y \leq 600 \tag{1}$$

$$x + 1.5y \leq 750 \tag{2}$$

$$x \quad\quad \leq 400 \tag{3}$$

$$x, \quad y \geq 0 \tag{4}$$

Start with the easiest constraint, the non-negativity constraint, which requires x and y to be non-negative. This constrains x and y to be in quadrant I (including the vertical axis and horizontal axis) of the Cartesian coordinate system. If non-negativity was the only constraint, then the feasible set would include all non-negative values for x and y, which is depicted graphically in Figure 1.1(a). Notice that this set includes all of quadrant I plus the x and y axis and is bounded from below, but unbounded from above, that is, $x, y \geq 0$.

Next consider the first structural constraint:

$$x + y \leq 600 \tag{1.1}$$

To show all points that satisfy this relation, start by graphing the line corresponding to the equation. Rewrite (1.1) using an equality constraint instead of a weak inequality, that is:

$$x + y = 600 \tag{1.2}$$

Equation (1.2) is called a **constraint line**. This line or equation gives all the values for x and y that lie on the border or **frontier** of this particular constraint. The simplest way to draw this graphically is to find the x and y intercepts and then connect the two by drawing a line. Using basic algebra, the x intercept of (1.2) can be found by setting y equal to zero:

$$1x + 1(0) = 600, \text{ or}$$

$$x = 600 \tag{1.3}$$

It is clear from equation (1.3) that the x intercept is 600. Repeating this procedure for y (set x equal to zero and solve (1.2) for y) shows that the y intercept is also 600. Drawing a line connecting the x and y intercepts gives the constraint line. Since the constraint is of the $\leq$ type, shading in the region on and below this frontier results in the feasible region for this particular constraint. Again, this area is the region where all

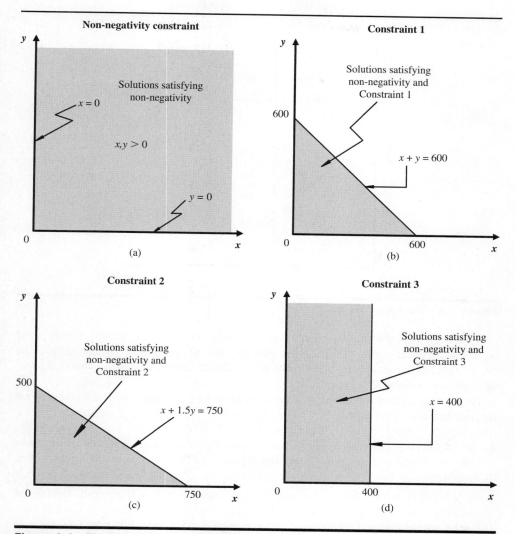

Figure 1.1 The four constraints for the maximization problem illustrated graphically.

solutions satisfy the first structural constraint. Figure 1.1(b) shows the feasible region for this constraint plus the non-negativity constraint.

Now consider the second structural constraint:

$$x + 1.5y \leq 750 \tag{1.4}$$

As in the previous case, we first make (1.4) an equality constraint, that is:

$$x + 1.5y = 750 \tag{1.5}$$

Then, solve for the x and y intercepts. The x intercept is 750 (i.e., solve $x + 1.5(0) = 750$ for x). The y intercept is $750/1.5 = 500$ (i.e., solve $1(0) + 1.5y = 750$ for y). Next, draw a line connecting the two intercepts. Finally, shade in all x and y points that lie on and below this line. Figure 1.1(c) shows the feasible region for this constraint and non-negativity.

The third and final structural constraint simply limits x to be no greater than 400. The constraint line is found by drawing a vertical line from the x axis at 400. Figure 1.1(d) shows the feasible region for this constraint plus non-negativity.

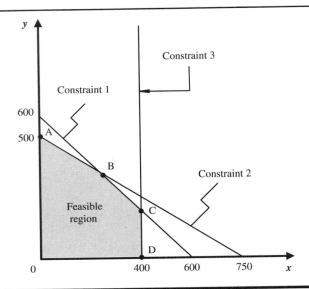

Figure 1.2   The feasible region.

In an LP model, the solution points need to satisfy all constraints simultaneously. That is, the full feasible region is found by superimposing all constraints into one graph. The feasible region is equal to the intersection of all feasible points. Figure 1.2 shows the feasible region for this LP problem.

There are several observations that can be made from examining the feasible region for this maximization problem:

1. A solution that does not violate any of the constraints is a feasible solution. For example, the origin and any point along the segment ABCD are feasible solutions to this problem.

2. If a set of feasible solutions for an LP model exists, then it will generally contain an infinite number of solutions. This is due to the assumption of divisibility (i.e., all variables are continuous). There are counter-examples to this rule; however, they are usually trivial cases.

3. The feasible region is a **convex** set. Intuitively, this means that if you pick any two points in this region and draw a line connecting them, all points lying on this line are also in this region. This condition will hold for any two points that lie within the feasible region. An important property of a convex set is that it will contain **extreme points**, which are the vertices or corners of the feasible region where one constraint intersects another. The origin is also considered to be an extreme point. In Figure 1.2, the extreme points are labeled as 0, A, B, C, and D.

4. An optimal solution, which will be discussed next, will always be found at one of the **extreme points** in the feasible region. In the case of multiple optimal solutions (discussed later on), one of the multiple optimal solutions will be an extreme point.

Finding the Optimal Solution

In proceeding to determine the optimal solution, choose an arbitrary level of profit for Z and identify all points associated with that level. For now, define the arbitrary level of profit generally as Z_a rather than a specific numeric value. Recall that from the objective function, profit is given by:

$$Z = 40x + 45y = Z_a \qquad (1.6)$$

Solving equation (1.6) for x or y gives the slope of the **iso-profit line**, or more generally called the **iso-contribution line**. An iso-contribution line shows all activity solutions that yield the same value for the objective function.

Solving (1.6) for y yields:

$$45y = Z_a - 40x, \text{ or}$$

$$y = 1/45\ Z_a - 8/9x \qquad (1.7)$$

This equation is the **slope-intercept form** of the iso-contribution line. The y intercept is $1/45$ and the slope is $-8/9$. We could also express the slope-intercept form in terms of solving for x. Figure 1.3 shows several values of Z_a plotted on a graph, along with the feasible region. The iso-contribution lines are parallel to one another since their slopes are the same. Also, if all objective function coefficients are positive, then the value of all iso-contribution lines will increase as they move out and to the right of the origin. Therefore, in maximization problems, the desired direction for iso-contribution lines is the northeast, north, or east.

Using these two properties, the optimal solution can now be found. Since the objective is to maximize Z, the optimal solution is found by moving the iso-contribution line out in the northeast (or north or east) direction as far as possible until only one point on the line touches (is tangent to) a point on the feasible region. The feasible point that lies on the highest Z_a line in this example is point B in Figure 1.3, which is the optimal solution. Hence, the optimal solution to this problem is $x^* = 300$, $y^* = 300$, and Z^* (the optimal value of the objective function) $= 25,500$ (i.e., $40(300) + 45(300)$). Note that

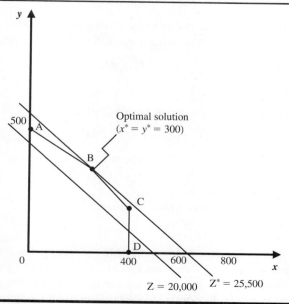

Figure 1.3   Iso-contribution lines and the optimal solution.

the optimal solution at point B is unique, in that all other solution combinations yielding $25,500 profit are infeasible.

To denote that the solution values are the optimal activities, asterisks ($*$) are used, that is,

$$x^* = 300, \; y^* = 300.$$

Not all structural constraints are **binding** at the optimal solution. A binding constraint is one where all of the resource endowment is utilized. In this example, the first two constraints are binding, but not the last constraint. To verify this, substitute the optimal activity values ($x^* = 300$, $y^* = 300$) into the constraints and compare the left-hand-side (LHS) values of these equations to their RHS values.

$$x^* + y^* \leq 600, \text{ or} \tag{1.8}$$

$$300 + 300 = 600 \text{ (binding)}$$

$$x^* + 1.5y^* \leq 750, \text{ or} \tag{1.9}$$

$$300 + 1.5(300) = 750 \text{ (binding)}$$

$$x^* \leq 400, \text{ or} \tag{1.10}$$

$$300 < 400 \text{ (nonbinding)}$$

We could also verify this for the two-activity case by inspecting the graphical solution.

Verifying the Solution: The Simultaneous Equations Approach

If we know which constraints are binding in the optimal solution, then we can determine the optimal solution by solving the constraints for x and y. In this example, since the first two constraints are binding, we can write them as equalities, that is:

$$x + y = 600 \tag{1.11}$$

$$x + 1.5y = 750 \tag{1.12}$$

Since there are two equations and two unknowns, they can be solved simultaneously to determine the values for the two unknown activities. Solving (1.11) for y yields:

$$y = 600 - x \tag{1.13}$$

Substituting (1.13) into (1.12) yields:

$$x + 1.5(600 - x) = 750, \text{ or}$$

$$x + 900 - 1.5x = 750, \text{ or}$$

$$0.5x = 150, \text{ or}$$

$$x^* = 300 \tag{1.14}$$

Substituting (1.14) into (1.13) determines y^*, that is:

$$y^* = 300 \tag{1.15}$$

The simultaneous approach is very handy for double-checking to see if the graphical solution is correct. The reader should always use this as a way of checking whether the graphical solution is correct or not. If a different answer is found than the graphical solution, then go back and re-do the graphs. This technique is also useful for determining exact solution values when the optimal value of an activity is not an integer such as $x^* = 22.29$.

To recap, use the following steps to find the optimal solution for two-activity maximization problems.

Step 1: Find the feasible region.

Step 2: Draw an **iso-contribution line**. An iso-contribution line is a set of solutions lying on a line whose objective function values are all the same. Note that all iso-contribution lines are parallel to one another, since their slopes are the same. Also, assuming that the objective function coefficients are positive, the value of iso-contribution lines increases as they move northeast from the origin.

Step 3: Continue to move the iso-contribution line away from the origin in a northeast direction until a further movement makes all points lying on the iso-contribution line infeasible. The optimal solution should occur at one of the extreme points along the feasibility region.

Step 4: Verify that the graphical solution is correct by using the simultaneous equations approach.

Some Comments on the Optimal Solution

Suppose that the objective function coefficient for x fell substantially from \$40 to \$10. How would this effect the optimal solution? The constraints would be unaffected since the profit parameters are not included in any constraint. However, the objective function would change to:

$$\text{Max: } Z = 10x + 45y.$$

This would change the slope of the iso-contribution line. The new iso-contribution line, in slope intercept form, is:

$$45y = Z_a - 10x, \text{ or}$$

$$y = 1/45Z_a - 2/9x \tag{1.16}$$

Note that the slope switches from $-8/9$ to $-2/9$. The optimal solution now occurs at point A in Figure 1.4 ($x^* = 0$, $y^* = 500$,). Notice that the first constraint is no longer binding, that is, substitute x^* and y^* into this constraint:

$$x^* + y^* = 0 + 500 = 500 < 600 \tag{1.17}$$

There are 100 units of the resource endowment that are not used now. What about the second constraint? This constraint is binding since

$$x^* + 1.5y^* = 0 + 1.5(500) = 750 = 750 \tag{1.18}$$

Finally, the last constraint is not binding since $x^* = 0$, that is:

$$x^* < 400 \tag{1.19}$$

This illustrates a very important condition stated earlier:
   An optimal solution to an LP problem will always occur at an extreme point of the feasible region.
Another important condition of LP which may be apparent from this example is the following:
   If the slope of the iso-contribution line is not the same as the slope of any of the constraints, then the optimal solution will be unique.
The term "unique" here means that there is one and only one optimal solution.

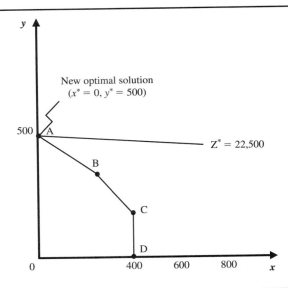

Figure 1.4 New optimal solution when profit for x decreases from \$40 to \$10 per unit.

Economic Interpretation

The economic interpretation of this profit maximization example comes from neoclassical economic theory of the firm. In this case, a firm is producing two outputs (x and y) and has a fixed amount of three resources (b_1, b_2, b_3) used to produce these outputs. The feasible region in this example is identical to the notion of a **production possibility set**, which is defined as any pair of output levels for x and y that is feasible to the firm given its technology and endowment of resources. It follows that the linear frontier to the feasible region (labeled ABCD in Figure 1.2) is the **production possibility frontier (PPF)**, which gives the maximum amount of y that can be produced, given each possible level of x, or vice versa.

The PPF will vary depending upon a firm's resource endowment level and technology. That is, an increase in b_1, b_2, and/or b_3 will cause the PPF to shift towards the northeast direction, and a decrease in b_1, b_2, and/or b_3 will cause the PPF to shift towards the origin. Likewise, an increase in the efficiency of converting inputs into outputs will cause the PPF to shift towards the northeast, and a decrease in efficiency will cause the PPF to shift towards the origin. Efficiency, in this context, means a reduction in the amount of each resource it takes to produce a unit of x and y. The term "technology," or phrase "technology of the firm," will sometimes be used to refer to a firm's technical coefficients.

Note that the PPF is concave to the origin, which means that as more and more of y is produced, the sacrifice in producing x becomes larger and larger. Likewise, increases in the production of x are accompanied by even larger sacrifices in the production of y. This notion is the same as the economic concept of increasing opportunity cost of one output in terms of the other. The only difference between the LP representation of the PPF and the neoclassical economic representation of the PPF is that the LP PPF does not result in a smooth and continuous frontier, whereas the neoclassical PPF is smooth and continuous.

In this example, it is assumed that the firm is a "price taker": that is, the firm cannot influence the price it receives for x or y by varying production. This is consistent with the assumption of perfect competition. The iso-contribution lines derived earlier are completely analogous to **iso-revenue lines** in economic theory. The slope of both of these lines is the price (or unit profit) of one output divided by the other output. As was shown earlier, to maximize total profit, the level of x and y produced should be chosen by the respective point of tangency between the iso-contribution line and the PPF. At this point, the slope of the iso-contribution line is equal to the ratio of the two prices (or unit profits). This corresponds to the economic notion of efficiency, namely, to maximize profit, the ratio of prices (or unit profits) must be equal to the rate of transformation of the two products, which is the same as the slope of the PPF. This is why the slope of the PPF is sometimes called the **marginal rate of product transformation**, as it gives how much of one product has to be sacrificed in order to produce one more unit of the other product.

Special Cases

There are three "special cases" that may arise in solving LP problems that should be noted. These special cases are: (1) unbounded solution, (2) no feasible solution, and (3) multiple optimal solutions. Each of these cases is discussed in the context of the graphical method below.

Unbounded Solution   An unbounded solution occurs whenever the feasible region is not constrained from above, which results in an infinite feasible region. For example, the following maximization problem is "unbounded from above,"

Max: $Z = 1x + 9y$

$\qquad\qquad\qquad\qquad\qquad\qquad\qquad\qquad\qquad\qquad\qquad\qquad\qquad$ (0)

s.t.:

$\qquad\qquad x, \quad y \geq 0$

$\qquad\qquad\qquad\qquad\qquad\qquad\qquad\qquad\qquad\qquad\qquad\qquad\qquad$ (1)

It should be clear that no finite optimal solution exists for this problem. The objective function value will consistently become larger with increases in x and/or y. Since both x and y are not bounded from above, Z will approach infinity as x and/or y approaches infinity.

Graphically, an unbounded solution for a maximization problem can be detected whenever the feasible region extends without limits upwards from the x and/or y axes assuming that the objective function coefficients are positive.

An unbounded solution is usually the result of leaving out one or more constraints. In any case, the problem needs to be reformulated by adding constraints that bound the feasible region in order to get a finite solution. Figure 1.5 illustrates several examples of unbounded feasible regions.

No Feasible Solution   The case of no feasible solution occurs whenever the feasible region is an empty set (a set containing no points), which is due to conflicting constraint specification. Obviously if the feasible region is empty, a feasible solution cannot be obtained. This case usually arises due to an error in specification. It may also arise when the decision maker is attempting to satisfy inconsistent constraints. For example,

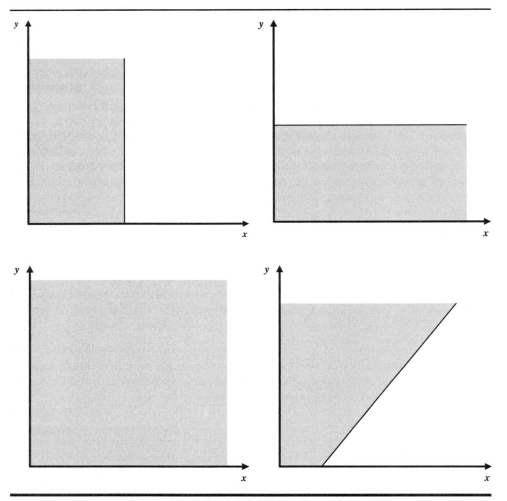

Figure 1.5   Four examples of LP models that result in unbounded feasible regions.

a decision maker might want to reduce costs, while at the same time triple output. The following model has no feasible solution:

Max: $Z = 1.5x + 10y$ (0)

s.t.:

$$1x + 2y \geq 100 \tag{1}$$

$$1x + 2y \leq 50 \tag{2}$$

$$x, \quad y \geq 0 \tag{3}$$

Figure 1.6 gives several specifications that result in no feasible solution.

Multiple Optimal Solutions   Multiple optimal solutions, or "alternative optimal solutions" as they are sometimes called, mean that there is more than one solution that is optimal, that is, "the optimal solution is not unique." This occurs whenever the slope of the

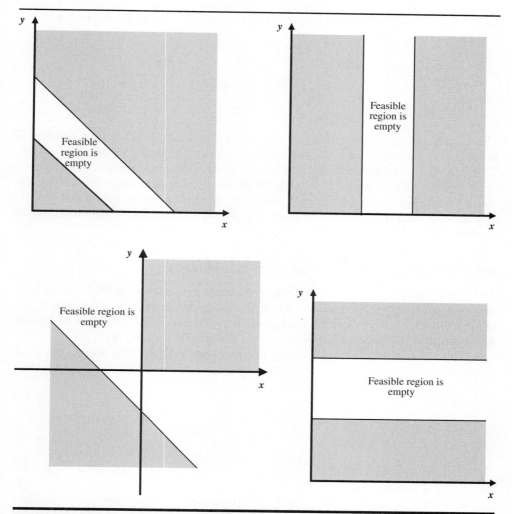

Figure 1.6   Four examples of LP models that result in no feasible solution.

iso-contribution line is the same as the slope of one of the line segments connecting two extreme points in the feasible regions. In this case, there will be an infinite number of optimal solutions; each point along this line segment is an optimal solution.

As an example, consider the following maximization problem:

Max: $Z = x + y$ $\qquad\qquad\qquad\qquad\qquad\qquad\qquad\qquad\qquad\qquad\qquad\qquad$ (0)

s.t.:

$$x + y \leq 100 \qquad\qquad\qquad\qquad\qquad (1)$$

$$x, \quad y \geq 0 \qquad\qquad\qquad\qquad\qquad (2)$$

In this case, the slope of the iso-contribution line and the slope of constraint (1) are both -1. Using the approach outlined earlier in this chapter, the reader can see that the iso-contribution line lies tangent to the entire line segment of constraint (1) rather than a single extreme point. Hence, there are an infinite number of solutions for this problem. The solution to this problem is:

$$x^* = 100 - y^*,$$

$$y^* = 100 - x^*,$$

$$Z^* = 100.$$

The existence of multiple optimal solutions is not a problem. In fact, such a situation is beneficial, as it gives the decision maker more flexibility in pursuing the optimal solution. That is, it offers the decision maker alternatives, which should be preferred. In such cases, alternative decision rules may be used by the decision maker in selecting one of the optimal solutions. For example, a decision maker may wish to choose an option that specializes in either x or y if the level of profitability is the same among a set of different optimal solutions.

An Example with Mixed Structural Constraints

The example presented earlier in this chapter had structural constraints that were all less-than-or-equal-to ($\leq$) restrictions. Not all maximization problems are like this; many contain equal-to ($=$) as well as greater-than-or-equal-to ($\geq$) constraints. This does not present a problem when solving such LP formulations. The same procedures apply.

For example, consider the following maximization problem with mixed structural constraints:

Max: $Z = 2x + 4y$ (0)

s.t.:

$$x + y \leq 100 \tag{1}$$

$$x \geq 25 \tag{2}$$

$$x - 0.5y = 0 \tag{3}$$

$$x, \quad y \geq 0 \tag{4}$$

Notice that not only does this example differ from the previous examples in that there are mixed constraints, but also in that there is a negative technical coefficient in the third constraint. There is no requirement that a_{ij} coefficients have to be non-negative. In this example, the negative technical coefficient (-0.5) is used because it forces x to equal $0.5y$, which can be seen by bringing $-0.5y$ to the RHS of the equation.

The first step in solving this problem graphically is to plot all constraints to determine the feasible region. Figure 1.7 displays the four constraints to this problem. Figure 1.7(a) shows the non-negativity restrictions on x and y. The first structural constraint is graphed in Figure 1.7(b).

Since it is a $\leq$ constraint, it is similar to the restrictions given in the previous example. The second structural constraint simply requires x to be at least as large as 25. This is demonstrated graphically in Figure 1.7(c). The last structural constraint requires that x be exactly 50% of the value for y. This is a very precise constraint, and unlike the two previous structural constraints, has a feasible region that is only a line segment following the equation:

$$x - 0.5y = 0.$$

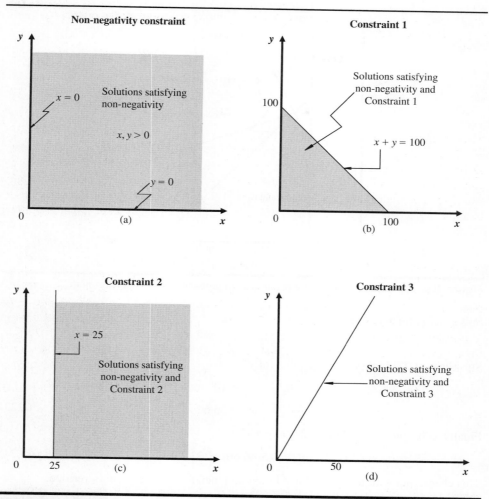

Figure 1.7   The four constraints for the maximization problem with mixed constraints illustrated graphically.

Superimposing all four constraints onto one another results in the feasible region for this problem, which is illustrated in Figure 1.8. The feasible region for this example is the thick line segment labeled AB. The optimal solution to this problem occurs at point A ($x^* = 33.33$, $y^* = 66.66$), where the iso-contribution line ($Z^* = 333.33$) is tangent to the northeastern boundary of the feasible region.

From this graph, it can be seen that constraints (1) and (3) are binding at extreme point A, while constraint (2) is not binding. To verify the optimal solution to this problem algebraically, use the simultaneous equation approach with the two binding constraints. That is, rewrite constraints (1) and (3) as equalities:

$$x + y = 100 \tag{1.20}$$

$$x - 0.5y = 0 \tag{1.21}$$

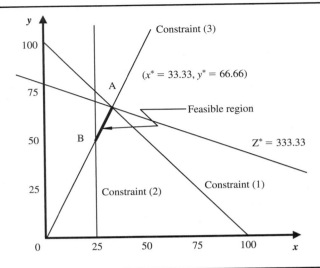

Figure 1.8 Optimal solution for the maximization problem with mixed constraints.

Solve (1.21) for x yields:

$$x = 0.5y \tag{1.22}$$

Substitute (1.22) into (1.20) and solve for y^*:

$$0.5y + y = 100, \text{ or}$$

$$y^* = 66.66 \tag{1.23}$$

Finally, substitute (1.23) into (1.22) to get x^*:

$$x^* = 0.5(66.66) = 33.33 \tag{1.24}$$

This checks out with the graphical solution and therefore the solution is verified.

1.6 THE GRAPHICAL APPROACH FOR SOLVING LINEAR PROGRAMMING MINIMIZATION PROBLEMS

The graphical approach to solving minimization problems is completely analogous to that of maximization. In this section, the reader will learn how to solve a two-activity minimization example. Since most of the concepts are analogous to the previous section, this section will emphasize the steps involved in the solution procedures. If any of the concepts are unclear, consult the previous section for more detail.

Consider the following LP problem written in general form.

$$\text{Min: } Z = 500x + 750y \tag{0}$$

s.t.:

$$\tfrac{1}{2}x + y \geq 50 \tag{1}$$

$$4/5x + 5/2y \geq 100 \tag{2}$$

$$x + y \geq 75 \tag{3}$$

$$x, \quad y \geq 0 \tag{4}$$

The optimal solution to minimization problems is found exactly as were found in the maximization problem using the following steps.

Step 1: Find the feasible region.

Step 2: Draw an **iso-cost line**. An iso-cost line is a set of solutions lying on a line whose objective function values are all the same. Similar to iso-contribution lines, all iso-cost lines are parallel to one another since their slopes are the same. Also, assuming that the objective function coefficients are positive, the value of iso-cost lines decreases as they move towards the origin.

Step 3: Continue to move the iso-cost line towards the origin until a further movement makes all points lying on the iso-cost line infeasible. Unlike the maximization problem, the preferred direction of Z is towards the origin because minimum values of Z are desired.

Step 4: Verify that the graphical solution is correct by using the simultaneous equations approach.

The feasible region for this problem is again determined by graphing the constraint line for each constraint and then finding the intersecting area of all constraints. Consider the first constraint:

$$1/2x + y \geq 50 \tag{1.25}$$

As in the maximization problem, derive the constraint line corresponding to (1.25). To do this, rewrite (1.25) as an equality:

$$1/2x + y = 50 \tag{1.26}$$

Then, compute the x and y intercepts:

$$x = 50/(1/2) = 100,$$

$$y = 50.$$

Because this is a $\geq$ constraint, the feasible region corresponding to (1) is all solutions lying on or above this constraint line (see Figure 1.9(b)).

The second constraint is:

$$4/5x + 5/2y \geq 100 \tag{1.27}$$

The constraint line for (1.27) is:

$$4/5x + 5/2y = 100 \tag{1.28}$$

The intercepts are:

$$x = 100/(4/5) = 125,$$

$$y = 100/(5/2) = 40.$$

The feasible region for constraint 2 is all points lying on or above the equation (1.28) (see Figure 1.9(c)).

The third constraint is:

$$x + y \geq 75 \tag{1.29}$$

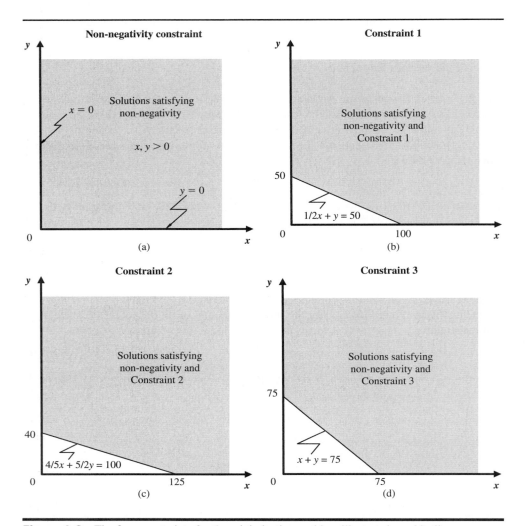

Figure 1.9   The four constraints for the minimization problem illustrated graphically.

The constraint line is:

$$x + y = 75 \tag{1.30}$$

The intercepts are:

$$x = 75, y = 75.$$

The feasible region for constraint 3 is all points lying on or above equation (1.30) (see Figure 1.9(d)).

   The feasible region to this problem is shown in Figure 1.10. The same observations made regarding the feasible region of a maximization problem apply to the minimization problem as well. That is:

1.   Any solution that does not violate the constraints is a feasible solution.

2.   The set of solutions in the feasible region is generally infinite, due to the assumption of perfect divisibility (i.e., all variables are continuous).

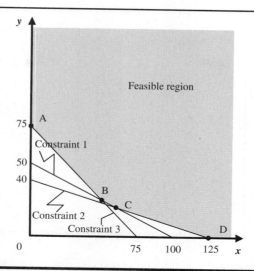

Figure 1.10 The feasible region.

3. The feasible region is a **convex** set. Recall that a convex set means that if one picks any two points in this region and draws a line connecting them, all points lying on this line are also in this region. This condition will hold for any two points that lie within the feasible region.

4. The optimal solution will always be found at one of the **extreme points** in the feasible region. In Figure 1.10, the extreme points are labeled as A, B, C, and D.

Unlike the maximization problem, the feasible region for this minimization problem does not include the origin. Furthermore, the structural constraints for this problem bound the feasible region from below, while it is unbounded from above. This is due to the fact that the constraints are $\geq$ type that represent some sort of minimum conditions on the activities.

The Iso-Cost Line

The iso-cost line is found by choosing an arbitrary objective function value, Z_a:

$$Z_a = 500x + 750y \text{ or in slope intercept form:}$$

$$y = 1/750 Z_a - 2/3x, \text{ or}$$

$$x = 1/500 Z_a - 3/2y.$$

The y and x intercepts are $Z_a/750$ and $Z_a/500$, respectively. Next, choose an arbitrary numeric value for Z_a. Letting $Z_a = \$40,000$, the intercepts are:

$$x = 80, y = 53.3.$$

This line is plotted in Figure 1.11. It is not feasible since all points lie outside the feasible region. After trying this several times, the optimal solution is found at

$$x^* = 50, y^* = 25, \text{ and } Z_a^* = 43,750.$$

Again, the same short-cut approach used in the maximization problem can be used in this case. That is, plot one Z_a line and determine visually which extreme point the optimal Z_a line will be tangent to.

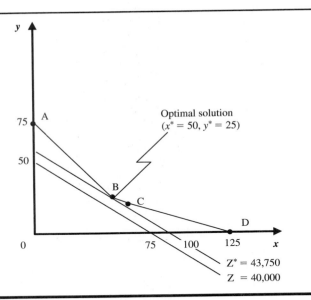

Figure 1.11   The iso-cost line and the optimal solution.

Which constraints are binding and which are nonbinding? You can answer this by inspecting the graphical solution, or you can substitute the optimal activities into the constraint. Plugging x^* and y^* into Constraint 1 yields:

$$1/2(50) + (25) = 50 \tag{1.31}$$

which is exactly equal to the RHS value, and hence the first structural constraint is binding.

Substituting x^* and y^* into Constraint 2 yields:

$$4/5(50) + 5/2(25) = 102.5 \tag{1.32}$$

which is greater than the RHS value of 100, and hence the second structural constraint is nonbinding. This minimum requirement was overachieved by 2.5 units.

Plugging x^* and y^* into Constraint 3 yields:

$$(50) + (25) = 75 \tag{1.33}$$

which is exactly equal to the RHS value, and hence the third structural constraint is binding.

The optimal (minimum) value of the objective function is:

$$Z^* = 500x^* + 750y^*, \text{ or}$$

$$Z^* = 500(50) + 750(25) = \$43,750.$$

Simultaneous Equations Approach

As was true for the maximization problem, the simultaneous equations approach can be used to solve the minimization problem once the binding constraints have been determined. In this case, Constraints 1 and 3 are binding. Therefore, they can be written as equalities:

$$1/2x + y = 50 \text{ (Constraint 1)} \qquad (1.34)$$

$$x + y = 75 \text{ (Constraint 3)} \qquad (1.35)$$

Solving (1.35) for y yields:

$$y = 75 - x \qquad (1.36)$$

Substituting (1.36) into (1.34) yields:

$$1/2x + 75 - x = 50, \text{ or}$$

$$1/2x = 25, \text{ or}$$

$$x^* = 50 \qquad (1.37)$$

Substituting (1.37) into (1.36) determines y^*, that is:

$$y^* = 25 \qquad (1.38)$$

In this case, the simultaneous equation method has verified that the graphical solution is indeed correct. Again, always use this approach to verify the graphical solution.

The Standard Form of the Model for Maximization and Minimization Problems

Recall the first maximization problem solved previously:

Max: $Z = 40x + 45y$ $\qquad (0)$

s.t.:

$$x + \quad y \le 600 \qquad (1)$$

$$x + 1.5y \le 750 \qquad (2)$$

$$x \qquad\quad \le 400 \qquad (3)$$

$$x, \quad y \ge 0 \qquad (4)$$

Another way to formulate the model is to use **slack variables** to represent idle capacity. Slack variables represent the difference between how much of the resource is available and how much is used. There should be one slack variable added to each structural constraint in the problem, excluding the non-negativity constraint. Therefore, in this problem there are three slack variables. Also, since slack variables measure the unused amount of each resource endowment in the optimal solution, each of the structural constraints is now stated in terms of an equality (=) rather than a weak inequality. For example, consider Constraint 1 of the problem above. To represent the amount of unused resource 1, define slack variable 1 (s_1), which equals:

$$s_1 = 600 - x - y.$$

Rearranging this equation to put the constant on the RHS yields:

$$x + y + s_1 = 600.$$

To include a slack variable for a $\le$ type constraint add it to the constraint and replace the $\le$ restriction with an equality restriction.

Since slack variables do not contribute to the objective function value, they are included as activities in the objective function with zero objective function coefficients. The model with slack variables is written as the following:

Max: $Z = 40x + 45y + 0s_1 + 0s_2 + 0s_3$ $\qquad\qquad$ (0)

s.t.:

$$x + y + 1s_1 \qquad\qquad = 600 \qquad\qquad (1)$$
$$x + 1.5y + \qquad 1s_2 \qquad = 750 \qquad\qquad (2)$$
$$x + \qquad\qquad 1s_3 = 400 \qquad\qquad (3)$$
$$x, \quad y, \quad s_1, \quad s_2, \quad s_3 \geq 0 \qquad\qquad (4)$$

This is called the **standard form** of the LP model. The difference between the general and standard forms of an LP model is that the standard form includes slack variables (and/or "surplus" variables, which are discussed later in this section) and structural equality constraints, while the general form uses weak inequality constraints and does not include slack (or surplus) variables. At the optimal solution, the values of the slack variables are found by solving equations (1) through (3), given the values for x^* and y^*,

$$s_1^* = 0, \ s_2^* = 0, \text{ and } s_3^* = 100.$$

The slack variables can now be used to distinguish whether a constraint is binding or not. A binding constraint is one where its slack variable is zero. A nonbinding constraint is one where its slack variable is positive.

Another way to write the standard form of the model is using **tableau form** as shown below.

Equation	x	y	s_1	s_2	s_3	b
(0)	40	45	0	0	0	—
(1)	1	1	1	0	0	600
(2)	1	1.5	0	1	0	750
(3)	1	0	0	0	1	400

The activities are arranged as columns, and the last column is the resource endowment (b). The first row contains the objective function coefficients. The rest of the rows correspond to the constraints of the LP problem. Finally, the non-negativity constraint is not included in the tableau, but is assumed. Alternatively, all zero coefficients could be left as blanks.

To illustrate the standard form of the model for minimization problems, consider the previous LP problem written in general form.

Min: $Z = 500x + 750y$ $\qquad\qquad$ (0)

s.t.:

$$1/2x + y \geq 50 \qquad\qquad (1)$$
$$4/5x + 5/2y \geq 100 \qquad\qquad (2)$$
$$x + y \geq 75 \qquad\qquad (3)$$
$$x, \quad y \geq 0 \qquad\qquad (4)$$

One difference between this minimization problem and the previous maximization problem has to do with the **standard form** of the problem. In the previous problem where all constraints were of the $\leq$ type, the standard form incorporated **slack variables** in order to rewrite the weak inequality restriction as an equality. Slack variables are used whenever the constraints are $\leq$ for maximization or minimization problems, implying some sort of maximum restriction on resource use: for instance, you cannot use more than 24 hours of your own labor per day to study. Whenever $\geq$ constraints are encountered, **surplus variables** are used in the standard form of the model in order to convert the relation from a weak inequality to an equality. Surplus variables represent the amount by which a minimum condition is overachieved. Unlike slack variables, a surplus variable is subtracted off each constraint, and an equality is used in the standard model. Surplus variables carry a negative sign in the constraint equations in order to guarantee that non-negativity holds for each surplus variable.[4] When a surplus variable is zero in the optimal solution (the constraint is binding), this means that the minimum condition has been exactly achieved. When a surplus variable in the optimal solution is greater than zero (the constraint is nonbinding), this means that a minimum condition has been more than achieved.

The standard form of the general model above is:

Min: $Z = 500x + 750y + 0s_1 + 0s_2 + 0s_3$ (0)

s.t.:

$$\frac{1}{2}x + y - s_1 \qquad\qquad = 50 \tag{1}$$
$$\frac{4}{5}x + \frac{5}{2}y \qquad - s_2 \qquad = 100 \tag{2}$$
$$x + y \qquad\qquad - s_3 = 75 \tag{3}$$
$$x, \quad y, \quad s_1, \quad s_2, \quad s_3 \geq 0 \tag{4}$$

In tableau form, this problem is:

Equation	x	y	s_1	s_2	s_3	b
(0)	500	750	0	0	0	—
(1)	1/2	1	−1	0	0	50
(2)	4/5	5/2	0	−1	0	100
(3)	1	1	0	0	−1	75

1.7 SENSITIVITY ANALYSIS WITH THE GRAPHICAL APPROACH

So far in this book, we have discussed the formulation and graphical solution techniques for small LP problems. To the decision maker, this is probably the least important aspect of LP. The decision maker wants to know how the results of your hard work relate to the problem at hand. What do the LP results tell the decision maker to do? This phase of a decision problem is called **analysis**.

Some analysis has already been performed in the previous section. For example, the optimal solutions derived in the maximization and minimization problems provide the decision maker with optimal strategies, assuming that all parameters in the problem have been correctly specified and the model adequately depicts the decision process. Unfortunately, LP models are models, not reality. Hence, we need to verify that the results of such models give plausible answers. In addition, "what if" questions on key parameters of the model are frequently asked. If they are not asked, they should be asked. The "what if" questions fall into a category of analysis called **sensitivity analysis**.

---

[4]The reader should verify why slack variables carry a positive coefficient and surplus variables have a negative coefficient in order to preserve non-negativity.

Sensitivity analysis is the examination of how changes in the parameters of an LP model (c_i, b_i, and a_{ij}) affect the optimal solution: that is, how sensitive the optimal solution is to changes in these parameters. There are two general uses of sensitivity analysis in economics. First, it is used in LP models as a means to conduct "what if" analyses of the problem. If two additional hired workers are added to the family farm, then how would the optimal solution on how much of x, y, and Z change? If the government increased its pollution control standards on pollutant A by 50%, how would the objective of adopting the lowest-cost method of reducing air pollution using two alternative abatement methods change? Sensitivity analysis is used to answer these types of questions. Second, and more important to economists, sensitivity analysis is used to derive output supply (and input demand)[5] functions. This is accomplished by holding all parameters in the model except for the price of one of the outputs constant. By varying this price, the model solution traces out a price-quantity schedule, which is the definition of supply.

Two types of sensitivity analysis are presented in this section: objective function coefficient and resource endowment sensitivity analysis.[6] Each will be discussed separately in the context of a maximization problem.

Objective Function Coefficients (C_i) Sensitivity Analysis

Consider the following simple example of a feed dealer that wants to maximize total profit (Z) from the weekly sale of two types of feed. The dealer receives a profit of $50 for each ton of feed 1 (x) sold and $60 for each ton of feed 2 (y) sold. Labor and machinery are the only two resources needed to manufacture x and y. The feed dealer faces the following technical coefficients and resource endowments in the production of x and y.

Resource Requirement

Resource	x	y	Resource Endowment	Units
Labor	1.66	1.00	500	Hours
Machinery	1.00	1.33	400	Hours

The LP problem is:

(Objective Function)   Max: $Z = 50x + 60y$ (0)

s.t.:

(Labor Constraint) $1.66x + 1.00y \leq 500$ (1)

(Machinery Constraint) $1.00x + 1.33y \leq 400$ (2)

(Non-negativity) $x, \quad y \geq 0$ (3)

The graphical solution to this problem is shown in Figure 1.12. The optimal solution is:

$$x^* = 219.41, y^* = 135.78, \text{ and } Z^* = \$19,117.30 \text{ (weekly profit)}.$$

---

[5]The case of input demand functions will not be considered in this section, but will be dealt with later in the book.

[6]While this section considers sensitivity analysis for the objective function and resource endowment parameters, it should be noted that sensitivity analysis can also be conducted on technical coefficients (a_{ij}).

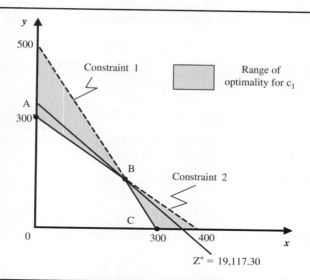

Figure 1.12 Optimal solution and range of optimality for the slope of the iso-contribution line.

Question: What range of values can the objective function coefficients on x and y be without changing the optimal solution (x^*, y^*)? The answer to this question is called the **range of optimality** for c_i.

As long as the slope of the iso-contribution line is between the slope of the two binding constraints, the optimal solution will always occur at extreme point B. This region is shaded in Figure 1.12. The region is equal to the area between the two binding constraints of the problem. Why is this the case? We can easily see graphically that if the iso-contribution line rotates around its point of tangency to B, then point B will remain optimal as long as the iso-contribution line remains within the shaded area. This is due to the fact that solution B still represents the highest achievable profit, given the structural constraints of the problem.

It is important to note that while the optimal activities will remain at $x^* = 219.41$, $y^* = 135.78$ if the slope of the iso-contribution line changes, but remains within this region, the value of the objective function at the optimal solution will change. Hence, the range of optimality for the objective function coefficients is defined by the activity values remaining unchanged, not the value of the objective function.

Why Calculate the Range of Optimality? If we know the range of optimality, then the problem does not have to be re-solved whenever the objective function coefficients are changed, as long as the iso-contribution line remains within this range. Also, knowledge of this range provides information as to how "sensitive" the model solution is with respect to parameters in the objective function, which is important as parameters are often estimates rather than exact measures.

Algebraic Solution

At extreme point B, both constraints are binding; hence they can be written as equalities:

$$1.66x + 1.00y = 500 \tag{1.39}$$

$$1.00x + 1.33y = 400 \tag{1.40}$$

Transforming (1.39) into slope-intercept form (solving for y) yields:

$$y = 500 - 1.66x \qquad (1.41)$$

In equation (1.41), 500 is the intercept, and -1.66 is the slope of the first constraint.
   Transforming (1.40) into slope-intercept form (solving for y) yields:

$$y = 300.75 - 0.75x \qquad (1.42)$$

In equation (1.42), 300.75 is the intercept, and -0.75 is the slope of the second constraint.
From (1.41) and (1.42), the following condition is derived:

Extreme point B will remain optimal if and only if:

$$-1.66 \leq \text{slope of iso-contribution line} \leq -0.75 \qquad (1.43)$$

Consider the following objective function in more general form:

$$\text{Max } Z = c_1x + c_2y.$$

Writing this in slope intercept form results in:

$$c_2y = Z - c_1x, \text{ or}$$
$$y = (1/c_2)Z - c_1/c_2x \qquad (1.44)$$

Thus, the slope of the iso-contribution line is $-c_1/c_2$.
   Substituting this into condition (1.43) yields:

$$-1.66 \leq -c_1/c_2 \leq -0.75, \text{ or multiply all sides by } -1:$$
$$0.75 \leq c_1/c_2 \leq 1.66 \qquad (1.45)$$

Condition (1.45) is the **range of optimality for c_1 and c_2**. To calculate the specific range
given the objective function coefficients, first hold c_2 at its initial level ($c_2 = 60$) and substitute into (1.45):

$$0.75 \leq c_1/60 \leq 1.66 \qquad (1.46)$$

Using the right-hand side of (1.46), we have

$$c_1/60 \leq 1.66, \text{ or}$$
$$c_1 \leq 99.6 \qquad (1.47)$$

Using the left-hand side of (1.46), we have

$$0.75 \leq c_1/60, \text{ or}$$
$$45 \leq c_1 \qquad (1.48)$$

Combining (1.47) and (1.48) yields the following condition for the range of optimality
for c_1:
   As long as $45 \leq c_1 \leq 99.6$, given $c_2 = 60$, the optimal solution will always occur at
extreme point B.
   What about the range for c_2? To answer this question, hold c_1 constant at its initial level,
that is, $c_1 = 50$ and substitute $c_1 = 50$ into (1.45), that is:

$$0.75 \leq 50/c_2 \leq 1.66 \qquad (1.49)$$

Using the right-hand side of (1.49), we have

$$50/c_2 \leq 1.66, \text{ or}$$

$$c_2 \geq 30.12 \tag{1.50}$$

Using the left-hand side of (1.49), we have

$$0.75 \leq 50/c_2, \text{ or}$$

$$c_2 \leq 66.67 \tag{1.51}$$

Combining (1.50) and (1.51) yields the following condition for the range of optimality for c_2:

As long as $30.12 \leq c_2 \leq 66.67$, given $c_1 = 50$, the optimal solution will always occur at extreme point B.

Deriving an Output Supply Curve

We can use this algebraic approach, which is called **parametric programming**, to generate this decision maker's supply function for either x or y. We will consider x. Strictly speaking, a supply schedule is a price-quantity relationship. However, the objective function coefficients for this problem are unit profits, not prices. For the sake of discussion, assume that these coefficients are now unit prices, not unit profits. This will not detract from the concepts that follow since this problem can easily be reformulated by disaggregating the objective function activities and coefficients in terms of marketing and production.

Holding the price for y constant at $60, we know from our analysis above that it is optimal for the feed dealer to sell 219.4 units of x when the price is between $45 and $99.60. In price-quantity space, this represents the dealer's supply schedule for the price range between $45 and $99.60 (see Figure 1.13).

What about a price below $45? If we resolved the problem using the same constraint set, but change the x objective function coefficient to $45, the new solution would occur at extreme point A in Figure 1.12 ($x^* = 0$, $y^* = 300$). Furthermore, parametric programming would reveal that as long as the price of x was within the range of minus infinity to $45, this solution would remain optimal. Thus, within the range of $0 to $45 it is optimal to supply 0 units of x.

Finally, if the price of x is above $99.60, the new solution would occur at extreme point C in Figure 1.12 ($x^* = 300$, $y^* = 0$). Based on these parametric programming results, the supply function for x is generated and presented in Figure 1.13. Figure 1.13 also shows the seller's supply function for y, which was derived in an analogous manner. It should be noted that these supply functions are for a single seller (in this example of a feed dealer) rather than for the entire market. To obtain the market supply function, we need to aggregate all individuals' supply functions within the market.

Note that the number of changes in optimal solutions is equal to the number of extreme points. This will always hold. Each optimal solution is called a **basis**. Each change in the optimal solution is called a **change in the basis**. We will discuss this more in the next chapter.

Resource Endowment (b_i) Sensitivity Analysis Sensitivity analysis can also be conducted on the resource endowments. If we change (increase or decrease) a resource endowment (holding constant all other parameters), what happens to the optimal solution? Resource endowment sensitivity analysis addresses this question.

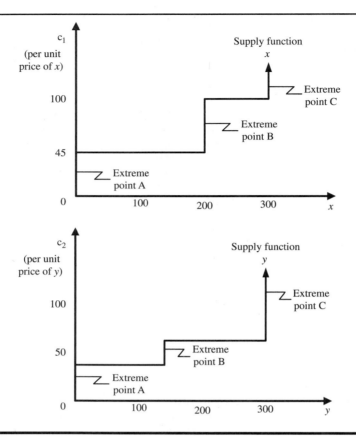

Figure 1.13   Supply functions for x and y.

We can demonstrate graphically this type of analysis. However, doing so can be very imprecise and can actually result in a wrong answer. This is due to the fact that one of the key results obtained from this type of analysis, "shadow prices," is defined by very small changes in resource endowments. The change in the feasible set due to very small changes in the b_i's is very difficult to see graphically, and hence will not be demonstrated here.

A **shadow price (SP)** gives the unit value of each resource to the objective of the problem. For example, the SP of labor shows how much one unit of labor increases profit (assuming the objective function is profit). Likewise, the SP of machinery gives how much one unit of machinery increases profit. Mathematically, an SP is defined as the change in the value of the objective function, given a one-unit change in the RHS parameter of a constraint, holding all other parameters constant.

$$\text{Shadow Price } b_1 \ (SP_{b1}) = (Z^{*\prime} - Z^{*}) / (b_1^{\prime} - b_1),$$

where:

$Z^{*\prime}$ = optimal value of the new solution when b_1 is changed by one unit

Z^{*} = optimal value of the original solution

$b_1^{\prime}$ = new level of resource 1 after the change

b_1 = original level of resource 1

What is the SP for a nonbinding constraint? It is zero because an increase in the resource endowment for such a constraint will only affect the amount of the slack variable and add nothing to the objective function since a slack variable's objective function coefficient is zero.

An Algebraic Way of Calculating Shadow Prices   In the optimal solution, both constraints are binding. These two binding constraints can be used with the simultaneous equation method to calculate SPs. First consider the SP for the labor constraint (SP_{b1}).

Increase b_1 one unit, for instance, from 500 to 501.[7] Now there is one more hour of labor. Then solve for x^*, y^*, and Z^* using the two binding constraints. The new constraint 1 expressed as an equality is:

$$1.66x + 1.00y = 501 \tag{1.52}$$

Constraint 2 expressed as an equality constraint is:

$$1.00x + 1.33y = 400 \tag{1.53}$$

Solve (1.52) for y,

$$y = 501 - 1.66x \tag{1.54}$$

Substitute (1.54) into (1.53) and solve for x:

$$1.00x + 1.33(501 - 1.66x) = 400, \text{ or}$$

$$x^* = 220.51 \tag{1.55}$$

Substitute (1.55) into (1.54) to determine y^*:

$$y = 501 - 1.66(220.51), \text{ or}$$

$$y^* = 134.95 \tag{1.56}$$

Substitute the new optimal values for the decision variables into Z to obtain the new Z^* (which will be denoted as $Z^{*\prime}$):

$$Z^{*\prime} = 50(220.51) + 60(134.95) = 19,122.50 \tag{1.57}$$

Recall that the original $Z^* = 19,117.30$. The formula for the SP is:

$$SP_{b1} = (Z^{*\prime} - Z^*)/ (b_1' - b_1), \text{ or}$$

$$SP_{b1} = (Z^{*\prime} - Z^*),$$

since $b_1' - b_1 = 1$.

Hence, in this example, the SP for labor is:

$$SP_{b1} = 19,122.50 - 19,117.30 = 5.20 \text{ per hour.}$$

If you added one more hour of labor, total profit would increase by \$5.20. The value of labor to the firm is worth \$5.20 per hour, given its current level of resource endowments.

---

[7]In this example the resource endowments were changed by increasing them by one unit in calculating SP. One could also compute SP by decreasing the b_i by one unit.

Next consider the SP for the machinery constraint (SP$_{b2}$). As before, first increase $b_2 = 400$ to $b'_2 = 401$. Then using the simultaneous equation approach Constraint 1 expressed as an equality is:

$$1.66x + 1.00y = 500 \tag{1.58}$$

The new Constraint 2 expressed as an equality constraint is:

$$1.00x + 1.33y = 401 \tag{1.59}$$

Solve (1.58) for y,

$$y = 500 - 1.66x \tag{1.60}$$

Substitute (1.60) into (1.59) and solve for x:

$$1.00x + 1.33(500 - 1.66x) = 401, \text{ or}$$

$$x^* = 218.58 \tag{1.61}$$

Substitute (1.61) into (1.60) to determine y^*: $y = 500 - 1.66(218.58)$, or

$$y^* = 137.16 \tag{1.62}$$

Substitute the new optimal values for the decision variables into Z to obtain the new $Z^{*\prime}$:

$$Z^{*\prime} = 50(218.58) + 60(137.16) = 19{,}158.43 \tag{1.63}$$

Recall that the original $Z^* = 19{,}117.30$. Hence, in this example, the SP for machinery is:

$$SP_{b2} = 19{,}158.43 - 19{,}117.30 = 41.13 \text{ per hour.}$$

If you added one more hour of machinery time, total profit would increase by \$41.13. The value of machinery to the firm is worth \$41.13 per hour, given its current level of resource endowments.

Sensitivity Analysis and Minimization Problems

Consider the minimization problem discussed earlier. Recall the LP problem was to:

Min: $Z = 500x + 750y$ $\hspace{8cm}$ (0)

s.t.:

$$1/2x + \quad y \geq 50 \tag{1}$$

$$4/5x + 5/2y \geq 100 \tag{2}$$

$$x + \quad y \geq 75 \tag{3}$$

$$x, \quad y \geq 0 \tag{4}$$

where:

x = usage of filters to reduce pollution (units = filters)

y = usage of cleansing additive to reduce pollution (units = additive)

The optimal solution is:

$$x^* = 50, y^* = 25, \text{ and } Z^* = \$43{,}750.$$

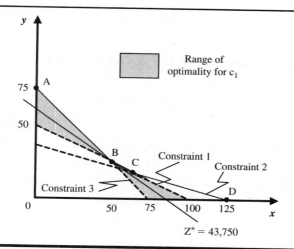

Figure 1.14   Optimal solution and range of optimality for the slope of the iso-cost line.

Figure 1.14 shows this optimal solution.

Question: What range of values can the cost of x and y be without changing the optimal solution? That is, what is the **range of optimality** for c_i? In Figure 1.14, the optimal solution is at extreme point B. As long as the slope of the iso-cost line is between the slope of constraint 1 and the slope of constraint 3 the optimal solution will always occur at extreme point B. This region is shaded in Figure 1.14. While the optimal decision variables will remain at $x^* = 50$ and $y^* = 25$ if the slope of Z changes but remains within this region, the value of the objective function will change.

Algebraic Solution

Note that at extreme point B, constraints 1 and 3 are binding. Solving constraint 1 for y yields:

$1/2x + y = 50$ (since this constraint is binding, use an equality relation), or

$$y = 50 - 1/2x \qquad (1.64)$$

In equation (1.64), 50 is the intercept of constraint 1 and $-1/2$ is its slope. Solving constraint 3 for y yields: $x + y = 75$ (since this constraint is binding, use an equality relation), or

$$y = 75 - x \qquad (1.65)$$

In equation (1.65), 75 is the intercept of constraint 3, and -1 is its slope. From equations (1.64) and (1.65), the following condition is derived:
   Extreme point B will remain optimal if and only if:

$$-1 \leq \text{slope of iso-cost line} \leq -1/2 \qquad (1.66)$$

Consider the objective function in more general form as:

$$\text{Min: } Z = c_1 x + c_2 y.$$

Writing this in slope-intercept form we have:

$$y = (1/c_2)Z - (c_1/c_2)x \tag{1.67}$$

Thus, the general slope of the iso-cost line will be $-c_1/c_2$. Substituting into condition (1.66) yields:

$$-1 \le -c_1/c_2 \le -1/2, \text{ or multiply by } -1:$$

$$1/2 \le c_1/c_2 \le 1 \tag{1.68}$$

To calculate the range of optimality, first hold the cost of cleansing additive (y) at its initial level, that is $c_2 = \$750$. Then we can calculate the range for c_1. Substituting $c_2 = 750$ into (1.68) yields:

$$1/2 \le c_1/750 \le 1 \tag{1.69}$$

Using the left-hand side of inequality (1.69):

$$c_1/750 \ge 1/2, \text{ or}$$

$$c_1 \ge 375 \tag{1.70}$$

Using the right-hand side of inequality (1.69):

$$c_1/750 \le 1, \text{ or}$$

$$c_1 \le 750 \tag{1.71}$$

Combining algebraic results (1.70) and (1.71) gives the following condition for c_1 for the range of optimality:

$$375 \le c_1 \le 750 \tag{1.72}$$

As long as the cost of x is between \$375 and \$750, given that the cost of y is \$750, the optimal solution will always occur at extreme point B.

Verify that repeating this procedure yields the similar condition for c_2 that:

$$500 \le c_2 \le 1,000 \tag{1.73}$$

Right-Hand-Side-Value (b_i)

Suppose the state changes its requirement on pollutant C from reducing emission by 75 units to reducing pollution by 74 units:

$$x + y \ge 75 \text{ (old constraint)},$$

$$x + y \ge 74 \text{ (new constraint)}.$$

By decreasing this requirement by one unit, the size of the feasibility region has been increased. SPs are calculated the same way as in maximization problems, except that decreasing the b_i's for the $\ge$ constraints yields a negative number. The negative SP means that if you reduce the RHS value by one unit, the objective function value will decrease by k units.

For example, to compute the SP for the third minimum condition (Constraint 3), substitute 74 for 75 into Constraint 3, and solve the two binding constraints (i.e., Constraint 1 and the new Constraint 3) simultaneously for $Z^{*'}$.

$$1/2x + y = 50 \quad \text{(Constraint 1)} \tag{1.74}$$

$$x + y = 74 \quad \text{(New Constraint 3)} \tag{1.75}$$

Solving (1.75) for y yields:

$$y = 74 - x \tag{1.76}$$

Substituting (1.76) into (1.74) yields the new x^*:

$$1/2x + (74 - x) = 50, \text{ or } 1/2x = 24, \text{ or}$$

$$x^* = 48 \tag{1.77}$$

Plugging (1.77) into (1.76) yields the new y^*:

$$y^* = 74 - 48 = 26 \tag{1.78}$$

Hence, the new solution is $(x^*, y^*, Z^*) = (48, 26, 43{,}500)$. The old solution was

$$(x^*, y^*, Z^*) = (50, 25, 43{,}750).$$

Using this information, the SP for the third minimum condition is:

$$SP_{b3} = 43{,}500 - 43{,}750 = -250 \tag{1.79}$$

If the minimum condition reflected by the third constraint is relaxed by one unit, this would decrease total pollution abatement costs to the firm by $250. Verify that the SPs for Constraints 1 and 2 of this problem are:

$$SP_{b1} = -500,$$

$$SP_{b2} = 0.$$

Interpret what these SPs mean in the context of this example on your own.

SUMMARY

The goals of this chapter were to (1) provide a general overview of what the components of an LP model are, (2) describe how to set up simple problems as LP problems, (3) show how to solve simple two-activity LP problems, and (4) introduce the notion of sensitivity analysis in LP. It was argued that the setting up of LP problems is more of an art than a science. The examples presented in the text will not be enough to make the reader an expert. This takes a lot of practice. Therefore, it is advised that you work through the section problems presented at the end of this chapter. The practice will help you become more familiar and comfortable with the topics covered in this chapter.

The reader should now be able to set up simple LP maximization and minimization models and solve them using the graphical approach. To recap, once the LP problem has been set up, the way to find the optimal solution is to perform the following steps:

1. Graph each constraint line by finding the two intercepts, drawing a line to connect them, and shading in the feasible region for each constraint.

2. Superimpose all constraints onto one graph and find the complete feasible region for the problem. The feasible region is equal to the intersection of all points that satisfy all the constraints.

3. Find the iso-contribution or iso-cost line by setting the objective function equal to an arbitrary numeric level, Z_a, and convert it to its slope-intercept form.

4. Plot the iso-contribution (or iso-cost) line for various levels of Z_a onto the feasible region graph. If for any value of Z_a more than one solution lies on the line, increase

(or decrease for minimization problems) the level of Z_a. The optimal solution is found by moving the iso-contribution line out to the right (or to the left for minimization problems) as far as possible until only one point on the line touches (is tangent to) a point on the feasible region. This holds for all cases except multiple optimal solutions.

5. Verify that your graphical solution is correct by using the simultaneous equations approach. This is done by solving for x and y using the binding structural constraints written as equations.

Three special cases were also presented. In the case of an **unbounded feasible region**, there is no finite solution. The **no feasible solution** case exists whenever the feasible region is an empty set. Finally, **multiple optimal solutions** occur whenever the slope of the iso-contribution line is the same as one of the line segments connecting two extreme points in the feasible region.

Once an optimal solution is found, it is generally recommended that sensitivity analysis be performed. Sensitivity analysis is the examination of how changes in the parameters of a LP model (c_i and b_i) affect the optimal solution. That is, how sensitive is the optimal solution to changes in these parameters? It was argued that since LP models are models not reality, the results of such models need to be verified to ensure that they give plausible answers. It is essential that you, as LP solvers ask and answer "what if" questions regarding key parameters of the model.

Two general uses of sensitivity analysis common in economics were discussed: (1) "what if" analyses of the problem and (2) derivation of output supply functions (this chapter illustrated supply not demand, which will be covered in a later section). In addition, two types of sensitivity analyses were examined: parametric programming on objective function coefficients and parametric programming on resource endowments. These concepts will be extremely useful later on in the book.

Unfortunately, two-variable LP problems are very unrealistic and are used only for teaching purposes. The real power of LP is found in much larger and more realistic applications. In order to solve larger LP problems, we need an alternative to the graphical approach. Fortunately, there exists an alternative approach called the simplex method, which we will begin to discuss in the next chapter.

EXERCISES

1. Set up the general form of the LP model for the following word problem.

   A cash grain farmer in Central Iowa has 600 acres of cropland available on which she plans to grow corn and soybeans in the spring of 2011. She has made some budgets, which take into account corn (x) and soybeans (y). The gross margin for corn is $40 per acre and for soybeans is $45 per acre. She has a maximum of 750 hours of tractor time available in the last half of May at the peak planting periods for both crops. It takes 1 hour per acre for field operations for corn (x) and 1.5 hours per acre for soybeans (y). The maximum acreage she can use for corn is 400 acres. Her sole objective is to select a cropping plan that will maximize net returns for this set of conditions in 2011.

2. Set up the general form of the LP model for the following word problem.

   A new pollution control law has been passed by the state legislature requiring manufacturers to reduce pollution by 20%. A local industrialist manufactures copper, which results in the emission of three pollutants into the air: A, B, and C. Under the state law, the industrialist is required to reduce A by at least 50 units, B by at least 100 units, and C by at least 75 units. There are two pollution abatement methods available to the industrialist: (1) to use filters (x) or (2) to use cleansing additive for fuel (y). For each filter that is used, the emission of the three pollutants can be reduced by 1/2, 4/5, and 1 units, respectively, for A, B, and C. For each ton of cleansing additive added to the fuel, the three pollutants can be reduced by 1, 5/2, and 1 units, respectively, for A, B, and C. The cost of one filter is $500, and the cleansing additive costs $750 per ton. Based on this information, the industrialist's objective is to adopt the lowest-cost method of reducing pollution according to the law.

3. Set up this decision problem as an LP model in general form.

You have been given $100,000 by a client to invest in the stock market. After some preliminary analysis, you find that there are five stocks that you want to consider for your client's investment portfolio. The price per share, expected annual rate of return, and risk index for each of the five stocks are summarized in the table below:

Stock	Price per Share ($)	Expected Annual Return ($)	Risk Index per Share ($)
Monsanto	85	15	0.15
Dean Foods	20	5	0.06
Kraft Foods	23	8	0.09
General Mills	52	10	0.12
Whole Foods	20	6	0.05

The risk index for each stock is your client's opinion of the riskiness of each investment. Your client has put a limit of 200 as the total amount of risk she will bear for the entire portfolio (total risk = risk index per share × number of shares). In addition, she does not want to invest in more than 500 shares of Monsanto stock. Finally, you cannot invest more than the $100,000 that your client has given to you. Your objective is to maximize the expected annual return for your client's portfolio. Assume that your commission does not figure into this exercise.

4. A steel factory that uses coal as its major source of energy causes three primary types of air pollution by releasing (1) particulate matter, (2) sulfur oxides, and (3) hydrocarbons. These three types of air pollution are caused by blast furnaces and open-hearth furnaces used in producing steel. The state has just passed a new clean air bill, which means that this factory must reduce its annual emission rate of particulates by 60 million pounds, sulfur oxides by 150 million pounds, and hydrocarbons by 125 million pounds. There are six pollution-abatement techniques (three for each type of furnace) that the factory can use to reduce air pollution. These six techniques, along with their per unit reduction for each pollutant, and their estimated annual cost per unit are listed below:

	Taller Smokestacks		Filters		Better fuels	
Pollutant	Blast Furnace	Open-Hearth Furnace	Blast Furnace	Open-Hearth Furnace	Blast Furnace	Open-Hearth Furnace
	(per unit reduction in million pounds)					
Particulate	12	9	25	20	17	13
Sulfur Oxides	35	42	18	31	56	49
Hydrocarbons	37	53	28	24	29	20
Cost/Unit	$80,000	$100,000	$70,000	$60,000	$110,000	$90,000

Assume that the factory's sole objective is to minimize the total cost of reducing emissions of these three pollutants to the new government standards by using any combination of the six pollution abatement techniques. Set this up as an LP problem.

5. A food firm is researching the profitability of introducing six new "healthy choice" food products (call them x, y, z, a, b, c). The firm currently has idle resource capacity on labor, machinery, and land of 100 hours, 300 hours, and 30,000 square feet, respectively. Hence, producing any or all of the new products will help solve the costs of excess capacity. The selling prices, total costs, and resource requirements for the production technology are summarized below.

| | New Product | | | | | |
Resource (Unit)	x	y	z	a	b	c
Labor (hours)	0.50	0.10	1.00	0.45	0.20	0.15
Machinery (hours)	1.00	0.45	3.50	1.00	1.10	2.00
Land (sq ft)	100	200	50	25	10	75
Unit Costs ($)	10	3	33	22	12	9
Unit Price ($)	12	4	36	23	15	11

Food products y and z are complements in the sense that for every unit of y produced and sold, 2 units of z must be produced and sold. Also, the firm requires that the amount of product c produced and sold be at least 50% of the total units of products a and b that are produced and sold. Set up an LP model that will result in a solution that maximizes total profit from the sale of any combination of these food products, subject to all constraints that were specified.

6. Now suppose that the firm in Exercise 5 can hire additional labor at $8 per hour. Reformulate Exercise 5 to allow for the firm to hire up to an additional 500 hours of labor. Note that now hired labor should be modeled as an activity in Exercise 5.

7. A farmer owns 500 acres of land, which is suitable for growing corn, soybeans, and sunflowers. His expectations are that the net profit from producing each crop is $55 per acre for corn, $60 per acre for soybeans, and $50 per acre for sunflowers. He and his family can supply 3,000 hours per year in performing all the farm operations necessary to grow these crops. In addition, he is endowed with the equivalent of 4,500 hours of tractor time necessary to grow these crops. Assume that the only resources necessary in crop production are land, labor, and tractor time.

| | Crop | | | |
Resource (Unit)	Corn	Soybeans	Sunflowers	Endowment
Land (acres)	1.0	1.0	1.0	500
Labor (hours)	0.4	0.2	0.3	3,000
Tractor (hours)	0.5	0.2	0.4	4,500

Assuming that the farmer's objective is to maximize total profits, what is the LP model for this exercise?

8. Now suppose that the farmer in Exercise 7 can rent an additional 100 acres of land at a cost of $15 per acre. Reformulate Exercise 7 to allow the farmer to rent up to 100

more acres of land. Note that now renting additional land should be modeled as an activity in Exercise 7.

9. An ice cream maker has hired you to help him decide next month's production schedule. He needs to determine the quantities of each flavor that should be produced based on the profitability of each flavor and several restrictions. He can produce six different flavors of ice cream: (1) super-super premium mocha chip, (2) super premium chocolate chocolate chip, (3) super premium Snickers bar crunch, (4) vanilla, (5) chocolate ice milk, and (6) Yuppie's Delight frozen yogurt. Each product is only available in quarts and has the following unit profits for the ice cream maker:

Product	Unit Profit ($/quart)
Super-super premium mocha chip	1.00
Super premium chocolate chocolate chip	0.75
Super premium Snickers bar crunch	0.88
Vanilla	0.43
Chocolate ice milk	0.50
Yuppie's Delight frozen yogurt	1.05

The total production capacity of the ice cream maker's plant is 10,000 gallons per month. He also knows that he can only sell 1,000 gallons of super-super premium mocha chip, and he must produce at least 2,500 gallons of chocolate ice milk for the local school district. Finally, because he is introducing Yuppie's Delight frozen yogurt and doesn't yet know the market for this product, he only wants to produce 500 gallons in the next month. Assuming he wishes to maximize profit and given these restrictions, formulate this decision problem as an LP model in general form.

10. A farmer has the following resource endowments: 1,000 acres of land, 1,500 hours of family labor, and $30,000 of capital investment. She can use these resources to grow the following crops: corn, sorghum, wheat, and soybeans. The farmer expects the following in terms of crop yields, prices, variable costs, and labor requirements.

Crop	Price ($/bushel)	Yield (bushel/acre)	Variable Cost ($/acre)	Labor Requirement (hours/acre)
Corn	2.75	120	250	3.25
Sorghum	2.65	100	200	3.00
Wheat	3.15	105	245	3.15
Soybeans	6.75	45	230	3.30

Also, the farmer can invest any part of her $30,000 to rent additional land at $100 per acre and hire additional labor at $6 per hour.

Assume that the farmer works to maximize net revenue (gross revenue minus variable costs) from the production of these four crops. Formulate this as an LP.

11. A college student on a tight budget wishes to plan a diet which will minimize his food expenditure while maintaining minimum nutritional requirements, according to the Recommended Daily Allowance (RDA). The student wants to plan his menu from

the following goods: hamburgers, hot dogs, salad, chicken, pizza, carrots, and cookies. The dietary information (in mg per pound) and cost of each of these foods is as follows:

Nutrient	Hamburger	Hotdog	Salad	Chicken	Pizza	Carrots	Cookies	RDA
Calories	2200	2100	500	700	2500	300	2600	2500
Calcium	100	200	400	300	475	400	150	80 mg
Protein	50	70	20	45	35	25	10	25 mg
Iron	25	15	30	10	5	15	20	15 mg
Cost/lb	2.50	2.00	1.75	3.00	5.00	2.25	3.50	

Furthermore, assume that the student wants to eat at least 0.25 pounds of cookies each day and will eat at most 0.50 pounds of carrots per day. Formulate an LP model that minimizes daily food expenditures while meeting the RDA and the other constraints given in this problem.

12. You have just been hired as an advertising manager for a generic advertising program for dairy farmers, Dairy Management, Inc. (DMI). DMI wants to conduct generic advertising to increase the demand for milk. DMI decided to consider both TV and radio, and wants you to do an analysis of how many TV and radio commercials to purchase for the month. You expect that one TV commercial will increase sales by 25,000 gallons, and one radio commercial will increase sales by 7,000 gallons. It costs $10,000 per TV commercial and $5,000 per radio commercial. Your boss tells you that you can't spend more than $200,000 on this project. Furthermore, the radio and TV stations tell you they have a combined maximum of 90 minutes for your commercials for the month. Each TV commercial takes 1 minute and each radio commercial takes 2 minutes to air. The boss tells you that he doesn't want more than 15 TV commercials because he gets sick of watching the same thing over and over again. The objective is to find the combination of TV (x) and radio (y) commercials that maximize the sale of milk.

   a. Set up this problem as an LP model.

   b. Write this problem in standard form (using slack variables) and in general form without slack variables.

   c. Graph the feasible region for this problem.

   d. Find the optimal solution for this problem.

13. Consider the following problem:

   Max: $Z = 15x + 10y$ (0)

   s.t.:

$$1x + 3/5y \leq 300 \tag{1}$$

$$1x + 1y \leq 400 \tag{2}$$

$$1x \quad\quad \leq 200 \tag{3}$$

$$x, \quad y \geq 0 \tag{4}$$

   Solve this problem using the graphical approach and the simultaneous equations approach to verify the graphical solution.

14. Write the following problem in standard form:

$$\text{Max: } Z = 15x + 10y \tag{0}$$

s.t.:

$$1x + 3/5y \leq 300 \tag{1}$$
$$1x + 1y \leq 400 \tag{2}$$
$$1x \qquad \leq 200 \tag{3}$$
$$x, \qquad y \geq 0 \tag{4}$$

Solve this problem using the graphical technique. Then derive the solution using the simultaneous equations approach. Show all your work.

15. What is the feasible region for the following problem?

$$\text{Max: } Z = 5x + 7y \tag{0}$$

s.t.:

$$x + y \leq 100 \tag{1}$$
$$5x + 5y \geq 500 \tag{2}$$
$$x, \qquad y \geq 0 \tag{3}$$

16. What is wrong with the following LP model? Explain.

$$\text{Max: } 5x \tag{0}$$

s.t.: $x < 100$ \hfill (1)

$\qquad x \geq 0$ \hfill (2)

17. For the following problem in general form, show graphically which constraints are binding and which are nonbinding. Clearly label your constraints.

$$\text{Max: } Z = 3x + 5y \tag{0}$$

s.t.:

$$x + y \leq 200 \tag{1}$$
$$0.25x + y \leq 100 \tag{2}$$
$$y \leq 50 \tag{3}$$
$$x, \qquad y \geq 0 \tag{4}$$

18. Write the following problem in standard form.

$$\text{Max: } Z = 35x + 15y \tag{0}$$

s.t.:

$$2x + 1/2y \leq 300 \tag{1}$$
$$1x + 1y \leq 500 \tag{2}$$
$$1y \leq 100 \tag{3}$$
$$x, \qquad y \geq 0 \tag{4}$$

Solve this problem using the graphical technique. Then derive the solution using the simultaneous equations approach. Show all your work.

19. A car manufacturing company produces an SUV (x) and a sedan (y). Long-term projections indicate an expected demand of at least 100 SUVs and 80 sedans each day. Because of limitations on production capacity, no more than 200 SUVs and 170 sedans can be made daily. To satisfy a shipping contract, a total of at least 200 cars must be shipped each day. If each SUV sold results in a $2,000 loss but each sedan produces a $5,000 profit, how many of each type should be made daily to maximize net profits?

20. Solve the following maximization problem:

Max: $Z = 10x + 12y$ (0)

s.t.:

$$x + \quad y \le 500 \tag{1}$$

$$x \qquad \le 250 \tag{2}$$

$$x - \quad y = 0 \tag{3}$$

$$x, \quad y \ge 0 \tag{4}$$

21. Consider the following LP problem:

Min: $Z = x + \quad y$ (0)

s.t.:

$$3.5x + \quad y \le 7 \tag{1}$$

$$-0.5x + \quad y \ge 1 \tag{2}$$

$$-8x + 10y \le 40 \tag{3}$$

$$x, \quad y \ge 0 \tag{4}$$

  a. Write this problem in standard form using slack and surplus variables.
  b. Graph the feasible region for this problem.
  c. Find the optimal solution for this problem.
  d. Use the simultaneous equations method to double check your graphical solution.

22. Write the following problem in standard form.

Min: $Z = 4x + 3y$ (0)

s.t.:

$$2x + \quad y \ge 10 \tag{1}$$

$$x + \quad y \ge 6 \tag{2}$$

$$x, \quad y \ge 0 \tag{3}$$

Solve this problem using the graphical technique. Also, derive the solution using the simultaneous equations approach. Show all your work.

23. Write the following problem in standard form.

Min: $Z = 20x + 15y$ (0)

s.t.:

$$x + y \geq 100 \tag{1}$$
$$3x + 2y \geq 250 \tag{2}$$
$$y \leq 90 \tag{3}$$
$$x, \quad y \geq 0 \tag{4}$$

24. Write the following problem in general form:

Max: $Z = 100x + 25y + 0s_1 + 0s_2 + 0s_3$ (0)

s.t.:

$$x + y + 1s_1 \qquad\qquad = 100 \tag{1}$$
$$y \quad - s_2 \qquad = 25 \tag{2}$$
$$x \qquad\quad - s_3 = 25 \tag{3}$$
$$x, \quad y, \quad s_1, \quad s_2, \quad s_3 \geq 0 \tag{4}$$

25. Solve the following LP problem with the graphical approach.

Max $Z = 3x + 3y$ (0)

s.t.:

$$4x + 2y \leq 70 \tag{1}$$
$$3x + 4y \leq 90 \tag{2}$$
$$x \qquad \leq 20 \tag{3}$$
$$x, \quad y \geq 0 \tag{4}$$

26. A small-scale poultry industry grows broilers, layers, and turkeys, and sells them at a profit of \$4, \$5, and \$6 respectively. The house is divided into three chambers separated by wooden bars to house the three kinds of birds. The house can accommodate no more than 45 birds. The labor time required for broilers and layers is 3 hours each. The turkeys require 4 hours of labor time. The house can grow a maximum of 20 broiler birds, and a maximum of 100 hours of labor are available. Formulate this problem as an LP model to maximize the total profit.

27. Solve the following problem using the graphical approach:

Min: $Z = x + 2y$ (0)

s.t.:

$$x + y \leq 100 \tag{1}$$
$$y \geq 45 \tag{2}$$
$$x - y = 0 \tag{3}$$
$$x, \quad y \geq 0 \tag{4}$$

28. Consider the following LP problem.

Max: $Z = 3x + 2.5y$ (0)

s.t.:

$$25/18x + 25/6y \leq 100 \tag{1}$$
$$2x + 2y \leq 60 \tag{2}$$
$$4x + 2y \leq 96 \tag{3}$$
$$x, \quad y \geq 0 \tag{4}$$

   a. Solve this problem using the graphical approach.

   b. Compute algebraically the optimal range for the objective function coefficient for x, that is, c_1.

   c. Compute algebraically the optimal range for the objective function coefficient for y, that is, c_2.

   d. Compute algebraically the SP for the resource endowment in constraint (1).

   e. Compute algebraically the SP for the resource endowment in constraint (2).

   f. Compute algebraically the SP for the resource endowment in constraint (3).

   g. Give one value for c_2 that would cause the optimal solution to contain multiple optimal solutions.

29. Consider the following LP problem.

Min: $Z = 100x + 100y$ (0)

s.t.:

$$x + 2y \geq 70 \tag{1}$$
$$20x + 10y \geq 500 \tag{2}$$
$$x + 55/9y \geq 110 \tag{3}$$
$$x + y \leq 160 \tag{4}$$
$$x, \quad y \geq 0 \tag{5}$$

   a. Solve this problem using the graphical approach.

   b. Compute algebraically the optimal range for the objective function coefficient for x, that is, c_1.

   c. Compute algebraically the optimal range for the objective function coefficient for y, that is, c_2.

   d. Compute algebraically the SP for constraint (1).

   e. Compute algebraically the SP for constraint (2).

   f. Compute algebraically the SP for constraint (3).

   g. Compute algebraically the SP for constraint (4).

   h. Give one value for c_2 that would cause this new optimal solution to contain multiple optimal solutions.

30. Solve the following LP problem graphically:

$$\text{Max: } Z = 3x + 3y \tag{0}$$

s.t.:

$$4x + 2y \le 70 \tag{1}$$

$$3x + 4y \le 90 \tag{2}$$

$$x \quad \le 20 \tag{3}$$

$$x, \quad y \ge 0 \tag{4}$$

31. A small Mexican food restaurant is open from 11:00 A.M. to 10:00 P.M. on weekdays. There are only two full-time employees, the chef and the owner. The waiters and waitresses are part-time, scheduled for 4-hour shifts. Due to variance in the arrival of customers throughout the day, the total number of full-time and part-time employees required and the wage-rate for part-time employees varies with the time of the day as follows:

Hour	Number of Employees Required	Wage-Rate for Part-Time Employees ($/hour)
11:00 A.M.–1:00 P.M.	6	8
1:00 P.M.–4:00 P.M.	4	9
4:00 P.M.–6:00 P.M.	5	9
6:00 P.M.–9:00 P.M.	10	10
9:00 P.M.–10:00 P.M.	8	8

The owner of the restaurant arrives at 11:00 A.M., works two hours, takes one hour off, and returns for three hours, takes another hour off and then works till 10:00 P.M. when the restaurant is closed. The chef also arrives at 12:00 A.M., works four hours, takes two hours off, and returns for another four hours. Develop a minimum-cost Monday schedule for the part-time employees. Set up the LP model in general form.

32. Consider the following LP model:

$$\text{Max: } Z = 15x + 10y \tag{0}$$

s.t.:

$$1x + 3/5y \le \quad 300 \tag{1}$$

$$1x + 1y \le \quad 400 \tag{2}$$

$$-1x \quad \ge -200 \tag{3}$$

$$x, \quad y \ge \quad 0 \tag{4}$$

a. Solve this problem graphically. Label all lines drawn on the graph and the axes.

b. What is the optimal solution to this problem?

c. Compute the range of optimality for c_1.

    d. Compute the range of optimality for c_2.

    e. What is the SP for constraint (1)? What is the economic interpretation of this number? Show all your work.

    f. What is the SP for constraint (2)? What is the economic interpretation of this number? Show all your work.

    g. What is the SP for constraint (3)? What is the economic interpretation of this number? Show all your work.

33. Consider the following minimization problem:

Min: $Z = 4x + 3y$ (0)

s.t.:

$$2x + 1y \geq 10 \qquad (1)$$
$$1x + 1y \geq 6 \qquad (2)$$
$$x, \quad y \geq 0 \qquad (3)$$

    a. Solve this problem using the graphical approach.

    b. Compute the range of optimality for c_1. How do you interpret this range?

    c. Compute the range of optimality for c_2. How do you interpret this range?

    d. What is the SP for constraint (1)? (Show your work.)

    e. What is the SP for constraint (2)? (Show your work.)

2

The Simplex Method to Solving Linear Programming Problems

Obviously most, if not all, real-world applications of linear programming (LP) involve more than two activities. Linear programming problems with thousands of activities and constraints are common. Hence, the graphical approach cannot be relied upon to solve realistic problems.

Instead, we rely on computers, which solve LP models using the **simplex method** (or **modified simplex method**).[1] The simplex method is an algebraic method, which systematically finds an optimal solution to the LP problem using iterative procedures. It is an iterative procedure because the simplex method uses basic steps that are repeated over and over again until an optimal solution is found by certain criteria. This chapter focuses on solving LP models using this technique. Understanding the simplex method provides an excellent basis for comprehending the logic behind many computer LP solvers, which are capable of solving large problems.

There are three objectives of this chapter. First, an overview is provided for solving simple maximization problems (with ≤ constraints) using the simplex technique. While a two-activity problem is used to illustrate the important concepts, all results are generalizable to more than two-activity applications. The second objective is to demonstrate how to use the simplex method for general maximization problems that include ≤, ≥, and = type constraints. The notion of "artificial variables," which are required to solve problems with ≥ and = constraints, is presented. Finally, the chapter concludes with a discussion of how to solve minimization problems with the simplex method. The fundamental goal of this chapter is to provide students with a sufficient knowledge of the simplex method to understand how computers solve LP problems and what the computer output means.

---

[1] There are other solution techniques for LP, but this chapter focuses on the simplex method.

2.1   THE SIMPLEX METHOD FOR A SIMPLE MAXIMIZATION PROBLEM

Consider the following maximization problem expressed in standard form with slack variables and equality constraints:

$$\text{Max: } Z = 35x_1 + 50x_2 + 0s_1 + 0s_2 + 0s_3 \tag{0}$$

s.t.:

$$x_1 + x_2 + s_1 \qquad\qquad = 1{,}000 \tag{1}$$

$$2.5x_1 + 0.75x_2 \qquad + s_2 \qquad = 1{,}500 \tag{2}$$

$$1.5x_2 \qquad\qquad + s_3 = 800 \tag{3}$$

$$x_1, \qquad x_2, \quad s_1, \quad s_2, \quad s_3 \geq 0 \tag{4}$$

Constraints (1) through (3) form a system of three linear equations with five variables.[2] Since this system has more variables than equations, it cannot be solved using the simultaneous equation approach. Instead, the simplex method uses an iterative procedure to get a solution for this system by assigning zeros to two variables, and then solving for the remaining three variables. More generally, when there are n variables and m constraints ($n > m$), then $n - m$ variables are set to zero, and the m constraints (equations) are solved for the remaining m variables. The solution to this is called a **basic solution**.

For example, if we let $x_1 = 0$ and $s_1 = 0$, then the above system becomes:

$$x_2 = 1{,}000 \tag{2.1}$$

$$0.75x_2 + s_2 = 1{,}500 \tag{2.2}$$

$$1.5x_2 + s_3 = 800 \tag{2.3}$$

From (2.1) we know that $x_2 = 1{,}000$. Substituting (2.1) into (2.2) results in the solution for s_2:

$$0.75(1{,}000) + s_2 = 1{,}500, \text{ or}$$

$$s_2 = 750.$$

Substituting (2.1) into (2.3) gives s_3:

$$1.5(1{,}000) + s_3 = 800, \text{ or}$$

$$s_3 = -700.$$

Hence, the basic solution when $x_1 = 0$ and $s_1 = 0$ is:

$$x_1 = 0, x_2 = 1{,}000, s_1 = 0, s_2 = 750, s_3 = -700.$$

In general, the $n - m$ variables set to zero are called **nonbasic variables** and the m (nonzero) variables are called **basic variables**. In this example, x_1 and s_1 are the nonbasic variables and x_2, s_2, and s_3 are the basic variables for this **basic solution**.

A basic solution can either be **feasible** or **infeasible**. A **basic feasible solution (BFS)** satisfies all constraints, including non-negativity. A **basic infeasible solution** violates at least one constraint. Is the above basic solution feasible or nonfeasible?

Fact: A BFS always occurs at an extreme point of the feasible region.

---

[2]Note that the term "variable" is used synonymously with the term "activity" in this book.

A BFS and an extreme point are one and the same. Since we know that an extreme point will be optimal (if any optimal solution exists), it seems reasonable to focus our attention on extreme points. The simplex method is based on this observation. It examines a sequence of BFSs, based on an iterative algorithm, until the optimal BFS is found.

The Simplex Tableau

The simplex method starts out by setting all "productive" activities to zero (i.e., the solution is the origin), and a simplex tableau is formed to do the first iteration. All nonslack and nonsurplus activities (i.e., the x_i's) will be referred to as "productive" activities in the discussion that follows. The first tableau is:

Basis	CB	x_1 35	x_2 50	s_1 0	s_2 0	s_3 0	b	b_i/a_{ij}
s_1	0	1	1	1	0	0	1,000	
s_2	0	2.5	0.75	0	1	0	1,500	
s_3	0	0	1.5	0	0	1	800	
Net Eval $(c_j - z_j)$	z_j							

Comments on Columns

1. The **basis** column includes all the basic variables. In the first iteration, the nonbasic variables are the productive activities ($x_1 = x_2 = 0$), and the basis therefore consists of the three slack variables s_1, s_2, and s_3.
2. The CB **column** contains the objective function coefficients for the basic variables. CB stands for the contribution of the current basis. Since the basic variables in the first iteration are all slack variables, $c_1, c_2,$ and $c_3 = 0$.
3. Columns x_1, x_2, s_1, s_2, and s_3 are the activities and slack variables to the problem. They include the basic and nonbasic variables.
4. The b column contains the right-hand-side (RHS) values (resource endowments) of the problem.
5. The b_i/a_{ij} column will be used to determine the pivot row, as will be explained later.
6. Note that the columns associated with the basic variables (s_1, s_2, and s_3 in this case) look like an identity matrix (1's on the diagonal and 0's in the off diagonal), for instance,

	s_1	s_2	s_3
s_1	1	0	0
s_2	0	1	0
s_3	0	0	1

Each of these columns is known as a **unit column** or **unit vector**. It is desirable to always have all basic variables forming unit vectors for the following reason:

When all basic variables are unit vectors, the solution for each basic variable is given by the value under the resource endowment b column associated with row i in the simplex tableau.

Comments on Rows

1. The first row under the activities row contains the objective function coefficients for the basic and nonbasic variables.

2. The next three rows correspond to the constraints of the problem. It is identical to the LP problem above, only expressed in tableau form.

3. The last two rows are called the z_j and $c_j - z_j$ rows.

The z_j and $c_j - z_j$ Rows

The z_j and $c_j - z_j$ rows provide a criterion for selecting which nonbasic variable, if any, should enter the next solution in order to increase the value of the objective function. There are two contrasting effects that bringing a nonbasic variable into the new basis will have on the value of the objective function.

1. **Direct Rate of Increase.** The objective function will increase at a rate of c_j per unit of x_j forced into the basis, where x_j is a nonbasic variable and c_j is its objective function coefficient.

2. **Indirect Rate of Decrease.** The objective function will decrease owing to a downward adjustment in the current basic variables due to bringing a nonbasic variable into the solution. The z_j row measures this indirect rate of decrease for each nonbasic variable.

The net effect of the direct rate of increase and the indirect rate of decrease in the objective function for each nonbasic variable is measured by the $c_j - z_j$ row.

Digression on the z_j Row

At first glance, it may sound counter-intuitive that bringing a nonbasic variable into the solution may result in a decrease in the value of the objective function. To see why this may occur, consider the somewhat analogous situation of adding a new marble to a bag of marbles that is already full. In order to make room for the new marble, an old marble has to be taken out, which, by itself, reduces the weight of the bag.

In the case of LP, forcing in a nonbasic variable requires a reduction in the value of current basic variables because scarce resources are now needed for the new variable, which competes with the old variables. To illustrate, consider the three linear equations of this example, where x_1 and x_2 are nonbasic and s_1, s_2, and s_3 are basic variables.

$$1x_1 + 1x_2 + 1s_1 = 1{,}000,$$

$$2.5x_1 + 0.75x_2 + 1s_2 = 1{,}500,$$

$$0x_1 + 1.5x_2 + 1s_3 = 800.$$

Now, solve each equation for the basic variables:

$$s_1 = 1{,}000 - 1x_1 - 1x_2,$$

$$s_2 = 1{,}500 - 2.5x_1 - 0.75x_2,$$

$$s_3 = 800 - 0x_1 - 1.5x_2.$$

Suppose that x_1 is forced into the solution. What happens to the current basic variables? s_1 will decrease from its current solution level of 1,000 at a rate of 1 per unit increase in x_1; s_2 will decrease from its current solution level of 1,500 at a rate of 2.5 per unit increase in x_1; and s_3 will decrease from its current solution level of 800 at a rate of 0 per unit increase in x_1.

Suppose that x_2 is forced into the solution. What happens to the current basic variables? s_1 will decrease from its current solution level of 1,000 at a rate of 1 per unit increase in x_2; s_2 will decrease from its current solution level of 1,500 at a rate of 0.75 per unit increase in x_2; and s_3 will decrease from its current solution level of 800 at a rate of 1.5 per unit increase in x_2.

These rates of decrease in basic variable levels are called **substitution coefficients**, as they indicate the tradeoff between current basic variables and how much they would have to decrease if a nonbasic variable were substituted into the system. The indirect rate of decrease (z_j row) measures this effect in economic terms by taking the product of the objective function coefficient for the basic variable and the substitution coefficient of the nonbasic variable. The general formula for this is:

$$z_j = \sum_{i=1}^{3} c_i a_{ij} \ (j=1,2,...,5).^3$$

Note that the appendix at the end of Chapter 3 provides a basic primer on summation notation and matrix operations.

While computing the indirect rate of decrease is only necessary for the nonbasic variables (z_1 and z_2 in this case), this measure is computed below for all variables:

$$z_1 = 0(1) + 0(2.5) \ + 0(0) \ \ = 0,$$
$$z_2 = 0(1) + 0(0.75) + 0(1.5) = 0,$$
$$z_3 = 0(1) + 0(0) \ \ \ \ + 0(0) \ \ = 0,$$
$$z_4 = 0(0) + 0(1) \ \ \ \ + 0(0) \ \ = 0,$$
$$z_5 = 0(0) + 0(0) \ \ \ \ + 0(1) \ \ = 0.$$

As you can see, the z_j values for the nonbasic variables, x_1 and x_2, are zero. In other words, no profit is given up by forcing x_1, or x_2 into the solution because the slack variables, which are the current basic variables, have objective function coefficients equal to zero.

A z_j value corresponding to the b column should also be computed. The formula for the RHS column for z_j is:

$$z_b = \sum_{i=1}^{3} c_i b_i.$$

Since the current solution values for the basic variables are contained in this column, the z_b value gives the value of the objective function for the current solution. In the initial tableau, z_b is equal to zero because only the slack variables are the basic variables.

In order to ascertain whether bringing a nonbasic variable into the solution will improve the subsequent solution, an examination of the net effect, that is, the $c_j - z_j$ row, is necessary. In essence, this row measures the gains minus the cost of making each nonbasic variable basic. Hence, if $c_j - z_j$ is positive for a nonbasic variable, then forcing it into the

---

[3]The technical coefficients, a_{ij}, form an $m \times n$ matrix with m rows and n columns. The subscript i references the row location and the subscript j references the column location in the matrix, for example, a_{32} is the element in the third row and second column.

next solution would be an improvement over the current solution. If it is zero or negative, then it would be an inferior move. The values for the $c_j - z_j$ row are calculated by simply subtracting z_j from c_j for each variable.

As before, while computing $c_j - z_j$ is only necessary for the nonbasic variables, this measure is computed below for all variables:

$$c_1 - z_1 = 35 - 0 = 35,$$

$$c_2 - z_2 = 50 - 0 = 50,$$

$$c_3 - z_3 = \ 0 - 0 = \ 0,$$

$$c_4 - z_4 = \ 0 - 0 = \ 0,$$

$$c_5 - z_5 = \ 0 - 0 = \ 0.$$

So the first tableau for this problem is:

First Tableau

Basis	CB	x_1 35	x_2 50	s_1 0	s_2 0	s_3 0	b	b_i/a_{ij}
s_1	0	1	1	1	0	0	1,000	1,000
s_2	0	2.5	0.75	0	1	0	1,500	2,000
s_3	0	0	1.5	0	0	1	800	533.33
z_j		0	0	0	0	0	0	
Net Eval ($c_j - z_j$)		35	50	0	0	0		

Improving Upon the Solution (Changing the Basis)

It is clear that the current solution can be improved since the values of the nonslack variables and total profit are zero. The criterion for selecting a new variable to enter the basis is:

Choose the nonbasic variable that yields the highest net contribution ($c_j - z_j$) value.

In this example, this variable is x_2 since $c_2 - z_2 = 50$. This seems logical since bringing in x_2 yields a higher contribution to profit than the other nonbasic variable x_1. The column containing the new variable, which is the x_2 column in this case, is called the **pivot column**. To add a new basic variable to the new basis means that one of the old basic variables has to be forced out of the basis, that is, has to become a nonbasic variable. To do this, the simplex method finds the basic variable that is the most restrictive in terms of constraining the problem and makes this variable nonbasic. This variable is determined by dividing all the RHS values by their respective positive, non-zero coefficients in the pivot column, x_2, that is, b_i/a_{i2} for i $= 1$, 2, and 3 and where the subscript "2" on a_{i2} indicates that x_2 is the pivot column. Note that if any row has a zero or negative a_{ij} coefficient in the pivot column, the ratio should not be computed for that row. Simply cross out that row from consideration. In this example, these ratios are:

$$b_1/a_{12} = 1,000/1 \quad = 1,000,$$

$$b_2/a_{22} = 1,500/(0.75) = 2,000,$$

$$b_3/a_{32} = 800/(1.5) \quad = 533.33.$$

The row with the smallest non-negative ratio represents the most restrictive row in the sense that it either requires more resources per activity, or has the least amount of resource endowment relative to the other rows. As such, the basic variable associated with this row becomes nonbasic. Since a slack variable is currently associated with this row (e.g., s_3 since it has the smallest non-negative ratio), s_3 becomes nonbasic (i.e., $s_3 = 0$) and is replaced in the new basis by the new entering variable, x_2. Intuitively, this represents the most restrictive constraint and hence making it nonbasic makes sense since $s_3 = 0$ implies it is a binding constraint. This row is called the **pivot row**. The element in the pivot column and the pivot row is called the **pivot element**. In this case, the pivot element is 1.5.

Sometimes there may be a tie among two variables which have the smallest b_i/a_{ij} ratio. If this occurs, then simply choose one at random to become the pivot row.

The Next Iteration

The new basis will be s_1, s_2, and x_2 which replaces s_3. The old s_3 row needs to be replaced with the new x_2 row. The problem is that unlike the old s_3 column, the x_2 column is not a unit vector, that is:

Basis	x_2
s_1	$1(a_{12})$
s_2	$0.75(a_{22})$
s_3	$1.5(a_{32})$

where a_{i2} (i = 1, 2, 3) are the original coefficients in the **A** matrix (a 3 × 5 matrix), x_2 column. The goal is to transform the a_{i2} coefficients so that they form a unit vector, that is:

Basis	x_2
s_1	$0(a_{12})$
s_2	$0(a_{22})$
x_2	$1(a_{32})$

where a_{i2} (i = 1, 2, 3) are the transformed coefficients in the **A** matrix, x_2 column. Recall that when all basic variables are unit vectors, their solution values are listed in the b column.

Digression on Two Facts About Matrix Algebra

1. Multiplying both sides of any row in a system of linear equations by a constant will not change the original solution. For example, multiplying both sides of the equation $[5x_1 + x_2 = 100]$ by 2 does not change the equation, that is:

$$2(5x_1 + x_2) = 2(100) \implies 5x_1 + x_2 = 100.$$

2. Replacing any row of a system of linear equations by the result of adding or subtracting a multiple of another row will not change the solution. Consider the following two equations:

$$x_1 + 5x_2 = 100 \tag{2.4}$$

$$7x_1 + 2x_2 = 200 \tag{2.5}$$

The solution is:

$$x_1 = 24.24, x_2 = 15.15.$$

Now multiply (2.5) by 0.5:

$$3.5x_1 + x_2 = 100 \qquad (2.6)$$

and add it to (2.4):

$$x_1 + 5x_2 + 3.5x_1 + x_2 = 100 + 100, \text{ or}$$

$$4.5x_1 + 6x_2 = 200 \qquad (2.7)$$

Solving (2.6) and (2.7) simultaneously yields

$$x_1 = 24.24, x_2 = 15.15,$$

the same as before.

These two elementary row operations are used in the simplex method.

To do this for the current example, the following procedures are used:

Step 1: Transform the old s_3 row to get a 1 for the x_2 parameter.

The new x_2 row in second tableau is created by dividing all a_{ij} and b_i coefficients in the pivot row (old s_3 row) by 1.5 in order to get a 1 coefficient in the x_2 column for a_{32}. That is, divide:

$$0x_1 + 1.5x_2 + 0s_1 + 0s_2 + \quad 1s_3 = 800 \qquad \text{(old } s_3 \text{ row)}$$

by 1.5 to get:

$$0x_1 + \quad 1x_2 + 0s_1 + 0s_2 + 0.67s_3 = 533.33 \qquad \text{(new } x_2 \text{ row)}$$

This results in each coefficient now being stated in terms of x_2 instead of s_3.

Step 2: Transform the old s_1 row to get a zero coefficient for a_{12}.

The new s_1 row in second tableau is created by transforming the old s_1 row to get a zero coefficient for a_{12}.

To do this, first multiply the new x_2 row by the negative of the a_{12} coefficient, which is -1 in this case. Then add the resulting row to the old s_1 row to get the new s_1 row, that is:

$$0x_1 - 1x_2 + 0s_1 + 0s_2 - 0.67s_3 = -533.33 \qquad \text{(new } x_2 \text{ row times } -1)$$

$$1x_1 + 1x_2 + 1s_1 + 0s_2 + \quad 0s_3 = 1{,}000 \qquad \begin{array}{l}\text{(old } s_1 \text{ row)}\\ \text{(add together)}\end{array}$$

$$\overline{}$$

$$1x_1 + 0x_2 + 1s_1 + 0s_2 - 0.67s_3 = \quad 466.67 \qquad \text{(new } s_1 \text{ row)}$$

Step 3: Transform the old s_2 row to get a zero coefficient for a_{22}.

The new s_2 row in the second tableau is created by transforming the old s_2 row to get a zero coefficient for a_{22}. To do this, first multiply the new x_2 row by the negative of the a_{22} coefficient, which is -0.75 in this case. Then add the resulting row to the old s_2 row to get the new s_2 row, that is:

$$0x_1 - 0.75x_2 + 0s_1 + 0s_2 - 0.5s_3 = -400 \qquad \text{(new } x_2 \text{ row times } -0.75)$$

$$2.5x_1 + 0.75x_2 + 0s_1 + 1s_2 + \quad 0s_3 = 1{,}500 \qquad \text{(old } s_2 \text{ row)}$$

$$\text{(add together)}$$

$$\overline{}$$

$$2.5x_1 + \quad 0x_2 + 0s_1 + 1s_2 - 0.5s_3 = 1{,}100 \qquad \text{(new } s_2 \text{ row)}$$

Using the information from steps 1, 2, 3, the second tableau becomes:

Second Tableau

		x_1	x_2	s_1	s_2	s_3		
Basis	CB	35	50	0	0	0	b	b_i/a_{ij}
s_1	0	1	0	1	0	−0.67	466.67	466.67
s_2	0	2.5	0	0	1	−0.50	1,100	440
x_2	50	0	1	0	0	0.67	533.33	—
z_j		0	50	0	0	33.5	26,666.67	
Net Eval ($c_j - z_j$)		35	0	0	0	−33.5		

Check to see that the new tableau has the elements of an identity matrix formed from the three columns of the basic variables s_1, s_2, and x_2. If not, an error has been made and should be corrected before proceeding to the next step. If all basic variables form an identity matrix, then calculate the z_j and $c_j - z_j$ values.

Stopping Rule: If the new $c_j - z_j$ values for all nonbasic variables are zero or negative, then stop since the solution is optimal.

Since the $c_1 - z_1$ entry is 35, which is positive, an optimal solution has not yet been found. Proceeding to the next iteration, the pivot column is:

$$\text{Pivot Column} = x_1.$$

Next Iteration

Divide all b_i elements by the positive coefficients in the pivot column, b_i/a_{i1} (note that the "1" subscript on a_{i1} represents that x_1 is the pivot column). Choose the row with the smallest resulting ratio as the pivot row. This is the row that is replaced by x_1 in the new basis. In this case, it is row s_2. Notice that the x_2 row is not considered here because it has a zero a_{ij} coefficient in the pivot column. The new basis in the third tableau will be s_1, x_1, and x_2.

Step 1: Transform the old pivot row replacing s_2 with x_1.
The new x_1 row in the third tableau is created by dividing all a_{ij} and b_i coefficients in the pivot row (old s_2 row) by 2.5 in order to get a 1 coefficient in the x_1 column for a_{21}. That is, divide:

$$2.5x_1 + 0x_2 + 0s_1 + 1s_2 - 0.5s_3 = 1,100 \qquad \text{(old } s_2 \text{ row)}$$

by 2.5 to get:

$$1x_1 + 0x_2 + 0s_1 + 0.4s_2 - 0.2s_3 = 440 \qquad \text{(new } x_1 \text{ row)}$$

Step 2: Transform s_1 row.
The new s_1 row in the third tableau is created by transforming the old s_1 row to get a zero coefficient for a_{11}. To do this, first multiply the new x_1 row by the negative of the a_{11} coefficient, which is −1 in this case. Then add the resulting row to the old s_1 row to get the new s_1 row, that is:

$$-1x_1 + 0x_2 + 0s_1 - 0.4s_2 + 0.2s_3 = -440 \qquad \text{(new } x_1 \text{ row times } -1)$$
$$1x_1 + 0x_2 + 1s_1 + 0s_2 - 0.67s_3 = 466.67 \qquad \text{(old } s_1 \text{ row)}$$
$$\overline{} \qquad \text{(add together)}$$
$$0x_1 + 0x_2 + 1s_1 - 0.4s_2 - 0.47s_3 = 26.67 \qquad \text{(new } s_1 \text{ row)}$$

Step 3: Transform x_2 row.

The new x_2 row is created by transforming the old x_2 row to get a zero coefficient for a_{31}.

Note that this coefficient is already zero. Hence, no transformation is needed.

Using the information from steps 1, 2, and 3, the third tableau becomes:

Third Tableau

Basis	CB	x_1	x_2	s_1	s_2	s_3	b	b_i/a_{ij}
		35	50	0	0	0		
s_1	0	0	0	1	−0.4	−0.47	26.67	
x_1	35	1	0	0	0.4	−0.20	440	
x_2	50	0	1	0	0	0.67	533.33	
z_j		35	50	0	14	26.5	42,066.67	
Net Eval ($c_j - z_j$)		0	0	0	−14	−26.5		

Check to see that the new tableau has the elements of an identity matrix formed from the three columns of the basic variables s_1, x_1, and x_2. Calculate the new $c_j - z_j$ values for the third tableau. Since no positive entries exist, we have found the optimal solution to the problem. Consequently, this iterative procedure can stop because no positive values are found in the $c_j - z_j$ row.

The optimal solution is given by the b column in the final simplex tableau. Reading down this column, the optimal solution is:

$$x_1^* = 440, \ x_2^* = 533.33, \ s_1^* = 26.67, \ s_2^* = 0, \ s_3^* = 0, \text{ and } Z^* = 42{,}066.67.$$

A flow chart of the simplex method for a maximization problem is presented in Figure 2.1, and the three simplex tableaus for this problem are presented in Figure 2.2. To recap, the simplex method starts by writing the LP problem in simplex tableau form. In this first tableau, the slack variables are basic (non-zero) and the nonslack variables, x_i's, are made nonbasic (zero). Next, calculate the z_j and $c_j - z_j$ rows. The column of coefficients associated with the highest $c_j - z_j$ value is called the pivot column. The rule for selecting a new basic variable is to choose the nonbasic variable that yields the highest positive net contribution value. If all nonbasic variables have nonpositive net contributions, then stop, as the current simplex tableau contains the optimal solution. If it is determined that a new basic variable should be added to the next tableau, then determine which of the old basic variables has to leave the next basis. This is done by dividing all b_i column values by the a_{ij} coefficients in the pivot column. The ratio with the smallest non-negative value represents the current basic variable that should be forced out of the subsequent basis. The row associated with this variable is called the pivot row. In a case where there is a tie between two variables having the smallest value, flip a coin to determine the pivot row.

Next, perform all the elementary row transformations required to make all new basic variables form unit vectors. This is somewhat cumbersome, but after some practice it becomes quite simple. Then, the new tableau can be written out, and the z_j and $c_j - z_j$ rows can be computed. Use the same criteria as before in selecting new basic activities, deleting old basic activities, and determining whether or not to stop.

2.2   THE SIMPLEX METHOD FOR MAXIMIZATION PROBLEMS: GENERAL CASE

In the previous section, the simplex method for solving LP models using $\le$ constraints was described. In this section, solution techniques using the simplex method for LP models containing $\ge$, $\le$, and $=$ types of constraints are discussed.

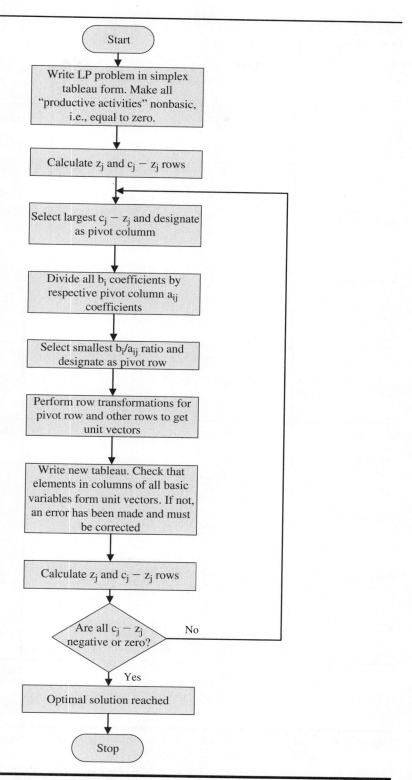

Figure 2.1   Flow chart of the simplex method for a maximization problem.

First Tableau

Basis	CB	x_1 35	x_2 50	s_1 0	s_2 0	s_3 0	b	b_i/a_{ij}
s_1	0	1	1	1	0	0	1,000	1,000
s_2	0	2.5	0.75	0	1	0	1,500	2,000
s_3	0	0	1.5	0	0	1	800	533.33
z_j		0	0	0	0	0	0	
Net Eval ($c_j - z_j$)		35	50	0	0	0		

Second Tableau

Basis	CB	x_1 35	x_2 50	s_1 0	s_2 0	s_3 0	b	b_i/a_{ij}
s_1	0	1	0	1	0	-0.67	466.67	466.67
s_2	0	2.5	0	0	1	-0.50	1,100	440
x_2	50	0	1	0	0	0.67	533.33	–
z_j		0	50	0	0	33.50	26,666.67	
Net Eval ($c_j - z_j$)		35	0	0	0	-33.50		

Third Tableau

Basis	CB	x_1 35	x_2 50	s_1 0	s_2 0	s_3 0	b	b_i/a_{ij}
s_1	0	0	0	1	-0.4	-0.47	26.67	
x_1	35	1	0	0	0.4	-0.20	440	
x_2	50	0	1	0	0	0.67	533.33	
z_j		35	50	0	14	26.50	42,066.67	
Net Eval ($c_j - z_j$)		0	0	0	-14	-26.50		

Figure 2.2   Three simplex tableaus for first maximization problem with pivot columns and rows highlighted, and pivot elements in bold.

Greater-Than-Or-Equal-To Constraints

Consider the following two-activity maximization problem:

Max: $Z = 500x_1 + 400x_2$ (0)

s.t.:

$$30x_1 + 50x_2 \leq 1500 \qquad (1)$$

$$10x_2 \leq 200 \qquad (2)$$

$$80x_1 + 50x_2 \leq 3000 \qquad (3)$$

$$10x_1 + 10x_2 \geq 250 \qquad (4)$$

$$x_1, \qquad x_2 \geq 0 \qquad (5)$$

This problem is different from the previous maximization problems discussed because one of the constraints is a $\geq$ restriction. It is totally permissible and indeed quite common for many maximization problems to include $\geq$ (and $=$) as well as $\leq$ types of constraints. Likewise, minimization problems do not necessarily have to include only $\geq$ constraints.

The above problem can be expressed in standard form as the following:

Max: $Z = 500x_1 + 400x_2 + 0s_1 + 0s_2 + 0s_3 + 0s_4$ (0)

s.t.:

$$30x_1 + 50x_2 + 1s_1 \qquad\qquad = 1{,}500 \qquad (1)$$

$$10x_2 \qquad + 1s_2 \qquad\qquad = 200 \qquad (2)$$

$$80x_1 + 50x_2 \qquad\quad + 1s_3 \qquad = 3{,}000 \qquad (3)$$

$$10x_1 + 10x_2 \qquad\qquad\qquad - 1s_4 = 250 \qquad (4)$$

$$x_1, \qquad x_2, \qquad s_1, \qquad s_2, \qquad s_3, \qquad s_4 \geq 0 \qquad (5)$$

To see why using the same simplex procedures discussed in the last section would cause problems for this example, let $x_1 = x_2 = 0$ and start the simplex method. Solving (1) through (4) yields:

$$x_1 = 0,\ x_2 = 0,\ s_1 = 1{,}500,\ s_2 = 200,\ s_3 = 3{,}000,\ s_4 = -250.$$

This solution is not feasible! Why? Because $s_4 = -250$, which violates the non-negativity restriction in (5). This problem is due to the $\geq$ in constraint (4). Since s_4 is a surplus rather than slack variable, the initial basic solution is not feasible. Hence, the previously described simplex method cannot be relied upon as a means to solve this problem. The problem is that the simplex method starts off at the origin, but in this case the origin is not part of the feasible region.

Artificial Variables   To solve this problem, **artificial variables** are used. Artificial variables (a_i) are "dummy" variables used to obtain an initial basic solution that is feasible when there are $\geq$ or $=$ constraints. They are dummy variables in the sense that they are included in the initial simplex tableau in order to get a feasible solution, but are forced out in subsequent tableaus because they do not have any real economic meaning. In the case of $\geq$ constraints, add an artificial variable, a_i, to the $\geq$ constraint (a_4, where the subscript value denotes the constraint number). The new constraint set becomes the following:

$$30x_1 + 50x_2 + 1s_1 = 1{,}500 \tag{1}$$

$$10x_2 + 1s_2 = 200 \tag{2}$$

$$80x_1 + 50x_2 + 1s_3 = 3{,}000 \tag{3}$$

$$10x_1 + 10x_2 - 1s_4 + 1a_4 = 250 \tag{4}$$

Letting $x_1 = x_2 = s_4 = 0$, the new solution is:

$$x_1 = 0,\ x_2 = 0,\ s_1 = 1{,}500,\ s_2 = 200,\ s_3 = 3{,}000,\ s_4 = 0,\ a_4 = 250.$$

This is now a BFS in a mathematical sense. However, in reality, it is still not feasible since constraint (4) in the general form of the model is still not satisfied since $s_4 = 0$. That is, this solution procedure is a mathematical means for using the simplex technique, but it does not make sense as a "real-world" solution. Therefore, a procedure needs to be devised that guarantees that no artificial variable will appear as a basic variable in the final solution. One approach that accomplishes this is called the **penalty approach**. This approach is very simple in that a very large cost is assigned to the artificial variable in the objective function. Rather than using a very large number, it is more convenient to denote this cost as m. This large penalty will have the effect of guaranteeing that the artificial variable will not be part of the optimal solution. The objective function becomes:

$$\text{Max: } Z = 500x_1 + 400x_2 + 0s_1 + 0s_2 + 0s_3 + 0s_4 - ma_4.$$

In larger models that contain more than one $\geq$ constraint, the procedure is the same. Add a surplus variable and an artificial variable for each $\geq$ constraint. Also, for each of these artificial variables in the objective function, assign its objective function coefficient a very large penalty m. In the case of maximization problems, the penalties are assigned by subtracting ma_i from the objective function. In the case of minimization problems, the penalties are assigned by adding ma_i to the objective function.

Which variables will be basic variables for the initial simplex tableau? The following rule will help determine the answer to this question:

The basic variables in the initial simplex tableau will all be slack and artificial variables that form unit vectors (i.e., have a coefficient of 1).

Using this rule, it is apparent that the basic variables for the initial tableau in this example are s_1, s_2, s_3, and a_4. The first tableau is presented below:

First Tableau

Basis	CB	x_1 500	x_2 400	s_1 0	s_2 0	s_3 0	s_4 0	a_4 $-m$	b	b_i/a_{ij}
s_1	0	30	50	1	0	0	0	0	1,500	50
s_2	0	0	10	0	1	0	0	0	200	–
s_3	0	80	50	0	0	1	0	0	3,000	37.5
a_4	$-m$	10	10	0	0	0	-1	1	250	25
z_j		$-10m$	$-10m$	0	0	0	m	$-m$	$-250m$	
$c_j - z_j$		$500+10m$	$400+10m$	0	0	0	$-m$	0		

Again notice that this first solution is feasible in the context of the simplex technique, but not in a real-world context. What happens to the artificial variable? In this case, the pivot column is x_1, and the pivot row is a_4. Hence, the artificial variable becomes nonbasic in the next tableau, which implies that the next iteration will be a "real" BFS. As was the case before, transformations are required to transform the new basic variable (x_1) for the next tableau into unit vectors.

Step 1: Divide old row a_4:

$$10x_1 + 10x_2 + 0s_1 + 0s_2 + 0s_3 - 1s_4 + 1a_4 = 250 \qquad \text{(old } a_4\text{)}$$

by 10 to get:

$$1x_1 + 1x_2 + 0s_1 + 0s_2 + 0s_3 - 0.1s_4 + 0.1a_4 = 25 \qquad \text{(new } x_1 \text{ row)}$$

Step 2: Create new s_1 row by multiplying new x_1 row by -30 ($-a_{11}$) and adding to the old s_1 row:

$$-30x_1 - 30x_2 + 0s_1 + 0s_2 + 0s_3 + 3s_4 - 3a_4 = -750 \qquad \text{(new } x_1 \text{ times } -30\text{)}$$

$$30x_1 + 50x_2 + 1s_1 + 0s_2 + 0s_3 + 0s_4 + 0a_4 = 1,500 \qquad \text{(old } s_1\text{)}$$

$$\overline{} \qquad \text{(add together)}$$

$$0x_1 + 20x_2 + 1s_1 + 0s_2 + 0s_3 + 3s_4 - 3a_4 = 750 \qquad \text{(new } s_1 \text{ row)}$$

Step 3: New s_2 row = old s_2 row since there is a zero coefficient in the x_1 column.

Step 4: Create new s_3 row by multiplying new x_1 row by -80 ($-a_{31}$) and adding it to the old s_3 row:

$$-80x_1 - 80x_2 + 0s_1 + 0s_2 + 0s_3 + 8s_4 - 8a_4 = -2,000 \qquad \text{(new } x_1 \text{ times } -80\text{)}$$

$$80x_1 + 50x_2 + 0s_1 + 0s_2 + 1s_3 + 0s_4 + 0a_4 = 3,000 \qquad \text{(old } s_3 \text{ row)}$$

$$\overline{} \qquad \text{(add together)}$$

$$0x_1 - 30x_2 + 0s_1 + 0s_2 + 1s_3 + 8s_4 - 8a_4 = 1,000 \qquad \text{(new } s_3 \text{ row)}$$

Using these results, the second tableau becomes:

Second Tableau

Basis	CB	x_1 500	x_2 400	s_1 0	s_2 0	s_3 0	s_4 0	a_4 $-m$	b	b_i/a_{ij}
s_1	0	0	20	1	0	0	3	-3	750	250
s_2	0	0	10	0	1	0	0	0	200	—
s_3	0	0	-30	0	0	1	8	-8	1,000	125
x_1	500	1	1	0	0	0	-0.1	0.1	25	—
	z_j	500	500	0	0	0	-50	50	12,500	
	$c_j - z_j$	0	-100	0	0	0	50	$-m-50$		

Notice that now the net contribution of a_4 is negative and the new solution is feasible. However, since not all net contributions are negative, this solution is not optimal, and it is necessary to proceed to the next iteration. Now the pivot column and pivot row are:

$$\text{Pivot Column} = s_4,$$
$$\text{Pivot Row} \quad = s_3.$$

The following transformations are needed for the third simplex tableau.

Step 1: Divide old row s_3

$$0x_1 - 30x_2 + 0s_1 + 0s_2 + 1s_3 + 8s_4 - 8a_4 = 1{,}000 \qquad \text{(old } s_3 \text{ row)}$$

by 8 to get:

$$0x_1 - 3.75x_2 + 0s_1 + 0s_2 + 0.125s_3 + 1s_4 - 1a_4 = 125 \qquad \text{(new } s_4 \text{ row)}$$

Step 2: Create the new s_1 row by multiplying the new s_4 row by -3 $(-a_{16})$ and adding to the old s_1 row:

$$
\begin{aligned}
0x_1 + 11.25x_2 + 0s_1 + 0s_2 - 0.375s_3 - 3s_4 + 3a_4 &= -375 \qquad \text{(new } s_4 \text{ times } -3\text{)} \\
0x_1 + 20x_2 + 1s_1 + 0s_2 + 0s_3 + 3s_4 - 3a_4 &= 750 \qquad \text{(old } s_1\text{)} \\
& \qquad \text{(add together)}
\end{aligned}
$$
$$\overline{}$$
$$0x_1 + 31.25x_2 + 1s_1 + 0s_2 - 0.375s_3 + 0s_4 + 0a_4 = 375 \qquad \text{(new } s_1 \text{ row)}$$

Step 3: New s_2 row = old s_2 row since there is a zero coefficient in the s_4 column.

Step 4: Create the new x_1 row by multiplying the new s_4 row by 0.1 $(-a_{46})$ and adding it to the old x_1 row:

$$
\begin{aligned}
0x_1 - 0.375x_2 + 0s_1 + 0s_2 + 0.0125s_3 + 0.1s_4 - 0.1a_4 &= 12.5 \qquad \text{(} s_4 \text{ times 0.1)} \\
1x_1 + 1x_2 + 0s_1 + 0s_2 + 0s_3 - 0.1s_4 + 0.1a_4 &= 25 \qquad \text{(old } x_1 \text{ row)} \\
& \qquad \text{(add together)}
\end{aligned}
$$
$$\overline{}$$
$$1x_1 + 0.625x_2 + 0s_1 + 0s_2 + 0.0125s_3 + 0s_4 + 0a_4 = 37.5 \qquad \text{(new } x_1 \text{ row)}$$

Using these results, the third tableau becomes:

Third Tableau

Basis	CB	x_1 500	x_2 400	s_1 0	s_2 0	s_3 0	s_4 0	a_4 $-m$	b	b_i/a_{ij}
s_1	0	0	31.25	1	0	−0.375	0	0	375	12
s_2	0	0	10	0	1	0	0	0	200	20
s_4	0	0	−3.75	0	0	0.125	1	−1	125	−
x_1	500	1	0.625	0	0	0.0125	0	0	37.5	60
	z_j	500	312.5	0	0	6.25	0	100	18,750	
	$c_j - z_j$	0	87.5	0	0	−6.25	0	−m		

Since not all net contributions are negative, this solution is not optimal, and we proceed to the next iteration.

$$\text{Pivot Column} = x_2,$$

$$\text{Pivot Row} \quad = s_1.$$

Step 1: Create the new x_2 row by dividing old s_1 row:

$$0x_1 + 31.25x_2 + \quad 1s_1 + 0s_2 - 0.375s_3 + 0s_4 + 0a_4 = 375 \qquad \text{(old } s_1 \text{ row)}$$

by 31.25 (a_{12}) to get:

$$0x_1 + \quad 1x_2 + 0.032s_1 + 0s_2 - 0.012s_3 + 0s_4 + 0a_4 = 12 \qquad \text{(new } x_2 \text{ row)}$$

Step 2: Create the new s_2 row by multiplying the new x_2 row by -10 ($-a_{22}$) and adding to the old s_2 row:

$$0x_1 - 10x_2 - 0.32s_1 + 0s_2 + 0.12s_3 + 0s_4 + 0a_4 = -120 \qquad \text{(new } x_2 \text{ times } -10\text{)}$$

$$0x_1 + 10x_2 + \quad 0s_1 + 1s_2 + \quad 0s_3 + 0s_4 + 0a_4 = \quad 200 \qquad \text{(old } s_2 \text{ row)}$$
$$\text{(add together)}$$

$$\overline{0x_1 + \quad 0x_2 - 0.32s_1 + 1s_2 + 0.12s_3 + 0s_4 + 0a_4 = \quad 80} \qquad \text{(new } s_2 \text{ row)}$$

Step 3: Create the new s_4 row by multiplying the new x_2 row by 3.75 ($-a_{32}$) and adding to the old s_4 row:

$$0x_1 + 3.75x_2 + 0.12s_1 + 0s_2 - 0.045s_3 + 0s_4 + 0a_4 = \quad 45 \qquad \text{(new } x_2 \text{ times } 3.75\text{)}$$

$$0x_1 - 3.75x_2 + \quad 0s_1 + 0s_2 + 0.125s_3 + 1s_4 - 1a_4 = 125 \qquad \text{(old } s_4 \text{ row)}$$
$$\text{(add together)}$$

$$\overline{0x_1 + \quad 0x_2 + 0.12s_1 + 0s_2 + \quad 0.08s_3 + 1s_4 - 1a_4 = 170} \qquad \text{(new } s_4 \text{ row)}$$

Step 4: Create the new x_1 row by multiplying the new x_2 row by $-.625$ ($-a_{42}$) and adding to the old x_1 row:

$$0x_1 - 0.625x_2 - 0.02s_1 + 0s_2 + 0.0075s_3 + 0s_4 + 0a_4 = -7.5 \qquad \text{(new } x_2 \text{ times } -0.625\text{)}$$

$$1x_1 + 0.625x_2 + \quad 0s_1 + 0s_2 + 0.0125s_3 + 0s_4 + 0a_4 = 37.5 \qquad \text{(old } x_1 \text{ row)}$$
$$\text{(add together)}$$

$$\overline{1x_1 + \quad 0x_2 - 0.02s_1 + 0s_2 + \quad 0.02s_3 + 0s_4 + 0a_4 = 30} \qquad \text{(new } x_1 \text{ row)}$$

Using these results, the fourth tableau becomes:

Fourth Tableau

Basis	CB	x_1 500	x_2 400	s_1 0	s_2 0	s_3 0	s_4 0	a_4 $-m$	b	b_i/a_{ij}
x_2	400	0	1	0.032	0	-0.012	0	0	12	
s_2	0	0	0	-0.32	1	0.12	0	0	80	
s_4	0	0	0	0.12	0	0.08	1	-1	170	
x_1	500	1	0	-0.02	0	0.02	0	0	30	
	z_j	500	400	2.8	0	5.2	0	0	19,800	
	$c_j - z_j$	0	0	-2.8	0	-5.2	0	$-m$		

This is the optimal solution because all $c_j - z_j$ values are negative. The optimal solution is given in the b column. The solution is:

$$x_1^* = 30,\ x_2^* = 12,\ Z^* = 19{,}800,\ s_1^* = 0,\ s_2^* = 80,\ s_3^* = 0,\ s_4^* = 170.$$

Equal-to Constraints

Consider the following LP problem:

Max: $Z = 7x_1 + 3x_2 + x_3$ (0)

s.t.:

$$x_1 + x_2 - 5x_3 = 775 \tag{1}$$

Plus other structural constraints and non-negativity.

There is now an equality constraint in the problem. In this case, an artificial variable (a_1) is necessary to create a BFS for the initial simplex tableau. Use the same procedures as those outlined for the $\geq$ constraints. The standard form of the model with the artificial variable is:

Max: $Z = 7x_1 + 3x_2 + x_3 - ma_1$ (0)

s.t.:

$$x_1 + x_2 - 5x_3 + 1a_1 = 775 \tag{1}$$

This can be solved with the simplex method in an identical fashion as before. As can be seen, the only difference between handling an equality constraint and a $\geq$ constraint is that the equality constraint has an artificial variable, but not a surplus variable associated with it.

Handling Negative Right-Hand-Side Values

At times constraints with negative RHS or b parameters are encountered. The problem with negative RHS parameters is that they violate the property of the tableau form that all RHS values be non-negative. Fortunately, there is an easy way to deal with this problem. Consider, for example:

$$-5x_1 + x_2 \leq -100 \tag{2.8}$$

This can be corrected by considering the following fact:

Multiplying both sides of a constraint or equation by -1 yields an identical constraint or equation.

Multiplying both sides of (2.8) by -1 results in:

$$5x_1 - x_2 \geq 100 \tag{2.9}$$

While multiplying both sides by a negative number reverses the sign of the inequality, constraint (2.9) remains mathematically identical to (2.8), and since (2.9) no longer has a negative b value, the simplex method can be used. Simply replace (2.8) with (2.9), add an artificial and surplus variable for (2.9), and solve via the simplex method.

2.3   THE SIMPLEX METHOD AND MINIMIZATION PROBLEMS

The following are two ways to approach the simplex method for minimization problems.

Approach 1: The first approach is the same as the one for maximization problems, except that the two rules for variable selection and stopping when the optimal solution is reached are reversed. First, the selection criterion for a new, nonbasic variable entering the solution is changed to the following:

Choose the nonbasic variable with the most negative $c_j - z_j$ value.

This makes sense, since $c_j - z_j$ gives the amount by which the objective function will change if one unit of a nonbasic variable is forced into the solution. In minimization problems the goal is to minimize the objective function value.

Second, the stopping criterion is now to halt the iterative simplex method whenever all the $c_j - z_j$ values are zero or positive. This should be clear as adding any nonbasic variable with a positive net contribution would make the subsequent objective function value higher.

Approach 2: The second approach is to convert the minimization problem to an equivalent "maximization" problem and solve using the same procedures as before. Consider the following fact:

Any minimization problem can be solved as a maximization problem and the result will be identical to the solution obtained by minimization. This is done by multiplying the objective function by -1 and maximizing.

To illustrate, consider the following minimization problem:

$$\text{Min: } Z = 10x_1 + 4x_2 \tag{0}$$

s.t.:

$$3x_1 + 2x_2 \geq 60 \tag{1}$$

$$7x_1 + 2x_2 \geq 84 \tag{2}$$

$$3x_1 + 6x_2 \geq 72 \tag{3}$$

$$x_1, \quad x_2 \geq 0 \tag{4}$$

The graphical solution to this problem is given in Figure 2.3. The iso-cost line is:

$$x_2 = 0.25Z_a - 2.5x_1 \tag{2.10}$$

The optimal solution is $(x_1^*, x_2^*) = (6, 21)$ and $Z^* = 144$. This problem could be equivalently formulated and solved by multiplying (0) by -1 and solving it as a maximization problem. That is, min $= -$max, or

$$\text{Max: } Z = -1(10x_1 + 4x_2) \tag{0}$$

s.t.:

$$3x_1 + 2x_2 \geq 60 \tag{1}$$

$$7x_1 + 2x_2 \geq 84 \tag{2}$$

$$3x_1 + 6x_2 \geq 72 \tag{3}$$

$$x_1, \quad x_2 \geq 0 \tag{4}$$

The objective function now is:

$$\text{Max: } Z = -10x_1 - 4x_2 \tag{0}$$

The iso-contribution line, in slope-intercept form is:

$$x_2 = -0.25Z_a - 2.5x_1 \tag{2.11}$$

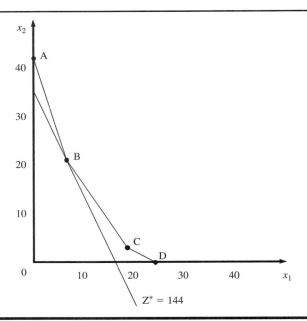

Figure 2.3   Graphical solution to the minimization problem.

Since constraints (1) through (4) are the same, the feasible region is the same. All that needs to be done is to find the "maximum" using the "iso-profit" line and the feasible region. The only difference between the "iso-profit" line given by (2.11) and the iso-cost line defined by (2.10) is that the intercept in (2.11) is negative while the intercept of (2.10) is positive. Let $Z_a = -200$. Then the intercepts are:

$$x_1 = -0.1(-200) = 20,$$

$$x_2 = -0.25(-200) = 50.$$

Let $Z_a = -144$. (Note $-144 \geq -200$.) Then intercepts are:

$$x_1 = -0.1(-144) = 14.4,$$

$$x_2 = -0.25(-144) = 36.$$

So the optimal solution using this approach is $(x_1^*, x_2^*) = (6, 21)$ and $Z^* = -144$ (see Figure 2.4). This illustrates that Min $= -$ Max, since $Z_{min} = Z_{max}$ (i.e., $144 = -(-144)$). This is not the same concept as "duality," which is covered in a later chapter. It is simply a procedure for converting a minimization problem to an equivalent maximization problem.

An Example

Consider the following four-activity minimization problem:

Min: $Z = 11x_1 + 12x_2 + 13x_3 + 9x_4$ (0)

s.t.:

$$1x_1 + 1x_2 + 1x_3 + 1x_4 \geq 100 \qquad (1)$$

$$2x_1 + 3x_2 + 1x_3 + 2x_4 \geq 250 \qquad (2)$$

$$x_1, \quad x_2, \quad x_3, \quad x_4 \geq 0 \qquad (3)$$

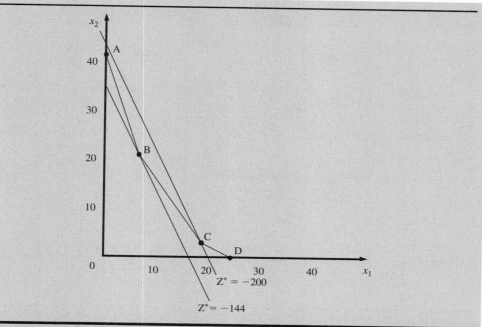

Figure 2.4   Graphical solution to the minimization problem using maximization of -1 times the objective function.

Using approach 1, the first tableau for this problem is:

First Tableau

Basis	CB	x_1	x_2	x_3	x_4	s_1	a_1	s_2	a_2	b	b_i/a_{ij}
		11	12	13	9	0	m	0	m		
a_1	m	1	1	1	1	-1	1	0	0	100	100
a_2	m	2	3	1	2	0	0	-1	1	250	83.33
	z_j	3m	4m	2m	3m	$-$m	m	$-$m	m	350m	
	$c_j - z_j$	11$-$3m	12$-$4m	13$-$2m	9$-$3m	m	0	m	0		

In this case, the pivot column is the column with the most negative $c_j - z_j$ value, which is column x_2. Since a_2 has the smallest b_i/a_{ij} ratio, it becomes the pivot row.

Next Iteration

Step 1: Create the new x_2 row by dividing the old a_2 row:

$$2x_1 + 3x_2 + \quad 1x_3 + \quad 2x_4 + 0s_1 + 0a_1 - \quad 1s_2 + \quad 1a_2 = 250 \qquad \text{(old } a_2 \text{ row)}$$

by 3 to get:

$$0.67x_1 + 1x_2 + 0.33x_3 + 0.67x_4 + 0s_1 + 0a_1 - 0.33s_2 + 0.33a_2 = 83.33 \qquad \text{(new } x_2 \text{ row)}$$

Step 2: Create the new a_1 row by multiplying new x_2 row by -1 and adding to the old a_1 row:

$$-0.67x_1 - 1x_2 - 0.33x_3 - 0.67x_4 - 0s_1 - 0a_1 + 0.33s_2 - 0.33a_2 = -83.33 \quad \text{(new } x_2$$
$$\text{row times } -1)$$

$$1x_1 + 1x_2 + \quad 1x_3 + \quad 1x_4 - 1s_1 + 1a_1 + \quad 0s_2 + \quad 0a_2 = 100 \quad \text{(old } a_1 \text{ row)}$$
$$\text{(add together)}$$

$$\overline{0.33x_1 + 0x_2 + 0.67x_3 + 0.33x_4 - 1s_1 + 1a_1 + 0.33s_2 - 0.33a_2 = 16.67} \quad \text{(new } a_1 \text{ row)}$$

Using these results, the second tableau becomes:

Second Tableau

		x_1	x_2	x_3	x_4	s_1	a_1	s_2	a_2		
Basis	CB	11	12	13	9	0	m	0	m	b	b_i/a_{ij}
a_1	m	0.33	0	0.67	0.33	-1	1	0.33	-0.33	16.67	25
x_2	12	0.66	1	0.33	0.67	0	0	-0.33	0.33	83.33	250
z_j		8 + 0.33m	12	4 + 0.67m	8 + 0.33m	$-m$	m	-4 + 0.33m	4 − 0.33m	999.96+ 16.67m	
$c_j - z_j$		3 − 0.33m	0	9 − 0.67m	1 − 0.33m	m	0	4 − 0.33m	1.33m − 4		

Since not all $c_j - z_j$ are positive, this tableau is not the optimal solution. The new pivot column is x_3 since it has the most negative $c_j - z_j$ value and the new pivot row is a_1.

Next Iteration

Step 1: Create the new x_3 row by dividing the old a_1 row:

$$0.33x_1 + 0x_2 + 0.67x_3 + 0.33x_4 - \quad 1s_1 + \quad 1a_1 + 0.33s_2 - 0.33a_2 = 16.66 \quad \text{(old } a_1 \text{ row)}$$

by 0.67 to get:

$$0.5x_1 + 0x_2 + \quad 1x_3 + 0.5x_4 - 1.5s_1 + 1.5a_1 + 0.5s_2 - 0.5a_2 = 25 \quad \text{(new } x_3 \text{ row)}$$

Step 2: Create new x_2 row by multiplying new x_3 row by -0.33 and adding to the old x_2 row:

$$-0.17x_1 + 0x_2 - 0.33x_3 - 0.17x_4 + 0.5s_1 - 0.5a_1 - 0.17s_2 + 0.17a_2 = -8.33 \quad \text{(new } x_3$$
$$\text{times } -0.33)$$

$$0.67x_1 + 1x_2 + 0.33x_3 + 0.67x_4 + \quad 0s_1 + \quad 0a_1 - 0.33s_2 + 0.33a_2 = 83.33 \quad \text{(old } x_2 \text{ row)}$$
$$\text{(add together)}$$

$$\overline{0.5x_1 + 1x_2 + \quad 0x_3 + 0.5x_4 + 0.5s_1 - 0.5a_1 - 0.5s_2 + 0.5a_2 = 75} \quad \text{(new } x_2 \text{ row)}$$

Using these results, the third tableau becomes:

Third Tableau

Basis	CB	x_1 11	x_2 12	x_3 13	x_4 9	s_1 0	a_1 m	s_2 0	a_2 m	b	b_i/a_{ij}
x_3	13	0.5	0	1	0.5	−1.5	1.5	0.5	−0.5	25	50
x_2	12	0.5	1	0	0.5	0.5	−0.5	−0.5	0.5	75	150
z_j		12.5	12	13	12.5	−13.5	13.5	0.5	−0.5	1,225	
$c_j - z_j$		−1.5	0	0	−3.5	13.5	m − 13.5	−0.5	m + 0.5		

Since not all $c_j - z_j$ are positive, this tableau is not the optimal solution. The new pivot column is x_4 since it has the most negative $c_j - z_j$ value and the new pivot row is x_3.

Next Iteration

Step 1: Create new x_4 row by dividing old x_3 row:

$$0.5x_1 + 0x_2 + 1x_3 + 0.5x_4 - 1.5s_1 + 1.5a_1 + 0.5s_2 - 0.5a_2 = 25 \qquad \text{(old } x_3 \text{ row)}$$

by 0.5 to get:

$$1x_1 + 0x_2 + 2x_3 + 1x_4 - 3s_1 + 3a_1 + 1s_2 - 1a_2 = 50 \qquad \text{(new } x_4 \text{ row)}$$

Step 2: Create new x_2 row by multiplying new x_4 row by −0.5 and adding to old x_2 row:

$$-0.5x_1 + 0x_2 - 1x_3 - 0.5x_4 + 1.5s_1 - 1.5a_1 - 0.5s_2 + 0.5a_2 = -25 \qquad \begin{array}{l}\text{(new } x_4 \text{ row}\\ \text{times } -0.5)\end{array}$$

$$0.5x_1 + 1x_2 + 0x_3 + 0.5x_4 + 0.5s_1 - 0.5a_1 - 0.5s_2 + 0.5a_2 = 75 \qquad \text{(old } x_2 \text{ row)}$$

_____ (add together)

$$0x_1 + 1x_2 - 1x_3 + 0x_4 + 2s_1 - 2a_1 - 1s_2 + 1a_2 = 50 \qquad \text{(new } x_2 \text{ row)}$$

Using these results, the fourth tableau becomes:

Fourth Tableau

Basis	CB	x_1 11	x_2 12	x_3 13	x_4 9	s_1 0	a_1 m	s_2 0	a_2 m	b	b_i/a_{ij}
x_4	9	1	0	2	1	−3	3	1	−1	50	
x_2	12	0	1	−1	0	2	−2	−1	1	50	
z_j		9	12	6	9	−3	3	−3	3	1,050	
$c_j - z_j$		2	0	7	0	3	m − 3	3	m − 3		

Since all $c_j - z_j$ are positive or zero, this tableau represents the optimal solution. The optimal solution is:

$$x_1^* = 0, \ x_2^* = 50, \ x_3^* = 0, \ x_4^* = 50, \ s_1^* = 0, \ a_1^* = 0, \ s_2^* = 0, \ a_2^* = 0, \text{ and } Z^* = 1,050.$$

This problem could have also been solved using the second approach, and an identical answer would have been found.

Special Cases

As was true in the graphical method, four special cases need to be mentioned for the simplex method. These are the cases of the unbounded solution, no feasible solution, multiple optimal solutions, and the special case of degeneracy.

Unbounded Solution   An unbounded solution exists for a maximization problem whenever it is possible for the value of the objective function to approach positive infinity. Such cases are recognized using the graphical approach whenever the feasible region extends to infinity in the x_1 and/or x_2 axes. An unbounded solution for a minimization problem occurs whenever it is possible for the value of the objective function to approach negative infinity.

Unbounded solutions are detected using the simplex method whenever the following occurs:

For a maximization problem, if a nonbasic variable's $c_j - z_j$ value is positive and its respective substitution coefficients (a_{ij}) are all nonpositive for any simplex tableau, then the solution is unbounded. For a minimization problem if a nonbasic variable's $c_j - z_j$ value is negative and its respective substitution coefficients are all nonpositive for any simplex tableau, then the solution is unbounded.

Why is an unbounded solution detected in this way? Recall that the substitution coefficients give the per unit decrease in the basic solution values for each unit increase in a nonbasic variable. If all of the substitution coefficients for a nonbasic variable are nonpositive, this implies that the solution values will actually increase (or not decrease) for every unit of the entering nonbasic variable being forced into the next solution. Hence, an unlimited amount of the new basic variable can be brought in without causing the existing basic variables to become zero or nonbasic. Therefore, for a maximization problem, the value of the objective function could increase forever at a rate of the net contribution ($c_j - z_j$) of the nonbasic variable since an infinite amount of it can be brought into the solution. This implies that the LP has no optimal solution because it is unbounded from above.

In the case of a minimization problem, if the $c_j - z_j$ value for the nonbasic variable is negative and if its substitution coefficients are all nonpositive, then an infinite amount of it could be brought into the solution, and the value of the objective function would approach negative infinity. Thus, this problem would have an unbounded solution. For either maximization or minimization problems, the conditions for being unbounded only occur for any nonbasic variable, not necessarily for the nonbasic variable that is in the pivot column.

No Feasible Solution   Recall from the previous chapter that no feasible solution occurs when the feasible region is empty. This is usually due to an inconsistency in the constraints. The case of no feasible solution is very easily recognized in the simplex approach with the use of the following rule:

If an artificial variable has a positive value in the final simplex tableau, then there is no feasible solution to the problem.

This should not be surprising since artificial variables are used to make infeasible solutions feasible, recognizing that they need to be made nonbasic in the final tableau. If artificial variables are basic in the final tableau, then you know that the solution is infeasible.

An infeasible solution can never occur for a problem that has all $\leq$ constraints because the origin will always be a feasible solution.

Multiple Optimal Solutions   Multiple optimal solutions occur whenever the slope of the iso-contribution line is the same as the slope of a line segment connecting two extreme points in the feasible region. The existence of alternative optima is recognized with the simplex method whenever the following occurs:

If any nonbasic variable in the final simplex tableau has a net contribution of zero (i.e., $c_j - z_j = 0$), then the optimal solution is not unique, but rather multiple optimal solutions exist.

Suppose that all nonbasic activities had zero net contributions. Then, if the simplex tableau continues to be iterated, a new solution will be found for the optimal activities, but each new solution would have the same objective function value because each new basic variable would have a net contribution of zero. This process could go on indefinitely.

Degeneracy   A final type of "special" solution is called **degeneracy**. Degeneracy is an optimal solution that is characterized by having at least one basic variable with a value of zero. Practically speaking, degeneracy is not usually a problem. However, degeneracy can theoretically be a problem if it results in "cycling." Cycling occurs whenever a degenerate basic variable is removed from one simplex tableau and brought back through the iterative process in a subsequent simplex tableau such that there is no improvement in the solution. Cycling therefore causes the possibility of the iterative simplex process becoming an infinite loop, implying that the stopping criterion is never satisfied.

There are methods designed to cure the potential problem of cycling. For example, a very small number can be added to or subtracted from the RHS or technical coefficients to cure degeneracy. However, degeneracy is usually more of a theoretical problem than one that is actually encountered in applications, and therefore it is not covered here in any detail.

SUMMARY

This chapter has focused on an algebraic technique used to solve LP problems. In order to fully understand the simplex method, it is recommended that the student rework the examples presented in this chapter and then try to answer the problems presented in the exercises. The more problems the student works through, the easier this technique becomes.

Solution techniques using the simplex technique for LP models containing $\leq$, $\geq$, and $=$ types of constraints for maximization and minimization problems were discussed in this chapter. The student should now be able to use this technique to solve any smaller (i.e., two- to five-activity) LP problems. Problems larger than this should be left to the computer as solving this type of problem can be dangerous to your health! A discussion of special cases of solutions including unbounded, multiple optimal, and no feasible solutions using the simplex approach was also presented. The case of degeneracy was examined, and it was argued that this is almost never a problem in applied LP.

EXERCISES

1. Consider the following LP maximization problem:

   Max: $Z = 90x_1 + 120x_2$ (0)

   s.t.:

$$1x_1 + 1x_2 \leq 200 \tag{1}$$
$$2x_1 + 3x_2 \leq 480 \tag{2}$$
$$1x_1 \qquad \leq 150 \tag{3}$$
$$x_1, \qquad x_2 \geq 0 \tag{4}$$

   a. If x_1 and x_2 are both equal to 0, then what are the solution values for the slack variables?
   b. If $x_1 = 90$ and $x_2 = 100$, then what are the solution values for the slack variables?
   c. Write out the initial simplex tableau for this problem.
   d. Solve the second simplex tableau for this problem.
   e. Solve the third simplex tableau for this problem.
   f. State the simplex criterion that indicates that an optimal solution has been reached.

2. An LP model is as follows:

   Min: $Z = 4x_1 + 39x_2 - 60x_3$ (0)

   s.t.:

$$x_1 + 2x_2 \qquad = 52 \tag{1}$$
$$3x_1 + \qquad 5x_3 \leq 36 \tag{2}$$
$$x_2 - 3x_3 \geq 3 \tag{3}$$
$$x_1, \quad x_2, \quad x_3 \geq 0 \tag{4}$$

   a. Transform the model into standard form.
   b. Solve the problem using the simplex method.
3. For the following LP:

   Max: $Z = 3x_1 + 4x_2$ (0)

   s.t.:

$$x_1 + 3x_2 \leq 8 \tag{1}$$
$$x_1 + x_2 \leq 4 \tag{2}$$
$$x_1, \qquad x_2 \geq 0 \tag{3}$$

   a. Formulate in standard form.
   b. Identify all basic solutions and decide if they are feasible or nonfeasible.
   c. Determine the optimal solution and what path the simplex method would follow to get to it.

4. Solve the following LP model using the simplex method:

Max: $Z = 7x_1 - 9x_2 - 4x_3 + 6x_4 + 12x_5$ (0)

s.t.:

$$1x_1 - 2x_2 + 2x_3 + 2x_4 + 1x_5 \leq 500 \quad (1)$$
$$2x_1 + 1x_2 + 3x_3 + 22x_4 \leq 900 \quad (2)$$
$$1x_5 \leq 200 \quad (3)$$
$$1x_1 \leq 350 \quad (4)$$
$$x_1, \quad x_2, \quad x_3, \quad x_4, \quad x_5 \geq 0 \quad (5)$$

Summarize the optimal solution (e.g., find the optimal productive and slack activity values, Z^*, and which constraints are binding).

5. Solve the following exercise using the simplex method:

Max: $Z = 5x + 3y$ (0)

s.t.:

$$2x + y \leq 20 \quad (1)$$
$$x + 2y \leq 36 \quad (2)$$
$$3x + y \leq 24 \quad (3)$$
$$x, \quad y \geq 0 \quad (4)$$

6. A cheese plant produces and sells three types of cheese: Cheddar, Monterey Jack, and Swiss. The sole objective is to maximize total profit from the production and sale of the three cheeses. The data below describes the production hours per unit in each of the three production operations required to produce each cheese and other data for the exercise.

Type of Cheese	Labor (hours/100 pounds)			Profit ($/100 pounds)
	1	2	3	
Cheddar	0.2	0.5	0.5	50.00
Monterey Jack	0.5	0.5	0.2	40.00
Swiss	1.0	0.3	0.2	70.00
Weekly Time Available (hours)	90	40	60	

a. Write this problem in standard form (including slack variables).
b. Solve this problem using the simplex method. Write each tableau on additional sheets of paper. Label tableau 1 as 1, tableau 2 as 2, etc. What is the optimal solution for this problem?

7. Consider the following model:

Max: $Z = 5x_1 + 6x_2 + 3x_3$ (0)

s.t.:

$$x_1 + x_2 + x_3 \geq 1,000 \quad (1)$$
$$x_1 - x_2 = 0 \quad (2)$$
$$x_1 + x_2 + x_3 \leq 2,000 \quad (3)$$
$$x_1, \quad x_2, \quad x_3 \geq 0 \quad (4)$$

a.   Set up this problem in standard form.
b.   Solve this problem using the simplex method.
c.   Report the optimal solution.

8.   Consider the following simplex tableau:

Basis	CB	x_1	x_2	x_3	s_1	s_2	s_3	s_4	b	b_i/a_{ij}
		3	4	2	0	0	0	0		
s_1	0	−0.33	0	0	1	−0.66	−1	0	2	
x_2	4	2	1	0	0	1	0	0	8	
x_3	2	−0.66	0	1	0	−0.33	1	0	2	
s_4	0	1.33	0	0	0	0.66	−2	1	2	
z_j										
$c_j - z_j$										

a.   Complete this simplex tableau including the z_j and $c_j - z_j$ rows, and the b_i/a_{ij} column.
b.   What is the total value of the objective function for this simplex tableau?
c.   Does this provide the optimal solution? If so, indicate what the solution is and how you decided it was optimal. If not, indicate how you decided and what the next variable introduced into the basis should be.
d.   What is the meaning of the $c_j - z_j$ value obtained in the x_1 column?

9.   Solve the following LP problem:

$$\text{Max } Z = 14x_1 + 5x_2 + 12x_3 + 9x_4 \tag{0}$$

s.t.:

$$4x_1 + 2x_2 + x_3 \geq 500 \tag{1}$$

$$5x_1 + 4x_4 \geq 2000 \tag{2}$$

$$x_2 + 2x_3 \leq 1000 \tag{3}$$

$$x_1, \quad x_2, \quad x_3, \quad x_4 \geq 0 \tag{4}$$

10. Solve the following LP model using the simplex method:

$$\text{Max: } Z = 9x_1 - 9x_2 - 4x_3 + 8x_4 + 14x_5 \tag{0}$$

s.t.:

$$2x_1 - 4x_2 + 1x_3 + 1x_4 + 2x_5 \leq 500 \tag{1}$$

$$3x_1 + 2x_2 + 2x_3 + 25x_4 \leq 900 \tag{2}$$

$$3x_5 \leq 200 \tag{3}$$

$$2x_1 \leq 350 \tag{4}$$

$$x_1, \quad x_2, \quad x_3, \quad x_4, \quad x_5 \geq 0 \tag{5}$$

Summarize the optimal solution (e.g., find the optimal productive and slack activity values, Z^*, and which constraints are binding).

11. Solve the following LP model using the simplex method:

Max: $Z = 25x_1 + 20x_2 - 2x_3 + 16x_4$ (0)

s.t.:

$$2x_1 + 8x_2 - 4x_3 + 6x_4 \leq 400 \qquad (1)$$

$$7x_1 + 3x_2 + 1x_3 + 15x_4 \leq 600 \qquad (2)$$

$$x_1, \quad x_2, \quad x_3, \quad x_4 \geq 0 \qquad (3)$$

Summarize the optimal solution (e.g., find the optimal productive and slack activity values, Z^*, and which constraints are binding).

12. A horticulturist is considering growing five types of flowers to sell in retail stores. The flowers are (1) roses, which net a profit of $2.00 per flower, (2) carnations, which net a profit of $0.75 per flower, (3) daisies, which net a profit of $0.35 per flower, (4) chrysanthemums (mums), which net a profit of $0.25 per flower, and (5) daffodils, which net a profit of $0.70 per flower. The horticulturist owns 9 acres of land, which is suitable for growing any of these five flowers. Assume the horticulturist faces the following production technology:

Unit Resource	Resource Requirement					Resource Endowment
	Roses	Carnations	Daisies	Mums	Daffodils	
Acres per Flower	0.001	0.0005	0.0003	0.0001	0.0004	9
Hours of Labor	0.100	0.0700	0.0600	0.0500	0.0700	1,200

Further assume that (1) at most, 2,500 roses can be grown and (2) the combination of daisies and mums cannot exceed 5,000 flowers. The sole objective is to maximize profits. Find the optimal solution to this problem using the simplex method. Write out each simplex tableau, and show all your work in arriving at each tableau. Clearly label each tableau.

13. A company manufacturing copper releases three pollutants into the air: A, B, and C. The state has just passed a law that mandates that the company's plant must reduce its emissions of A, B, and C by at least 15, 10, and 20 units, respectively. The company can reduce its emissions of A, B, and C by using three pollution abatement processes: x_1, x_2, and x_3. Use of 1 unit of x_1 will reduce pollutants A, B, and C by 1, 1, and 2 units, respectively. Use of 1 unit of x_2 will reduce pollutants A, B, and C by 3, 0, and 1 units, respectively. That is, only pollutants A and C can be reduced by using x_2; pollutant B cannot be reduced by using x_2. Finally, use of 1 unit of X_3 will reduce pollutants A, B, and C by 0, 2, and 0 units, respectively. That is, only pollutant B can be reduced under pollution control device x_3. Each unit of x_1, x_2, and x_3 costs the company $4, $2, and $3, respectively. The objective of the company is to install the least expensive combination of pollution control devices (x_1, x_2, and x_3) that also satisfies the state's requirement that A, B, and C be reduced by at least 15, 10, and 20 units, respectively.

  a. State this problem in standard form.

  b. Restate the objective function of this problem in such a way that it is technically a maximization problem, with a solution identical to a minimization problem given the constraints stated in part a.

c. Solve this problem using the simplex method.

d. What is the optimal solution to this problem?

14. Consider the following LP problem:

$$\text{Max: } Z = 4x_1 + 2x_2 - 3x_3 + 5x_4 \tag{0}$$

s.t.:

$$2x_1 - 1x_2 + 1x_3 + 2x_4 \geq 50 \tag{1}$$

$$3x_1 - 1x_3 + 2x_4 \leq 80 \tag{2}$$

$$1x_1 + 1x_2 + 1x_4 = 60 \tag{3}$$

$$x_1, \quad x_2, \quad x_3, \quad x_4 \geq 0 \tag{4}$$

a. Write the first tableau for this problem in standard form including artificial variables.

b. Solve this problem using the simplex method.

15. Solve the following maximization problem using the simplex method:

$$\text{Max: } Z = 10x + 5y + 15v \tag{0}$$

s.t.:

$$x + y + v \leq 1{,}000 \tag{1}$$

$$y = 100 \tag{2}$$

$$x, \quad y, \quad v \geq 0 \tag{3}$$

16. A dairy farmer's cows need three nutrients (A, B, and C) to subsist and produce milk each day. Each cow must receive the equivalent of 100 units of nutrient A, 200 units of nutrient B, and 50 units of nutrient C in order to maximize milk output. The farmer can use any combination of three feeds (f_1, f_2, and f_3) in meeting these minimum requirements. The local feed dealer sells all three feeds, which have the following cost per pound and nutrient equivalents (for A, B, and C) per pound.

Feed	Nutrient Content Per Pound			Cost ($/pound)
	A	B	C	
f_1	5	22	3	0.25
f_2	10	25	2	0.50
f_3	7	12	5	0.27

Assume that the farmer's objective is to minimize the cost per cow of buying any combination of these three feeds that satisfy the daily nutrient requirements of the cow.

a. Write the standard form of this LP problem.

b. Solve this problem using the simplex method.

17. Solve the following LP model using the simplex method:

Max: $Z = 10x_1 + 8x_2 - 2x_3 + 6x_4$ (0)

s.t.:

$$5x_1 + 3x_2 - 4x_3 + 2x_4 \leq 200 \quad (1)$$

$$6x_1 + 5x_2 + 1x_3 + 10x_4 \leq 300 \quad (2)$$

$$x_1, \quad x_2, \quad x_3, \quad x_4 \geq 0 \quad (3)$$

Summarize the optimal solution (e.g., find the optimal productive and slack activity values, Z^*, and which constraints are binding).

18. Consider the following problem:

Max: $Z = 10a + 20b - 5c + 15d$ (0)

s.t.:

$$1a + 2b + 3c + 4d \leq 1{,}000 \quad (1)$$

$$1a - 2b - 2c \qquad \leq -500 \quad (2)$$

$$4c - 7d = -100 \quad (3)$$

$$a, \quad b, \quad c, \quad d \geq 0 \quad (4)$$

a. Reformulate this problem so that the negative RHS values are eliminated from the constraints.

b. Write the first simplex tableau for this problem using the reformulated problem from part a.

19. A farmer owns 500 acres of land, which are suitable for growing corn, soybeans, and sunflowers. His expectations are that the net profit from producing each crop is: $55 per acre for corn, $60 per acre for soybeans, and $50 per acre for sunflowers. He and his family can supply 3,000 hours per year in performing all the farm operations necessary to grow these crops. In addition, he is endowed with the equivalent of 4,500 hours of tractor time necessary to grow these crops. Assume that the only resources necessary in crop production are land, labor, and tractor time. The technology is summarized below.

Resource (Unit)	Crop			Resource Endowment
	Corn	Soybeans	Sunflowers	
Land (acres)	1.0	1.0	1.0	500
Labor (hours)	0.4	0.2	0.3	3,000
Tractor (hours)	0.5	0.2	0.4	4,500

Assuming that the farmer's objective is to maximize total profits, what is the LP model for this problem? Solve this problem using the simplex method.

20. A farmer has the following resource endowments: 1,000 acres of land, 1,500 hours of family labor, and $30,000 for capital investment. She can use these resources to grow corn, wheat, and soybeans. The farmer expects the following in terms of crop yields, prices, variable costs, and labor requirements:

Crop	Price ($1 bushel)	Yield (bushel/acre)	Variable Cost ($/acre)	Labor Requirement (hours/acre)
Corn	2.75	120	250	3.25
Wheat	2.65	100	200	3.00
Soybeans	6.75	45	230	3.30

Assume that the farmer works to maximize net revenue (gross revenue minus variable costs) from the production of these three crops. Formulate this as an LP problem, and solve it using the simplex method.

21. Explain what is wrong with the following problem:

$$\text{Max: } Z = 10x + 7y \tag{0}$$

s.t.:

$$1x + 2y \le 100 \tag{1}$$

$$10x + 20y \ge 2{,}000 \tag{2}$$

$$x, \quad y \ge 0 \tag{3}$$

22. Write the following LP problem in general form:

$$\text{Max: } Z = 5k + 10l + 9a + 20f + 0s_1 + 0s_2 - Ma_2 - Ma_3 + 0s_4 - Ma_4 \tag{0}$$

s.t.:

$$1k + 1l + 1a + 1f + 1s_1 \qquad\qquad\qquad = 1{,}000 \tag{1}$$

$$1k + 1l + 1a + 1f - \qquad 1s_2 + a_2 \qquad\qquad = 1{,}000 \tag{2}$$

$$1k + \qquad\qquad\qquad\qquad a_3 \qquad = 200 \tag{3}$$

$$1a - \qquad\qquad\qquad 1s_4 + 1a_4 = 150 \tag{4}$$

$$k, \quad l, \quad a, \quad f, \quad s_1, \quad s_2, \quad a_2, \quad a_3, \quad s_4, \quad a_4 \ge 0 \tag{5}$$

23. What is wrong with the following model?

$$\text{Max: } Z = 10a + 15c + 5e \tag{0}$$

s.t.:

$$a + \quad c + \quad e \ge 100 \tag{1}$$

$$a - \quad c \qquad = 0 \tag{2}$$

$$a, \quad c, \quad e \ge 0 \tag{3}$$

How would you tell from the final simplex tableau that there is a problem with this model?

24. You have just been hired as an advertising manager for a generic advertising program for dairy farmers, Dairy Management, Inc. (DMI). DMI wants to conduct generic

advertising to increase the demand for milk. DMI has decided to consider both TV and radio, and wants you to do an analysis of how many TV and radio commercials to purchase for the month. You expect that one TV commercial will increase sales by 25,000 gallons and one radio commercial will increase sales by 7,000 gallons. It costs $10,000 per TV commercial and $5,000 per radio commercial. You have a budget of $200,000 for this project. Furthermore, the radio and TV stations have a combined maximum of 90 minutes for your commercials for the month. Each TV commercial takes 1 minute, and each radio commercial takes 2 minutes to air. Marketing research indicates that milk consumers do not want more than 15 TV commercials because they find watching the same commercial multiple times irritating. The objective is to find the combination of TV (x) and radio (y) commercials that maximize the sale of milk. Solve this problem using the simplex method.

3

Sensitivity Analysis Using the Simplex Method and Duality

In the previous chapter, the reader learned how to solve linear programming (LP) models using the simplex method. In this chapter, simplex-based sensitivity analysis is examined for both maximization and minimization problems. As was argued in the first chapter, sensitivity analysis is an essential element of any analysis because it provides answers to "what if" types of questions. More important to economists, sensitivity analysis can be used to derive output supply and input demand functions from an LP model.

This chapter also examines the notion of **duality**, which is the presentation of the same problem in two different ways. It cannot be emphasized enough that duality is one of the most important concepts in mathematical programming and is also an important topic in economics. Duality is fundamental to sensitivity analysis particularly with respect to shadow prices (SP). Recall that an SP gives the value of the change in the objective function given a one unit change in a constraint's right-hand-side (RHS) value. The SP is also fundamental to solution algorithms for mathematical programming, as well as LP, because it provides two possible formulations of the problem, which yield identical solutions. Obviously this adds computational efficiency, since the easier of the two problems may be solved.

There are three objectives of this chapter. The first objective is to learn how to use the simplex method to conduct sensitivity analysis. A solid understanding of simplex-based sensitivity analysis will enable a better understanding of how to interpret sensitivity reports from LP software. Second, several properties, concepts, and examples of duality are explained. A better understanding, as well as appreciation for the importance of duality as it relates to LP will be gained upon completion of this section. Finally, a brief overview of how to solve LP problems with Solver is presented. Solver is an add-in software optimization package that is used in conjunction with Excel, and can be used to solve linear and nonlinear programming models. The ability to set up, solve, and analyze LP models with Solver will be gained upon completion of this chapter.

In the duality section of this chapter, there is some use of summation and matrix notation to present LP models more compactly. This chapter includes an appendix that provides a basic primer on summation notation and matrix operations. For readers that are unfamiliar with the use of matrix and summation notation, read the appendix prior to this section because knowledge of this notation is necessary for understanding the presentation on duality. In addition, some models in future chapters will use both summation and matrix notation; therefore, readers not possessing a basic knowledge of this notation or needing their knowledge refreshed should refer to this appendix.

3.1   SIMPLEX-BASED SENSITIVITY ANALYSIS FOR MAXIMIZATION PROBLEMS

Sensitivity analysis can be performed using the final simplex tableau, which gives the optimal basic feasible solution (BFS). The information contained in the final tableau allows us to compute SPs for the constraints, and to do sensitivity analysis for objective function coefficient values and resource endowments.

Objective Function Coefficients (c_i) Sensitivity Analysis

Recall that the range of optimality refers to that range of objective function coefficient values such that the optimal solution will not change. This can be computed for nonbasic and for basic variables using the final simplex tableau.

Consider the following maximization problem expressed in standard form with slack variables and equality constraints:

$$\text{Max: } Z = 35x_1 + \quad 50x_2 + 0s_1 + 0s_2 + 0s_3 \tag{0}$$

s.t.:

$$x_1 + \quad x_2 + s_1 \qquad\qquad = 1{,}000 \tag{1}$$

$$2.5x_1 + 0.75x_2 \qquad + s_2 \qquad = 1{,}500 \tag{2}$$

$$1.5x_2 \qquad\qquad + s_3 = 800 \tag{3}$$

$$x_1, \qquad x_2, \quad s_1, \quad s_2, \quad s_3 \geq 0 \tag{4}$$

The final simplex tableau is:

Basis	CB	x_1 35	x_2 50	s_1 0	s_2 0	s_3 0	b	b_i/a_{ij}
s_1	0	0	0	1	−0.4	−0.47	26.67	
x_1	35	1	0	0	0.4	−0.2	440	
x_2	50	0	1	0	0	0.66	533.33	
z_j		35	50	0	14	26.33	42,066.67	
Net Eval ($c_j − z_j$)		0	0	0	−14	−26.33		

The range of optimality for this simplex tableau is defined based on the optimal solution condition: that is, $c_j − z_j \leq 0$. That is, find the range of objective function coefficients such that $c_j − z_j \leq 0$, for all values of j (meaning basic and nonbasic variables).

Finding the Range of Optimality for a Basic Variable

Finding the **range of optimality** for a basic variable is done by replacing its numeric objective function coefficient in the final tableau with the more general c_j. For example, the range of optimality for the basic variable x_1 is determined by replacing $c_1 = 35$ with c_1 in the final tableau, which results in the following:

		x_1	x_2	s_1	s_2	s_3	
Basis	CB	c_1	50	0	0	0	b
s_1	0	0	0	1	-0.4	-0.47	26.67
x_1	c_1	1	0	0	0.4	-0.2	440
x_2	50	0	1	0	0	0.66	533.33
	z_j	c_1	50	0	$0.4c_1$	$-0.2c_1 + 33.33$	$440c_1 + 26{,}666.5$
Net Eval $(c_j - z_j)$		0	0	0	$-0.4c_1$	$0.2c_1 - 33.33$	

It is now well known that this tableau will be the optimal solution if and only if:

$$c_j - z_j \leq 0 \text{ for all } j \tag{3.1}$$

Using condition (3.1), the following must hold:

$$-0.4c_1 \leq 0, \text{ and} \tag{3.2}$$

$$0.2c_1 - 33.33 \leq 0 \tag{3.3}$$

In order for (3.2) to hold,

$$c_1 \geq 0 \tag{3.4}$$

since it is being multiplied by a negative coefficient. In order for (3.3) to hold,

$$c_1 \leq 166.67 \tag{3.5}$$

(i.e., solve (3.3) for c_1). Combining conditions (3.4) and (3.5) provides the range of optimality for c_1, that is:

$$0 \leq c_1 \leq 166.67 \tag{3.6}$$

Verify that when $c_1 = 0$ or $c_1 = 166.67$, the result is that $c_j - z_j \leq 0$ for all j.

Finding the Range of Optimality for c_{S2}, a Nonbasic Variable

What about the range of optimality for a nonbasic variable? You may use the exact same procedure as that for a basic variable to obtain this range. Consider the nonbasic variable s_2 for this example. Replacing the 0 objective function coefficient on s_2 with c_{S2} in the final simplex tableau yields:

		x_1	x_2	s_1	s_2	s_3		
Basis	CB	35	50	0	c_{S2}	0	b	b_i/a_{ij}
s_1	0	0	0	1	-0.4	-0.47	26.67	
x_1	35	1	0	0	0.4	-0.2	440	
x_2	50	0	1	0	0	0.66	533.33	
	z_j	35	50	0	14	26.33	42,066.67	
Net Eval $(c_j - z_j)$		0	0	0	$c_{S2} - 14$	-26.33		

Condition (3.1) will now hold if and only if:

$$c_{S2} - 14 \leq 0, \text{ or}$$

$$c_{S2} \leq 14 \qquad (3.7)$$

Since there is no lower bound (LB) on what c_{S2} would have to be in order to satisfy condition (3.1), it is convention that the range of optimality for c_{S2} be written as:

$$-\infty \leq c_{S2} \leq 14 \qquad (3.8)$$

A relevant question is: Should one be concerned with the range of optimality for a nonbasic variable? The answer becomes clear when looking at larger models, where it is quite common to have some productive activities (the x_i's) as nonbasic variables in the optimal solution. In such cases, sensitivity analysis can be used to determine how much the objective function coefficients of the nonbasic productive activities need to change before the basis changes. A research example determining what minimum price is necessary to make a new energy crop, velvet beans, economically viable for Alabama cotton–corn farmers to produce is illustrated in Chapter 4.

For example, suppose that a food company may sell up to 100 different products (x_i, i = 1,...,100). Given the product prices (p_i) and all the structural constraints, suppose that the optimal solution includes 90 of these products as basic variables. An important question to the analyst is: What would it take for the 10 nonbasic products to become profitable enough to become basic? The answer to this question is found by determining the range of optimality for the nonbasic variable objective function coefficients, which are product prices for this example.

Resource Endowments (b) Sensitivity Analysis

Shadow prices are easy to calculate based on the final simplex tableau. They are found in the z_j row. Again, consider the final simplex tableau:

		x_1	x_2	s_1	s_2	s_3	
Basis	CB	35	50	0	0	0	b
s_1	0	0	0	1	-0.4	-0.47	26.67
x_1	35	1	0	0	0.4	-0.2	440
x_2	50	0	1	0	0	0.66	533.33
z_j		35	50	0	14	26.33	42,0665.67
Net Eval ($c_j - z_j$)		0	0	0	-14	-26.33	

The SPs for the three structural constraints (which are all $\leq$ type constraints) to this maximization problem are equal to the z_j value given in the final tableau for each of the slack variables. Hence, in this example:

$$SP_{b1} = 0,$$

$$SP_{b2} = 14,$$

$$SP_{b3} = 26.33.$$

For $\geq$ structural constraints in a maximization problem, the SPs are equal to the negative of the z_j value for each surplus variable associated with the $\geq$ constraint in the final tableau. Finally, for equality structural constraints in a maximization problem, the SPs are equal to the z_j value associated with each = constraint in the final tableau.

Range of Feasibility for Less-than-or-Equal-to Constraints

The **range of feasibility** is the range of values of b_i for which it can vary without causing any basic variables in the current solution to become infeasible (e.g. negative). In addition, this concept can also be interpreted as the range for b_i values where its SP do not change. The discussion that follows is applicable for $\leq$ structural constraints.

For example, suppose b_2 in the original problem (constraint (2)) was increased from 1,500 to 2,000. Will the current basis still yield a feasible basic solution? If so, then we know that given its SP of 14, the value of the objective function will increase by 14(500) = 7,000. To find the new solution, use the final simplex and the following formula:

$$\text{New Solution} = \text{Old Solution} + \Delta b_i \, (s_i \text{ Column})$$

where Δ means "change in," and s_i is the column of a_{ij} coefficients in the final tableau for the slack variable associated with constraint i. For this example, the new solution is:

$$\begin{bmatrix} -173.33 \\ 640 \\ 533.33 \end{bmatrix} = \begin{bmatrix} 26.67 \\ 440 \\ 533.33 \end{bmatrix} + 500 \begin{bmatrix} -0.4 \\ 0.4 \\ 0 \end{bmatrix}.$$

Clearly when b_2 is increased by 500 units the new solution is not feasible since $s_1 = -173.33$. What if b_2 was increased from 1,500 to 1,560? Then:

$$\text{New Solution} = \text{Old Solution} + \Delta b_2 \, (s_2 \text{ Column}) \text{ or}$$

$$\begin{bmatrix} 2.67 \\ 464 \\ 533.33 \end{bmatrix} = \begin{bmatrix} 26.67 \\ 440 \\ 533.33 \end{bmatrix} + 60 \begin{bmatrix} -0.4 \\ 0.4 \\ 0 \end{bmatrix}.$$

For this level of increase in b_2, the new solution is feasible. More generally, to find the range for which b_2 can vary without making the current solution infeasible, manipulate the formula above by solving for the permissible range of values for b_2, which satisfy that the new solution is still non-negative. In this case, solve the following for Δb_2:

$$\begin{bmatrix} s_1 \\ x_1 \\ x_2 \end{bmatrix} = \begin{bmatrix} 26.67 \\ 440 \\ 533.33 \end{bmatrix} + \Delta b_2 \begin{bmatrix} -0.4 \\ 0.4 \\ 0 \end{bmatrix}.$$

To be feasible, s_1, x_1, and $x_2 \geq 0$. Hence, the RHS of the above system of equations can be rewritten as (3.9), (3.10), and (3.11), and this range can be solved algebraically:

$$26.67 + \Delta b_2(-0.4) \geq 0 \qquad\qquad (3.9)$$

$$440 + \Delta b_2(0.4) \geq 0 \qquad\qquad (3.10)$$

$$533.33 + \Delta b_2(0) \geq 0 \qquad\qquad (3.11)$$

Solving each inequality separately for Δb_2 results in:

$$\Delta b_2 \leq 66.67 \tag{3.12}$$

$$\Delta b_2 \geq -1{,}100 \tag{3.13}$$

$$\text{Can't be solved for } \Delta b_2 \tag{3.14}$$

The range of feasibility for b_2 is given by the following condition, which satisfies both (3.12) and (3.13) simultaneously:

$$-1{,}100 \leq \Delta b_2 \leq 66.67 \tag{3.15}$$

or adding the upper limit of 66.67 and the lower limit of $-1{,}100$ to the current b_2 level of 1,500, the range of feasibility is:

$$400 \leq b_2 \leq 1{,}566.67 \tag{3.16}$$

In other words, as long as the change in b_2 is between $-1{,}100$ and 66.67, the current optimal basis will remain optimal. As a separate exercise, determine the range of feasibility for the other b_i values for this problem.

3.2   SIMPLEX-BASED SENSITIVITY ANALYSIS FOR MINIMIZATION PROBLEMS

Consider the following minimization problem:

Min: $Z = 11x_1 + 12x_2 + 13x_3 + 9x_4$ $\tag{0}$

s.t.:

$$1x_1 + 1x_2 + 1x_3 + 1x_4 \geq 100 \tag{1}$$

$$2x_1 + 3x_2 + 1x_3 + 2x_4 \geq 250 \tag{2}$$

$$x_1, \quad x_2, \quad x_3, \quad x_4 \geq 0 \tag{3}$$

The final tableau for this problem is:

Basis	CB	x_1	x_2	x_3	x_4	s_1	a_1	s_2	a_2	b	b_i/a_{ij}
		11	12	13	9	0	m	0	m		
x_4	9	1	0	2	1	-3	3	1	-1	50	
x_2	12	0	1	-1	0	2	-2	-1	1	50	
z_j		9	12	6	9	-3	3	-3	3	1,050	
$c_j - z_j$		2	0	7	0	3	$m-3$	3	$m-3$		

The range of optimality for this simplex tableau is defined based on the optimal solution condition, that is, $c_j - z_j \geq 0$. In other words, find the range of the objective function coefficients such that $c_j - z_j \geq 0$, for all values of j (meaning basic and nonbasic variables).

As was true in the maximization problem, finding the range of optimality for a basic variable is done by replacing its numeric objective function coefficient in the final tableau with the more general c_i. For example, the range of optimality for the basic variable x_2 is determined by replacing $c_2 = 12$ with c_2 in the final tableau, which results in the following:

Basis	CB	x_1 11	x_2 c_2	x_3 13	x_4 9	s_1 0	a_1 m	s_2 0	a_2 m	b	b_i/a_{ij}
x_4	9	1	0	2	1	-3	3	1	-1	50	
x_2	c_2	0	1	-1	0	2	-2	-1	1	50	
z_j		9	c_2	$18 - c_2$	9	$-27 + 2c_2$	$27 - 2c_2$	$9 - c_2$	$-9 + c_2$	1,050	
$c_j - z_j$		2	0	$-5 + c_2$	0	$27 - 2c_2$	$m - 27 + 2c_2$	$c_2 - 9$	$m + 9 - c_2$		

The following condition is used to derive the range of optimality:

$$c_j - z_j \geq 0 \text{ for all } j \tag{3.17}$$

Using condition (3.17), the following must hold:

$$-5 + c_2 \geq 0 \tag{3.18}$$
$$27 - 2c_2 \geq 0 \tag{3.19}$$
$$m - 27 + 2c_2 \geq 0 \tag{3.20}$$
$$c_2 - 9 \geq 0, \text{ and} \tag{3.21}$$
$$m + 9 - c_2 \geq 0 \tag{3.22}$$

In order for (3.18) to hold,

$$c_2 \geq 5 \tag{3.23}$$

In order for (3.19) to hold,

$$c_2 \leq 13.5 \tag{3.24}$$

In order for (3.20) to hold,

$$c_2 \geq 13.5 - 0.5m \tag{3.25}$$

In order for (3.21) to hold,

$$c_2 \geq 9 \tag{3.26}$$

In order for (3.22) to hold,

$$c_2 \leq m + 9 \tag{3.27}$$

First consider all the $\geq$ conditions in (3.23) through (3.27), that is, conditions (3.23), (3.25), and (3.26). Choose the most restrictive of these conditions as the LB for the range of optimality. By "most restrictive," we mean that it is a subset of the other conditions.[1] Condition (3.23) is more restrictive than (3.25) since if we let m approach infinity, then condition (3.25) implies that $c_2 \geq$ negative infinity. Condition (3.26) is more restrictive than condition (3.23) since (3.26) restricts c_2 to a smaller range on the number line than (3.23). Hence, the LB on c_2 for its range of optimality is condition (3.26), $c_2 \geq 9$.

---

[1] For example, the condition $x \geq 13$ is more restrictive than the condition $x \geq 0$ because the first is a subset of the second. Put differently, the condition $x \geq 13$ comprises a smaller range of real numbers than the condition $x \geq 0$.

Now consider all the $\leq$ conditions in (3.23) through (3.27), that is, conditions (3.24) and (3.27). Choose the most restrictive of these conditions as the upper bound (UB) for the range of optimality. Condition (3.24) is clearly more restrictive because if we let m approach infinity, condition (3.27) implies that $c_2 \leq \infty$. Hence, the UB is condition (3.24), $c_2 \leq 13.5$. Combining conditions (3.24) and (3.26) yields the following range of optimality for c_2:

$$9 \leq c_2 \leq 13.5 \tag{3.28}$$

What about the range of optimality for a nonbasic variable? The exact same procedure can be followed for the basic variable to obtain this range. Consider the nonbasic variable x_1 for this example. Replacing the 11 objective function coefficient on x_1 with c_1 in the final simplex tableau yields:

Basis	CB	x_1	x_2	x_3	x_4	s_1	a_1	s_2	a_2	b	b_i/a_{ij}
		c_1	12	13	9	0	m	0	m		
x_4	9	1	0	2	1	−3	3	1	−1	50	
x_2	12	0	1	−1	0	2	−2	−1	1	50	
z_j		9	12	6	9	−3	3	−3	3	1,050	
$c_j - z_j$		$c_1 - 9$	0	9	0	3	m − 3	3	m − 3		

Now the final basis will remain optimal if and only if:

$$c_1 - 9 \geq 0, \text{ or}$$

$$c_1 \geq 9 \tag{3.29}$$

Since there is no UB on what c_1 would have to be in order to satisfy condition (3.29), it is convention that the range of optimality for c_1 be written as:

$$9 \leq c_1 \leq \infty \tag{3.30}$$

Right-Hand-Side Sensitivity Analysis

Shadow prices are easy to calculate based on the final simplex tableau. They are found in the z_j row. Again, consider the final simplex tableau:

Basis	CB	x_1	x_2	x_3	x_4	s_1	a_1	s_2	a_2	b	b_i/a_{ij}
		11	12	13	9	0	m	0	m		
x_4	9	1	0	2	1	−3	3	1	−1	50	
x_2	12	0	1	−1	0	2	−2	−1	1	50	
z_j		9	12	6	9	−3	3	−3	3	1,050	
$c_j - z_j$		2	0	7	0	3	m − 3	3	m − 3		

The SP for the two structural constraints (which are both $\geq$ type constraints) to this minimization problem are equal to the z_j value given in the final tableau for each of the artificial variables. Hence, in this example:

$$SP_{b1} = 3,$$

$$SP_{b2} = 3.$$

More generally, SPs are found in the z_j row of the final simplex tableau for all types of LP problems. The table below presents a summary of SPs for minimization and maximization problems involving all three types of structural constraints.

Type of Constraint	SP for Either a Maximization or Minimization Problem
$\leq$	z_j value of slack variable associated with that constraint
$\geq$	z_j value of artificial variable associated with that constraint
$=$	z_j value of artificial variable associated with that constraint

Range of Feasibility for Greater-than-or-Equal-to and Equal-to Constraints

The following formula is used to calculate the range of feasibility for $\geq$ or $=$ structural constraints:

$$\text{New Solution} = \text{Old Solution} + \Delta b_i\,(a_i \text{ column}),$$

where Δ means "change in," and the a_i column refers to the column of a_{ij} coefficients in the final simplex tableau for the artificial variable associated with constraint i. Use this formula again by setting it $\geq$ zero to derive the range of values for Δb_i to satisfy non-negativity.

For example, consider the RHS value for constraint (1) in the minimization problem on the previous page. In this case, the following conditions are derived:

$$x_4 = 50 + 3\Delta b_1,$$

$$x_2 = 50 - 2\Delta b_1.$$

To be feasible, x_4 and x_2 must be non-negative. Rewriting the system of equations above as individual inequalities yields:

$$50 + 3\Delta b_1 \geq 0 \tag{3.31}$$

$$50 - 2\Delta b_1 \geq 0 \tag{3.32}$$

Solving each inequality for Δb_1 results in:

$$\Delta b_1 \geq -16.67 \tag{3.33}$$

$$\Delta b_1 \leq 25 \tag{3.34}$$

The range of feasibility for b_1 is thus given by the following condition:

$$-16.67 \leq \Delta b_1 \leq 25 \tag{3.35}$$

or expressing it in terms of b_1 instead of in terms of Δb_1:

$$83.33 \leq b_1 \leq 125 \tag{3.36}$$

As an exercise, determine the range of feasibility of b_2 for this problem.

3.3   DUALITY

This section examines the notion of duality. For those that are unfamiliar with the use of matrix and summation notation, read the appendix prior to this section because knowledge of this notation is necessary for understanding the presentation on duality.

All LP problems exist in pairs. For every LP problem there exists another LP problem whose formulation is different, but whose solution gives identical results to the original problem. This important notion is called **duality**.

The Relationship Between the Primal and Dual Problems

Duality is the formulation of one problem in two different ways. To illustrate, it is well known that under certain conditions, the solution to a problem that maximizes profits subject to technology constraints is identical to the solution to a problem that minimizes costs subject to equilibrium conditions (e.g., marginal costs = marginal revenue).

To understand the notion of duality, it is useful to begin with the definition of the two problems that result in identical solutions. As a matter of convention, the original LP problem will be called the **primal problem**, which has the following characteristic:

If the original problem is a minimization problem, then the primal is a minimization problem; and if the original problem is a maximization problem, then the primal is a maximization problem.

For example, consider the following maximization problem, whose primal problem is:

$$\text{Max: } Z = c_1 x_1 + c_2 x_2 \tag{0}$$

s.t.:

$$a_{11} x_1 + a_{12} x_2 \leq b_1 \tag{1}$$

$$a_{21} x_1 + a_{22} x_2 \leq b_2 \tag{2}$$

$$:: \quad ::: \qquad\qquad :$$

$$a_{m1} x_1 + a_{m1} x_2 \leq b_m \tag{m}$$

$$x_1, \qquad x_2 \geq 0 \tag{m+1}$$

The essence of duality is that for every primal problem there exists a **dual problem**. The dual problem is an alternative way of expressing the primal problem, and it has the following relationships with the primal problem:

1. The number of activities of the dual problem will equal the number of structural constraints (not including non-negativity) in the primal problem. In the example above, the dual problem will have n activities since the primal problem has n structural constraints.

2. The number of structural constraints of the dual problem will equal the number of activities in the primal problem. In this example, the dual problem will have two structural constraints because the primal problem has two activities.

3. The objective function coefficients for the dual problem are the RHS (b_i's) values in the primal problem. In this example, the dual problem's objective function coefficients will be the coefficients $b_1, b_2, \ldots, bn$ from the primal problem.

4. The RHS values for the dual problem correspond to the objective function coefficients of the primal problem. Hence, for this example the dual problem's RHS values will be c_1 and c_2.

5. If all activities in the primal maximization problem are non-negative and all constraints are $\leq$ restrictions, then all activities in the dual minimization problem must also be non-negative and all constraints will be $\geq$ restrictions.

6. The collection of technical coefficients (a_{ij}) in the dual problem is the same as the collection of technical coefficients in the primal, except that the dual problem uses the **transpose** of this matrix of coefficients, that is, the row elements become the column elements and vice versa. In this example, this implies that all the coefficients a_{ij} in the dual are equal to a_{ji} in the primal, for $i = 1, \ldots, m$; and $j = 1,2$.

7. The activities in the dual problem, which will be denoted as y_i to distinguish the dual from the primal problem's activities (i.e., the x_i's), are called the **dual variables**. The meaning of these variables will be discussed later in this section.

Using these facts, the dual problem for this example can be constructed:

Min: $Z = b_1 y_1 + b_2 y_2 + \cdots + b_n y_n$ $\qquad\qquad$ (0)

s.t.:

$$a_{11} y_1 + a_{21} y_2 + \cdots + a_{n1} y_n \geq c_1 \qquad\qquad (1)$$

$$a_{12} y_1 + a_{22} y_2 + \cdots + a_{n2} y_n \geq c_2 \qquad\qquad (2)$$

$$y_1, \qquad y_2, \qquad \cdots, \qquad y_n \geq 0 \qquad\qquad (3)$$

The dual variables (y_i) in the dual problem give the marginal value of the RHS values associated with the primal problem. Does this sound familiar? It should! This is precisely the definition of a SP. Hence, the solution to the dual problem directly provides a solution for all the SPs for the constraints in the primal problem.

The relationship between the primal and dual problem can be stated more compactly by using matrix notation. Let the m × 2 matrix **A** be the technical coefficients of the primal problem. Let the m × 1 vector b represent all the RHS (the b_i's) values of the primal problem. Similarly, let the vectors **c** (1 × 2) and **x** (2 × 1) represent the objective function coefficients and the activities, respectively, in the primal problem. Using this more compact notation, the primal problem (for a maximization problem) can be expressed generally as:

Max: $Z = \mathbf{cx}$ $\qquad\qquad$ (0)

s.t.:

$$\mathbf{Ax} \leq \mathbf{b} \qquad\qquad (1)$$

$$\mathbf{x} \geq 0 \qquad\qquad (2)$$

The dual problem then becomes:

Min: $V = \mathbf{b'y}$ $\qquad\qquad$ (0)

s.t.:

$$\mathbf{A'y} \geq \mathbf{c'} \qquad\qquad (1)$$

$$\mathbf{y} \geq 0 \qquad\qquad (2)$$

where **y** is a row vector of dual variables and **A'** is the transpose of matrix **A**.

Additional Properties of Duality

Duality is completely symmetric in that every element of the primal is contained in the dual. The following are several common properties of duality that illustrate this symmetry. The example above of the primal and dual problem is used for this illustration.

Weak Duality Property   Let $x\sim = (x_1\sim, x_2\sim)$ be a 1×2 vector of solution values that represents a feasible solution for the primal problem (which is a maximization problem), and let $y\sim = (y_1\sim, y_2\sim, \ldots, y_n\sim)$ be an $n \times 1$ vector that represents a feasible solution for its dual problem. The weak duality property states that the following condition will always hold:

$$c_1 x_1\sim + c_2 x_2\sim \leq b_1 y_1\sim + b_2 y_2\sim + \cdots + b_n y_n\sim$$

or using matrix notation,

$$cx\sim \leq b'y\sim.$$

This means that the value of the objective function for any feasible solution to a primal problem will never be greater than the value of the objective function for its dual counterpart problem. If the primal problem is a minimization problem, then the reverse is true, that is, $\leq$ becomes $\geq$ for this property.

Strong Duality Property   Let $x^* = (x_1^*, x_2^*)$ be the optimal solution for the primal problem, and let $y^* = (y_1^*, y_2^*, \ldots, y_n^*)$ be the optimal solution for its dual problem. The strong duality property states that the following condition will always hold at optimality:

$$c_1 x_1^* + c_2 x_2^* = b_1 y_1^* + b_2 y_2^* + \cdots + b_n y_m^*,$$

or using matrix notation,

$$cx^* = b'y^*.$$

This simply means that the optimal value of the objective function for a primal problem is equal to the optimal value of the objective function for its dual problem.

Complementary Optimal Solution Property   The final simplex tableau for the primal property generates not only an optimal solution for the primal activities (x_i^*), but also a "complementary" solution for the dual problem activities (y_i^*), where these optimal dual variables are the SPs found in the primal problem's z_i row for the slack variables.

Symmetry Property   The dual of the dual problem is the primal problem. This property is easily proven. Let the dual problem be defined in matrix form as follows:

(3.37) Min: $V = b'y$ $\qquad\qquad\qquad\qquad\qquad\qquad\qquad\qquad\qquad\qquad$ (0)

s.t.:

$$A'y \geq c' \qquad\qquad\qquad\qquad\qquad (1)$$

$$y \geq 0 \qquad\qquad\qquad\qquad\qquad (2)$$

This problem is identical to the following:

(3.38) Max: $-V = -b'y$ $\qquad\qquad\qquad\qquad\qquad\qquad\qquad\qquad\qquad$ (0)

s.t.:

$$-A'y \leq -c' \qquad\qquad\qquad\qquad\qquad (1)$$

$$y \geq 0 \qquad\qquad\qquad\qquad\qquad (2)$$

That is, (3.38) was obtained by multiplying the objective function and constraint set of (3.37) by -1 and then maximize it. Hence, problem (3.38) is identical to problem (3.37) and both are by definition the dual problem. Now, what is the dual of this dual problem? The dual of (3.38) is:

(3.39) Min: $-Z = -\mathbf{c}x$ (0)

s.t.:

$$-\mathbf{A}x \geq -\mathbf{b} \tag{1}$$

$$x \geq 0 \tag{2}$$

Note that the transpose of a transpose is the original matrix, for instance, $[\mathbf{A}']' = \mathbf{A}$. By performing the same conversion of problem (3.39) that was done to transform problem (3.37) to (3.38), the net result is:

(3.40) Max: $Z = \mathbf{c}x$ (0)

s.t.:

$$\mathbf{A}x \leq \mathbf{b} \tag{1}$$

$$x \geq 0 \tag{2}$$

Since (3.40) is by definition the primal problem, this proves that the dual of the dual is the primal problem.

The Relationship Between the Primal and Dual Solutions

Previously, the following maximization problem was solved using the simplex method.

Max: $Z = 35x_1 + 50x_2$ (0)

s.t.:

$$x_1 + x_2 \leq 1{,}000 \tag{1}$$

$$2.5x_1 + 0.75x_2 \leq 1{,}500 \tag{2}$$

$$1.5x_2 \leq 800 \tag{3}$$

$$x_1, \quad x_2 \geq 0 \tag{4}$$

The final simplex tableau was:

		x_1	x_2	s_1	s_2	s_3	
Basis	CB	35	50	0	0	0	b
s_1	0	0	0	1	-0.4	-0.47	26.67
x_1	35	1	0	0	0.4	-0.2	440
x_2	50	0	1	0	0	0.66	533.33
z_j		35	50	0	14	26.33	42,066.67
Net Eval $(c_j - z_j)$		0	0	0	-14	-26.33	

The optimal solution therefore is:

$$(x_1^*, x_2^*, s_1^*, s_2^*, s_3^*) = (440, 533.33, 26.67, 0, 0) \text{ and } z^* = 42{,}066.67.$$

In addition, the SPs for the three resources are:

$$(SP_{b1}, SP_{b2}, SP_{b3}) = (0, 14, 26.33).$$

Now, consider solving the dual of this problem. The dual problem is:

Min: $Z = 1,000y_1 + 1,500y_2 + 800y_3$ (0)

s.t.:

$$1y_1 + \quad 2.5y_2 + \quad 0y_3 \geq 35 \tag{1}$$
$$1y_1 + \quad 0.75y_2 + \quad 1.5y_3 \geq 50 \tag{2}$$
$$y_1, \qquad y_2, \qquad y_3 \geq 0 \tag{3}$$

In standard form with surplus and artificial variables this dual problem is:

Min $= 1,000y_1 + 1,500y_2 + 800y_3 + 0s_1 + 0s_2 + ma_1 + ma_2$ (0)

s.t.:

$$1y_1 + \quad 2.5y_2 + \quad 0y_3 - 1s_1 \qquad + 1a_1 \qquad = 35 \tag{1}$$
$$1y_1 + \quad 0.75y_2 + \quad 1.5y_3 \qquad - 1s_2 \qquad + 1a_2 = 50 \tag{2}$$
$$y_1, \qquad y_2, \qquad y_3, \qquad s_1, \qquad s_2, \qquad a_1, \qquad a_2 \geq 0 \tag{3}$$

Note that the two m objective function coefficients for the two artificial variables are positive rather than negative. This is due to the fact that for minimization problems, the penalties are positive. To solve this using the simplex tableau method, either solve as a minimization problem and reverse the $c_j - z_j$ rule, or multiply the objective function by -1 and solve as a maximization problem. Consider using the reverse $c_j - z_j$ rule approach. Then the first tableau becomes:

First Tableau

Basis	CB	y_1 1,000	y_2 1,500	y_3 800	s_1 0	s_2 0	a_1 m	a_2 m	b	b_i/a_{ij}
a_1	m	1	2.5	0	−1	0	1	0	35	14
a_2	m	1	0.75	1.5	0	−1	0	1	50	66.6
	z_j	2m	3.25m	1.5m	−m	−m	m	m	85m	
	$c_j - z_j$	$1,000 - 2m$	$1,500 - 3.25m$	$800 - 1.5m$	m	m	0	0		

While the subsequent simplex tableaus are listed below, the student should independently work through the simplex method to make sure the same tableaus are derived.

Second Tableau

Basis	CB	y_1 1,000	y_2 1,500	y_3 800	s_1 0	s_2 0	a_1 m	a_2 m	b	b_i/a_{ij}
y_2	1,500	0.4	1	0	−0.4	0	0.4	0	14	—
a_2	m	0.7	0	1.5	0.3	−1	−0.3	1	39.5	26.3
	z_j	$600 + 0.7m$	1,500	1.5m	$-600 + 0.3m$	−m	$600 - 0.3m$	m	$21,000 +$	
	$c_j - z_j$	$400 - 0.7m$	0	$800 - 1.5m$	$600 - 0.3m$	m	$1.3m - 600$	0	$39.5m$	

Third Tableau

Basis	CB	y_1 1,000	y_2 1,500	y_3 800	s_1 0	s_2 0	a_1 m	a_2 m	b
y_2	1,500	0.4	1	0	−0.4	0	0.4	0	14
y_3	800	0.47	0	1	0.2	−0.67	−0.2	0.67	26.33
z_j		976	1,500	800	−440	−533.33	440	533.33	42,066.67
$c_j - z_j$		24	0	0	440	533.33	m − 440	m − 533.33	

An optimal solution has been reached since $c_j - z_j \geq 0$ for all j. The optimal solution is:

$$(y_1^*, y_2^*, y_3^*) = (0, 14, 26.33) \text{ and } z^* = 42,066.67.$$

In addition, the SPs for the two constraints are:

$$(SP_{b1}, SP_{b2}) = (440, 533.33).$$

Comparing the final tableau of the dual with the final tableau of the primal illustrates several additional very important and useful properties of the final simplex tableaus for the primal and dual problems.[2] These properties are listed below.

Property 1　If the dual problem has an optimal solution, then the primal problem has an optimal solution and vice versa.

Furthermore, the value of the objective function of the dual and primal problems will be identical. For instance, if the optimal value is $42,066.67 when the dual problem is solved, then the solution for the primal problem is also $42,066.67.

Property 2　The optimal values of the productive activities for the primal problem are given by the absolute value of the SPs from the dual solution. Also, the optimal values of the primal slack variables are given by the absolute value of the $c_j - z_j$ values for the y_i variables in the final simplex tableau of the dual problem.

For this example, the absolute value of the z_j values of the surplus variables in the final tableau of the dual problem are 440 and 533.33, which are identical to x_1^* and x_2^*, respectively, in the primal. The absolute value of the $c_j - z_j$ values of the dual variables in the final tableau of the dual problem are 26.67, 0, and 0, which are identical to s_1^*, s_2^*, and s_3^*, respectively, in the primal.

Property 3　If the primal problem is unbounded from above, then the dual problem will not have a feasible solution.

Property 4　If the dual problem is unbounded from below, then the primal problem will not have a feasible solution.

---

[2]The signs in the relationship between the primal and dual solution may be different in some cases. This is the reason for using absolute values.

Property 5   The range of optimality on objective function coefficients for the primal problem is the range of feasibility on RHS values for the dual problem, and vice versa.

The complete primal solution can be obtained from the dual, and vice versa. These properties enable either the use of the primal or dual LP problem to get the same solution and sensitivity analysis results, which is important since the dual can be easier to solve than the primal, and vice versa. Generally speaking, the problem that has the fewest structural constraints will be the easiest to solve with the simplex method. For example, if the primal problem is a three-activity problem with 10 structural constraints, then the dual problem will only have three constraints (with 10 activities) and probably will be easier to solve computationally.

Figure 3.1 summarizes some of these properties by illustrating a simple example.

The Normal Form of a Linear Programming Model

Most of the models encountered thus far have been in **normal form**. For a **maximization problem**, normal form means that the objective function is maximized subject to all structural constraints being the $\leq$ type plus the non-negativity constraint on the activities. For a **minimization problem**, normal form means that the objective function is minimized subject to all structural constraints being the $\geq$ type plus the non-negativity constraint on the activities. A **mixed linear programming model** for either a maximization or a minimization problem means that there is at least one $\geq$ and at least one $\leq$ type of constraint in the problem.

Normal form is important in duality theory because it is much easier to find the dual of a problem expressed in normal form than in mixed form. Thus, the first step in deriving the dual for a primal problem is to check whether the primal is in normal form. If it is not, the primal should be converted into normal form before the dual problem is constructed. To illustrate, consider the following minimization problem with mixed constraints:

Min: $Z = 200x_1 + 175x_2$ (0)

s.t.:

$$3x_1 + 5x_2 \geq 1000 \tag{1}$$

$$1x_1 \leq 800 \tag{2}$$

$$30x_1 - 5x_2 = 30 \tag{3}$$

$$x_1, \quad x_2 \geq 0 \tag{4}$$

Before formulating the dual to this problem, convert this primal problem into normal form. For a minimization problem, normal form requires that all structural constraints be $\geq$ constraints. Therefore, constraints (2) and (3) in this problem need to be transformed. The following steps are recommended for this example.

Step 1:   Convert $\leq$ constraint in (2) to a $\geq$ constraint by multiplying (2) by -1, which results in the following:

$$-1x_1 \geq -800 \tag{2$'$}$$

Step 2:   Convert the equality in (3) to two constraints, that is:

$$30x_1 - 5x_2 \leq 30 \tag{3a}$$

$$30x_1 - 5x_2 \geq 30 \tag{3b}$$

Note: These two constraints imply an equality.

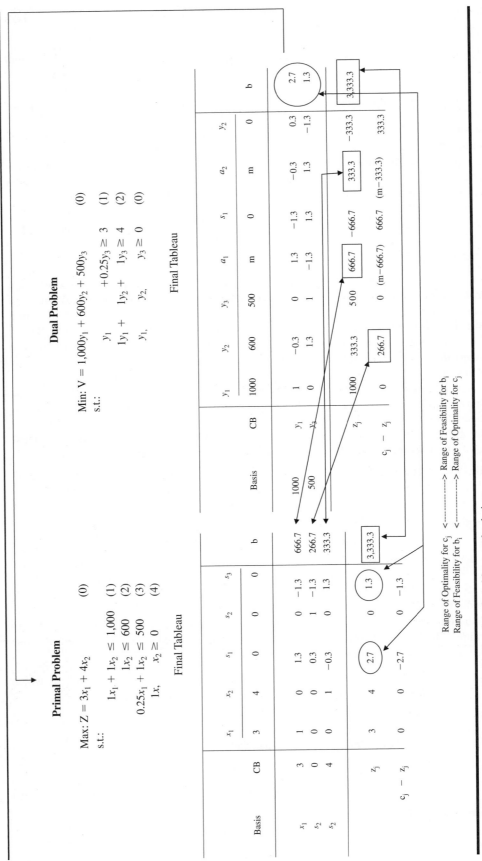

Figure 3.1 Relationships between primal and dual problems and solutions.

Step 3: Convert (3a) into a $\geq$ constraint by multiplying (3a) by -1, that is:

$$-30x_1 + 5x_2 \geq -30 \tag{3a}'$$

Using (0), (1), (2)', (3a)', and (3b), the equivalent primal problem in normal form is:

Min: $Z = 200x_1 + 175x_2$ $\hfill (0)$

s.t.:

$$3x_1 + 5x_2 \geq 1{,}000 \tag{1}$$

$$-1x_1 \qquad \geq -800 \tag{2}'$$

$$-30x_1 + 5x_2 \geq -30 \tag{3a}'$$

$$30x_1 - 5x_2 \geq 30 \tag{3b}$$

$$x_1, \qquad x_2 \geq 0 \tag{4}$$

Now, the dual to this problem will be much easier to solve. The dual to this problem is:

Max: $V = 1000y_1 - 800y_2 - 30y_3 + 30y_4$ $\hfill (0)$

s.t.:

$$3y_1 - 1y_2 - 30y_3 + 30y_4 \leq 200 \tag{1}$$

$$5y_1 + 0y_2 + 5y_3 - 5y_4 \leq 175 \tag{2}$$

$$y_1, \qquad y_2, \qquad y_3, \qquad y_4 \geq 0 \tag{3}$$

The Economic Intuition Behind Duality

Consider a farmer who produces and sells five crops: corn (a), soybeans (b), oats (c), wheat (d), and sunflowers (e). The resource requirements, resource endowments, and net selling prices of each crop are summarized below.

Resource (unit)	Crop Resource Requirement					Resource Endowment
	a	b	c	d	e	
Labor (hours)	1.5	1.3	1.0	0.8	0.3	2,700
Land (acres)	1	1	1	1	1	2,500
Machinery (hours)	0.5	0.4	0.3	0.2	0.1	1,000
Net profit/acre	$500	$400	$450	$350	$300	

The primal problem therefore is:

Max: $Z = 500a + 400b + 450c + 350d + 300e$ $\hfill (0)$

s.t.:

$$1.5a + 1.3b + 1.0c + 0.8d + 0.3e \leq 2{,}700 \tag{1}$$

$$1.0a + 1.0b + 1.0c + 1.0d + 1.0e \leq 2{,}500 \tag{2}$$

$$0.5a + 0.4b + 0.3c + 0.2d + 0.1e \leq 1{,}000 \tag{3}$$

$$a, \qquad b, \qquad c, \qquad d, \qquad e \geq 0 \tag{4}$$

Consider the dual to this problem from a buyer's point of view. That is, suppose that a buyer was interested in buying the farmer's resources to go into the farm business. The buyer needs to submit a bid for each of the three resources used to produce the five crops. Let the dual variables y_1, y_2, and y_3 represent the buyer's bids for the fixed supply of labor, land, and machinery, respectively. Furthermore, the buyer is interested in making the lowest possible bid to acquire each resource. Since the fixed supply of labor, land, and machinery is 2,700 hours, 2,500 acres, and 1,000 hours, the objective function from the buyer's perspective is:

$$\text{Min: } V = 2{,}700y_1 + 2{,}500y_2 + 1{,}000y_3 \tag{0}$$

Obviously, if there were no constraints on the buyer's bids, then the optimal solution would be to offer \$0 for each resource. However, the farmer would probably not be interested in selling resources for nothing. A resource will only be sold if the farmer can receive an amount that is at least as much as the resource's value in producing the five crops. The value of these resources is reflected in the net profit of all five crops.

For example, for each acre of corn that is produced and sold, the farmer currently makes a profit of \$500. This implies that the bid prices for the three resources, when multiplied by their respective resource requirements and summed up, must be at least \$500. This restriction must be true for the profitability of all five crops. Hence, for crop e, the bid prices for the three resources, when multiplied by their respective resource requirements and summed up, must be at least \$300. Mathematically, this is reflected by constraints (1) through (5) below.

$$1.5y_1 + 1.0y_2 + 0.5y_3 \geq 500 \tag{1}$$

$$1.3y_1 + 1.0y_2 + 0.4y_3 \geq 400 \tag{2}$$

$$1.0y_1 + 1.0y_2 + 0.3y_3 \geq 450 \tag{3}$$

$$0.8y_1 + 1.0y_2 + 0.2y_3 \geq 350 \tag{4}$$

$$0.3y_1 + 1.0y_2 + 0.1y_3 \geq 300 \tag{5}$$

The meaning of these five constraints can also be explained by economic theory. Recall that according to neoclassical economic theory, it is optimal to produce up to a point where marginal cost of producing each crop is equal to the marginal revenue for each crop. It is actually more accurate to state that economic theory requires that marginal costs be $\geq$ marginal revenue in order to have an optimal allocation of resources. In constraints (1) through (5), the marginal costs of producing each crop are given by the left-hand-sides (LHS) and the marginal revenues from each crop are given by the RHSs. For example, the marginal cost of producing crop c is: $1.0y_1 + 1.0y_2 + 0.3y_3$.

Optimization therefore requires that constraints (1) through (5) all be satisfied. Note that any of these constraints will be binding (i.e., marginal cost will equal marginal revenue) if the solution indicates that it is optimal to sell a positive amount of the crop. Hence, if $a^* > 0$, then constraint (1) will be binding; this is also true for b^*, c^*, d^*, and e^*.

Now consider an example, where the primal problem is a minimization problem, such as pollution abatement. Assume a new law requires a factory to reduce its emissions of three contaminants—methane (M), nitrous oxide (NO), and carbon dioxide (CO_2)—by 100 million pounds, 200 million pounds, and 300 million pounds per year, respectively. The plant has two devices it can use to cut emissions of each pollutant: scrubbers (s), which cost \$500,000 per device, and large smokestacks (ls), which cost \$2 million per device. One installed scrubber will remove 50 million pounds of M, 75 million pounds of NO, and 125 million pounds of CO_2. One taller smokestack will remove 100 million pounds of M, 150 million pounds of NO, and 175 million pounds of CO_2. Therefore, the primal for this problem is:

$$\text{Min: } Z = 500,000s + 2,000,000ls \tag{0}$$

s.t.:

$$50,000,000s + 100,000,000ls \geq 100,000,000 \tag{1}$$

$$75,000,000s + 150,000,000ls \geq 200,000,000 \tag{2}$$

$$125,000,000s + 175,000,000ls \geq 300,000,000 \tag{3}$$

$$s, \qquad\qquad ls \geq 0 \tag{4}$$

Consider the dual to this problem from a seller's point of view. That is, suppose that a seller was interested in selling the reductions in the three pollutants (M, NO, and CO_2) to the factory. The seller needs to submit an offer to sell each of the three minimum reductions in M, NO, and CO_2. Let the dual variables y_1, y_2, and y_3 represent the seller's offers for the fixed minimum reductions in M, NO, and CO_2, respectively. Furthermore, the seller is interested in making the highest possible offer for the reduction in each pollutant. Since the fixed minimum reductions for M, NO, and CO_2 are 100 million, 200 million, and 300 million, respectively, the objective function from the seller's perspective is:

$$\text{Max: } V = 100,000,000y_1 + 200,000,000y_2 + 300,000,000y_3.$$

Obviously, if there were no constraints on the seller's offers, then the optimal solution would be to offer an infinite amount for the reduction in each pollutant. However, the factory would not be interested in buying the seller's services for such a hefty price. The factory owner will only be interested in buying the seller's service if the factory owner can pay an amount that is less-than-or-equal to the cost of installing scrubbers and/or large smokestacks to meet the minimum reductions. The cost of s and ls is reflected in the net unit cost of installing scrubbers and large smokestacks.

For example, each scrubber (s) that is installed costs $500,000. This implies that the offer prices for the three minimum reductions, when multiplied by their respective unit reductions (y_i) and summed up, must be no more than $500,000. Likewise, the cost of installing each large smokestack is $2,000,000. This implies that the seller's offer prices for the minimum reductions, when multiplied by their respective unit reductions and summed up, must be no more than $2,000,000. Hence, mathematically, the dual to this problem is:

$$\text{Max: } V = 100,000,000y_1 + 200,000,000y_2 + 300,000,000y_3 \tag{0}$$

s.t.:

$$50,000,000y_1 + 75,000,000y_2 + 125,000,000y_3 \leq 500,000 \tag{1}$$

$$100,000,000y_1 + 150,000,000y_2 + 175,000,000y_3 \leq 2,000,000 \tag{2}$$

$$y_1, \qquad\qquad y_2, \qquad\qquad y_3 \geq 0 \tag{3}$$

3.4   SOLVING LINEAR PROGRAMMING PROBLEMS USING SOLVER

Microsoft Excel has become a ubiquitous tool for data analysis in business. It can also be used for modeling and solving optimization problems. This is accomplished through an add-in called Risk Solver Platform for Education (generally hereafter as Solver), which is available for free for those who purchase this textbook.[3] Solver offers an implementation of the Simplex Algorithm, along with several other optimization algorithms all within the context of Excel spreadsheets. This enables users to take advantage of Excel equations and features, including graphically displaying model results and customization of mathematical programming models using Visual Basic with Applications (VBA).

This section provides a walkthrough of building and solving an LP in Excel with the most recent version of Solver. Supplementary materials related to this example and all the other examples in this textbook are available online at **www.wiley.com/college/kaiser**. These materials include the initial model set-up, the solved model, and the related sensitivity analyses are provided to help students and instructors learn the art of modeling using spreadsheets.

Step 1: Set up an LP in spreadsheet form.

Consider the following LP problem:

$$\text{Max: } Z = 10x + 5y + 9z - 3a + 7b \tag{0}$$

s.t.:

$$2.5x + y + z + a + b \le 100 \tag{1}$$

$$z \le 70 \tag{2}$$

$$a \ge 25 \tag{3}$$

$$z - b \ge 0 \tag{4}$$

$$x, \quad y, \quad z, \quad a, \quad b \ge 0 \tag{5}$$

This problem can be set up in an Excel spreadsheet as displayed in Figure 3.2. Cells B2 through F2 (referred hereafter simply as Cells B2:F2) are reserved to hold the values of the decision variables. In this case, they are left blank, implying a zero starting point for the algorithm. This works fine with LPs; however, with nonlinear algorithms the initial values of the decision variables should be feasible. The objective function coefficients are typed into Cells B3:F3. The coefficients of the constraints are in the block of cells from B5:F8 and the RHS values are in Cells I5:I9. The formulas to calculate the total values of the objective function and constraints in Cells G3 and G5:G9 will be covered shortly.

Next, **range names** are set up, which assign certain blocks of cells a meaningful identifier. The use of range names is not required; however, they are highly recommended as they make setting up the problem in Solver easier and also help individuals not involved with the development of the model interpret the model results. Range names offer several advantages in organizing a spreadsheet and working with formulas. Range names can be set using the Name Box next to the Formula Box in Excel, or through the Define Name button on the Formulas ribbon. Here, for instance, Cells B2:F2 are named DecisionVars. The full list of defined range names is displayed in Figure 3.2. Range names can be used in place of the row and column identifiers so that now it states "Z = SUMPRODUCT(DecisionVars, ObjFuncCoef)" instead of the more difficult to interpret version of this equation "G3 = SUMPRODUCT(B2:F2, B3:F3)."

The **sumproduct** function can be used to set up formulas for the objective function and constraints. Sumproduct is used in Excel to perform vector multiplication of two ranges. For

---

[3]See the Preface for a more complete discussion of using Solver in the context of other available mathematical programming software.

	A	B	C	D	E	F	G	H	I
1		x	y	z	a	b			
2	Decision Variables						Objective Function		
3	Objective Function Coefficients	10	5	9	-3	7	0		
4							Total		RHS
5	Constraints	2.5	1	1	1	1	0	≤	100
6		0	0	1	0	0	0	≤	70
7		0	0	0	1	0	0	≥	25
8		0	0	1	0	-1	0	≥	0
9									
10	DecisionVars	=Sheet1!$B$2:$F$2							
11	LHS	=Sheet1!$G$5:$G$8							
12	ObjFuncCoef	=Sheet1!$B$3:$F$3							
13	RHS	=Sheet1!$I$5:$I$8							
14	Z	=Sheet1!$G$3							

Figure 3.2 The model in spreadsheet form.

	G
1	
2	**Objective Function**
3	=SUMPRODUCT(DecisionVars,B3:F3)
4	**Total**
5	=SUMPRODUCT(DecisionVars,B5:F5)
6	=SUMPRODUCT(DecisionVars,B6:F6)
7	=SUMPRODUCT(DecisionVars,B7:F7)
8	=SUMPRODUCT(DecisionVars,B8:F8)
9	

Figure 3.3 The formulas used in the model.

example, typing =SUMPRODUCT(B1:B3, C1:C3) in a cell produces the same result as typing =B1*C1 + B2*C2 + B3*C3. Using the previously defined name ranges, we can set our objective function cell to =SUMPRODUCT(DecisionVars, ObjFuncCoef). Likewise, each constraint total in the G column will be the sum product of the coefficients in that row and the decision variables. G5, for instance, will be =SUMPRODUCT(DecisionVars, B5:F5). Once G5 is defined, it can be copied and pasted down the column. In this case, the reference to the constraint coefficients will move, but the DecisionVars reference will stay put. The formulas used in the model are displayed in Figure 3.3. As a final touch, labels, borders, and shading have been included to make the model easier to interpret for others who might need to work with it or have access to the results.

The model is now set up in Excel and ready to be solved. At this point, you should review the input data and formulas, and try a few test values to make sure that everything is working properly before proceeding to define the model in Solver. Sometimes functions will display error codes starting with hash symbols (#) instead of the expected value as the result. Usually this indicates that there is something wrong with the formula that needs to be fixed. It can sometimes be a hassle to track down the source of the error, but there are some tools that make the **troubleshooting** process easier.

To display the functions instead of the results, hold down the CONTROL key and press the tilde ($\sim$) key. Use the same key combination to toggle back to displaying the numerical results. This makes it possible to review and compare multiple equations at once. The "Trace Precedents" and "Trace Dependents" buttons on the Formula tab on the Ribbon are also useful in resolving formula errors. Selecting a formula and clicking Trace Precedents displays arrows pointing to all of the cells that the cell's formulas reference, while the Trace Dependents displays arrows pointing to all of the other cells with formulas referencing that cell. The use of this feature can help an analyst review the entire model to ensure that it has the proper flow from the initial information given through the decision variables and constraints to the desired objective function.

Other common errors in Excel include the following:

- If a cell displays "#######" it means that the cell is too small to display the values in the cell. In most cases, this does not affect the model as the underlying values still exist; it just makes reading the model on the screen or in printed form difficult. To correct this problem, adjust the column width to display the numerical results.

- The "#VALUE!" error usually means that a function is receiving the wrong type of data, for instance, a range of cells containing text used as an argument for the =SUM() function.

- If "#NUM!" is displayed it means that there is some sort of numerical error in the calculation. An example of this could be $=(-1)^\wedge.5$, which would return the square root of -1, an imaginary number. To correct this problem, correct the formula.

- The error message "#REF!" is usually caused when the cell or range that a formula references is deleted by moving or copying over. To correct this problem, update the equation with the proper reference cells.

Step 2: Define the model with Solver.

A model can be defined and run in Solver in several different ways. We will focus on using the Solver Options and Model Specification box, which shows up on the right side of Excel, which is available on the Risk Solver Platform tab in Excel's Ribbon (Figure 3.4).[4] The Risk Solver Platform has four tabs: Model, Platform, Engine, and Output. Starting on the Model tab, the center pane lists Sensitivity, Optimization, Simulation, and Decision Tree icons. This is where the objective function, decision variables, and constraints all need to be defined within the model. First, to define the objective function, expand the Optimization Model options by clicking on the small plus sign next to Optimization. Select (by clicking on) Objective under the Optimization Icon, then select the objective function cell (G3) in the spreadsheet, and finally select the Add Button (a green plus sign below the tabs at the top of the Model Specification box), which adds it to the model as a maximization objective.

The objective function is set to be maximized by default, and it can be changed by clicking on the objective in the Model Specification box. This will bring up the Change Objective dialog box as shown in Figure 3.5. This allows the selection of an objective of Max, Min, or Value Of, which will respectively attempt to Maximize, Minimize, or achieve a predefined value for the objective function. Decision variables can likewise be

---

[4]If the Risk Solver Platform tab is not on the Ribbon it is likely not installed properly. Refer to the Risk Solver documentation for further troubleshooting.

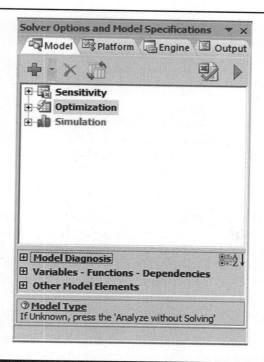

Figure 3.4   The Solver Options and Model Specification box.

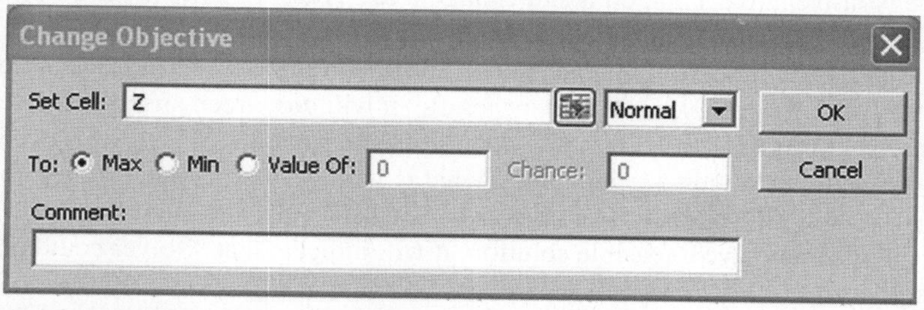

Figure 3.5   The Change Objective dialog box.

defined by selecting Variables, highlighting the DecisionVars cells and clicking the Add Button.

The three types of constraints, $\geq$, $\leq$, or $=$, have to be defined separately. To define the $\geq$ constraints, select Constraints in the Model Specification box, highlight the first two LHS totals, Cells G5 and G6, and click the Add Button. This will bring up the Change Constraint Dialog box as shown in Figure 3.6. Make sure the center pull-down box displays "$<=$", click inside the Constraint box, highlight the two corresponding

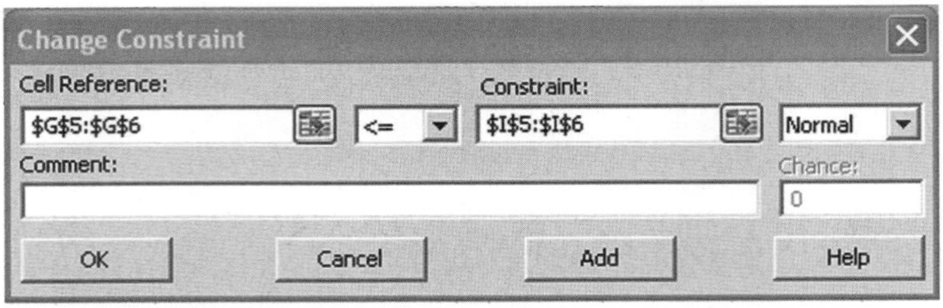

Figure 3.6   The Change Constraint dialog box.

RHS values in the spreadsheet (Cells I5 and I6), and click the Add button in the Change Constraint dialog box. This will add the $\leq$ constraints to the model and clear the Cell Reference and Constraint boxes so that the $\geq$ constraint can be added. Click inside the Cell Reference box and highlight the LHS totals for the second two constraints, G7 and G8. Then click inside the Constraint box and highlight the RHS values for the second two constraints, I7 and I8. Finally, in the pull-down menu between the two, change the "$<=$" to "$=>$" and click OK. Equality constraints can be added in a similar fashion through the pull-down menu.

Now the LP is defined in Solver, and the Model Specification box should look like Figure 3.7. If there are any errors, items can be changed by double-clicking on an item and making the necessary changes in the resultant dialog box, or deleted by selecting the item and clicking on the big red "x" next to the add button. Before running the model, click on the Engine tab to check some important settings. Immediately underneath the tabs is a dropdown box that specifies which engine is being used. Make sure that "Standard LP/Quadratic Engine" is selected. Also, under the General section, if it says "False" next to Assume Non-Negative, click on it and change it to "True." Finally, to have Solver solve the LP, click on the Output tab and click the green arrow.

A message bar at the bottom of the Model Specification box will indicate whether the model is successfully optimized. If it is successful, it will turn green and say "Solver found a solution. All constraints and optimality conditions are satisfied." If it is not successful, it will turn red and give a message indicating what is wrong. For example, if the problem is unbounded, it will indicate that "The objective (Set Cell) values do not converge." Also, if the problem does not have a feasible solution, it will indicate that "Solver could not find a feasible solution."

In the example presented above, Solver successfully solved the problem with an objective function value of 590. The optimum values of the decision variables are displayed in the DecisionVars cells. The final spreadsheet is displayed in Figure 3.8. Now that the LP has been optimized there are several useful reports that Solver can generate. These can be found under the Reports button on the Risk Solver Platform ribbon, under the Optimization menu, as shown in Figure 3.9.

The **Answer Report**, which provides the optimal values for all decision variables and the objective function, is presented in Figure 3.10. All optimal values are listed under the column labeled "Final Value." In this example, the optimal activities and objective function value are:

$$x^* = 0, \, y^* = 0, \, z^* = 70, \, a^* = 25, \, b^* = 5, \, z^* = 590.$$

The answer report also provides information on whether the constraints are binding as well as slack values for unbinding constraints.

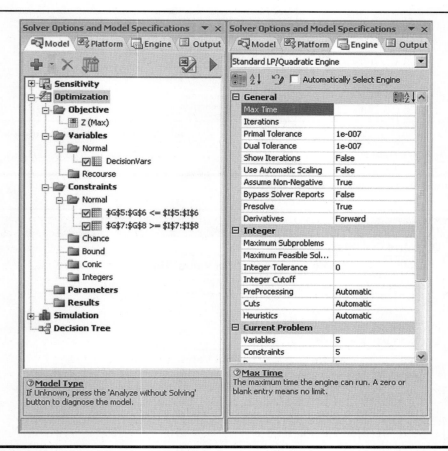

Figure 3.7   The complete model and options.

	A	B	C	D	E	F	G	H	I
1		x	y	z	a	b			
2	Decision Variables	0	0	70	25	5	Objective Function		
3	Objective Function Coefficients	10	5	9	-3	7	590		
4							Total		RHS
5	Constraints	2.5	1	1	1	1	100	≤	100
6		0	0	1	0	0	70	≤	70
7		0	0	0	1	0	25	≥	25
8		0	0	1	0	-1	65	≥	0
9									
10	DecisionVars	=Sheet1!$B$2:$F$2							
11	LHS	=Sheet1!$G$5:$G$8							
12	ObjFuncCoef	=Sheet1!$B$3:$F$3							
13	RHS	=Sheet1!$I$5:$I$8							
14	Z	=Sheet1!$G$3							

Figure 3.8   The solution to the model.

The **Sensitivity Report**, listed in Figure 3.11, provides information on the shadow values, ranges of optimality for all activities, and range of feasibility for all RHS values. The range of optimality is presented in the middle portion of Figure 3.11 under the heading "Decision Variable Cells." The numbers under the column heading "Final Value" list the

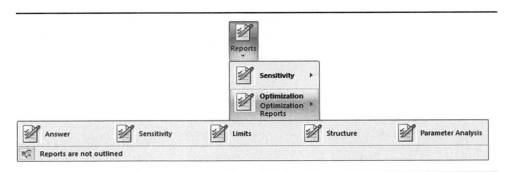

Figure 3.9   Accessing the analysis reports.

	A	B	C	D	E	F	G
1		Objective Cell (Max)					
2		Cell	Name	Original Value	Final Value		
3		Z	Z	0	590		
4							
5							
6		Decision Variable Cells					
7		Cell	Name	Original Value	Final Value		
8		$B$2	Decision Variables x	0	0		
9		$C$2	Decision Variables y	0	0		
10		$D$2	Decision Variables z	0	70		
11		$E$2	Decision Variables a	0	25		
12		$F$2	Decision Variables b	0	5		
13							
14		Constraints					
15		Cell	Name	Cell Value	Formula	Status	Slack
16		$G$5	Constraints Total	100	$G$5<=$I$5	Binding	0
17		$G$6	Total	70	$G$6<=$I$6	Binding	0
18		$G$7	Total	25	$G$7>=$I$7	Binding	0
19		$G$8	Total	65	$G$8>=$I$8	Not Binding	65

Figure 3.10   The answer report.

optimal values for the activities. The numbers in the "Objective Coefficient" column are the original objective function coefficients for the problem. The numbers under the "Allowable Increase" and Allowable Decrease" columns give the amount that the objective function coefficient can be increased or decreased without changing the optimal activity level. For example, the allowable increase for x is 7.5, which means as long as x's objective function coefficients are less than 17.5 (i.e., $10 + 7.5$), x^* will remain 0 in the optimal solution. The allowable decrease in this example is 1E+30, which is a large number and a proxy for infinity, and it means if the objective function coefficient is decreased by infinity, x^* will remain 0 in the optimal solution.

	A	B	C	D	E	F	G	H
1	Objective Cell (Max)							
2		Cell	Name	Final Value				
3		Z	Z	590				
4								
5	Decision Variable Cells							
6				Final	Reduced	Objective	Allowable	Allowable
7		Cell	Name	Value	Cost	Coefficient	Increase	Decrease
8		$B$2	Decision Variables x	0	-7.5	10	7.5	1E+30
9		$C$2	Decision Variables y	0	-2	5	2	1E+30
10		$D$2	Decision Variables z	70	0	9	1E+30	2.0
11		$E$2	Decision Variables a	25	0	-3	10.0	1E+30
12		$F$2	Decision Variables b	5	0	7	2.0	2.0
13								
14	Constraints							
15				Final	Shadow	Constraint	Allowable	Allowable
16		Cell	Name	Value	Price	R.H. Side	Increase	Decrease
17		$G$5	Constraints Total	100	7	100	65	5
18		$G$6	Total	70	2	70	5	32.5
19		$G$7	Total	25	-10	25	5	25
20		$G$8	Total	65	0	0	65	1E+30

Figure 3.11   The sensitivity report.

The SP or dual values for each constraint are listed under the column labeled "SP" in the bottom portion of Figure 3.11 called "Constraints." In this example, the SPs are:

$$SP_{b1} = 7, SP_{b2} = 2, SP_{b3} = -10, SP_{b4} = 0.$$

The range of feasibility is provided in Figure 3.11 in the "Allowable Increase" and "Allowable Decrease" columns. For example, for the first constraint, the RHS value of 100 can be increased by 65 to 165, or decreased by 5 to 95 without changing the current SP of 7.

The last item in Figure 3.11 is a column labeled "Reduced Cost." The reduced cost is relevant for each activity that is currently zero, and provides a measure of how much the objective function would change if one unit of the variable were forced into the solution. For example, if one unit of x were forced into the current solution, the objective function would decrease by 7.5 to 582.5.

The **Limits Report** is presented in Figure 3.12. The Limits Report re-solves the model to find the lower and upper bounds that each variable can take while satisfying the constraints and holding all other variables constant. For example, if an optimal value for a variable is 10, and the lower and upper bounds are 5 and 15, then the variable could take on any value between 5 and 15, and still all the same constraints would be satisfied as with the optimal value 10. In Limits Report, you can also find the value of the target cell for both upper and lower bound values for the variable. For example, if the optimal value where a decision variable is 10 leads to a profit of $10,000, Limits Report specifies how much the objective function value would fall if the decision variable were set to, say, 5.

The Risk Solver Platform has additional powerful tools, including Parameter Analysis, which can do more powerful sensitivity analysis beyond the limits of what's available through the Sensitivity Report, as well as built-in probabilistic simulation tools. Some of these will be covered in later chapters.

	Cell	Objective Name	Value		Lower Limit	Objective Result		Upper Limit	Objective Result
z	z		590						

	Cell	Decision Variable Name	Value		Lower Limit	Objective Result		Upper Limit	Objective Result
	$B$2	Decision Variables x	0		0	590		0	590
	$C$2	Decision Variables y	0		0	590		0	590
	$D$2	Decision Variables z	70		5	5		70	590
	$E$2	Decision Variables a	25		25	590		25	590
	$F$2	Decision Variables b	5		0	555		5	590

Figure 3.12   The limits report.

SUMMARY

This chapter focused on simplex-based sensitivity analysis, duality, and software for solving LP problems. Sensitivity analysis can be performed using information from the final simplex tableau, which provides the optimal feasible solution. The information contained in the final tableau includes the SPs and sensitivity analysis that can be performed on the objective function coefficients and resource endowments. Finding the range of optimality for a basic or nonbasic variable is done by replacing its numeric objective function coefficient in the final tableau with the more general c_j, and solving the optimality condition that $c_j - z_j \leq 0$ for all j. The range of feasibility can be found given the range of values of b_i. This range indicates how much variable values can change without causing any basic variables in the current solution to become infeasible (e.g. negative). The SPs for all structural constraints are also contained in the final simplex tableau. Sensitivity analysis for both maximization and minimization problems was also presented.

This chapter also provided an overview of duality, which is one of the most important concepts in mathematical programming, as well as an important topic in economics. Duality is simply the formulation of one problem in two different ways. While the attention given to this topic has been rather limited, the reader should now have a basic understanding of the important relationships between the primal and dual problems and solutions. Duality is powerful for two reasons. First, it is fundamental to sensitivity analysis since the solution of the primal problem also contains a "complementary" dual solution, which gives the SPs for the primal problem's resources. Second, duality adds tremendous computational efficiency to solution techniques. Since identical results may be obtained from two different problems, the reader may solve the easier of the two (which generally will be the problem with fewer constraints).

Finally, the Solver software package that will be used in this textbook to solve linear and nonlinear programming was introduced. A five-activity example was presented to demonstrate how to set up an LP in Excel and solve it with Solver.

REFERENCES

Gass, Saul I. (1985) *Linear Programming*, 5th ed. New York, NY: McGraw-Hill, Inc.

Messer, Kent D., and William L. Allen III. (2010) "Applying Optimization and the Analytic Hierarchy Process to Enhance Agricultural Preservation Strategies in the State of Delaware." Agricultural and Resource Economics Review 39(3): 442–456.

APPENDIX: SUMMATION AND MATRIX NOTATION

A.1 Summation Notation

So far in this book, the general form of a maximization problem has been expressed as follows:

$$\text{Max: } Z = c_1 x_1 + c_2 x_2 + \cdots + c_n x_n \tag{0}$$

s.t.:

$$a_{11} x_1 + a_{12} x_2 + \cdots + a_{1n} x_n \leq b_1 \tag{1}$$

$$a_{21} x_1 + a_{22} x_2 + \cdots + a_{2n} x_n \leq b_2 \tag{2}$$

$$\vdots \; \vdots \; \vdots \; \vdots \; \vdots \hspace{6cm} \vdots$$

$$a_{m1} x_1 + a_{m2} x_2 + \cdots + a_{mn} x_n \leq b_m \tag{m}$$

$$x_1, \quad x_2, \cdots \quad\quad x_n \geq 0 \tag{m+1}$$

Note that the subscript on the single subscripted activities (x_i) refers to the activity number, the subscript on the single subscripted RHS variables (b_i) refers to the RHS value for constraint i, and the subscripts on the double subscripted technical coefficients (a_{ij}) refers to the amount of resource i required of the jth activity (put differently, i refers to the row and j refers to the column for a_{ij}).

Whenever there is a set of mathematical equations that have a logical sequence of subscripts, these equations can be expressed in a more compact manner by using summation notation. For example, consider the following equation:

$$y = b_1 x_1 + b_2 x_2 + b_3 x_3 + b_4 x_4 + b_5 x_5 + b_6 x_6 \tag{A.1}$$

In this case, equation (A.1) has a logical sequence of subscripts because each pair of b and x variables contain the same subscript, beginning with 1 and increased in increments of 1 until the number 6 is reached. Using these characteristics of equation (A.1), it can be expressed as follows:

$$y = \text{the sum over i from i equal 1 to i equal 6 of } b_i x_i \tag{A.2}$$

This is precisely the verbal interpretation of summation notation. That is, we will define

$$y = \sum_{i=1}^{6} b_i x_i \tag{A.3}$$

to mean, "the sum from 1 to 6 of the product $b_i x_i$." The summation notation used in (A.3) is equivalent to the full and complete mathematical expression in (A.1).

EXAMPLES

The following examples will make this type of notation easier to understand.

Example 1.

Consider the following equation:

$$y = b_1x_1 + b_2x_2 + b_3x_3 \tag{A.4}$$

Using summation notation, (A.4) can be written as:

$$y = \sum_{i=1}^{3} b_i x_i \tag{A.5}$$

Example 2.

The following is an example of double subscripted variables:

$$y = b_{11}x_{11} + b_{12}x_{12} + b_{13}x_{13} + b_{21}x_{21} + b_{22}x_{22} + b_{23}x_{23} + b_{31}x_{31} \\ + b_{32}x_{32} + b_{33}x_{33} \tag{A.6}$$

Using summation notation, (A.6) becomes:

$$y = \sum_{i=1}^{3} \sum_{j=1}^{3} b_{ij} x_{ij} \tag{A.7}$$

In this case, start with the first subscript (i) set to 1, and the second subscript (j) is varied from 1 to 3; then i is set to 2 and j is again varied from 1 to 3, and so on. Whenever there are more than one summation signs, the left-most summation sign is varied only after the ones on its right have been varied for their entire range.

Example 3.

Consider the following system of equations:

$$y_1 = b_{11}x_1 + b_{12}x_2 + b_{13}x_3 + b_{14}x_4 \tag{A.8}$$

$$y_2 = b_{21}x_1 + b_{22}x_2 + b_{23}x_3 + b_{24}x_4 \tag{A.9}$$

$$y_3 = b_{31}x_1 + b_{32}x_2 + b_{33}x_3 + b_{34}x_4 \tag{A.10}$$

$$y_4 = b_{41}x_1 + b_{42}x_2 + b_{43}x_3 + b_{44}x_4 \tag{A.11}$$

This system can be expressed in two different ways using summation notation. First, all four equations can be written, that is:

$$y_1 = \sum_{j=1}^{4} b_{1j} x_j \tag{A.12}$$

$$y_2 = \sum_{j=1}^{4} b_{2j} x_j \qquad (A.13)$$

$$y_3 = \sum_{j=1}^{4} b_{3j} x_j \qquad (A.14)$$

$$y_4 = \sum_{j=1}^{4} b_{4j} x_j \qquad (A.15)$$

The second way is to use subscripts for the equation number, which makes the model even more compact, that is:

$$y_i = \sum_{j=1}^{4} b_{ij} x_j \quad (\text{for } i = 1,2,3, \text{ and } 4) \qquad (A.16)$$

In this case, the ith subscript refers to each of the four equations, and the jth subscript is what is being summed.

Using this information, the general form of the LP model depicted in equations (0) through (m+1) of the original problem can be expressed as:

$$\text{Max: } Z = \sum_{i=1}^{n} c_i x_i \qquad (0)$$

s.t.:

$$\sum_{j=1}^{n} a_{ij} x_j \leq b_i \quad \text{for } i = 1, \ldots, m \qquad (1)$$

$$x_j \geq 0 \quad \text{for } j = 1, \ldots, n \qquad (2)$$

The somewhat longer way of expressing the constraints individually can also be done, as with the first approach in Example 3.

A.2 Basic Matrix Operations and Notation

A **matrix** is a rectangular array of numbers. For example, define matrix **A** to be:

$$\mathbf{A} = \begin{bmatrix} 4 & 6 & 10 \\ 5 & 9 & 9 \\ 1 & 2 & 3 \\ 7 & 9 & 12 \end{bmatrix}.$$

In this example, $\mathbf{A}$ is a 4×3 array of numbers because it has 4 rows and 3 columns. More generally, a matrix's **dimension** is given by the number of rows it has by the number of columns it has, for instance, an $m \times n$ matrix has m rows and n columns. Each number in the matrix is called an **element** of the matrix. Consider the more general $m \times n$ matrix $\mathbf{A}$, where:

$$\mathbf{A} = \begin{bmatrix} a_{11} & a_{12} & a_{13} & \cdots & a_{1n} \\ a_{21} & a_{22} & a_{23} & \cdots & a_{2n} \\ a_{31} & a_{32} & a_{33} & \cdots & a_{3n} \\ \vdots & \vdots & \vdots & \vdots & \vdots \\ a_{m1} & a_{m2} & a_{m3} & \cdots & a_{mn} \end{bmatrix}$$

The elements that have the same subscript numbers for i and j (i.e., $i = j$) are called the **diagonal elements** of matrix $\mathbf{A}$. In this example, $a_{11}, a_{22}, a_{33}, \ldots$ are the diagonal elements. All other elements whose subscript numbers for i and j are not the same are called the off-diagonal elements.

A **vector** is a special type of matrix. Like a matrix, a vector is an array of numbers with the special characteristic that it either has only one row, or only one column. A row vector has one row and m columns, while a column vector has one column and n rows. Vector $\boldsymbol{x}$ below is an example of a row vector ($1 \times n$) and vector $\mathbf{k}$ is an example of a column vector ($m \times 1$).

$$\boldsymbol{x} = \begin{bmatrix} a_{11} & a_{12} & a_{13} & \cdots & a_{1n} \end{bmatrix},$$

$$\mathbf{k} = \begin{bmatrix} a_{11} \\ a_{21} \\ a_{31} \\ \vdots \\ a_{m1} \end{bmatrix}$$

Basic Matrix Operations　　While a matrix does not have a numeric value, matrices can be added, subtracted, and multiplied. The following discussion focuses on these simple matrix operations.

Matrix Addition. The elements of an $m \times n$ matrix $\mathbf{A}$ can be added to the elements of another $m \times n$ matrix $\mathbf{B}$ to form a new $m \times n$ matrix $\mathbf{C}$ by adding each a_{ij} element in $\mathbf{A}$ with each a_{ij} element in $\mathbf{B}$. The only requirement for matrix addition is that all matrices to be summed have the same row–column dimension. For example, let

$$\mathbf{A} = \begin{bmatrix} 2 & -5 & 10 \\ -9 & 7 & 22 \end{bmatrix}, \text{ and}$$

$$\mathbf{B} = \begin{bmatrix} 11 & 1 & -12 \\ 10 & 2 & 44 \end{bmatrix}.$$

Since both **A** and **B** have the same dimension (2×3), they can be added together to form a new 2×3 matrix **C**:

$$\mathbf{A} + \mathbf{B} = \mathbf{C}, \text{ or}$$

$$\mathbf{A} + \mathbf{B} = \begin{bmatrix} 2+11 & -5+1 & 10-12 \\ -9+10 & 7+2 & 22+44 \end{bmatrix} = \begin{bmatrix} 13 & -4 & -2 \\ 1 & 9 & 66 \end{bmatrix} = \mathbf{C}.$$

Matrix Subtraction. The same rule holds for matrix subtraction as was true for matrix addition. Two matrices, **A** and **B**, with dimensions m $\times$ n may be subtracted by taking the difference between each similar element (a_{ij}) for both matrices. Using the above two matrices **A** and **B**,

$$\mathbf{A} - \mathbf{B} = \mathbf{C}, \text{ or}$$

$$\mathbf{A} - \mathbf{B} = \begin{bmatrix} 2-11 & -5-1 & 10+12 \\ -9-10 & 7-2 & 22-44 \end{bmatrix} = \begin{bmatrix} -9 & -6 & 22 \\ -19 & 5 & -22 \end{bmatrix} = \mathbf{C}.$$

Scalar Multiplication. A **scalar** is simply a constant number (i.e., a 1×1 matrix). To multiply a scalar, k, by an m $\times$ n matrix, multiply each element of the matrix by k. For example, suppose **A** is a 3×2 matrix as shown below.

$$\mathbf{A} = \begin{bmatrix} 5 & 4 \\ 6 & 3 \\ 2 & 1 \end{bmatrix}.$$

Then, k multiplied by **A** yields the following:

$$k\mathbf{A} = k\begin{bmatrix} 5 & 4 \\ 6 & 3 \\ 2 & 1 \end{bmatrix} = \begin{bmatrix} 5k & 4k \\ 6k & 3k \\ 2k & 1k \end{bmatrix}.$$

If k = 2, then k**A** equals

$$k\mathbf{A} = \begin{bmatrix} 10 & 8 \\ 12 & 6 \\ 4 & 2 \end{bmatrix}.$$

Matrix Multiplication. Matrix multiplication is a bit more complicated than the previous elementary operations. Multiplication of two matrices **A** and **B** is permissible if and only if the two matrices have the following characteristic regarding their dimensions:

The number of columns in matrix **A** must be equal to the number of rows in matrix **B** in order to perform matrix multiplication.

Hence, an m $\times$ n matrix **A** can be multiplied by another matrix **B** if and only if **B** has m rows. The only exception to this condition is scalar multiplication. If this condition holds, then multiplication of **A** and **B** can be done to form a new matrix **C**, which has the following characteristic regarding its dimensions:

The new matrix **C** formed by the multiplication of an m $\times$ n matrix **A** with an n $\times$ t matrix **B** will have an m $\times$ t dimension.

For example, if **A** is 12 × 15, and **B** is 15 × 3, then the new matrix **C** = **AB** will be a 12 × 3 matrix. If **A** and **B** are compatible for multiplication, then the following rule is used for multiplication to obtain the new matrix **C**:

The new element in the ith row and jth column of **C** (i.e., element a_{ij}) is equal to the summation of the product of each element in row i of matrix **A** multiplied by each element in column j of matrix **B**. Mathematically speaking, if **A** is m × n, and **B** is n × t, then the result of their product is a new m × t matrix **C** whose element for any row i and column j is equal to:

$$\mathbf{C} = \mathbf{AB} = \left[\sum_{k=1}^{n} a_{ik} b_{kj} \right].$$

For example, let **A** and **B** be the following:

$$\mathbf{A} = \begin{bmatrix} 5 & 10 \\ 20 & 25 \\ 10 & 4 \end{bmatrix} \quad \mathbf{B} = \begin{bmatrix} 2 & 1 \\ 3 & 4 \end{bmatrix}.$$

Since **A** is a 3 × 2 matrix and **B** is a 2 × 2 matrix, you may multiply the two matrices to get a new 3 × 2 matrix **C**:

$$\mathbf{C} = \mathbf{AB} = \begin{bmatrix} 5 & 10 \\ 20 & 25 \\ 10 & 4 \end{bmatrix} \begin{bmatrix} 2 & 1 \\ 3 & 4 \end{bmatrix} = \begin{bmatrix} 5 \times 2 + 10 \times 3 & 5 \times 1 + 10 \times 4 \\ 20 \times 2 + 25 \times 3 & 20 \times 1 + 25 \times 4 \\ 10 \times 2 + 4 \times 3 & 10 \times 1 + 4 \times 4 \end{bmatrix} \text{ or}$$

$$\mathbf{C} = \begin{bmatrix} 40 & 45 \\ 115 & 120 \\ 32 & 26 \end{bmatrix}.$$

Note that unlike ordinary multiplication of two variables, the direction of the multiplication for matrices does matter. In this case, you can multiply **A** times **B**, but not **B** times **A**. Verify this on your own.

Transpose of a Matrix. The **transpose** of an m × n matrix **A** (denoted by **A′**) is a simple operation by which a new matrix **A′** is formed by interchanging the row elements in **A** by its column elements, and the column elements in **A** are replaced by its row elements, such that the matrix is now n × m. That is, if **A** is defined as:

$$\mathbf{A} = \begin{bmatrix} a_{11} & a_{12} & a_{13} & a_{14} & \dots & a_{1n} \\ a_{21} & a_{22} & a_{23} & a_{24} & \dots & a_{2n} \\ a_{31} & a_{32} & a_{33} & a_{34} & \dots & a_{3n} \\ \vdots & \vdots & \vdots & \vdots & \dots & \vdots \\ a_{m1} & a_{m2} & a_{m3} & a_{m4} & \dots & a_{mn} \end{bmatrix}, \quad (\mathbf{A} \text{ is an m} \times \text{n matrix})$$

then the transpose of $\mathbf{A}$ ($\mathbf{A}'$) is defined as:

$$\mathbf{A}' = \begin{bmatrix} a_{11} & a_{21} & a_{31} & a_{41} & \cdots & a_{m1} \\ a_{12} & a_{22} & a_{32} & a_{42} & \cdots & a_{m2} \\ a_{13} & a_{23} & a_{33} & a_{43} & \cdots & a_{m3} \\ \vdots & \vdots & \vdots & \vdots & \cdots & \vdots \\ a_{1n} & a_{2n} & a_{3n} & a_{4n} & \cdots & a_{mn} \end{bmatrix}, \quad (\mathbf{A}' \text{ is an } n \times m \text{ matrix})$$

Because of the multiplication requirement that the number of columns in $\mathbf{A}$ must equal the number of rows in $\mathbf{B}$ to multiply $\mathbf{A}$ times $\mathbf{B}$, the transpose property is very useful. For example, suppose that $\mathbf{A}$ is a 10×12 matrix and $\mathbf{B}$ is a 14×12 matrix. Clearly, you cannot multiply $\mathbf{A}$ times $\mathbf{B}$ or multiply $\mathbf{B}$ times $\mathbf{A}$ because their dimensions are not compatible for matrix multiplication. However, $\mathbf{A}$ can be multiplied by the transpose of $\mathbf{B}$ because $\mathbf{A}$ is a 10×12 and $\mathbf{B}'$ is a 12×14 matrix. The product, $\mathbf{AB}'$ is therefore permissible.

Using Matrix Notation to Express a Linear Programming Model   Recall that the general form of the model has been expressed thus far in the course as:

$$\text{Max: } Z = c_1x_1 + c_2x_2 + \cdots + c_nx_n \tag{0}$$

s.t.:

$$a_{11}x_1 + a_{12}x_2 + \cdots + a_{1n}x_n \le b_1 \tag{1}$$

$$a_{21}x_1 + a_{22}x_2 + \cdots + a_{2n}x_n \le b_2 \tag{2}$$

$$\vdots \qquad \vdots \qquad \vdots \quad \vdots \quad \vdots \qquad\qquad \vdots$$

$$a_{m1}x_1 + a_{m2}x_2 + \cdots + a_{mn}x_n \le b_m \tag{m}$$

$$x_1, \qquad x_2, \qquad \cdots \qquad x_n \ge 0 \tag{m+1}$$

Now consider the following matrices and vectors. Let:
$\mathbf{c} = (c_1 \, c_2 \, c_3 \, \ldots \, c_n)$ be a $1 \times n$ row vector of objective function coefficients;
$\mathbf{x} = (x_1 \, x_2 \, x_3 \, \ldots \, x_n)$ be a $m \times 1$ column vector of activities;

$$\mathbf{A} = \begin{bmatrix} a_{11} & a_{12} & \cdots & a_{1n} \\ a_{21} & a_{22} & \cdots & a_{2n} \\ \vdots & \vdots & \cdots & \vdots \\ a_{m1} & a_{m2} & \cdots & a_{mn} \end{bmatrix}$$

be an $m \times n$ matrix of technical coefficients; $\mathbf{b} = (b_1 \, b_2 \, b_3 \, \ldots \, b_m)$ be an $m \times 1$ column vector of RHS values.

Using these matrices and vectors, the original LP problem can be expressed much more compactly as:

$$\text{Max: } Z = \mathbf{cx} \tag{0}$$

s.t.:

$$\mathbf{Ax} \le \mathbf{b} \tag{1}$$

$$\mathbf{x} \ge 0 \tag{2}$$

The procedures for solving LP problems with the simplex or revised simplex methods can be presented using matrix algebra. While this approach was not used in this chapter, the interested reader can refer to a more general book on LP, such as Gass (1985).

EXERCISES

1. Write the dual to this LP problem in general form.

$$\text{Max: } Z = 4x_1 + 3x_2 \tag{0}$$

s.t.:

$$x_1 \qquad \geq 4 \tag{1}$$
$$x_2 \geq 1 \tag{2}$$
$$x_1 + x_2 \geq 24 \tag{3}$$
$$x_1 + x_2 \leq 120 \tag{4}$$
$$x_1, \quad x_2 \geq 0 \tag{5}$$

2. Consider the following LP problem:

$$\text{Min: } Z = 5x_1 + 3x_2 + 2x_3 \tag{0}$$

s.t.:

$$x_1 + x_2 \qquad \geq 50 \tag{1}$$
$$2x_2 + x_3 \geq 350 \tag{2}$$
$$3x_1 - \qquad 2x_3 \leq 25 \tag{3}$$
$$4x_1 + x_2 + x_3 \geq 33 \tag{4}$$
$$x_1, \quad x_2, \quad x_3 \geq 0 \tag{5}$$

   a.   What is the dual to this primal problem?
   b.   Which would be easier to solve: the primal or the dual problem?

3. Write the dual of the following primal problem, and solve it using the simplex method.

$$\text{Max: } Z = 4x_1 + 2x_2 \tag{0}$$

s.t.:

$$-x_1 - x_2 \leq -3 \tag{1}$$
$$-x_1 - x_2 \leq 2 \tag{2}$$
$$x_1, \quad x_2 \geq 0 \tag{3}$$

4. Solve the dual of the following problem with the simplex method:

Min: $Z = 12x_1 + 5x_2 + 20x_3$ (0)

s.t.:

$$x_1 - x_2 + x_3 \geq 140 \qquad (1)$$
$$2x_1 + x_3 \geq 160 \qquad (2)$$
$$2x_2 - x_3 \geq 15 \qquad (3)$$
$$x_1, \quad x_2, \quad x_3 \geq 0 \qquad (4)$$

5. The final simplex tableau for a maximization problem is:

		x_1	x_2	s_1	s_2	s_3	s_4	
Basis	CB	10	9	0	0	0	0	b
x_2	9	0	1	30/16	0	−21/16	0	252
s_2	0	0	0	−15/16	1	5/32	0	120
x_1	10	1	0	−20/16	0	30/16	0	540
s_4	0	0	0	−11/32	0	9/64	1	18
	z_j	10	9	70/16	0	111/16	0	7,668
	$c_j - z_j$	0	0	−70/16	0	−111/16	0	

 a. Calculate the range of optimality for the profit contribution of x_1, that is, the c_1 coefficient.

 b. Calculate the range of optimality for the profit contribution of x_2, that is, the c_2 coefficient.

 c. Calculate the range of feasibility for b_1.

 d. Calculate the range of feasibility for b_2.

6. Consider the following LP problem, and its final simplex tableau:

Max: $Z = 2x + 3y + w$ (0)

s.t.:

$$x + 4y + w \leq 300 \qquad (1)$$
$$3x + 2y - w \leq 150 \qquad (2)$$
$$-y + w \leq 100 \qquad (3)$$
$$x, \quad y, \quad w \geq 0 \qquad (4)$$

Final simplex tableau:

		x	y	w	s_1	s_2	s_3	
Basis	CB	2	3	1	0	0	0	b
w	1	0	0	1	0.2142	−0.0714	0.7143	125
y	3	0	1	0	0.2142	−0.0714	−0.2857	25
x	2	1	0	0	−0.0714	0.3571	0.4286	75
	z_j	2	3	1	0.7143	0.4285	0.7143	350
	$c_j - z_j$	0	0	0	−0.7143	−0.4285	−0.7143	

Based on the final simplex tableau, find the following:

a. The optimal solution.

b. The SPs for each constraint.

c. The range of optimality for the objective function coefficients.

d. The range of feasibility for each RHS value.

7. A company imports electronic components that are used to assemble two different types of computer modules for tractors. One model, A, generates a profit contribution of $50 per unit whereas the other, B, generates a profit contribution of $40 per unit. For next week's production, a maximum of 150 hours (b_1) of assembly time can be made available. Each unit of A requires three hours of assembly time, and each unit of B requires five hours. In addition, the company currently has in inventory 20 display units used in B; thus, no more than 20 units of B can be assembled. Finally, only 300 (b_2) square meters of warehouse space can be made available next week. Each unit of A requires eight square meters, and each unit of B requires five square meters.

a. Construct an LP problem to find the optimal production for next week to maximize profit.

b. Find the dual of the LP problem from part a.

c. Solve either the primal or the dual problem using the simplex method.

d. Calculate the range of feasibility for b_1.

e. Calculate the range of optimality for the profit contribution of part a.

8. A dairy farmer's cows needs three nutrients (A, B, and C) to subsist and produce milk each day. Each cow must receive the equivalent of 100 units of nutrient A, 200 units of nutrient B, and 50 units of nutrient C in order to maximize milk output. The farmer can use any combination of three feeds (f_1, f_2, and f_3) in meeting these minimum requirements. The local feed dealer sells all three feeds, which have the following cost per pound and nutrient equivalents (for A, B, and C) per pound.

Feed	Nutrient Content Per Pound			Cost ($/pound)
	A	B	C	
f_1	5	22	3	0.25
f_2	10	25	2	0.50
f_3	7	12	5	0.27

Assume that the farmer's objective is to minimize the cost per cow of buying any combination of these three feeds that satisfies the daily nutrient requirement of the cows.

a. Solve this problem using the simplex method.

b. Your optimal solution should indicate that no amount of feed f_2 should be purchased and fed to the farmer's cows. By how much should feed f_2's current price of $0.50 per pound decrease in order for the optimal solution to change?

c. Your optimal solution should indicate that some amount of feed f_3 should be purchased and fed to the farmer's cows. By how much should feed f_3's current price of $0.27 per pound increase in order for the optimal solution to change?

d. What is the range of feasibility for the minimum requirement for nutrient A?

e.   What is the range of feasibility for the minimum requirement for nutrient B?

f.   What is the range of feasibility for the minimum requirement for nutrient C?

g.   How would you interpret these ranges for this problem?

h.   What are the SPs for each constraint of this problem? What is their economic interpretation for this problem?

9.   Consider the following model:

$$\text{Max: } Z = 5x_1 + 6x_2 + 3x_3 \tag{0}$$

s.t.:

$$x_1 + x_2 + x_3 \geq 1{,}000 \tag{1}$$

$$x_1 - x_2 \qquad = 0 \tag{2}$$

$$x_1 + x_2 + x_3 \leq 2{,}000 \tag{3}$$

$$x_1, \quad x_2, \quad x_3 \geq 0 \tag{4}$$

a.   Solve this problem using the simplex method.

b.   Report the SPs for each constraint.

c.   Derive the range of optimality for all objective function coefficients.

d.   Derive the range of feasibility for all RHS values.

10.   A farmer owns 500 acres of land, which is suitable for growing corn, soybeans, and sunflowers. His expectations are that the net profit from producing each crop is: $55 per acre for corn, $60 per acre for soybeans, and $50 per acre for sunflowers. He and his family can supply 3,000 hours per year in performing all the farm operations necessary to grow these crops. In addition, he is endowed with the equivalent of 4,500 hours of tractor time necessary to grow these crops. Assume that the only resources necessary in crop production are land, labor, and tractor time; the technological relationships are summarized below.

Resource (unit)	Crop			Resource Endowment
	Corn	Soybeans	Sunflowers	
Land (acres)	1.0	1.0	1.0	500
Labor (hours)	0.4	0.2	0.3	3,000
Tractor (hours)	0.5	0.2	0.4	4,500

Assume that the farmer's objective is to maximize total profits.

a.   Solve this problem using the simplex method.

b.   Derive and graph the output supply functions for each crop. Output supply in this context is the relationship between acres produced and net profitability of the crop.

c.   Derive and graph the input demand functions for land and labor.

11. Consider the following LP problem:

$$\text{Min: } Z = 10x + 20y + 5z \tag{0}$$

s.t.:

$$1x + 1y + 1z \geq 100 \tag{1}$$
$$1z \leq 50 \tag{2}$$
$$1x \qquad\qquad \leq 25 \tag{3}$$
$$x, \quad y, \quad z \geq 0 \tag{4}$$

    a.   Solve this problem using the simplex method.

    b.   Would the optimal solution change if the objective function coefficient for x were increased from 10 to 19? Explain.

    c.   If the RHS of constraint (3) were increased from 25 to 60, would there be a change in the basis? Explain.

    d.   By how much would the value of the objective function change if the RHS of constraint (1) was increased by one unit?

    e.   Is the solution (found in the final simplex tableau in part a) unique, or is it part of a set of multiple optimal solutions? Explain.

12. Consider the following LP model:

$$\text{Min: } Z = c_1 x_1 + c_2 x_2 + c_3 x_3 + \cdots + c_m x_m \tag{0}$$

s.t.:

$$a_{11} x_1 + a_{12} x_2 + a_{13} x_3 + \cdots + a_{1m} x_m \geq b_1 \tag{1}$$
$$a_{21} x_1 + a_{22} x_2 + a_{23} x_3 + \cdots + a_{2m} x_m \geq b_2 \tag{2}$$
$$a_{31} x_1 + a_{32} x_2 + a_{33} x_3 + \cdots + a_{3m} x_m \geq b_3 \tag{3}$$
$$\vdots \qquad \vdots \qquad \vdots \qquad \vdots \qquad \vdots \qquad\qquad \vdots$$
$$a_{n1} x_1 + a_{n2} x_2 + a_{n3} x_3 + \cdots + a_{nm} x_m \geq b_n \tag{n}$$
$$x_1, \quad x_2, \quad x_3, \quad \cdots, \quad x_m \geq 0 \tag{n+1}$$

    a.   Rewrite this problem using summation signs.

    b.   Write the standard form of this model using summation notation.

    c.   Write the standard form of this model using matrix notation. Again, define any new matrices or vectors used in your model.

13. Consider the following LP problem:

$$\text{Min: } Z = c_1x_1 + c_2x_2 + c_3x_3 + c_4x_4 + c_5x_5 \tag{0}$$

s.t.:

$$a_{11}x_1 + a_{12}x_2 + a_{13}x_3 + a_{14}x_4 + a_{15}x_5 \geq b_1 \tag{1}$$

$$a_{21}x_1 + a_{22}x_2 + a_{23}x_3 + a_{24}x_4 + a_{25}x_5 \geq b_2 \tag{2}$$

$$a_{31}x_1 + a_{32}x_2 + a_{33}x_3 + a_{34}x_4 + a_{35}x_5 \geq b_3 \tag{3}$$

$$a_{41}x_1 + a_{42}x_2 + a_{43}x_3 + a_{44}x_4 + a_{45}x_5 \geq b_4 \tag{4}$$

$$a_{51}x_1 + a_{52}x_2 + a_{53}x_3 + a_{54}x_4 + a_{55}x_5 \geq b_5 \tag{5}$$

$$a_{61}x_1 + a_{62}x_2 + a_{63}x_3 + a_{64}x_4 + a_{65}x_5 \geq b_6 \tag{6}$$

$$a_{71}x_1 + a_{72}x_2 + a_{73}x_3 + a_{74}x_4 + a_{75}x_5 \geq b_7 \tag{7}$$

$$a_{81}x_1 + a_{82}x_2 + a_{83}x_3 + a_{84}x_4 + a_{85}x_5 \geq b_8 \tag{8}$$

$$a_{91}x_1 + a_{92}x_2 + a_{93}x_3 + a_{94}x_4 + a_{95}x_5 \geq b_9 \tag{9}$$

$$x_1, \quad x_2, \quad x_3, \quad x_4, \quad x_5 \geq 0 \tag{10}$$

a. Write this problem in general form using summation signs.
b. Write this problem in general form using matrix notation.
c. Write this problem in standard form (with slack variables) using summation signs.
d. Write this problem in standard form using matrix notation.
e. Write the dual to this problem using any form you want.
f. Which would probably be easier to solve: the primal or the dual? Why?

14. Write the dual to the following primal problem:

$$\text{Min: } Z = c_1x_1 + c_2x_2 + c_3x_3 + c_4x_4 \tag{0}$$

s.t.:

$$a_{11}x_1 + a_{12}x_2 + a_{13}x_3 + a_{14}x_4 \geq b_1 \tag{1}$$

$$a_{21}x_1 + a_{22}x_2 + a_{23}x_3 + a_{24}x_4 \geq b_2 \tag{2}$$

$$a_{31}x_1 + a_{32}x_2 + a_{33}x_3 + a_{34}x_4 \geq b_3 \tag{3}$$

$$a_{41}x_1 + a_{42}x_2 + a_{43}x_3 + a_{44}x_4 \geq b_4 \tag{4}$$

$$a_{51}x_1 + a_{52}x_2 + a_{53}x_3 + a_{54}x_4 \geq b_5 \tag{5}$$

$$x_1, \quad x_2, \quad x_3, \quad x_4 \geq 0 \tag{6}$$

15. Consider the following LP problem:

Max: $Z = 4x_1 + 2x_2 - 3x_3 + 5x_4$ (0)

s.t.:

$$2x_1 - 1x_2 + 1x_3 + 2x_4 \geq 50 \tag{1}$$

$$3x_1 - 1x_3 + 2x_4 \leq 80 \tag{2}$$

$$1x_1 + 1x_2 + 1x_4 = 60 \tag{3}$$

$$x_1, \quad x_2, \quad x_3, \quad x_4 \geq 0 \tag{4}$$

   a.   Write this problem in its equivalent normal form.

   b.   Write the dual to the primal problem in part a.

   c.   Solve either the primal or the dual problem using the simplex method.

16. Consider the following LP problem:

Max: $Z = 7x_1 - 1x_2 + 10x_3$ (0)

s.t.:

$$1x_1 + 1x_2 + 1x_3 \leq 500 \tag{1}$$

$$5x_1 - 1x_2 \leq 0 \tag{2}$$

$$1x_1 \leq 400 \tag{3}$$

$$1x_2 \leq 400 \tag{4}$$

$$1x_3 \leq 400 \tag{5}$$

$$1x_1 + 1x_2 \leq 450 \tag{6}$$

$$x_1, \quad x_2, \quad x_3 \geq 0 \tag{7}$$

   a.   What is the dual to this primal problem?

   b.   Which do you think would be easier to solve, the primal or the dual problem?

17. Consider the following LP problem:

Min: $Z = 5x_1 + 4x_2$ (0)

s.t.:

$$1x_1 + 1x_2 \geq 100 \tag{1}$$

$$-5x_1 + 3x_2 \geq 1 \tag{2}$$

$$x_1, \quad x_2 \geq 0 \tag{3}$$

   a.   Solve this problem using the graphical approach.

   b.   Write the dual to this primal problem.

   c.   Solve the dual problem using the graphical approach.

18. Consider the following model:

$$\text{Min: } Z = c_1 x_1 + c_2 x_2 + c_3 x_3 + c_4 x_4 + c_5 x_5 \tag{0}$$

s.t.:

$$a_{11} x_1 + a_{12} x_2 + a_{13} x_3 + a_{14} x_4 + a_{15} x_5 \geq b_1 \tag{1}$$

$$a_{21} x_1 + a_{22} x_2 + a_{23} x_3 + a_{24} x_4 + a_{25} x_5 \geq b_2 \tag{2}$$

$$a_{31} x_1 + a_{32} x_2 + a_{33} x_3 + a_{34} x_4 + a_{35} x_5 \geq b_3 \tag{3}$$

$$a_{41} x_1 + a_{42} x_2 + a_{43} x_3 + a_{44} x_4 + a_{45} x_5 \geq b_4 \tag{4}$$

$$a_{51} x_1 + a_{52} x_2 + a_{53} x_3 + a_{54} x_4 + a_{55} x_5 \geq b_5 \tag{5}$$

$$a_{61} x_1 + a_{62} x_2 + a_{63} x_3 + a_{64} x_4 + a_{65} x_5 \geq b_6 \tag{6}$$

$$a_{71} x_1 + a_{72} x_2 + a_{73} x_3 + a_{74} x_4 + a_{75} x_5 \geq b_7 \tag{7}$$

$$a_{81} x_1 + a_{82} x_2 + a_{83} x_3 + a_{84} x_4 + a_{85} x_5 \geq b_8 \tag{8}$$

$$a_{91} x_1 + a_{92} x_2 + a_{93} x_3 + a_{94} x_4 + a_{95} x_5 \geq b_9 \tag{9}$$

$$x_1, \quad x_2, \quad x_3, \quad x_4, \quad x_5 \geq 0 \tag{10}$$

a. Write the dual of this problem in general form using summation signs.
b. Write the primal of this problem in matrix notation (define all matrices and vectors).
c. Write the dual of this problem in general form using matrix notation.
d. Which would be easier to solve (take fewer iterations), the primal or the dual? Explain.

19. Consider the following model:

$$\text{Min: } Z = 1.1 x_1 + 5.1 x_2 \tag{0}$$

s.t.:

$$x_1 + x_2 \geq 500 \tag{1}$$

$$x_1 \geq 325 \tag{2}$$

$$x_2 \geq 125 \tag{3}$$

$$x_1 + x_2 \leq 750 \tag{4}$$

$$x_1, \quad x_2 \geq 0 \tag{5}$$

a. Write the dual of this problem. (*Hint*: Put primal into normal form first).
b. Solve the primal problem using the simplex method. Show all your work.
c. Solve the dual problem using the simplex method. Show all your work.
d. Summarize the similarities between the two optimal solutions.

20. Consider the following model:

$$\text{Max: } Z = -3x_1 + 2x_2 + 3x_3 - 9x_4 \tag{0}$$

s.t.:

$$x_1 \qquad\qquad\qquad \le 1{,}000 \tag{1}$$

$$x_2 \qquad\qquad \le 1{,}000 \tag{2}$$

$$x_3 \qquad \le 1{,}000 \tag{3}$$

$$x_4 \le 1{,}000 \tag{4}$$

$$x_1 + x_2 + x_3 + x_4 \le 3{,}000 \tag{5}$$

$$x_1 \qquad\qquad\qquad \ge 250 \tag{6}$$

$$x_4 \ge 250 \tag{7}$$

$$x_1 - \qquad\qquad x_4 = 0 \tag{8}$$

$$x_1, \quad x_2, \quad x_3, \quad x_4 \ge 0 \tag{9}$$

Write the dual to this problem.

21. A farmer has 5,000 acres of land, 1,000 hours of labor, and 2,000 hours of tractor time to grow two crops: corn and wheat. He expects that he can earn \$250 per acre for corn and \$200 per acre for wheat. It takes 0.25 hours per acre in labor to grow corn and 0.15 hours per acre in labor to grow wheat. Finally, it takes 0.30 hours per acre of tractor time to grow corn and 0.15 hours per acre of tractor time to grow wheat. The farmer's objective is to maximize total net income, given his production constraints. State the dual to this problem intuitively (without equations). Think of the dual problem from a buyer's perspective, someone who wants to buy the resources to the farm. Give enough detail to fully explain the economic interpretation of the dual problem using only words.

22. Give the dual to the following primal problem:

$$\text{Min: } Z = 5x_1 + 2x_2 \tag{0}$$

s.t.:

$$3x_1 + 1x_2 \ge 125 \tag{1}$$

$$2x_1 \qquad = 75 \tag{2}$$

$$1x_2 \le 500 \tag{3}$$

$$5x_1 \qquad \le 1{,}000 \tag{4}$$

$$x_1, \quad x_2 \ge 0 \tag{5}$$

23. You have just been hired as an advertising manager for a generic advertising program for dairy farmers, Dairy Management, Inc. (DMI). DMI wants to conduct generic advertising to increase the demand for milk. DMI decided to consider TV, radio, print, and outdoor advertising and wants you to determine how much money should be allocated for each type of media for the next month. You expect that one TV commercial will increase sales by 25,000 gallons; one radio commercial will increase sales by 7,000 gallons; one print advertisement will increase sales by 3,500 gallons; and one billboard will increase sales by 5,000 gallons. It costs $10,000 per TV commercial, $5,000 per radio commercial, $1,000 per print advertisement, and $3,500 per billboard. Your boss tells you that you can't spend more than $500,000 on this project. Furthermore, the radio and TV stations tell you they have a combined maximum of 45 minutes for your commercials for the month. Each TV commercial takes 1 minute, and each radio commercial takes 0.5 minutes to air. The boss tells you that he doesn't want more than 25 TV commercials because he gets sick of watching the same thing over and over again. The objective is to find the combination of TV, radio, print, and billboard advertisements that maximizes milk sales. Solve this problem with Solver. Report and analyze the solution.

24. A steel factory that uses coal as its major source of energy causes three major types of air pollution by releasing: (1) particulate matter, (2) sulfur oxides, and (3) hydrocarbons. These three types of air pollution are caused by blast furnaces and open hearth furnaces used in producing steel. The state has just passed a new clean air bill, which means that this factory must reduce its annual emission rate of particulate matter by 60 million pounds, sulfur oxides by 150 million pounds, and hydrocarbons by 125 million pounds. There are six pollution abatement techniques (three for each type of furnace) that the factory can use to reduce air pollution. These six techniques, their per unit reduction for each pollutant, and their estimated annual cost per unit are listed below:

	Taller Smokestacks		Filters		Better Fuels	
Pollutant	Blast Furnace	Open-Hearth Furnace	Blast Furnace	Open-Hearth Furnace	Blast Furnace	Open-Hearth Furnace
	(per unit reduction in million pounds)					
Particulate	12	9	25	20	17	13
Sulfur Oxides	35	42	18	31	56	49
Hydrocarbons	37	53	28	24	29	20
Cost/Unit	$80,000	$100,000	$70,000	$60,000	$110,000	$90,000

Assume that the factory's sole objective is to minimize the total cost of reducing emissions of these three pollutants to the new government standards by using any combination of the six pollution abatement techniques. Set this problem up as a linear program, and solve it with Solver. Report and analyze the solution.

25. A food firm is researching the profitability of introducing six new "healthy choice" food products (call them x, y, z, a, b, c). The firm currently has idle resource capacity on labor, machinery, and land of 100 hours, 300 hours, and 30,000 square feet, respectively. Hence, producing any or all of the new products will help solve the costs of excess capacity. The selling prices, total costs, resource requirements, and endowments for the production technology are summarized on the next page.

Resource (Unit)	New Product					
	x	y	z	a	b	c
Labor (hours)	0.50	0.10	1.00	0.45	0.2	0.15
Machinery (hours)	1.00	0.45	3.50	1.00	1.10	2.00
Land (sq ft)	100	200	50	25	10	75
Unit Costs ($)	10	3	33	22	12	9
Unit Price ($)	12	4	36	23	15	11

Food products y and z are complements in the sense that for every unit of y produced and sold, 2 units of z must be produced and sold. Also, the firm requires that the amount of product c produced and sold be at least 50% of the total units of products a and b that are produced and sold. Set up a LP model that will result in a solution that maximizes total profit from the sale of any combination of these food products, subject to all constraints that were specified (use Solver).

26. An ice cream maker has hired you to help him decide next month's production schedule. He needs to determine the quantities of each flavor that should be produced based on the profitability of each flavor and several restrictions. He has the capability of producing six different flavors of ice cream: (1) super-super premium mocha chip, (2) super premium chocolate chocolate chip, (3) super premium snickers bar crunch, (4) vanilla, (5) chocolate ice milk, and (6) Yuppie's Delight frozen yogurt. Each product is only available in quarts and has the following unit profits for the ice cream maker:

Product	Unit Profit ($/quart)
Super-Super Premium Mocha Chip	1.00
Super Premium chocolate chocolate chip	0.75
Super Premium Snickers Bar Crunch	0.88
Vanilla	0.43
Chocolate Ice Milk	0.50
Yuppie's Delight Frozen Yogurt	1.05

The total production capacity of the ice cream maker's plant is 10,000 gallons per month. He also knows that he can only sell 1,000 gallons of super-super premium mocha chip, and he must produce at least 2,500 gallons of chocolate ice milk for the local school district. Finally, because he is introducing Yuppie's Delight frozen yogurt and doesn't know the market for this product, he only wants to produce 500 gallons in the next month. Assuming he wishes to maximize profit, given these restrictions, formulate this decision problem as an LP model in general form. Solve it using Solver, and report and analyze the optimal solution.

4

Farm-Level Linear Programming Models

Modern farming is a complicated business. Farmers have to make a lot of complex production and marketing decisions throughout the year. For example, in crop agriculture, decisions need to be made about which crops will be grown, how the soil will be prepared, how much land to rent, how much labor to employ, and the optimal timing of these operations. Marketing decisions involve how and when to sell the harvested output throughout the marketing year. Linear programming (LP) is an excellent tool for assisting farmers in this decision making and is widely used in agriculture.

Agriculture is one of the principal economic sectors that uses LP modeling. Many land grant universities, through their cooperative extension programs, offer numerous types of LP models to assist farmers in their decision-making process. Such models tend to be developed for the characteristics of the region, but also allow farmers to input characteristics of their own farms as LP parameters.

The primary goal of this chapter is to provide a detailed overview of several types of LP models that have been used to assist farmers and used by researchers to address problems within the agricultural sector.

The purposes of this chapter are six-fold:

1. To illustrate several types of LP models used to assist farmers in making production and marketing decisions.

2. To introduce the topic of sequencing constraints, which guarantee that basic operations incurring costs are performed and in the proper sequence.

3. To demonstrate how to disaggregate activities into individual operations in the production process.

4. To discuss how to validate and calibrate a mathematical programming model, which applies not only to farm LPs, but to all applications of mathematical programming.

5. To illustrate how time may be incorporated into these basic models to make them dynamic, multiperiod models.

6. To examine two research applications that demonstrate the usefulness of LP modeling in agricultural economics.

The second, third, fourth, and fifth objectives apply to LP models for any type of enterprise, not just farming. It should be noted that some of the parameters in the agricultural examples that follow are hypothetical and are only used for illustrative purposes.

4.1   STATIC MODELS OF A CROP FARM

The term **static** means that the element of time is not included. It is useful to begin the discussion of agricultural applications with static models since they are the simplest. After several static models are presented, the discussion proceeds to the more realistic dynamic model.

A Simple Model

Consider a decision problem for a crop farmer who owns 600 acres of tillable land. Farmer Pat wants to decide what combination of corn, soybeans, and wheat to produce in order to maximize net revenue (NR = total revenue minus variable costs). In order to produce corn, soybeans, and wheat, Farmer Pat has to perform the following field operations: plow (*pl*) and disk (*d*) all land, plant corn (*pc*), plant soybeans (*ps*), plant wheat (*pw*), harvest corn (*hc*), harvest soybeans (*hs*), and harvest wheat (*hw*). On the marketing side, we will assume that the farmer sells the entire harvested crop at harvest time. Farmer Pat is also constrained by the amount of family labor that is available to perform all operations. It is assumed that Pat's family can contribute up to 1,700 hours of "endowed" labor per year. However, Pat can also hire up to 900 hours of additional "hired" labor at $6.00 per hour.

Farmer Pat has the following expectations regarding labor requirements (hours per acre) and variable costs for each operation.[1]

Operation	Labor (hour/acre)	Variable cost ($/acre)	Other
Plowing (*pl*)	0.60	10	
Disking (*d*)	0.50	10	
Plant Corn (*pc*)	0.45	60	
Plant Soybeans (*ps*)	0.45	45	
Plant Wheat (*pw*)	0.30	30	
Harvesting Corn (*hc*)	1.48	100	
Harvesting Soybeans (*hs*)	1.00	50	
Harvesting Wheat (*hw*)	1.00	40	
Hired Labor (*hl*) ($/hour)		6	
Corn Price at Harvest ($/bushel)			2.60
Soybean Price at Harvest ($/bushel)			6.35
Wheat Price at Harvest ($/bushel)			3.70
Corn Yield (bushel/acre)			135
Soybean Yield (bushel/acre)			45
Wheat Yield (bushel/acre)			65

---

[1]These and other parameters in this chapter are based on farm conditions in the late 1980s and are not reflective of current market conditions.

The simplest model aggregates all the production and marketing operations into three activities: corn, soybean, and wheat output. The model is used to maximize net revenue by choosing the amount of corn, soybeans, and/or wheat to produce and sell subject to land, family labor, and hired labor constraints:

$$\text{Max: } Z = 171.00c + 170.80s + 150.50w - 6.00hl \text{ [units = NR/acre]} \tag{0}$$

s.t.:

$$c + \quad s + \quad w \qquad\qquad \le 600 \quad \text{(units = acres)} \tag{1}$$

$$3.03c + \quad 2.55s + \quad 2.40w - \quad hl \le 1700 \text{ (units = hours)} \tag{2}$$

$$hl \le 900 \quad \text{(units = hours)} \tag{3}$$

$$c, \quad s, \quad w, \quad hl \ge 0 \quad \text{(units = acres)} \tag{4}$$

The objective function coefficients represent net revenue per acre. They are computed from the information given in the table above, where:

$$c_c = (2.60)(135) - (10 + 10 + 60 + 100) = 171.00,$$

$$c_s = (6.35)(45) \quad - (10 + 10 + 45 + 50) \quad = 170.80,$$

$$c_w = (3.70)(65) \quad - (10 + 10 + 30 + 40) \quad = 150.50.$$

The first constraint is simply a land constraint requiring the combined amount of corn, soybeans, and wheat produced and sold to be 600 acres or less.

The second constraint is a labor constraint that limits the amount of labor used in producing corn, soybeans, and wheat to not exceed 1,700 hours. The technical coefficient for the corn operations in the second constraint is equal to the sum of the technical coefficients for all corn operations, that is, 0.60 (plowing) + 0.50 (disking) + 0.45 (plant corn) + 1.48 (harvest corn) = 3.03. For soybeans, the technical coefficient is calculated the same way: that is, 0.60 (plowing) + 0.50 (disking) + 0.45 (plant soybeans) + 1.00 (harvest soybeans) = 2.55. The technical coefficient for wheat is calculated similarly: that is, 0.60 (plowing) + 0.50 (disking) + 0.30 (plant wheat) + 1.00 (harvest wheat) = 2.40.

The optimal solution to this problem is:

$$c^* = 354.2, \ s^* = 245.8, \ w^* = 0, \ hl^* = 0, \ s_1{}^* = 0, \ s_2{}^* = 0, \ s_3{}^* = 900, \ Z^* = 102,551.$$

The shadow prices (SP) for land and labor are $169.74/acre and $0.41/hour, respectively. This suggests that Farmer Pat should be willing to pay up to $169.74 per acre, and $0.41 per hour for an additional hour of hired labor.

The range of optimality for the objective function coefficient for per unit profitability of corn (c_c), soybeans (c_s), and wheat (c_w) are:

$$170.80 \le c_c \le 173.68,$$

$$168.12 \le c_s \le 171.00,$$

$$-\infty \le c_w \le 170.74.$$

These ranges are very narrow, suggesting that the current solution is quite sensitive to the estimate of the unit profitability for each crop. In this case, further sensitivity analysis in the form of varying the objective function coefficients and examining the corresponding solutions is highly recommended. Later on in this chapter, this type of sensitivity analysis in the form of deriving output supply functions will be demonstrated.

The range of feasibility for land, family labor, and hired labor are:

$$561.1 \leq land \leq 666.67$$

$$1{,}530 \leq family\ labor \leq 1{,}818$$

$$0 \leq hired\ labor \leq \infty.$$

Recall that these are the ranges for which the respective SPs will hold. Also, these ranges are used to estimate input demand functions for each of the resources. This will be demonstrated later in this chapter.

Obviously this model does not provide the farmer with much information since it aggregates all of the operations (plowing, disking, planting, and harvesting) into one activity for each crop.

A More Disaggregated Model

Consider the following model, where all field operations are explicitly incorporated. Table 4.1 illustrates the LP tableau for this problem. The activities in this tableau are defined as follows:

pl = plowing

d = disking

pc = plant corn

ps = plant soybeans

pw = plant wheat

hc = harvesting corn

hs = harvesting soybeans

hw = harvesting wheat

hl = hired labor

hsc = harvest sale of corn

hss = harvest sale of soybeans

hsw = harvest sale of wheat

The objective function (0) differs from the previous problem in that the individual field operations and marketing activities are now explicit in the formulation. For each operation, the objective function coefficients are the variable costs (per acre) and are entered as negative numbers since they are costs. Note that all production operations are expressed in units of acres. The marketing activity objective function coefficients are the prices (per bushel) for the selling activities in the model and are entered as positive numbers since they represent contributions to net revenue.

The first constraint is a land restriction that limits total amount of acres planted to corn, soybeans, and wheat to not exceed 600 acres. The second constraint limits the amount of labor used in all the operations to not exceed 1,700 hours of family labor plus any additional hired labor, which costs $6.00 per hour. Notice that there is also a hired labor constraint restricting total amount of hired labor to not exceed 900 hours. This type of constraint is needed when there are labor shortages in the area, which is often common.

Table 4.1   Linear Programming Tableau for Static Crop Farm Problem

pl	d	pc	ps	pw	hc	hs	hw	hl	hsc	hss	hsw			
−10	−10	−60	−45	−30	−100	−50	−40	−6	2.6	6.35	3.7			(0)
		1	1	1								≤	600	(1)
0.6	0.5	0.45	0.45	0.3	1.48	1	1	−1				≤	1,700	(2)
					−135				1			≤	0	(3)
						−45				1		≤	0	(4)
							−65				1	≤	0	(5)
								1				≤	900	(6)
−1	1											≤	0	(7)
	−1	1	1	1								≤	0	(8)
		−1			1							≤	0	(9)
			−1			1						≤	0	(10)
				−1			1					≤	0	(11)
pl	d	pc	ps	pw	hc	hs	hw	hl	hsc	hss	hsw	≥	0	(12)

In order to make costs and revenues comparable, you must convert the marketing activities, which are currently expressed in \$/bushel, to gross revenue per acre. The third set of constraints in Table 4.1—equations (3), (4) and (5)—make this transformation. The first of these constraints is for corn:

$$-135hc + 1hsc \leq 0.$$

This is easy to interpret if we rewrite it as:

$$1hsc \leq 135hc.$$

This simply means that the number of bushels of corn sold at harvest must not exceed the number of acres of corn harvested times its yield per acre (135 bushels). In other words, if the optimal solution is to plant and harvest 100 acres of corn, then the total amount of corn marketed cannot exceed 13,500 bushels. An identical interpretation holds for the soybeans and wheat output constraints. Through these three constraints, the measurement units in the objective function are consistent. The value of the objective function will be expressed in dollars of total net revenue.

Finally, the **sequencing constraints** (constraints (7) through (11) in Table 4.1) assure the proper order of the field operations, as well as guarantees that all operations are performed. Notice that without these constraints, the activities that incur costs would not end up being performed, which is not an unrealistic assumption. Constraint (7) is a plow-before-disk constraint (*pl/d*), which requires that before an acre of land is disked, it must be plowed:

$$-1pl + 1d \leq 0 \text{ implies } 1d \leq 1pl.$$

Constraint (8) is a disk-before-planting constraint (*d/p*), which requires that before an acre of land is planted, it must be disked:

$$-1d + 1pc + 1ps + 1pw \leq 0 \text{ implies } 1pc + 1ps + 1pw \leq 1d.$$

Constraints (9), (10), and (11) require that in order to harvest an acre of corn (or an acre of soybeans, or wheat), an acre must have been planted with corn (or soybeans, or wheat):

$$-1pc + 1hc \leq 0 \text{ implies } 1hc \leq 1pc,$$

$$-1ps + 1hs \leq 0 \text{ implies } 1hs \leq 1ps,$$

$$-1pw + 1hw \leq 0 \text{ implies } 1hw \leq 1pw.$$

One of the first things to check from the solution of these types of problems is whether the sequencing constraints have been formulated correctly. To do this, make sure that all acres of land that are harvested and sold have also been plowed, disked, and planted. You can see that this is the case here by examining the optimal solution, which is:

$$pl^* = 600, \, d^* = 600, \, pc^* = 354.2, \, ps^* = 245.8, \, pw^* = 0,$$
$$hc^* = 354.2, \, hs^* = 245.8, \, hw^* = 0, \, hl^* = 0, \, hsc^* = 47,812.5,$$
$$hss^* = 11,062.5, \, hsw^* = 0, \, Z^* = \$102,539.$$

That is, the farmer will plow and disk 600 acres of land. The farmer will then plant 354.2 acres of corn and 245.8 acres of soybeans. In the fall, all acres planted with corn and soybeans will be harvested and sold on the cash market. The farmer's expected net revenue is $102,539.

The important SPs are:

$$SP_{land} = 169.42, \, SP_{family\ labor} = 0.52.$$

This implies that the farmer should be willing to pay up to $169.42 for another acre of land and up to $0.52 for another hour of family labor. The SP of hired labor is zero because this constraint is not binding.

Note that sequencing constraints are correct since 600 acres of land are plowed, then disked, then planted, and finally harvested.

The crop output constraints are binding, and their SPs are 2.60, 6.35, and 3.70, respectively for corn, soybeans, and wheat. These SPs are simply the market prices per bushel that were given for the three crops.

The SPs for the sequencing constraints vary from 10.31 to 250.23. Consider the plant-before-harvest sequencing constraints. For corn, the SP for this constraint is 250.23 and is interpreted as follows. The sequencing constraint is:

$$-pc + hc \leq 0, \text{ or}$$

$$hc \leq pc.$$

This means that for each acre of corn harvested, you need to plant 1 acre of corn. Now increase the RHS from 0 to 1, that is,

$$hc \leq pc + 1.$$

This means that the farmer gets one "free" acre of corn harvested that does not have to be planted, plowed, or disked. The farmer does, however, have to pay the variable cost of harvesting and additional labor costs. Hence, the SP is equal to:

(corn price)(corn yield) − (harvest variable cost) − (harvesting corn
technical coefficient)($SP_{family\ labor}$), or

$$(2.60)(135) - 100 - (1.48)(0.52) = 250.23.$$

Notice that the subtracted labor cost is a derived cost based on the family labor SP and the additional amount of labor required to harvest the "free" acre of corn. The SPs for soybeans and wheat plant-before-harvest sequencing constraints are derived the same way and are equal to:

$$(6.35)(45) - 50 - (1)(0.52) = 235.23,$$

$$(3.70)(65) - 40 - (1)(0.52) = 199.98.$$

Output Supply Functions

An output supply function gives the relationship between optimal quantity and price of each crop, holding all other parameters of the problem constant. For example, the corn output supply function maps optimal corn acreage for different corn prices, holding constant all other prices, costs, technical coefficients, and resources endowments. To derive an output supply function, use the range of optimality with a simple iterative procedure.

For instance, to derive the corn supply function, prepare a two-column table that lists the range of optimality for the corn price in one column and optimal corn acreage in the other. For the current solution, the range of optimality for the corn price is

$$2.60 \leq c_c \leq 2.62.$$

The current optimal corn acreage is $c^* = 354.2$. As long as the price of corn is between $2.60 and $2.62 per bushel, 354.2 acres of corn should be produced and sold (note that you could also express quantity in terms of bushels marketed, which would be equivalent). Therefore, the first entry in the supply function table is:

Corn Price Range of Optimality	Optimal Corn Acreage
2.60 to 2.62	354.2

What happens to corn acreage if the corn price is decreased slightly below the lower limit for its range of optimality? The answer to this question will give the next entry in the supply function table. When resolving the model using a corn price of $2.59, optimal corn acreage falls to zero. Not surprisingly, the range of optimality associated with this solution is

$$-\infty \leq c_c < 2.60.$$

As long as the price of corn is below $2.60, no corn should be produced and sold. Entering the new range of optimality and optimal corn acreage into the supply function table yields:

Corn Price Range of Optimality	Optimal Corn Acreage
0 to 2.59	0
2.60 to 2.62	354.2

Since the lower bound (LB) of zero has been reached for optimal corn acreage, you need not consider further solutions that decrease the corn price below its original price. Now consider what happens if the price of corn is increased above the upper bound (UB) for its initial range of optimality, which in this case is $2.62. When resolving the model using a corn price of $2.63, optimal corn acreage increases to 600 acres. The new range of optimality for this solution is:

$$2.62 < c_c \leq \infty.$$

As long as the corn price is higher than $2.62, the optimal solution is to only grow corn on the farmer's land. Adding the new range of optimality and corn acreage to the supply function table yields:

Corn Price Range of Optimality	Optimal Corn Acreage
0 to 2.59	0
2.60 to 2.62	354.2
2.63 to ∞	600

All the information necessary to graph an output supply function for corn is now complete since the lower range for the price is zero and the upper range is infinity. Figure 4.1 displays the output supply function for corn. In applications that are larger than this, the number of "kinks" in the supply schedule increases.

Input Demand Functions

An input demand function gives the relationship between the level (amount) of the resource endowment and its SP, holding all other parameters of the problem constant. For example, the land input demand function maps the SP of land for different levels of land endowments, holding constant all other prices, costs, technical coefficients, and resource endowments. To derive an input demand function, use the range of feasibility with a simple iterative procedure.

For instance, to derive the land input demand function, again prepare a two-column table that lists the range of feasibility for the land endowment in one column and the land SP in the other. For the current solution, the range of feasibility for the land endowment is:

$$561.06 \leq land \leq 666.67.$$

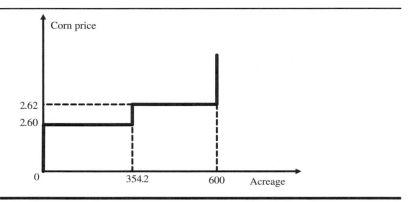

Figure 4.1   Corn output supply function.

The current SP for land is 169.42. As long as the amount of land endowed is between 561.06 and 666.67 acres, the land SP will be $169.42 per acre. Therefore, the first entry in the input demand function table is:

Land Range of Feasibility	Land SP
561.06 to 666.67	169.42

What happens to the land SP if the land endowment is decreased slightly below the lower limit for its range of feasibility? The answer to this question will give the next entry in the input demand function table. When resolving the model using a land endowment of 560 acres, the SP for land rises to $171.00 per acre. In this case, the range of feasibility is:

$$0 \le land \le 561.06.$$

As long as the amount of land is less than 561.06 acres, the SP for land will be $171.00. Entering the new range of feasibility and SP into the input demand function table yields:

Land Range of Feasibility	Land SP
0 to 561.06	171.00
561.06 to 666.67	169.42

Since the LB of zero has been reached for the range of feasibility, you need not consider further solutions that decrease land acreage below its original level. Now consider what happens if land acreage is increased above the UB for its initial range of feasibility, which in this case is 666.67. When resolving the model using total acreage equal to 667, the SP declines to 155.45. The new range of feasibility for this solution is:

$$666.67 < land \le 1,019.60.$$

Adding the new range of feasibility and SP to the input demand function table yields:

Land Range of Feasibility	Land SP
0 to 561.05	171.00
561.06 to 666.67	169.42
666.68 to 1,019.60	155.45

Since the land constraint is still binding, you must re-solve the problem again setting land acreage above the new upper limit from the range of feasibility, which in this case is above 1,019.60 acres. Resolving the model with acreage set at 1,020 acres results in the land constraint no longer binding. Hence, the SP falls to zero and the range of feasibility is:

$$1,019.60 < land \le \infty.$$

Adding this to the input demand function table yields:

Land Range of Feasibility	Land SP
0 to 561.05	171.00
561.06 to 666.67	169.42
666.68 to 1,019.60	155.45
1,019.61 to ∞	0

All the information necessary to graph an input demand function for land is now complete since the LB for the range of feasibility in the table above is zero and the UB is infinity. Figure 4.2 displays the input demand function for land. On your own, derive the input demand function for labor for this problem.

Discussion

There are many modifications you could make in this model to account for various farming practices. For example, many farmers practice crop rotation where they grow corn on last year's soybean acreage and soybeans on last year's corn acreage. To accommodate for crop rotation, you could add maximum corn and soybeans constraints. The RHS parameters for each constraint would be the amount of acres devoted to corn and soybeans consistent with crop rotation patterns. Alternatively, you could require corn acreage to equal soybean acreage by adding the following constraint:

$$pc - ps = 0.$$

There are also many other resource constraints that farmers are confronted with that should be added to the model. For example, in addition to labor, farmers have machine time constraints. Suppose the farmer owns one tractor (for planting) and one combine (for harvesting). You should then add to the model a tractor-time and a combine-time constraint. The technical coefficients for these constraints would be the hours per acre that it takes to plant corn and soybeans with the tractor, and harvest corn and soybeans with the combine. The RHS-values would be the number of tractors (or combines) times the number of hours available.

How are the technical coefficients and RHS values for farm models derived? Technical coefficients are based on machinery size and can be generated from computer programs

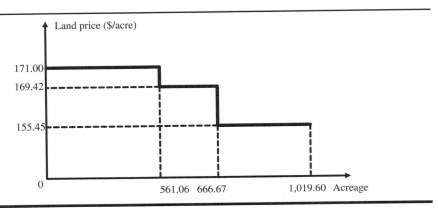

Figure 4.2   Land input demand function.

written by farm management experts (such programs are often available from farm management professors and extension associates in departments of agricultural economics). A general formula for machinery technical coefficients is:

$$\text{Field rate (acres per hour)} = (mph)(ft)(e)/8.25, \text{ where}$$

mph is the speed of the machine in miles per hour, *ft* is the width covered by the machine measured in feet, *e* is the efficiency of the machine, and 8.25 is the width-to-area conversion factor and is used when width is measured in feet. Labor technical coefficients are usually based on a percentage of the machinery coefficients, for instance, 110 percent of machinery rates.

Right-hand-side parameters for labor and machinery time are based on the number of available machines and persons, and field time availability, which depends on weather and soil conditions, that is, rainfall, temperature, and soil type. Often much of this data are available from historical daily data from county agricultural experiment stations. For example, if you wanted an estimate of available field time for a two-week period, May 1 to May 15, you could take the average number of field hours (hours that farmers could actually be performing operations in the field) over the past three years from the local agricultural experiment station. Suppose that amounted to 84 hours, and the farm owned two tractors, then total tractor time availability for this two week period would be 2(84) = 168 hours.

This model is also very naive from a marketing point of view. In reality, there are many other marketing strategies available to farmers than simply selling the entire crop at harvest time. For example, one strategy is to include on-farm storage of crops, which enables the farmer to store the crop at harvest when prices are typically low and sell the crop from storage at a later date when the prices are higher. This could be done by defining additional marketing strategies that require storage, adding a maximum storage constraint, and modifying the crop output constraints so that the additional marketing activities are also included in these constraints. Other strategies would be to include futures market activities like hedging or forward marketing. This would be done by adding additional marketing activities to the objective function along with their expected prices as objective function coefficients.

In this example, the optimal solution does not tell the farmer anything about timing because it is a static model. When should the field be plowed and disked, when should the crops be planted and harvested, and so on? In reality, the timing of such operations is critical information for farmers in order to make sound management decisions. Hence, it is desirable to look at "dynamic" models so that time, as well as operations, can be explicitly accounted for in the model.

4.2   A MULTIPLE-YEAR MODEL

Multiple-year farm models are useful in planning longer-term operations. A crucially important decision for a person deciding whether or not to go into farming is what type of enterprises to pursue. Suppose a person has inherited 90 acres of land that is suitable for fruit and vegetable production, and is close to a major city for direct sales to consumers via a close-by farmers' market. The major decision for this person is whether to grow a perennial crop, apples, versus two annual crops: organic tomatoes and organic lettuce. When deciding what to produce, the person is not interested in simply looking at revenue streams for the coming year, but rather wants to take a longer-term view. In this case, a multiple-year model is necessary.

To illustrate such a model, consider a four-year planning period. The person can grow organic tomatoes and lettuce in years 1, 2, 3, and 4 and expect to receive positive unit profits for each crop each year. If the future farmer decides to put in apple trees, negative unit profits will be incurred in year 1 since there is no revenue from the harvest in the first year, but costs are still incurred. However, positive profits are expected in years 2, 3, and 4. Assume the following profit per acre expectations for each crop for each year:

Year	Apples (*ap*)	Tomatoes (*tm*)	Lettuce (*lt*)
1	−800	500	600
2	750	600	500
3	950	700	550
4	999	450	600

Also assume that the unit profits for each crop for each year are discounted,[2] so the objective function is to maximize discounted net revenue for the farmer over a four-year period.

There are two resources that the farmer controls: land (90 acres) and family labor (a total of 60 person weeks available from the entire family). Suppose that the apple operation requires 0.6 person weeks per acre per year, organic tomatoes requires 0.8 person weeks per acre per year, and organic lettuce requires 0.9 person weeks per acre per year in labor. In the interest of diversification, the farmer wants to grow a minimum of 10 acres of apples, 10 acres of tomatoes, and 10 acres of lettuce in year 1; however, there are no minimum requirements for years 2, 3, and 4. The LP tableau corresponding to this problem is presented in Table 4.2.

The objective function in (0) is to maximize discounted profit (r) over the four-year period. Constraint (1) is a definitional constraint that defines total profit for the four-year period as equaling the sum of profits in years 1 through 4 (r_1, r_2, r_3, and r_4). Likewise, constraints (2) through (5) define yearly profits for each of the four years from the production and sale of apples (ap_i), tomatoes (tm_i), and lettuce (lt_i), where i = 1, 2, 3, 4. Constraints (6) through (13) are structural constraints for land availability (90 acres) and labor availability (60 weeks) for each of the four years. Unlike annual crops such as tomatoes and lettuce, apples require apple trees to be planted, and it takes a minimum of one year before they bear fruit. Hence, the model needs a set of constraints that guarantee that the number of acres devoted to apple production in each year is the same since this production decision, once made, is more permanent than annual crop decisions. Constraints (14) through (16) assure this by requiring apple acreage in year 1 to be the same as in year 2 (constraint 14), year 2 acreage to be the same as year 3 (constraint 15), and year 3 acreage to be the same as year 4 (constraint 16). Constraints (17) through (19) are the 10-acre minimum conditions for apples, tomatoes, and lettuce in year 1. Finally, (20) is the non-negativity restriction for all activities except for r_1, which is allowed to be negative since if the model chooses a solution with significant apple acreage in year 1, negative profits are allowable.

The optimal solution to this problem yields a total discounted profit of $184,607 with $r_1 = -\$32,333$, $r_2 = \$63,000$, $r_3 = \$78,000$, and $r_4 = \$75,940$. In year 1, the farmer should devote 60 acres to apple production, 10 acres to tomato production, and 17.78 acres to lettuce production. In the first year, all but 2.22 acres are used because the labor constraint is binding with a SP of 666.66: that is, one more additional week of labor would yield $666.66 in extra profit. In years 2 and 3, 60 acres are devoted to apple production, and the remaining 30 acres to tomato production. Both land and labor are binding constraints in year 2, and have SPs of 496 and 130, respectively. The SPs for land and labor in year 3 are 875 and 0. In year 4, 60 acres of apples and 26.67

---

[2]One could discount the stream on expected profits each year by the following net present value (NPV) formula: $NPV = \Sigma r_t/(1 + i)^t$, where r_t is the net cash flow at time t, i is the discount rate, and t is the time period.

Table 4.2 Linear Programming Tableau for Multiyear Fruit–Vegetable Farm Problem

r	r₁	r₂	r₃	r₄	ap₁	tm₁	lt₁	ap₂	tm₂	lt₂	ap₃	tm₃	lt₃	ap₄	tm₄	lt₄			
r	r_1	r_2	r_3	r_4	ap_1	tm_1	lt_1	ap_2	tm_2	lt_2	ap_3	tm_3	lt_3	ap_4	tm_4	lt_4			
1																			(0)
−1	1	1	1	1													=	0	(1)
	−1				−800	500	600										=	0	(2)
		−1						750	600	500							=	0	(3)
			−1								950	700	550				=	0	(4)
				−1										999	450	600	=	0	(5)
					1	1	1										≤	90	(6)
					0.6	0.8	0.9										≤	60	(7)
								1	1	1							≤	90	(8)
								0.6	0.8	0.9							≤	60	(9)
											1	1	1				≤	90	(10)
											0.6	0.8	0.9				≤	60	(11)
														1	1	1	≤	90	(12)
														0.6	0.8	0.9	≤	60	(13)
					−1			1									=	0	(14)
								−1			1						=	0	(15)
											−1			1			=	0	(16)
					1												≥	10	(17)
						1											≥	10	(18)
							1										≥	10	(19)
																	≥	0	(20)

acres of lettuce are grown. The SPs for land and labor in the last year are 0 and 666.66, respectively.

If perennial crops like apples have values that are positive at the end of the last year of the problem, then a **terminal value** must be computed and included in the analysis. In this example, an apple tree would still be valuable at the end of the fourth year. One way to compute a terminal value is to calculate the discounted stream of future profits that the apple tree would have, and then add that value to the year 4 unit profit coefficient. In this way, the future value of this asset is explicitly accounted for.

4.3   CROP-LIVESTOCK ENTERPRISES[3]

So far, only crop enterprises have been examined. Another common farm enterprise is a joint crop and livestock operation such as a dairy farm. The basic difference between these enterprises and crop-only farms is that the former consumes part (or all) of what it produces by feeding it to the livestock. Of course, some of the crop can still be sold in the market place.

To illustrate how LP can be used to assist these farmers, consider the second (disaggregated) static farm problem presented in this chapter. You could easily extend this to a dynamic model using identical logic, but it is not done here in the interest of space. Suppose that this crop farmer wants to evaluate whether or not it would be profitable to add up to 60 dairy cows to the farm. Assume that the cost per cow is $500, and each cow can produce 15,000 pounds of milk per year which sells for $0.12 per pound. Because a joint crop and dairy farm requires more labor than the crop farm, assume that the amount of family labor is 5,000 hours per year and the farmer can still hire up to an additional 900 hours of labor at $6.00 per hour.

To maximize milk production, dairy cows must be fed a certain diet. Assume that each cow requires the following annual diet, which will be produced entirely on the land owned by the farmer:

1.   Concentrate made out of 35 bushels of corn grain mixed with 9 bushels of soybeans

2.   1.22 tons of hay

3.   0.6 tons of forage from pasture

4.   7.9 tons of corn silage

Therefore, in addition to corn grain, soybeans, and wheat, the farmer must now grow pastureland, hay, and corn silage. Assume that the farmer will grow only as much pastureland and corn silage that is necessary to feed the herd, but may grow additional hay beyond that needed for the herd. Any excess hay that is grown can be sold at harvest for $75.00 per ton. The farmer expects the following variable costs and yields for each of these additional crops (assume all other parameters for the other crops are as defined previously).

Crop	Variable Cost	Yield	Technical Coefficient
Plant corn silage	60		0.45
Plant hay	15		0.25
Plant pastureland	15	2.3	0.25
Harvest corn silage	65	4.0	1.00
Harvest hay	16	2.5	0.50

---

[3]This problem, solution, and corresponding sensitivity analysis are shown in the Chapter 4 supplemental materials available at www.wiley.com/college/kaiser.

Table 4.3   Linear Programming Tableau for Static Crop–Livestock Farm Problem

	pl	d	pc	pcs	psb	pw	ph	pp	hc	hcs	hh	hsb	hw	cow	hl	hsc	hsh	hss	hsw	ms		
(0)	−10	−10	−60	−60	−45	−30	−15	−15	−100	−65	−16	−50	−40	−500	−6	2.6	75	6.35	3.7	0.12		
(1)	0.6	0.5	0.45	0.45	0.45	0.3	0.25	0.25													≤	600
(2)									1.48	1	0.5	1	1	65	−1	0	0	0	0	0	≤	5,000
(3)														1		1					≤	60
(4)									−135					35				1			≤	0
(5)														9					1		≤	0
(6)												−45								1	≤	0
(7)								−2.3					−65	−15,000							≤	0
(8)															1						≤	900
(9)										−4				0.6							≤	0
(10)											−2.5			7.9							≤	0
(11)														1.22			1				≤	0
(12)	−1	1							1			1									≤	0
(13)		−1	−1							1			1								≤	0
(14)				−1							1										≤	0
(15)					−1																≤	0
(16)						−1															≤	0
(17)	−1						−1														≤	0
(18)																					≤	0
(19)																					≥	0
	pl	d	pc	pcs	psb	pw	ph	pp	hc	hcs	hh	hsb	hw	cow	hl	hsc	hsh	hss	hsw	ms		

The mathematical formulation of this problem is presented in tableau form in Table 4.3. The activities for this problem are denoted as:

$$pl = \text{plowing}$$
$$d = \text{disking}$$
$$pc = \text{plant corn}$$
$$pcs = \text{plant corn silage}$$
$$psb = \text{plant soybeans}$$
$$pw = \text{plant wheat}$$
$$ph = \text{plant hay}$$
$$pp = \text{plant pasture}$$
$$hc = \text{harvesting corn}$$
$$hcs = \text{harvesting corn silage}$$
$$hh = \text{harvesting hay}$$
$$hsb = \text{harvesting soybeans}$$
$$hw = \text{harvesting wheat}$$
$$cow = \text{number of cows purchased}$$
$$hl = \text{hired labor}$$
$$hsc = \text{harvest sale of corn}$$
$$hsh = \text{harvest sale of hay}$$
$$hss = \text{harvest sale of soybeans}$$
$$hsw = \text{harvest sale of wheat}$$
$$ms = \text{milk sales}$$

The objective function (0) differs from before only by the addition of several activities for the dairy operation, that is, plant corn silage (pcs), plant hay (ph), plant pasture (pp), harvesting corn silage (hcs), harvesting hay (hh), number of dairy cows to purchase (cow), harvest sale of hay (hsh), and sale of milk (ms). The land (1) and labor (2) constraints have been modified to include technical coefficients for these additional activities, where appropriate. Constraint (3) is a herd size limitation which restricts the farmer from purchasing more than 60 cows. This is followed by the crop output constraints, (4) to (7). Notice that the corn and soybean crop output constraints have been modified so that the amount of crop sold at harvest must not exceed the amount harvested times its respective yield minus the amount required for feeding cows, that is,

$$-135hc + 35cow + 1hsc \leq 0 \text{ (for corn)},$$

$$-45hsb + 9cow + 1hss \leq 0 \text{ (for soybeans)}.$$

To see this more clearly, rearrange the corn output constraint as:

$$35cow + 1hsc \leq 135hc.$$

This means the number of cows added times their annual requirement of 35 bushels of corn, plus the bushels of corn sold at harvest cannot exceed the number of acres harvested to corn times its yield, 135 bushels per acre.

The wheat output constraint is the same as before, since it is assumed that wheat is not used as feed for the cows.

The next constraint, (7), is a cow output constraint that restricts the total pounds of milk sales to be $\leq$ 15,000 pounds times the number of cows that are purchased. Recall that each cow can produce 15,000 pounds of milk per year. Constraint (8) restricts the total amount of hired labor to 900 hours or less. This is followed by three constraints relating pasture, corn silage, and hay yields per acre to cow feed intake requirements. For example, the first constraint, (9), is:

$$-2.3pp + 0.6cow \leq 0, \text{ or}$$

$$0.6cow \leq 2.3pp.$$

This constraint requires that the farmer must plant enough pasture to satisfy the feeding requirements of the herd, given a yield of 2.3 tons per acre for pasture. The same interpretation holds for the other two constraints.

Finally, the sequencing constraints are similar to those presented earlier, but now they are modified to account for the dairy operation. Examine these on your own to see how they differ from the crop-only farm.

The optimal solution to this problem shows that establishing a dairy operation will be profitable to the farmer. The solution suggests that 54.52 cows[4] should be purchased and milked, and in order to have enough feed for the herd, the farmer must produce 14.1 acres of corn grain, 107.7 acres of corn silage, 26.6 acres of hay, and 14.2 acres of pasture. While no wheat is produced, the remaining 437.4 acres of land is used to grow and sell soybeans. Based on the farmer's expectations, the net revenue from this operation is $123,244, which is more than $20,000 higher than the crop-only solution.

The SPs for the resource endowments show that the implicit values of the farmer's land and labor in this example are $155.47 per acre and $5.99 per hour. Since the maximum herd size constraint is not binding, its respective SP is zero. If you wanted to see what a positive SP for cows is, one way to make this constraint binding is to increase the amount of family labor from 5,000 hours to say 10,000 hours. In this case, the herd constraint becomes binding and its SP is $380.60 per cow. That is, adding an additional cow would increase net revenue by $380.60. The new constraint could then be used to derive an input demand function for cows, which would provide useful information for the farmer, particularly in deciding how many cows to purchase. Derivation of output supply functions for corn grain, hay, soybeans, wheat, and milk, and of input demand functions for land, labor, and other fixed resources would also be useful information to the farmer.

4.4 DYNAMIC MODELS

The term **dynamic** here means a model that incorporates time. However, while uncertainty is usually incorporated into dynamic models, it will be ignored in this model. Later in the book, the concept of risk and uncertainty will be covered. The multiple-year model presented in a previous section is actually one type of dynamic model since time was explicit considered. In this section, a within-year model with discrete time periods disaggregated across a crop production and marketing year is pressented. In other words, it is a model useful for planning over a one-year time horizon.

---

[4]Obviously, the purchase of 0.52 of a living dairy cow is unrealistic. How to handle integer solutions will be discussed in Chapter 7.

A One-Year Model With Discrete Time Periods[5]

Suppose that the farmer provides the following information regarding time periods and the production process (assume that the 1,700 hours are divided as displayed in the table below):

Period	Available Labor (hours)	Plow	Disk	Planting Corn	Planting Beans	Planting Wheat	Harvesting Corn	Harvesting Beans	Harvesting Wheat	Harvest Sales Corn	Harvest Sales Beans	Harvest Sales Wheat
1 Mar 15–May 9	283	X	X	X		X						
2 May 10–May 23	283	X	X	X	X	X						
3 May 24–Jun 6	283		X		X							
4 Sep 13–Sep 26	283						X					
5 Sep 27–Oct 17	283	X					X	X	X			
6 Oct 18–Nov 7	285	X					X		X	X	X	X

Corn and soybean yields are influenced by planting and harvesting dates (wheat is not). The corn and soybean yields by their respective planting and harvesting dates are given below. For this variety of corn, it is better to plant early and harvest late in order to obtain the highest yield: for instance, planting period 1 and harvesting period 6 gives the highest corn yield. The same is true for soybeans in this example.

Harvest Period	Yield (bushels/acre) by Planting Date			
	Corn		Soybeans	
	1	2	2	3
4	—	—	35	40
5	140	130	55	50
6	150	120	—	—

Assume that plowing may take place in the spring prior to disking and/or in the fall after harvest (e.g., periods 5 and 6). This assumption serves as a link between an annual model and a longer-run model. The only difference between this model and the last model is that now there is an activity for each operation and each time period. For example, rather than having just one plowing activity (pl), now the model has four plowing activities, one for period 1 (pl_1), one for period 2 (pl_2), one for period 5 (pl_5), and one for period 6 (pl_6).

The mathematical formulation of this problem is:

$$\text{Max: } Z = -10pl_5 - 10pl_6 - 10pl_1 - 10pl_2 - 10d_1 - 10d_2 - 10d_3$$
$$- 60pc_1 - 60pc_2 - 45ps_2 - 45ps_3 - 30pw_1 - 30pw_2 - 100hc_{15}$$
$$- 100hc_{16} - 100hc_{25} - 100hc_{26} - 50hs_{24} - 50hs_{25} - 50hs_{24}$$
$$- 50hs_{35} - 40hw_{15} - 40hw_{16} - 40hw_{25} - 40hw_{26} + 2.60hsc$$
$$+ 6.35hss + 3.70hsw - 6hl_1 - 6hl_2 - 6hl_3 - 6hl_4$$
$$- 6hl_5 - 6hl_6 \tag{0}$$

s.t.:

Land Constraint

$$1pc_1 + 1pc_2 + 1ps_2 + 1ps_3 + 1pw_1 + 1pw_2 \le 600 \tag{1}$$

[5]This problem, solution, and corresponding sensitivity analysis are shown in the Chapter 4 supplemental materials available at www.wiley.com/college/kaiser.

Labor Constraints

Period 1: $0.60pl_1 + 0.50d_1 + 0.45pc_1 + 0.30pw_1 - 1hl_1 \leq 283$ (2)

Period 2: $0.60pl_2 + 0.50d_2 + 0.45pc_2 + 0.45ps_2 + 0.30pw_2$
$- 1hl_2 \leq 283$ (3)

Period 3: $0.50d_3 + 0.45ps_3 - 1hl_3 \leq 283$ (4)

Period 4: $1.00hs_{24} + 1.00hs_{34} - 1hl_4 \leq 283$ (5)

Period 5: $0.60pl_5 + 1.48hc_{15} + 1.48hc_{25} + 1.00hs_{25} + 1.00hs_{35}$
$+ 1.00hw_{15} + 1.00hw_{25} - 1hl_5 \leq 283$ (6)

Period 6: $0.60pl_6 + 1.48hc_{16} + 1.48hc_{26} + 1.00hw_{16} + 1.00hw_{26}$
$- 1hl_6 \leq 285$ (7)

Output Constraints

Corn: $- 140hc_{15} - 150hc_{16} - 130hc_{25} - 120hc_{26} + 1hsc \leq 0$ (8)

Soybeans: $- 35hs_{24} - 55hs_{25} - 40hs_{34} - 50hs_{35} + 1hss \leq 0$ (9)

Wheat: $- 65hw_{15} - 65hw_{16} - 65hw_{25} - 65hw_{26} + 1hsw \leq 0$ (10)

Hired Labor Maximum Constraints

Period 1: $1hl_1 \leq 150$ (11)

Period 2: $1hl_2 \leq 150$ (12)

Period 3: $1hl_3 \leq 150$ (13)

Period 4: $1hl_4 \leq 150$ (14)

Period 5: $1hl_5 \leq 150$ (15)

Period 6: $1hl_6 \leq 150$ (16)

Plow Before Disk Sequencing Constraints

Period 1: $- pl_5 - pl_6 - pl_1 + d_1 \leq 0$ (17)

Period 2: $- pl_5 - pl_6 - pl_1 - pl_2 + d_1 + d_2 \leq 0$ (18)

Period 3: $- pl_5 - pl_6 - pl_1 - pl_2 + d_1 + d_2 + d_3 \leq 0$ (19)

Disk Before Plant Sequencing Constraints

Period 1: $- d_1 + pc_1 + pw_1 \leq 0$ (20)

Period 2: $- d_1 - d_2 + pc_1 + pc_2 + ps_2 + pw_1 + pw_2 \leq 0$ (21)

Period 3: $- d_1 - d_2 - d_3 + pc_1 + pc_2 + ps_2 + ps_3 + pw_1 + pw_2 \leq 0$ (22)

Plant Corn Before Harvest Corn Sequencing Constraints

Period 1: $- pc_1 + hc_{15} + hc_{16} \leq 0$ (23)

Period 2: $- pc_2 + hc_{25} + hc_{26} \leq 0$ (24)

Plant Soybeans Before Harvest Soybeans Sequencing Constraints

$$\textbf{Period 2: } - ps_2 + hs_{24} + hs_{25} \leq 0 \tag{25}$$

$$\textbf{Period 3: } - ps_3 + hs_{34} + hs_{35} \leq 0 \tag{26}$$

Plant Wheat Before Harvest Wheat Sequencing Constraints

$$\textbf{Period 1: } - pw_1 + hw_{15} + hw_{16} \leq 0 \tag{27}$$

$$\textbf{Period 2: } - pw_2 + hw_{25} + hw_{26} \leq 0 \tag{28}$$

$$\textbf{Non-negativity}: \text{All Activities} \geq 0 \tag{29}$$

The formulation of this problem has a similar interpretation as before, except now the activities are disaggregated to include time periods. The single numeric subscript for each activity refers to the time period the operation occurs in, for instance, pw_2 is the wheat planting activity in period 2. The two subscripts for the harvesting activity give the time period in which the crop was planted and the time period in which it is harvested. For example, hc_{25} means harvest an acre of corn that was planted in period 2 and is harvested in period 5.

Constraint (1) is the land restriction and is similar to the previous model except that activities are further disaggregated by time period. Constraints (2) through (7) are the labor constraints for the six time periods. Each time period corresponds to the activities that are permitted in that time period. The output constraints (8) to (10) are similar to the previous model except the crop yield coefficients are now disaggregated by planting and harvesting dates.

Note that the sequencing constraints appear to be the most different. The first set of sequencing constraints require that prior to disking an acre of land, you must plow an acre of land. Plowing can take place in either the fall of the previous year (periods 5 and 6), or in the spring prior to disking (periods 1 and 2). That is, it is assumed that the plowing activities begin in periods 5 and 6 of the previous year and may carry through into periods 1 and 2 of the current crop year. Plowing is the between-year linkage in the model.

Consider the first constraint in this set, plow before disk, period 1. This restricts all period 1 disking from exceeding all the land that was plowed in the previous year (periods 5 and 6) and in period 1. The next constraint for period 2 restricts combined disking in periods 1 and 2 from exceeding combined plowing in periods 5, 6, 1, and 2. The same logic applies to the last of these constraints for period 3.

The second set of sequencing constraints requires that before an acre of land is planted with corn, soybeans, or wheat, it must be disked. These three specific constraints do just that, assuring that the operations are performed in the correct order by time period.

Finally, the last set of sequencing constraints require that in order to harvest an acre of corn, soybeans, or wheat, you must have planted an acre of corn, soybeans, or wheat.

The optimal solution is that 167 acres of corn and 433 acres of soybeans be produced and sold. The sequencing constraints are constructed appropriately since all the field operations take place and are done in proper order. Plowing takes place in periods 1, 2, and 6, while all disking takes place in periods 1 and 2. Note that the amount of disking in period 1 and/or 2 never exceeds the amount of acres plowed prior or concurrent to it. All corn is planted in period 1, while all soybeans are planted in period 2. The farmer makes extensive use of hired labor in periods 1, 2, and 5 when the majority of operations must be performed.

What about the SPs for each resource? The Solver sensitivity analysis shows that the SP for land is 191.82. The SPs for labor give interesting information. These SPs range from zero (periods 1, 2, 3 and 4) to \$27.13 per hour for period 5. This suggests that the labor supply for period 5 operations is the tightest. Look at the rest of the SPs and make your own economic interpretations.

As was the case before, it is recommended that detailed sensitivity analysis be conducted for this problem. At a minimum, output supply functions for the three crops should be derived in order to see how sensitive output decisions are to various prices. It is also interesting to compare the price responsiveness of the crops to one another. Is corn more price elastic than wheat or soybeans? Additionally, you should generate input demand functions for land and labor. The labor demand functions would be disaggregated by the six time periods when field operations take place.

You could expand this model to include additional resources that farmers use, including important constraints for their use. Machinery constraints would be important for U.S. crop farmers. In developing countries, animal power (instead of machinery) would be an important resource constraint to add to the problem.

4.5   MODEL VALIDATION

Model validation is one of the most important steps in any quantitative modeling, including mathematical programming. Model validation is the process of determining how well the model represents the real world. Are the model outcomes consistent with reality? This section, which is based exclusively on McCarl and Spreen (2003), examines several systematic methods for model validation. While this section is contained in the farm LP chapter, model validation applies to all remaining chapters on both linear and nonlinear programming models.

McCarl and Spreen (2003) argue that model validation is basically a subjective process since the modelers themselves are the ones who choose the way the model is to be judged. Modelers select the validity tests, which variables to examine, the thresholds for passing the tests, and so on. However, while the subjective nature of validation may give rise to bad models passing validation tests, at the very least, the process of validation reveals model strengths and weaknesses.

The authors present two general validation approaches: **validation by construct** and **validation by results**. Each is discussed separately below.

Validation by construct is always a part of good modeling. Validation by construct means that the correct procedures were used in building the model. For example, the model is built on the basis of economic theory and is consistent with expert opinion. All model coefficients have a sound basis, coming from other reputable sources, sound scientific models, and/or accurate data. All model constraints are imposed on the basis of real-world limitations in the decision process. To be fully transparent for validation by construct, it is essential that the modeler openly document and make available all the steps involved in building the model and determining the model coefficients. In this way, other researchers can judge the validity of the model. McCarl and Spreen (2003) point out that the major flaw of validation by construct is that this approach assumes rather than tests model validity. Hence, while validation by construct is a useful tool in model building, a more rigorous validation procedure is validation by results.

Validation by results compares values of variables generated by the model to actual observations in the real world. McCarl and Spreen (2003) outline five steps involved in this validation process:

Step 1: Real-world observations are collected to compare against model results.

Step 2: Validation experiment is designed.

Step 3: Experiment is conducted with the model to obtain solutions.

Step 4: Statistical tests are used to measure how closely model solutions conform to actual observations.

Step 5: Model is judged valid or not valid.

Real-world observations are used to compare with both model inputs and outputs. For example, in the crop models presented in this chapter, model inputs include crop prices, costs, yields, and available field time. The analyst needs to have relative certainty that these model inputs are reflective of reality. Model outputs include optimal crop mix and marketing solutions. Comparing optimal crop mix and marketing solutions from the crop model with actual acreage and marketing outcomes in the location to be modeled would be an obvious test of validating the model.

There is a variety of validation experiments that could be used in model validation. McCarl and Spreen (2003) present several general types. The most common is the **prediction experiment**, which involves solving the model and comparing solutions directly with real-world outcomes. For example, how close is the optimal crop mix from a farm LP model to real-world crop mix in the location being studied? The second experiment is called a **change experiment**, which examines whether the model can correctly predict changes in key variables. In order to do this, you need observed data for different situations. For example, you could look at two different time periods that have different crop prices and crop mixes, and use the two sets of data with the model to see how well the model solutions coincide with actual changes in the real world. The third experiment discussed by McCarl and Spreen (2003) is called a **tracking experiment**, which is similar to the change experiment, but rather than looking at two different situations, tracking looks at a greater number of situations over time. For example, the analyst could look at the real-world data from eight consecutive quarters and compare it with model solutions for the same time intervals.

In addition to these, McCarl and Spreen (2003) discuss three other validation experiments: **feasibility**, **quantity**, and **price experiments**. Under the feasibility experiment, the model is solved by setting the decision variables to their observed levels to determine whether the solution is feasible. This is done by adding equality constraints to the model that restrict the variables to equal their real-world observed level. There is also a **dual feasibility experiment**, which involves testing whether the observed SPs from the primal problem are feasible in the dual problem. Under the quantity experiment, the output supply (or input demand) term in the objective function is dropped, but an output equality constraint is added, setting output to the real-world value. Then, the corresponding SPs for the output constraint are compared with real-world market prices. In the input demand version, the input term is dropped from the objective function, while an equality constraint is added that sets the input level equal to observed levels. Then, the corresponding SPs for inputs can be compared with the actual market prices for the inputs. The price experiment is only conducted for price endogenous models (covered in a later chapter), and it entails setting objective function coefficients at real-world levels, then comparing the solution values for optimal quantities to real-world quantity levels.

McCarl and Spreen (2003) identify a systematic process for conducting the validation experiment involving the following steps:

Step 1: Depending on type of experiment, adjust the variables, equations, constraints, and data for the model.

Step 2: Solve the model.

Step 3: Examine the model solutions to determine whether they are infeasible, unbounded, or optimal. If infeasible, find cause of infeasibility, then proceed to Step 5 below. If unbounded, then place an UB and continue to Step 6 below. If optimal, then measure how close solution is to real-world levels (see discussion on statistical tests below).

Step 4: Assuming the model solution is sufficiently close to real-world values, then either judge the model to be valid for application, or do additional "higher-level" validation experiments (one or more of the experiments described above).

Step 5: Assuming the model fails the validation experiment, diagnose the model by looking at whether there are errors in any data, or the objective function is properly specified, or the model structure provides an accurate depiction of the real-world decision environment.

Step 6: Correct the model. This will obviously depend upon the problem at hand, as well as the type of experiment being conducted. Consider, for example, the case of a corn–soybean farmer and the prediction validation experiment. Assume that the solution indicates that all the acreage is being grown to soybeans, whereas in reality the crop mix found in the location is 65% corn and 35% soybeans. Determine from the model structure why this is occurring. Is the price (and/or costs) for corn lower (higher) than it actually is causing soybeans to be over-represented in the model solution? Is the price (and/or costs) for soybeans higher (lower) than it actually is causing soybeans to be over-represented in the model solution? Are the RHS values for field day availability lower than in reality, thus making the more time-consuming corn production less feasible? Look thoroughly at all the possible reasons why corn is being understated by the model solutions.

Step 7: Assuming these steps fail to lead to a valid model, then either abandon the model, use the model but point out its deficiencies, limit the scope of the validation tests to fewer variables, or use a different validation test to see if it passes.

There are numerous evaluation criteria for establishing how well the model solutions approximate reality. McCarl and Spreen (2003) discuss the use of regression analysis, which is performed by regressing model solutions on a constant and observed values. In this case, a perfect fit would be indicated by a constant value of zero and a slope coefficient of one. Other statistical measures include simple correlation coefficients between observed and model results, as well as means, standard deviations, and mean absolute deviations.

An Example

Consider the dynamic corn–soybean model presented earlier in this chapter. The optimal solution to this model involved producing 27.8% of the farmer's crop acreage to corn and 72.2% to soybeans. In reality, this region (Southern Minnesota) devotes more cropland to corn than to soybeans. For instance, in 2007 and 2008, the average crop mix for Minnesota was 56.5% corn and 43.5% soybeans. In validating the model based on the prediction experiment, a key concern would be that the model results in too much soybeans and too little corn being grown. Why might that be the case?

There are at least three possibilities for corn production being understated in the model solution. First, the price of corn relative to the price of soybeans used in the model may be lower than the actual relative prices. In fact, that is the case here. The price of corn and soybeans used in the model were $2.60 and $6.35 per bushel, respectively, which yields a relative corn–soybean price ratio of 0.409. The actual average corn–soybean price ratio has been higher in recent years. For example, the average ratio in 2007 and 2008 was 0.42. In other words, the actual corn–soybean price ratio is 2.8% (i.e., 0.42/0.409) higher than the one used in the model. Second, the cost of corn relative to soybeans used in the model may be higher than the actual relative costs. Again, that is the case here. The ratio of corn to soybean costs used in the model is equal to 1.684, while recent relative costs in the region were substantially lower, that is, 1.297. In other words, the model overstates corn costs relative to soybean costs by 29.8% (i.e., 1.685/1.297). Finally, the model corn yields relative to soybean yields may be lower than what is actually observed in this region. Again, this is the case here. The ratio of corn to soybean yields used in the model was 2.27, while in 2007 and 2008, that ratio averaged 3.85.

If the model is re-solved by inflating the corn price by 2.8%, inflating soybean costs by 29.8%, and making the ratio of corn to soybean yields equal to 3.85, the new solution gives an optimal crop mix of 49% corn and 51% soybeans, which is much closer to the real-world percentage. With these adjustments, the model actually wants to grow more corn than this, but is limited by the amount of labor the farm has in the six production periods. If the amount of labor is increased, then the model will grow even more corn. Hence, making this adjustment to the model will produce results that are even closer to real-world observations on crop mix. These sorts of adjustments in the model that make the results more reflective of reality are sometimes called **model calibration**.

Any sound research should include model validation and calibration. The use of validation by construct is important for constructing a sound model. But even sound models will usually produce initial results that deviate from real-world observations. In these cases, it is useful to perform model validation and calibration similar to the example above.

4.6   RESEARCH APPLICATION: CROP FARM MODEL

There have been many research applications using crop LP models for solving real-world problems. For example, LP models have been used in examining agricultural-environmental problems involving "best management practices" by farms. For instance, what would the environmental impacts of a large dairy farm in an environmentally sensitive region be if the farm followed best management practices rather than actual practices? Crop LP models have been used to examine the economics of new farm technologies such as reduced tillage, new crop varieties, and irrigation technologies to answer whether such new technologies are more profitable than existing ones. Another area of use of these models has been the economics of sustainable agriculture in developed and developing countries. Here, LP models have been used to determine the most optimal sustainable practices for agriculture. Linear programming models have been used extensively to formulate least-cost feed rations for livestock, optimal cropping patterns for individual farms, optimal marketing plans for selling farm output, and countless other applications.

To illustrate the usefulness of the crop model, consider the following application based on a previous student's research project that deals with an important policy question facing grain farmers (Watanabe, 1988). While federal crop subsidy programs have changed since the late 1980s, this out-of-date study is still an excellent example of how a farmer could use LP to answer an important research question: is it profitable to participate in federal commodity programs? There are benefits to participation, such as price supports and deficiency payments, as well as costs, such as forgone production on acreage that is required to be set aside as part of the participation requirements. In this application, the commodity is rice, and a farm-level LP model is used to analyze the problem. This problem is examined for the two most important rice producing regions in the United States: Arkansas and California.

Rice was designated as one of the original seven commodities covered by the Agricultural Adjustment Act of 1933. This Act established a price support and deficiency payment program for rice. Participation in this program, which is voluntary, requires that rice producers set aside some of their production. In return, they are guaranteed a minimum price if market prices fall below that price. Hence, it offers greater price stability and higher expected prices.

The rice program has affected the prices received by rice producers, their incomes, the costs and values of resources used in rice production, and rice growers' productions planning process. Rice producers have higher program participation rates than producers of other commodities (85–95 percent in recent years) and government payments to producers have accounted for 42 percent of gross income from rice, contributing significantly to producers' welfare.

The traditionally high program participation rates of rice producers may be explained by the fact that rice production is so capital-intensive that resources may be more fixed than for other crops. Irrigation systems, land leveling, the construction of levees around fields, and harvesting equipment may not easily be adjusted from season to season. Also, as the rice industry has been facing market prices below target prices since 1981 due to rising production capacity, weak foreign demand, and hefty supplies and stocks, the rice program now seems to play a more important role in a rice grower's decision-making process compared to when the first deficiency payments were paid for the 1976 crop.

The objective of this study is to examine whether a profit-maximizing rice farmer is better or worse off participating in the government's rice program. The impacts of risk and uncertainty, while important, are ignored in this deterministic model. In a later chapter, we will explore how to include risk in crop models.

The total U.S. rice crop is produced by six states: Arkansas, Louisiana, Mississippi, Missouri, Texas, and California. Rice production costs and, as a consequence, the returns to its producers vary widely among the regions due mainly to the differences in operating characteristics, production practices, and types of rice produced. According to "Economic Indicators of the Farm Sector: Costs of Production" published by the U.S. Department of Agriculture (USDA), cash expenses were lowest in the non-Delta area of Arkansas (the major rice production region) and highest in California (the second-largest, producing 20 percent of the total U.S. rice) in 1986. Medium/short-grain varieties predominate in California, where yields are 50 percent higher than those of long-grain growers in other regions. However, the average price for medium/short-grain rice is 10–15 percent lower, so the higher California costs are not necessarily balanced by higher receipts. For this reason, Arkansas (non-Delta) and California were selected as different cases of rice production among the four rice production regions (the others are the Mississippi River Delta and Gulf coast) cited in the USDA report mentioned above.

The principal alternative crops in Arkansas are soybeans and cotton. In California, a number of alternatives are similarly important: hay, sugar beets, vegetables, wheat, and feed grains. All of these alternative crops in California are irrigated. However, only one-fifth of the soybean area is irrigated in Arkansas. In order to simplify the crop mix problem, soybeans were chosen as the alternative crop in Arkansas and wheat was chosen as the alternative crop for California.

To formulate LP models for both regions, data for production costs and farm prices and other relevant information were collected from the USDA report. Production costs, yields, and farm prices are the averages of those in 1984–1986, respectively, for each region, and data for the Arkansas model is displayed in Table 4.4, and for the California model in Table 4.5.

Table 4.4   Data for Arkansas Rice–Soybean Crop Farm Model

Operation	Labor (hour/acre)	Variable Cost ($/acre)
Plowing (*pl*)	0.35	4.00
Disk rice (*dr*)	0.40	43.60
Disk soybeans (*ds*)	0.25	2.00
Plant rice (*pr*)	0.30	52.00
Plant soybeans (*ps*)	0.25	10.00
Fertilize rice (*fr*)	0.20	32.80
Fertilize soybeans (*fs*)	0.20	16.50
Irrigate rice (*ir*)	0.10	35.40
Harvest rice (*hr*)	0.60	38.50
Harvest soybeans (*hs*)	0.36	12.00
Rice harvest price (*sr*, $/bushel)	7.01	
Soybean harvest price (*ss*, $/bushel)	4.97	
Rice target price ($/bushel)	11.66	
Soybean loan rate ($/bushel)	5.02	
Rice yield (bushel/acre)	50.00	
Soybean yield (bushel/acre)	30.00	

Table 4.5   Data for California Rice–Wheat Crop Farm Model

Operation	Labor (hour/acre)	Variable Cost ($/acre)
Plowing (*pl*)	0.35	29.30
Disk rice (*dr*)	0.40	47.00
Disk wheat (*dw*)	0.25	15.00
Plant rice (*pr*)	0.20	35.00
Plant wheat (*pw*)	0.20	40.60
Fertilize rice (*fr*)	0.20	71.30
Fertilize wheat (*fw*)	0.20	54.00
Irrigate rice (*ir*)	0.15	33.30
Irrigate wheat (*iw*)	0.10	11.30
Harvest rice (*hr*)	0.60	70.80
Harvest wheat (*hw*)	0.28	21.00
Rice harvest price (*sr*, $/bushel)	6.23	
Wheat harvest price (*sw*, $/bushel)	4.02	
Rice target price ($/bushel)	11.66	
Wheat loan rate ($/bushel)	4.38	
Rice yield (bushel/acre)	73.00	
Wheat yield (bushel/acre)	73.00	

In Arkansas, the average land endowment of rice growers (land operated = land owned plus land rented in minus land rented out) is 950 acres, and total labor of 1,600 hours is available to perform all of the production operations. The model is static and assumes that each production operation occurs at once and therefore the optimal solution does not tell us anything about timing. Also, it is assumed that there is only one type of marketing activity for each crop, namely, to sell at harvest when not participating in the program. This assumption was made due to the fact that farm prices for rice are generally higher during the first five months of the marketing year, that is, around and immediately after harvest. Finally, it is assumed that there is no risk aversion by the farmer, that is, the farmer is risk neutral or profit maximizing.

In California, the average land endowment of rice growers is 1,150 acres, and a total labor of 2,000 hours is available to perform all of the production operations shown. The assumptions made to formulate this model are the same as those in the Arkansas model. In addition, for wheat, which has deficiency payments associated with acreage reductions, the 1987 target price is used as support price.

With respect to the support price, the target price is used for rice, and the loan rate is used for soybeans since these prices are actually received by farmers if they participate in the respective commodity programs. To be eligible for program benefits, rice producers must reduce their planting, but acreage reductions are not required as a condition of eligibility for price support loans under the Food Security Act of 1985. The support prices and acreage reduction levels used in this model are for 1987 (35% for rice, 0% for soybeans). Hence, if the market price is below the target (or support) price, the government price replaces it as a parameter in the model.

The LP tableau for the nonparticipating and participating Arkansas farm is shown in Tables 4.6 and 4.7 (for brevity the California LP tableau is not shown, but is similar to the Arkansas model). The activities are denoted as follows:

Table 4.6   Linear Programming Tableau for Nonparticipating Arkansas Crop Farm

	pl	*dr*	*ds*	*pr*	*ps*	*fr*	*fs*	*ir*	*hr*	*hs*	*sr*	*ss*			
Max	−4	−43.6	−2	−52	−10	−32.8	−16.5	−35.4	−38.5	−12	7.01	4.97			(0)
s.t.:															
Land				1	1								≤	950	(1)
Labor	0.35	0.4	0.25	0.3	0.25	0.2	0.2	0.1	0.6	0.36			≤	1,600	(2)
RI Yield									−50		1		≤	0	(3)
SB Yield										−30		1	≤	0	(4)
pl/d	−1	1	1										≤	0	(5)
dr/pr		−1		1									≤	0	(6)
pr/fr				−1		1							≤	0	(7)
fr/ir						−1		1					≤	0	(8)
ir/hr								−1	1				≤	0	(9)
ds/ps		−1			1								≤	0	(10)
ps/fs					−1		1						≤	0	(11)
fs/hs							−1			1			≤	0	(12)
Non-neg	*pl*	*dr*	*ds*	*pr*	*ps*	*fr*	*fs*	*ir*	*hr*	*hs*	*sr*	*ss*	≥	0	(13)

Table 4.7   Linear Programming Tableau for Participating Arkansas Crop Farm

	pl	dr	ds	pr	ps	fr	fs	ir	hr	hs	sr	ss			
Max	−4	−43.6	−2	−52	−10	−32.8	−16.5	−35.4	−38.5	−12	11.66	5.02			(0)
s.t.:															
Land				1.35	1								≤	950	(1)
Labor	0.35	0.4	0.25	0.3	0.25	0.2	0.2	0.1	0.6	0.36			≤	1,600	(2)
RI Yield								−50		1			≤	0	(3)
SB Yield									−30		1		≤	0	(4)
pl/d	−1	1	1										≤	0	(5)
dr/pr		−1		1									≤	0	(6)
pr/fr				−1		1							≤	0	(7)
fr/ir						−1		1					≤	0	(8)
ir/hr								−1	1				≤	0	(9)
ds/ps			−1		1								≤	0	(10)
ps/fs					−1		1						≤	0	(11)
fs/hs							−1			1			≤	0	(12)
Non-neg	pl	dr	ds	pr	ps	fr	fs	ir	hr	hs	sr	ss	≥	0	(13)

pl = plow

dr = disk for rice

ds = disk for soybeans

pr = plant rice

ps = plant soybeans

fr = fertilize rice

fs = fertilize soybeans

ir = irrigate rice

hr = harvest rice

hs = harvest soybeans

sr = sell rice

ss = sell soybeans

Net revenue is maximized subject to constraints for land, labor, and crop output and sequencing restrictions. Notice that the 35 percent reduction in acreage as required for the participating case in Table 4.7 is accomplished by putting a coefficient of 1.35 instead of 1 for the rice planting activity.

The optimal results for Arkansas are as follows. If not participating in the commodity programs, the farmer should plow 950 acres of land, disk 482.4 acres for rice planting and

467.6 acres for soybean planting, and plant rice and soybeans to the disked acreage. After planting occurs, fertilizer should be applied to the rice and soybean acreage, and the rice should be irrigated (there is no irrigation for soybeans). In the fall, all rice and soybeans are harvested with 24,120.37 cwt of rice sold at $7.01 per cwt and 14,027.78 bushels of soybeans sold at $4.97 per bushel. The net revenue for the nonparticipating Arkansas farm is $118,473. The important SPs are 1.20 for land and 73.33 for labor, implying that the farmer should be willing to pay up to $1.20 for another acre of land and up to $73.33 for another hour of labor. The extremely low SP of land relative to labor tells us that labor is a much scarcer resource than land for this particular problem.

If the Arkansas rice farmer participates in the government programs, all resources should be devoted to rice production. In this case, 703.7 acres of land (the remaining acreage is set aside) are plowed, disked, planted, fertilized, irrigated, and harvested for rice production. After harvest, 35,185.18 cwt. of rice is sold at the higher government price of $11.66 per cwt. Net revenue for this participating farmer is $265,085. Therefore, even though total acreage planted is reduced by 35 percent, the farmer is significantly better off by $146,612 through participating in the program. The SP for land is $279.04 for land, which is $277.84 higher than the case of nonparticipation. This implies that if the farmer added one more acre of land, net revenue would increase by $279.04, and therefore the farmer should be willing to pay up to $279.04 for another acre of land. The SP for labor is zero since not all 1,600 hours are used. Notice that the SP for land is significantly larger than in the nonparticipation case, which illustrates how commodity programs become capitalized into land values.

In California, if not participating in the commodity programs, the rice farmer will plow 1,150 acres of land, and disk 794.23 acres for rice planting and 355.77 acres for wheat planting. After planting, fertilizer and irrigation is applied to the rice and wheat acreage. At harvest (rice in fall and wheat in early summer), all acreage planted to rice and wheat are harvested and sold at 57,987.84 cwt. for rice at $6.23 per cwt. and 25,971.16 bushels of wheat at $4.02 per bushel. Net revenue for the nonparticipating California farm is $176,999. The important SPs are 198.98 for land and 88.13 for labor. This implies that the farmer should be willing to pay up to $198.98 for another acre of land and up to $88.13 for another hour of labor.

Similar to the Arkansas farmer, if the California rice farmer participates in the government programs, then all resources should be devoted to rice production. That is, 851.85 acres are plowed, disked, planted, fertilized, irrigated, and harvested to rice. After harvest, 62,185.18 cwt of rice is sold for $11.66 per cwt. Net revenue for the participating California farm is $480,853. Therefore, even though the farmer is required to reduce total acreage planted by 35 percent, the farmer is significantly better off by $303,854 by participating in the program. The SP for land is 418.13 for land, which is 2.1 times higher than the case of nonparticipation. The SP for labor is zero since not all 2,000 hours are used.

The results of this research indicate that government payments to rice growers, which have been set very high relative to market prices since the 1981 Act, contribute significantly to their revenue. The results also explain the extremely high participation rates in the rice commodity program. This is particularly true in California, where rice production costs are higher and farm prices for rice are lower than the southern states, since the rice program does not distinguish between the various types of rice. The results also imply significant profitability of rice production if farmers participate in the program. A USDA study shows that the relative economic advantage of producing rice is evident when returns including government payments are compared across major crops produced in rice growing regions. These results partly explain why rice growers tend to be generally heavier program participants than other grain producers.

4.7   RESEARCH APPLICATION: ECONOMIC FEASIBILITY OF AN ENERGY CROP FOR A SOUTH ALABAMA COTTON–PEANUT FARM

The U.S. government has encouraged the development of alternative fuel sources as a means to become more energy self-sufficient. Crop residues have the potential of displacing the equivalent of over 12 percent of petroleum imports or 5 percent of electricity consumption in the United States (Gallagher et al., 2003). While currently corn-based ethanol has been the main bio-fuel employed in the United States, there are other fuel crops that could be used to produce ethanol. However, farmers will only grow such crops if it is profitable to do so.

The research study summarized here used an LP model to determine the minimum price that would make production of velvet beans, a potential energy crop, profitable for a cotton–peanut farm in Alabama (Frank et al., 2004). The biomass from velvet beans can be used to produce fuel, and they are desirable in a rotation with other crops because they have positive effects on soil and enhance the yield of other crops. However, Alabama farmers tend to grow cotton and peanuts, which are currently more profitable than velvet beans.

In an effort to promote bio-fuels, the federal government offers several subsidies to firms to produce alternative fuels. This has the potential to increase the demand for crops like velvet beans, which would raise the farm price and encourage its expansion. To determine what the minimum price would be to make it economically attractive for the typical cotton and peanut Alabama farm, Frank et al. (2004) developed a farm-level LP model along with crop enterprise budgets. The data for the model came from a variety of sources including an experimental station (crop yields), and cost data from the Alabama Cooperative Extension Service. Because there were no enterprise data for velvet beans, the authors assumed a yield of 7 tons per acre and that velvet beans were intercropped with sorghum. The authors used data from the Alabama Cooperative Extension Service and the Alabama Agricultural Experiment Station to construct an enterprise budget for it.

The LP model used by Frank et al. (2004) is small and very simple, yet it is an innovative application of LP to a very important topic. The objective function maximizes profit over a three-year period. The objective function is:

$$\text{Max: } Z = 84.43vb_1 + 34.82pt_1 + 152.51ct_1 + 103.25ptvbrot_2 + 103.64ptctrot_2 \\ + 187.40ptvbrot_3 + 194.69ptctrot_3,$$

where:

vb_1 = velvet bean acres, year 1

pt_1 = peanut acres, year 1

ct_1 = cotton acres, year 1

$ptvbrot_2$ = peanut–velvet bean rotation, year 2

$ptctrot_2$ = peanut–cotton rotation, year 2

$ptvbrot_3$ = peanut–velvet bean rotation, year 3

$ptctrot_3$ = peanut–cotton rotation, year 3

The objective function coefficients are the net revenues, per acre, for each crop. The first constraint in the LP is a land constraint, which limits the total acreage for all these activities to not exceed 1,000 acres, which is the average size for cotton-peanut farms in Southeastern Alabama, that is,

$$vb_1 + pt_1 + ct_1 + ptvbrot_2 + ptctrot_2 + ptvbrot_3 + ptctrot_3 \leq 1{,}000.$$

The second constraint in the LP model is a velvet bean–peanut rotation constraint, that is,

$$-vb_1 + ptvbrot_2 \leq 0.$$

This constraint limits the peanuts acreage in year 2 under this rotation strategy to not exceed the acreage grown for velvet beans in year 1.

The third constraint is a cotton–peanut rotation constraint, that is,

$$-ct_1 + ptctrot_2 \leq 0.$$

This constraint limits the peanuts acreage in year 2 under this rotation strategy to not exceed the acreage grown for cotton in year 1.

The fourth constraint is a velvet bean–peanut second year rotation constraint, that is,

$$-0.5vb_1 + ptvbrot_3 \leq 0.$$

This constraint limits peanut acreage under this rotational strategy in year 3 to not exceed 50% of the velvet bean acreage in year 1. Similarly, the last constraint is added for the rotational strategy of cotton and peanuts in year 3, that is,

$$-0.5ct_1 + ptctrot_3 \leq 0.$$

The first run of the model is solved using net revenues for velvet beans assuming a base price of $30.00 per ton. Then, the model is resolved parametrically by increasing this price while holding all other parameters in the model constant.

The results indicate that when the price of velvet beans is less than $41.00 per ton, the optimal solution is a rotation of cotton in year 1, cotton in year 2, and peanuts in year 3, with profits equaling $166,568. If the price of velvet beans is between $41.00 and $44.00 per ton, then the optimal rotation for the farm is velvet beans in year 1, velvet beans in year 2, and peanuts in year 3, with profits ranging from $169,418 to $183,418. Finally, if the price of velvet beans is higher than $44.00 per ton, then the optimal rotation is to grow velvet beans in all three years, and profit is above $188,000.

Frank et al.'s (2004) results are similar to a previous study examining the potential of switch grass as a bio-fuel. De La Torre Ugarte et al. (2003) found that at a price of $40 per dry ton, 42 million acres of idled pasture or Conservation Reserve Program acres would be converted to biomass production of switch grass. Frank et al.'s (2004) findings suggest that velvet beans will not be economically attractive for Alabama farmers until the farm price reaches $41.00 per ton of dry biomass. Until the price reaches this level, farmers in this region will not consider a rotation other than cotton-cotton-peanuts. The price will have to be greater than $45.00 per ton of dry biomass before farmers in this region would switch to a velvet beans-only rotation. The authors conclude that depending upon distance to the power plant, and pending federal subsidies, this level of price is possible in the near future.

One shortcoming of this research is that the sensitivity analysis for the price of velvet beans assumes that the prices of other competing crops would not change. However, it is likely that if the price of velvet beans changed, so would the price of competing crops, probably in the same direction. Hence, you need to bear in mind the assumption of changing one price without the other prices changing when interpreting these results.

SUMMARY

This chapter examined farm-level LP models. The purposes of this chapter were to (1) illustrate several basic farm LP models that could be used to assist farmers in making production and marketing decisions; (2) illustrate how time may be incorporated into these basic models both in the context of within-year decisions and multiple-year decisions; (3) demonstrate how to disaggregate production activities by operation; (4) discuss how to validate and calibrate mathematical programming models; and (5) illustrate two examples of these models used to research real-world problems. These types of models are used frequently by agricultural economists.

The chapter began with a discussion of static models that ignore the timing of operations. The discussion began with a simple and highly aggregate crop model, which was later disaggregated to include the various field operations required to grow crops. We then discussed how sensitivity analysis could be used to derive output supply and input demand functions. The output supply function for a specific crop maps optimal crop acreage for different crop prices, holding constant all other crop prices, costs, technical coefficients, and resources endowments. The range of optimality was used to derive the output supply, and an iterative procedure that sets the price above and below the limits of the range and resolves the problem to get the new optimal acreage. An input demand function gives the relationship between the level (amount) of the resource endowment and its SP, holding all other parameters of the problem constant. The range of feasibility was used to derive the input demand functions using a similar iterative procedure.

In reality, the timing of field operations is just as critical to farmers as the choice of crop mix in making sound management decisions. Hence, the chapter also examined "dynamic" models so that time, as well as operations, can be explicitly accounted for in the model. An annual, multiperiod crop model with discrete time periods was presented, along with the concept of sequencing constraints. Sequencing constraints insure that all operations occur and are in proper order. A multiyear model for annual and perennial crops was also presented.

Another common farm enterprise is a joint crop and livestock operation such as dairy farms. The basic difference between these enterprises and crop-only farms is that the former consumes part (or all) of what it produces by feeding it to the livestock. Of course, some of the crop can still be sold in the market place. This chapter demonstrated how to extend the crop farm model into a dairy and crop farm model.

Determining how well the model reflects reality is called model validation. Two types of calibration were examined: validation by construct and validation by results. The first is implied in sound development of the model, which includes thorough and open documentation of data and all model coefficients. The second is much more rigorous testing of the model, comparing it with real-world results. An example was presented to show how to diagnose and calibrate a model through validation to get more realistic results.

Finally, two research examples were examined. The first developed a farm-level LP model to examine the benefits and costs of participating in farm commodity programs. A case study for rice was examined, and the results indicated substantial benefits for participation in the form of increased net farm revenue. The second example looked at the economics of velvet beans as a potential energy crop to be added to a cotton-peanut farm in Alabama. The results showed what the minimum output price would need to be for velvet beans to enter the optimal solution for the farm.

REFERENCES

De La Torre Ugarte, G. D., Shapouri, H., Walsh, M. E., & Slinsky, S. P. (2003, February). The economic impacts of bioenergy crop production on U.S. agriculture. Washington, D.C.: USDA.

Frank, E. T., Duffy, P., Taylor, C. R., Bransby, D., Runge, M., & Rodriguez-Kabana, R. (2004, February 14–18). Economic feasibility of an energy crop on a south Alabama cotton-peanut farm. Selected paper at the Southern Agricultural Economics Association annual meeting, Tulsa, Oklahoma.

Gallagher, P., Dikeman, M., Fritz, J., Wailes, E., Gauther, W., & Shapouri, H. (2003). Biomass from crop residues: Cost and supply estimates. *Agricultural Economic Report* 819. U.S. Department of Agriculture, Office of Chief Economist, Office of Energy Policy and New Uses.

McCarl, B., & Spreen, T. (2003) *Applied Mathematical Programming Using Algebraic Systems.* [online] Unpublished monograph. Available: http://agecon2.tamu.edu/people/faculty/mccarl-bruce/books.htm. [Revised 10 July 2003].

Watanabe, S. (1988). Should rice farmers participate in commodity programs? Course project, Introduction to Mathematical Programming, Department of Applied Economics and Management, Cornell University.

EXERCISES

1. A farmer can grow three crops on 1,000 acres of land: corn, soybeans, and wheat. In producing these three crops, the farmer has to: (1) plow the land (*pl*), (2) plant corn (*pc*), (3) plant soybeans (*ps*), (4) plant wheat (*pw*), (5) harvest corn (*hc*), (6) harvest soybeans (*hs*), (7) harvest wheat (*hw*), (8) sell the corn after harvest (*sc*), (9) sell the soybeans after harvest (*ss*), and (10) sell the wheat after harvest (*sw*). The farmer must plow the land prior to planting and must plant the crops prior to harvesting the crops. In addition to the land endowment of 1,000 acres, the farmer has a total of 1,200 hours available to perform all of the above production operations. The farmer does not want to plant more than 400 acres to corn for soil conservation reasons. The farmer's expectations regarding the labor requirements (hours per acre) and variable costs for each operation, as well as expected price and yield (bushels per acre) at harvest for the three crops are presented below:

Operation	Labor Requirement (hours/acre)	Variable Cost ($/acre)	Crop Yields and Prices
Plowing (*pl*)	0.40	4.00	
Plant Corn (*pc*)	0.39	114.00	
Plant Soybeans (*ps*)	0.30	80.00	
Plant Wheat (*pw*)	0.30	78.00	
Harvest Corn (*hc*)	0.60	48.00	120
Harvest Soybeans (*hs*)	0.30	17.00	40
Harvest Wheat (*hw*)	0.28	10.00	70
Corn Price ($/bushel)		2.90	
Soybean Price ($/bushel)		5.75	
Wheat Price ($/bushel)		3.00	

Assume that the farmer's objective is to maximize net revenue from corn, soybeans, and wheat production.

a. Write the LP tableau for this problem.

b. There are many assumptions regarding the decision process this farmer follows in this model. List three of these assumptions.

c. Solve this exercise using Solver. Summarize the optimal solution and SPs. List three factors (other than relative crop prices) that influence the optimal crop mix in this exercise.

2. For the previous exercise, derive and graph the output supply functions for corn, soybeans, and wheat. Compare the price responsiveness of the three supply functions.

3. For the previous exercise, derive and graph the input demand functions for land and labor. Compare the price responsiveness of the two input demand functions.

4. In addition to the parameters laid out in the first exercise, suppose that the farmer provides the following information regarding time periods and the production process (assume that the 1,200 hours are divided as displayed in the table below):

Period	Available Labor (hours)	Planting Plow	Planting Corn	Planting Beans	Planting Wheat	Harvesting Corn	Harvesting Beans	Harvesting Wheat	Harvest Sales Corn	Harvest Sales Beans	Harvest Sales Wheat	Post-Harvest Sales Corn	Post-Harvest Sales Beans	Post-Harvest Sales Wheat
1 Mar 15–May 9	175	X	X		X									
2 May 10–May 23	175	X	X	X	X									
3 May 24–Jun 6	150			X										
4 Sep 13–Sep 26	150					X	X							
5 Sep 27–Oct 17	275	X				X	X	X	X					
6 Oct 18–Nov 7	275	X				X		X	X		X			
7 Jan 27	–											X	X	X

In addition, the farmer has a 10,000 bushel on-farm storage facility in which any combination of corn, soybeans, and wheat can be stored. There are now six instead of three marketing activities. Any portion of the crop can be sold at harvest for the following prices: $2.90 per bushel of corn, $5.75 per bushel of soybeans, and $3.02 per bushel of wheat (note that the wheat harvest price has been changed). Assume that the stored crop can be sold on January 27 at the following prices: $3.10 per bushel of corn, $6.30 per bushel of soybeans, and $3.20 per bushel of wheat. Again, the farmer has 1,000 acres of land and will not grow more than 400 acres of corn. Corn and soybean yields are influenced by planting and harvesting dates, but wheat is not. The corn and soybean yields by their respective planting and harvest dates are given below.

Harvest Period	Corn 1	Corn 2	Soybeans 2	Soybeans 3
	Yield (bushel/acre) by Planting Date			
4	–	–	40	30
5	100	115	50	15
6	135	110	–	–

Assume that plowing may take place in the spring prior to planting and/or in the fall after harvest (e.g. periods 5 and 6).

a. Write out this problem as an LP model assuming maximization of net revenue. You may do so in general form or in tableau form (without slack variables).

b. Solve this problem in Solver. Summarize the optimal solution and SPs.

5. For the previous exercise, derive and graph the input demand functions for land and storage. In which periods is labor the most scarce?

6. For the previous exercise, how would hired labor be incorporated as activities into this model?

7. A farmer can plant two crops in a 450-acre plot of land. One acre of land can produce 2.7 bushels of crop A and 2.4 bushels of crop B. Crop A requires $30 for plowing, disking, and harvesting altogether per bushel, and crop B requires $40 for plowing, disking, and harvesting per bushel (the cost includes labor for both crops). Now the same

crops could be sold at $120 per bushel and $180 per bushel respectively. The family has 15,000 total hours available, and crop A requires 40 hours per acre, and crop B requires 60 hours per acre. The family can arrange 10,000 hours for $6.00 hour as additional labor. As for the availability of seeds, crop A could be planted for 300 acres, and crop B could be planted for 200 acres. If the farmer has $30,000 to invest in his crops, what is his maximum potential profit?

8. Consider an organic chili pepper farm in New Mexico. They can grow a combination of jalapenos, habaneros, and poblanos. For inputs, they use land, labor, and organic fertilizer of which they have endowments of 350 acres, 500 hours, and 300 bags, respectively. They have a total working budget of $45,000. Based on the following production information, use Solver to calculate the farm's output supply and input demand functions for all factors and products.

	Labor (hours/acre)	Cost ($/acre)	Fertilizer (bags/acre)	Yields/Prices ($)
Plowing	0.75	4		
Plant Jalapenos	0.40	110	0.70	
Plant Habaneros	0.45	75	0.75	
Plant Poblanos	0.35	95	0.70	
Harvest Jalapenos	0.30	20		100 bushels/acre
Harvest Habaneros	0.45	20		95 bushels/acre
Harvest Poblanos	0.40	30		85 bushels/acre
Jalapeno Price				4.00/bushel
Habanero Price				3.75/bushel
Poblano Price				4.75/bushel

9. A farmer has 100 acres of land to plant tomatoes, radishes, and lettuce. She can sell tomatoes at $2.00 per lb., radishes at $2.00 per lb., and $1.00 for lettuce heads. The fertilizer costs are $50.00 for tomatoes and radish each and $25.00 for lettuce. A total of 400 man days of labor are available at $20.00 per man day. The other variable costs and labor costs are tabulated below.

	Tomato	Radish	Lettuce
Price per lb	2	2	1
Yield per acre	2,000	1,000	3,000
Fertilizer cost	50	50	25
Labor days required	5	5	4

Formulate this problem as an LP model to maximize the profit. Also find the input demand function for labor and land constraints.

10. ABC Company produces feedstuff for chicken. According to national standards, the feedstuff produced should meet nutritional criteria.

	Crude Protein	Coarse Fiber	Lysine	Calcium	Phosphor	Salt
National Criteria	135 ~145g/kg	<50g/kg	≥5.6g/kg	23~40g/kg	4.6~6.5g/kg	3.7g/kg

The company produces feedstuff from several raw materials including corn, wheat, wheat bran, DL-met, and salt. The nutrition fact of different raw materials is listed in the following table.

	Price ($/kg)	Crude Protein (g/kg)	Coarse Fiber (g/kg)	Lysine (g/kg)	Calcium (g/kg)	Phosphor (g/kg)	Salt (g/kg)
Corn	0.68	78	16	2.3	0.7	0.30	
Wheat	0.72	114	22	3.4	0.6	0.34	
Wheat Bran	0.23	142	95	6.0	0.3	10.00	
Rice Bran	0.22	117	72	6.5	1.0	13.00	
DL-met	23.00						
Bone-meal	0.56				300	140	
Calcium Carbonate	1.12				400		
Salt	0.42						1,000

How would you mix those raw materials, according to national standards, to produce 1kg mixed feedstuff with least cost?

11. Use the dynamic crop farm model presented in this chapter to derive an output supply function for corn, soybeans, and wheat.

12. Use the dynamic crop farm model presented in this chapter to derive an input demand function for land and for labor by period.

13. With respect to the dynamic crop farm model presented in this chapter, what would be some useful and relevant parameters to vary for sensitivity analysis?

14. Make up your own production schedule by time period to make the example of the static crop-livestock farm presented in this chapter. Formulate the LP problem and solve it using Solver.

15. Consider the dynamic crop farm problem in this chapter. Modify this problem to include the following. The farmer has an on-farm storage capacity of 20,000 bushels for corn, soybeans, and wheat (assume they can be stored together). The farmer has decided to evaluate another marketing strategy in addition to harvest sales. Specifically, the farmer expects the net price of corn, soybeans, and wheat in March of the following year to be $3.05, $7.00, and $4.50 per bushel respectively. Include these three marketing strategies in the model (remember the storage constraint).

    a. Solve this exercise using Solver, and report the solution.

    b. For each of the six marketing strategies, derive a supply function in tabular and graphical form.

16. Consider the dynamic crop farm problem in this chapter. Modify this problem to include the following. The farmer can rent up to 100 acres of land from his neighbor, Mr. Bishop, at a cost of $140 per acre. The farmer can also rent up to another 200 acres of land from another neighbor, Mr. Nichols, at a cost of $165 per acre. Add these rental possibilities to the model. Solve this exercise using Solver, and report the solution.

17. Consider the dynamic crop farm problem in this chapter. Modify this problem to include the following. In reality, corn must be dried to a 15% moisture content to be sold or stored. Suppose that corn has the following moisture contents, which vary depending upon when it is planted and harvested:

Harvest Period	Planting Period	
	1	2
5	17%	21%
6	13%	18%

Note that corn harvested with moisture contents of 15% or below does not need to be dried. Assume that it costs $0.02 per bushel per percentage point to dry corn. Then, for example, if corn is harvested with a 20% moisture content and a yield of 150 bushels per acre, the per acre drying cost is:

$$\text{Drying cost/acre} = (20 - 15)\,(0.02)\,(150) = \$15$$

In the model, calculate the corn drying costs per acre for each combination of planting and harvesting periods, and add these costs to the corn harvesting variable costs. Solve this exercise using Solver, and report the solution.

18. List and explain 10 factors that influence the optimal crop mix for the dynamic crop farm problem presented in this chapter.

19. A farmer owns 1,000 acres of cropland suitable for growing corn, soybeans, sorghum, and wheat. The expected at-harvest price for each of these crops is: $2.50 per bushel for corn, $6.35 per bushel for soybeans, $2.75 per bushel for sorghum, and $3.10 per bushel for wheat. The farmer expects the following yields for each crop: 135 bushels per acre for corn, 45 bushels per acre for soybeans, 95 bushels per acre for sorghum, and 100 bushels per acre for wheat. In order to grow each crop, the farmer has to plow the land, plant each of the crops, and harvest each of the crops. The following summarizes the labor requirements and variable cost for each operation:

Production Operation	Labor Requirement (hours/acre)	Variable Cost ($/acre)
Plow	0.50	10
Plant Corn	0.30	75
Plant Soybeans	0.30	35
Plant Sorghum	0.30	45
Plant Wheat	0.30	40
Harvest Corn	1.35	50
Harvest Soybeans	0.65	20
Harvest Sorghum	0.75	25
Harvest Wheat	0.60	44

Assume that the farmer is endowed with 5,000 hours of family labor and can hire up to an additional 600 hours of hired labor at a cost of $5.00 per hour. The farmer can rent up to 200 acres of land from a neighbor at a cost of $100 per acre. On the other hand, the farmer can also rent out up to 500 acres of her own 1,000 acres to another neighbor, who is willing to pay $100 per acre.

Suppose that this crop farmer wants to evaluate whether or not it would be profitable to add up to 60 dairy cows to her farm. Assume that the cost per cow is $500, and each cow can produce 15,000 pounds of milk per year which sells for $0.12 per pound. Assume that each cow requires the following annual diet, which will be produced entirely on the land owned by the farmer: (1) a concentrate made out of 35 bushels of corn mixed with 9 bushels of soybeans, and (2) 1.22 tons of hay. Therefore, in addition to corn, soybeans, sorghum, and wheat, the farmer must now grow hay. Assume that any excess hay that is grown beyond the need of the dairy herd can be sold at harvest for $75.00 per ton. The farmer expects a yield of 2.5 tons per acre for hay, a cost of $15 per acre for planting hay, and $16.00 per acre for harvesting hay. Also, it takes 0.25 hours per acre for planting hay, 0.50 hours per acre for harvesting hay, and 65 hours per cow for milking cows.

a. Formulate the LP model for this problem.

b. Solve it with Solver.

c. Derive an output supply function for milk, corn, and soybeans.

d. Derive an input demand function for land and another for family labor.

20. Consider the example of the multiyear apple, tomato, and lettuce farm presented in this chapter. Show how you would modify this problem to account for the following:

a. The option of renting an additional 50 acres from a neighboring farmer and being able to hire up to 25 weeks of labor from immigrant workers for $12.00 per hour.

b. Derive an output supply function for apples.

21. Consider the example of the multiyear apple, tomato, and lettuce farm presented in this chapter. How would you modify this problem to account for a crop rotation strategy between tomatoes and lettuce? Crop rotation means that you grow tomatoes where lettuce was grown the previous year, and grow lettuce where tomatoes were grown the previous year.

22. Consider the example of the crop-livestock farm presented in this chapter. Formulate this problem in Solver and derive the solution. Next, using sensitivity analysis in Solver, derive an output supply function for cows.

23. Suppose you were interested in a farm family in a developing country that sold part of its harvest each year for needed currency, but mainly relied on the farm to meet its own food needs. The farm grows beans, corn, and lettuce for sale and its own consumption. Suppose the limiting production resources are land and labor, and it is endowed with 10 acres of land and 5,000 hours of family labor. It faces the following per acres resource requirements:

Resource	Beans	Corn	Lettuce
Land	1	1	1
Labor	1,000	1,300	1,400

For every acre, the farmer can produce 60 bushels of beans, 80 bushels of corn, and 1,000 head of lettuce. It can receive the following unit profits if the produce is sold on the market: $8 per bushel of beans, $6 per bushel of corn, and $2 per head of lettuce. In addition, the farm-family needs to produce enough to have 50 bushels of beans, 100 bushels of corn, and 365 head of lettuce for their own consumption. Formulate an LP model that maximizes profit from the sale of produce while meeting the constraints outlined in this exercise including the minimum consumption constraints. Solve the exercise using Solver.

24. For the previous exercise, derive a supply function for beans, lettuce, and corn.

25. For the previous exercise, derive an input demand function for land and labor.

5

Transportation and Assignment Models for Food and Agricultural Markets

A frequent application of linear programming (LP) is transportation decision problems. Transportation problems are a special class of LP models known as **network optimization** problems. Transportation problems involve determining how to move a product in the most efficient way given certain constraints such as available supply (plant capacity) from each production location and prevailing demand at each consumption destination. Generally, the objective is to minimize transportation costs, but other objectives may be modeled, such as maximization of the profits net of transportation costs.

Transportation costs are particularly important for agricultural commodities. Most agricultural commodities are bulky, and many are highly perishable. Consequently, transportation costs are relatively high for many agricultural commodities. As a result, food manufacturers and agricultural cooperatives are interested in designing farm assembly routes that are as efficient as possible. Linear programming models are often used to help design efficient routes.

In the next section, the transportation problem is presented as an LP problem. The standard assumptions of the LP model are discussed, and a simple example is used to illustrate the basic structure of such models. In addition, the general model involving **n** supply and **m** demand nodes is described along with several modifications in the basic transportation model. Next, the **transshipment model** is presented, which extends the transportation model to include warehouse and/or processing activities that represent intermediate nodes in the network. This is especially important for agricultural commodities, since most involve some form of processing or transformation along the marketing chain. A special case of the transportation model, called the **assignment problem**, is then described. The assignment model, which is quite similar to the transportation model, involves matching resources with tasks in the most efficient way possible. The chapter concludes with a research application of a large transportation LP model involving the production and distribution system of the U.S. dairy industry.

This chapter will not cover other special-purpose solution procedures for network optimization problems discussed elsewhere. Interested readers should consult any standard management science or operations research text to learn more about these special-purpose solution procedures, which are more efficient than the simplex method for solving specific transportation problems.

5.1   GENERAL TRANSPORTATION MODEL

The following assumptions are standard for the transportation model:

1.   One good (x) is produced and sold in different geographic locations.
2.   There are n **supply nodes** where x is produced. Denote a supply node as $*$.
3.   There are m **demand nodes** where x is consumed. Denote a demand node as $+$.
4.   Supply and demand at each location are **known** and **fixed**.
5.   The objective is to **minimize total transportation** costs for shipment flows from supply to demand nodes.
6.   The only costs assumed in this problem are **transportation costs**; that is, the product has already been produced.
7.   The amount of total supply is $\geq$ the amount of total demand.

Later on in this section, modifications are made in the model in order to relax some of these assumptions.

Any supply node, i, can sell to any demand node, j. The model determines the flow of x from supply node i to each demand node j based on the network of transportation costs and the quantities supplied and demanded.

To illustrate the potential movements of x, let n = m = 2. The diagram below shows these potential movements. The lines connecting the supply and demand nodes are called **arcs**. An arc gives the possible direction of shipments from a supply origin to a demand destination. For example, supply in location 1, denoted as (1^*) in this diagram, can be shipped to demand locations $(+1)$ and/or $(+2)$. Likewise, supply in location 2, denoted as (2^*) in this diagram, can be shipped to demand locations $(+1)$ and/or $(+2)$. This type of diagram is called a **network** because it shows the relationship between supply and demand areas connected by transportation distances.

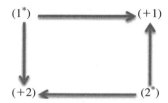

Consider the case of n supply origins and m demand destinations. Let:

   i = the ith supply node; i = $1, \ldots, $ n,

   j = the jth demand node; j = $1, \ldots, $ m,

   c_{ij} = the unit transportation costs from supply i to demand j (e.g., C_{36} is the unit transportation cost from supply location 3 to demand location 6),

x_{ij} = the amount of product shipped from supply origin i to demand destination j (e.g., x_{19} is the amount of good x that is transported from supply location 1 to demand location 9),

s_i = supply (fixed plant capacity) at supply origin i,

d_j = demand (fixed consumption level) at demand location j.

This problem can be formulated as an LP problem as follows:

$$\text{Min: } Z = \sum_{i=1}^{n} \sum_{j=1}^{m} c_{ij} x_{ij} \tag{0}$$

s.t.:

$$\sum_{j=1}^{m} x_{ij} \le s_i, \quad i = 1, \ldots, n \tag{1}$$

$$\sum_{i=1}^{n} x_{ij} \ge d_j, \quad j = 1, \ldots, m \tag{2}$$

$$x_{ij} \ge 0, \quad i = 1, \ldots, n; \quad j = 1, \ldots, m \tag{3}$$

The objective function (0) is the total cost of transportation from all supply origins to all final demand destinations in the model. The c_{ij} parameters are the unit transportation costs for shipping the output from origin i to destination j. The x_{ij} activities represent the level of shipments from origin i to destination j.

Constraint (1) requires that the amount shipped from any supply origin i to the demand destinations not exceed the available supply from supply origin i. For example, if supply origin 4 has 400 units of product to distribute, then this constraint requires that the shipment out of origin 4 to the demand destinations not exceed 400 units. There are n of these constraints, which is equal to the number of supply nodes in the network.

Constraint (2) requires that the amount shipped to any demand destination j from the supply origins not be less than the fixed demand for demand destination j. It is therefore implicitly assumed that all demand destinations always receive at least as much product as is demanded. For example, if demand destination 2 consumes 500 units of the product, then this constraint requires that at least 500 units be shipped to destination 2 from the supply origins. An equality constraint could also be constructed so that $s = d$. There are m of these constraints, which is equal to the number of demand nodes in the network.

Constraint (3) requires non-negativity on all activities, x_{ij}.

It is clear that the structure of the LP transportation model is quite simple. Regardless of how large the problem is in terms of the number of supply and demand nodes, the basic structure of the problem is to minimize total shipment costs subject to all demand nodes being satisfied and all supply nodes not shipping more than their fixed capacity.

An Example[1]

A large food manufacturer produces premium ice cream (x) in three different supply plants in the United States. The firm sells their ice cream to four different demand

---

[1]This problem, solution, and corresponding sensitivity analysis are shown in the Chapter 5 supplemental materials available at www.wiley.com/college/kaiser.

destinations in the U.S. The supply origins, demand destinations, and unit transportation costs of shipping ice cream are summarized below.[2]

Supply nodes: Boston, Milwaukee, San Diego.

Demand nodes: Orlando, San Antonio, St. Paul, Seattle.

Fixed Weekly Supplies (in 1,000 gallons)

Location	1 Boston	2 Milwaukee	3 San Diego
Supply	125	75	100

Fixed Weekly Demands (in 1,000 gallons)

Location	1 Orlando	2 San Antonio	3 St. Paul	4 Seattle
Demand	50	75	100	50

Unit Transportation Costs ($/1,000 gallons)

From/To	1 Orlando	2 San Antonio	3 St. Paul	4 Seattle	Total Supply
1 Boston	25	50	40	125	125
2 Milwaukee	55	65	25	75	75
3 San Diego	90	45	75	45	100
Total Demand	50	75	100	50	

Using this information, the LP tableau, which minimizes total transportation costs can be constructed as follows:

	x_{11}	x_{12}	x_{13}	x_{14}	x_{21}	x_{22}	x_{23}	x_{24}	x_{31}	x_{32}	x_{33}	x_{34}	
Obj (Min)	25	50	40	125	55	65	25	75	90	45	75	45	RHS
Supply 1	1	1	1	1									≤ 125
Supply 2					1	1	1	1					≤ 75
Supply 3									1	1	1	1	≤ 100
Demand 1	1				1				1				≥ 50
Demand 2		1				1				1			≥ 75
Demand 3			1				1				1		≥ 100
Demand 4				1				1				1	≥ 50
Non-neg	x_{11}	x_{12}	x_{13}	x_{14}	x_{21}	x_{22}	x_{23}	x_{24}	x_{31}	x_{32}	x_{33}	x_{34}	≥ 0

The formulation of this problem in Solver is presented in Figure 5.1. The optimal solution to this problem is:

$$x_{11} = 50, x_{12} = 25, x_{13} = 25, x_{23} = 75, x_{32} = 50, x_{34} = 50, Z^* = \$9,875,$$
$$\text{and all other } x_{ij} = 0$$

$$SP_{sup1} = 0, SP_{sup2} = -15, SP_{sup3} = -5, SP_{dem1} = 25, SP_{dem2} = 50,$$
$$SP_{dem3} = 40, SP_{dem4} = 50$$

---

[2]All costs, supply, and demand for this example are fictitious and not based on real-world data.

	A B	C	D	E	F
1	**Decision Variables** Miami		Houston	Minneapolis	Portland
2	New York	30	25	15	0
3	Chicago	0	0	75	0
4	Los Angeles	0	50	0	40
5					
6	**Costs**	Miami	Houston	Minneapolis	Portland
7	New York	20	40	35	120
8	Chicago	50	60	20	70
9	Los Angeles	80	35	70	40
10					
11	**Objective**				
12	Min Total Costs	$ 6,975			
13					
14	**Constraints**				
15	**Demand**	Total		RHS	
16	Miami	30	≥	30	
17	Houston	75	≥	75	
18	Minneapolis	90	≥	90	
19	Portland	40	≥	40	
20	**Supply**				
21	New York	70	≤	100	
22	Chicago	75	≤	75	
23	Los Angeles	90	≤	90	
24					

Figure 5.1   Solver formulation for ice cream transportation problem.

Summary and Interpretation of Optimal Solution   There are several steps that are quite useful in summarizing and interpreting the optimal solution for a transportation model. These are summarized below.

Step 1: Draw a map of the supply origins and demand destination nodes. In this map, specify the fixed demands, fixed supplies, and transportation costs. Figure 5.2 displays an example of such a map for this problem. In Figure 5.2 the fixed supplies and demands are given by the [ ] type brackets, and the amount imported or exported is given by the ( ) type parentheses.

Step 2: Draw the optimal shipments arcs and values from the supply origins to the demand destinations. Check to see that none of the supply and demand constraints are violated. Note: for optimal shipments with zero values, the arcs and values should not be drawn, otherwise the map will become very cluttered. In Figure 5.2, the optimal shipment levels are listed aside the arc lines.

Step 3: Interpret the shadow prices (SP) on the supply and demand constraints. The map constructed in steps 1 and 2 will be very useful in doing this.

Supply Shadow Prices   Boston Supply: The SP equals zero because the total supply of 125 units at the Boston plant is not fully utilized; only 100 units are shipped.

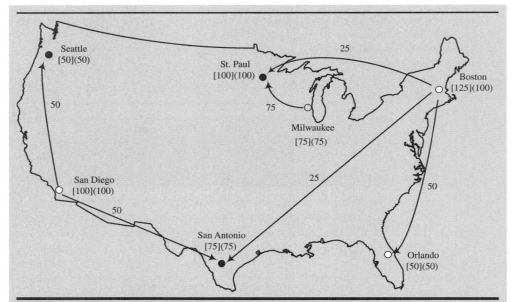

Figure 5.2   Map of supply and demand nodes for the ice cream example.

Consequently, increasing the 125 available units by one to 126 units will add zero to the optimal value of the objective function, which in this case is total transportation cost.

Milwaukee Supply: The SP equals -15. This means that if the plant capacity in Milwaukee is increased by one more unit, total transportation costs would decrease by $15. Why?

If the Milwaukee supply is increased by one unit, it will be shipped to St. Paul, which causes total transportation costs to increase by $20. However, if one more unit is shipped to St. Paul from Milwaukee, then one less unit will be shipped from Boston to St. Paul, which will cause total transportation costs to decrease by $35. The net result is $20 - 35 = -15$.

San Diego Supply: The SP equals -5. This means that if the plant capacity in San Diego is increased by one more unit, total transportation costs would decrease by $5. Why?

If San Diego's supply is increased by one unit, then the additional unit would be shipped to San Antonio rather than Seattle since Seattle's demand is fixed and its sole supplier is San Diego. The additional unit shipped from San Diego to San Antonio adds $35 to total transportation costs. However, San Antonio would then require one less unit from Boston, which would lower total transportation costs by $40. The net result is $35 - 40 = -5$.

Demand Shadow Prices   Orlando Demand: The SP equals 20. This means that if demand in Orlando increases by one unit then the total transportation cost would increase by $20. Why?

Boston would ship an additional unit (from its slack of $30) to Orlando at a cost of $20.

San Antonio Demand: The SP equals 40. This means that if demand in San Antonio increases by one unit then the total transportation costs would increase by $40. Why?

Boston would ship an additional unit (from its slack of $30) to San Antonio at a cost of $40.

St. Paul Demand: The SP equals 35. This means that if demand in St. Paul increases by one unit then the total transportation cost would increase by $35. Why?

Boston would ship an additional unit (from its slack of $30) to St. Paul at a cost of $35.

Seattle Demand: The SP equals 45. This means that if demand in Seattle increases by one unit then the total transportation costs would increase by $45. Why?

The derivation of this SP is more complicated. If Seattle's demand increased by one unit, then ice cream would be shipped from San Diego at a cost of $40 per unit. However, one less unit would be shipped from San Diego to San Antonio, which means a savings of $35. San Antonio would then receive an additional unit from Boston at a cost of $40. Thus, the change in the objective function would be equal to $40 - 35 + 40 = 45$.

Sensitivity Analysis

Most decision problems are not complete without some sort of sensitivity analysis. In this problem, as well as all transportation problems, the actual solution will depend upon the key parameters in the model. In these applications, the key parameters will be the unit transportation cost estimates (c_{ij}), the level of plant capacity for each supply origin i (b_i), and the level of demand in each demand destination j (b_j). Hence, it follows that basic sensitivity analysis should focus on "what if" questions regarding these parameters. What if plant capacity in origin i was increased from A to Z? What if the demand in destination j decreased from Y to W? What if the unit transportation costs for several arcs in the network increased or decreased? And so on.

Another general rule regarding sensitivity analysis is to use this type of analysis on parameter estimates that are suspected to be the least accurate. For example, suppose that exact estimates on the amount of ice cream produced in each supply origin and the amount demanded from each demand destination are available, while good estimates on some or all of the unit transportation costs are not available. In this case, the sensitivity analysis should focus on the objective function coefficients in order to ascertain how sensitive the model is to the estimates of c_{ij}. If the model is not very sensitive to these estimates, that is, the results do not change much, then the results are more credible. On the other hand, if the results do change significantly with changes in the c_{ij} parameters, then great caution should be taken in using the model results to make decisions unless better estimates of c_{ij} can be obtained.

Another type of sensitivity analysis is to modify the problem to add and/or delete new and old supply and demand locations to see if the firm can construct a better distributional network. With respect to supply, it may not be possible to consider building a lot of new plants or close down existing plants. But, this type of analysis might be useful to a firm considering adding one or two new plants and/or closing down one or two old plants. Such analysis would provide insights into efficiency gains due to new plant investment. Analogous examples for this type of sensitivity analysis for demand exist as well, but are left to the reader for further consideration.

5.2   EXTENSIONS OF THE MODEL

There are several extensions of the basic transportation model. Three of these are briefly listed and described below.

1. Maximizing Revenue Rather than Minimizing Transportation Costs

A manager might be interested in maximizing revenue or profit net of the transportation costs rather than simply minimizing costs. To modify the model, first replace the objective function with:

$$\text{Max: } Z = \sum_{i=1}^{n} \sum_{j=1}^{m} c_{ij} x_{ij} \tag{0}$$

where c_{ij} is net revenue per unit of the good shipped, that is, total revenue minus variable production costs minus transportation costs.

This can be maximized subject to the same constraints above. It should be noted, however, that since this is a maximization model, the total supply and demand should be equal in order for all the constraints to be binding and make sense. That is, if total supply is greater than total demand, one of the regions will receive more shipments than it demands, since the objective function is being maximized. If demand is indeed fixed, then it would not make sense to have additional flows to these markets. On the other hand, if demand is not fixed, then use a formulation with total supply being greater than demand.

2. Incorporating Route Constraints

Suppose only k amount of good x can be shipped from Seattle to Boston. To account for this, simply add the following constraint to the original problem:

$$x_{32} \leq k,$$

where k is the amount of the shipment constrained by the routing constraint.

3. Incorporating Unacceptable Routes into the Network

Suppose that, due to construction, several key roads are closed, which makes it physically impossible to ship by truck any amount of a good from supply i to demand j. In this case, simply remove the activity from the problem for which the route is unavailable. For example, suppose that the manager says it is impossible to ship ice cream from San Diego to Seattle. Then simply delete decision variable x_{34} from the objective function and constraint set.

An Example

Consider the following hypothetical transportation problem in international trade. The United States produces more oranges than it consumes, and exports to England, Germany, Canada, Japan, and China. The two states that orange exports come from are California and Florida. Suppose that on an annual basis, each region has the following fixed supply and demand for oranges:

Fixed Annual Supplies (in millions of tons):

Location	1 Florida	2 California
Supply	750	1,000

Fixed Annual Demand (in millions of tons):

Location	1 England	2 Germany	3 Canada	4 Japan	5 China
Demand	400	300	300	250	500

The following is the net revenue (net of all variable costs including transportation costs) for shipping a ton of oranges to and from the various locations:

From/To	1 England	2 Germany	3 Canada	4 Japan	5 China
1 Florida	1,000	950	1,100	750	800
2 California	800	850	1,200	1,375	1,500

Consider the following trade policy scenarios:

1. Free trade—there are no restrictions on U.S. exports to any countries.
2. Import quota—same as free trade, except Japan puts an import quota on U.S. orange exports of no more than 150 million tons.
3. Fixed rate tariff—same as free trade, except China puts a fixed rate tariff of $200 per ton of U.S. oranges.
4. Trade embargo—same as free trade, except the United States places a complete orange export embargo on China.

A transportation LP model could be used to examine the revenue impacts of the four trade policies on U.S. orange exports. Consider the free trade scenario first, which can be used as a baseline to compare the trade restriction scenarios. The problem is:

$$\text{Max: } Z = 1{,}000x_{11} + 950x_{12} + 1{,}100x_{13} + 750x_{14} + 800x_{15} + 800x_{21}$$
$$+ 850x_{22} + 1{,}200x_{23} + 1{,}375x_{24} + 1{,}500x_{25} \tag{0}$$

s.t.:

$$x_{11} + x_{12} + x_{13} + x_{14} + x_{15} \leq 750 \tag{1}$$

$$x_{21} + x_{22} + x_{23} + x_{24} + x_{25} \leq 1{,}000 \tag{2}$$

$$x_{11} + x_{21} \geq 400 \tag{3}$$

$$x_{12} + x_{22} \geq 300 \tag{4}$$

$$x_{13} + x_{23} \geq 300 \tag{5}$$

$$x_{14} + x_{24} \geq 250 \tag{6}$$

$$x_{15} + x_{25} \geq 500 \tag{7}$$

$$x_{ij} \geq 0 \quad i=1, 2; j=1, \ldots, 5 \tag{8}$$

To model the second trade scenario, simply change the right-hand-side (RHS) value of constraint (6) from 250 to 150. However, we need to make an assumption regarding what happens to the 100 million tons of oranges that were previously exported to Japan. For this problem, assume that they can be shipped to any of the other markets without impacting unit net revenues. In reality, unit net revenues would likely decline in the region receiving the additional oranges.

To model the third trade scenario, modify the free trade objective function by changing the objective function coefficients on unit profits from Florida to China from 800 to 600 (i.e., $800 - 200 = 600$) and from California to China from 1,500 to 1,300 (i.e., $1{,}500 - 200 = 1{,}300$).

To model the fourth trade scenario (ban on Chinese exports), eliminate constraint (7) from the free trade model so that there are no Florida and California exports to China. Also, eliminate the two activities for trade flows to China, x_{15} and x_{25}. An assumption again needs to be made about where the 500 million tons of oranges previously shipped to China would go. Assume here it is free to go to any other market without changing unit net revenues.

The optimal solution to the free trade problem is: Florida should sell 400 million tons of oranges to England, 300 million tons to Germany, and 50 million tons to Canada. California should sell 500 million tons of oranges to China, 250 million tons to Japan, and 250 million tons to Canada. The total net export revenue to the U.S. orange industry is $2,133,750.

In the second Japanese import quota scenario, the optimal solution is: Florida should sell 400 million tons of oranges to England, 300 million tons to Germany, and 50 million tons to Canada. California should sell 600 million tons of oranges to China, 150 million tons to Japan, and 250 million tons to Canada. The total net export revenue to the U.S. orange industry is $2,146,250. Note that in this case, China picks up an additional 100 million tons of oranges from California, and total industry net revenue actually increases by 0.6 percent due to Japan's import quota. In reality, revenue would likely fall in this case because in order for China to increase its imports from California by 100 million tons, unit net revenues would need to decline.

In the third scenario, where the Chinese impose a $200 per ton fixed-rate tariff, Florida should sell 400 million tons of oranges to England, 300 million tons to Germany, and 50 million tons to Canada. California should sell 500 million tons of oranges to China, 250 million tons to Japan, and 250 million tons to Canada. These trade flows are identical to the free trade case. However, in this case, total net export revenue to the U.S. orange industry falls to $2,033,750, which is 4.7 percent lower than the free trade scenario.

The optimal solution to the fourth scenario, that is, the ban of exports to China, is that Florida should sell 400 million tons of oranges to England, 300 million tons to Germany, and 50 million tons to Canada. California should sell 750 million tons to Japan, and 250 million tons to Canada. The total net export revenue to the U.S. orange industry is $2,071,250. Note that in this case, Japan picks up an additional 500 million tons of oranges from California, and total industry net revenue declines by 2.9 percent due to the export ban to China. In reality, revenue would likely fall by more than this because in order for Japan to increase its imports from California by 500 million tons, unit net revenues would need to decline.

This type of model would be useful for examining different types of trade barriers and their impacts on industry revenues.

5.3 THE TRANSSHIPMENT MODEL

In reality, distributional systems often include not only plant-to-retail routes, but also intermediate nodes as well. For example, supermarket chains usually ship products from their point of creation to warehouses rather than directly to the supermarkets. Supermarkets then receive shipments of goods from the warehouses. The warehouse in this network is called a **transshipment node**. Transshipment nodes do not need to be physical storage areas, as is the case in this example. They may also be processing centers. For example, raw milk is produced on the farm and then is delivered to processing plants. These plants, in turn, process the raw milk into fluid products, soft dairy products, or hard dairy products. From there, the processed milk products are shipped to supermarkets. In this case, the transshipment nodes not only serve as a storage function, but also as a processing function.

The Transshipment Warehouse Model[3]

In general, a transshipment model is any transportation model involving an intermediate node in the distributional network. In this class of models, shipments are permitted between any pair of nodes, for instance, shipments can occur between supply and demand locations, between supply and transshipment locations, or between transshipment and demand locations. A transportation model that includes shipments within nodes can also be designed, but this will not be considered here. Figure 5.3 shows an example of shipment between nodes.

Consider the transshipment problem corresponding to Figure 5.3. The objective function is to minimize transportation costs. The possible shipments, as indicated in Figure 5.3, are:

x_{13} (Supply 1 to intermediate location 3),

x_{14} (Supply 1 to intermediate location 4),

x_{23} (Supply 2 to intermediate location 3),

x_{24} (Supply 2 to intermediate location 4),

x_{35} (Intermediate location 3 to demand 5),

x_{36} (Intermediate location 3 to demand 6),

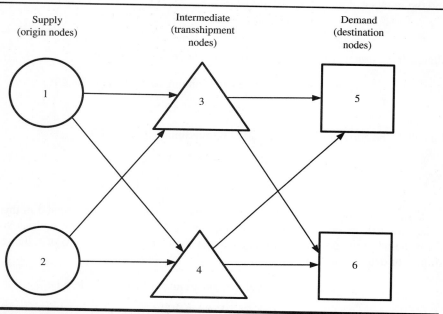

Figure 5.3 Network diagram of transshipment problem ($n=m=k=2$).

---

[3] This problem, solution, and corresponding sensitivity analysis are shown in the Chapter 5 supplemental materials available at www.wiley.com/college/kaiser.

x_{45} (Intermediate location 4 to demand 5),

x_{46} (Intermediate location 4 to demand 6).

Let c_{ij} be the transportation costs for each shipment from node i to node j. Then the objective function is:

$$\text{Min: } Z = c_{13}x_{13} + c_{14}x_{14} + c_{23}x_{23} + c_{24}x_{24} + c_{35}x_{35} + c_{36}x_{36}$$
$$+ c_{45}x_{45} + c_{46}x_{46}$$

There are three sets of constraints, including (1) supply constraints, (2) transshipment constraints, and (3) demand constraints. The first set of constraints is identical to that of the transportation problem that limits the amount shipped from each supply origin to not exceed available supply.

Suppose supply 1 = 300 and supply 2 = 700. Then the first two constraints are:

$$x_{13} + x_{14} \leq 300,$$

$$x_{23} + x_{24} \leq 700.$$

The constraints for the transshipment nodes require that the amount shipped in equals the amount shipped out. This assumes that the capacity for each transshipment node can handle any or all supplies coming in from the supply nodes. For transshipment node 3, the number of units shipped in is:

$$x_{13} + x_{23},$$

and the number of units shipped out is:

$$x_{35} + x_{36}.$$

Hence, to guarantee that units shipped in equals units shipped out

$$x_{13} + x_{23} = x_{35} + x_{36}, \text{ or rearranging to be suitable for LP}$$

$$x_{13} + x_{23} - x_{35} - x_{36} = 0.$$

Likewise, the corresponding constraint on transshipment node 4 is:

$$x_{14} + x_{24} = x_{45} + x_{46}, \text{ or}$$

$$x_{14} + x_{24} - x_{45} - x_{46} = 0.$$

Finally, the last set of constraints is on the demand nodes, because demand in this model is constructed by satisfying the transshipment nodes, not the supply nodes. Assuming demand node 5 requires 600 units and demand node 6 requires 400 units of x, then the constraint for demand node 5 is:

$$x_{35} + x_{45} = 600, \text{ and the constraint for demand node 6 is:}$$

$$x_{36} + x_{46} = 400.$$

Given the transportation costs, c_{ij}, this model can be solved.

For example, suppose there are two supply nodes, two transshipment nodes, and two demand nodes. Also, assume supply can be shipped to either transshipment node, or demand node. A diagram of this problem is given in Figure 5.4. The possible

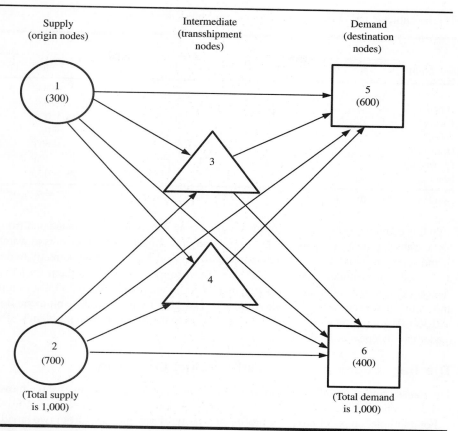

Figure 5.4   Network diagram of transshipment example.

shipments between any two pairs of nodes and the unit transportation costs are defined below:

x_{13} = Supply 1 to transshipment 3, unit transportation cost = \$40,

x_{14} = Supply 1 to transshipment 4, unit transportation cost = \$50,

x_{15} = Supply 1 to demand 5, unit transportation cost = \$90,

x_{16} = Supply 1 to demand 6, unit transportation cost = \$125,

x_{23} = Supply 2 to transshipment 3, unit transportation cost = \$40,

x_{24} = Supply 2 to transshipment 4, unit transportation cost = \$50,

x_{25} = Supply 2 to demand 5, unit transportation cost = \$75,

x_{26} = Supply 2 to demand 6, unit transportation cost = \$85,

x_{35} = Transshipment 3 to demand 5, unit transportation cost = \$40,

x_{36} = Transshipment 3 to demand 6, unit transportation cost = \$95,

x_{45} = Transshipment 4 to demand 5, unit transportation cost = \$85,

x_{46} = Transshipment 4 to demand 6, unit transportation cost = \$50.

The LP tableau and optimal solution to the problem are given below.

	x_{13}	x_{14}	x_{15}	x_{16}	x_{23}	x_{24}	x_{25}	x_{26}	x_{35}	x_{36}	x_{45}	x_{46}	b	Shadow Price (SP)
Obj. (Min)	40	50	90	125	40	50	75	85	40	95	85	50		
Supply 1	1	1	1	1									≤300	0
Supply 2					1	1	1	1					≤700	−5
Trans 3	−1				−1				1	1			=0	−40
Trans 4		−1				−1					1	1	=0	−40
Demand 5			1				1		1		1		≥600	80
Demand 6				1				1		1		1	≥400	90
Non-neg	x_{13}	x_{14}	x_{15}	x_{16}	x_{23}	x_{24}	x_{25}	x_{26}	x_{35}	x_{36}	x_{45}	x_{46}	≥0	
Solution	300	0	0	0	0	0	300	400	300	0	0	0	$Z^* = 80{,}500$	

In this example, shipments are occurring both from supply to final demand and from supply to transshipment to demand nodes. Supply origin 1 ships all 300 of its units to warehouse 3, and supply origin 2 ships 300 units directly to demand 5 and 400 units directly to demand 6. Transshipment node 3 receives 300 units from supply 1 and ships them to demand 5. Hence, demand 5 receives 600 units in total (300 from warehouse 3 and 300 from supply 2), and demand 6 receives 400 from supply 2. The cost of the optimal shipment flows is $80,500. The interpretation of the SPs is analogous to the transportation problem's SPs. The reader can independently determine how each SP is derived.

The Transshipment Model with Product Conversion

The transshipment model is also very useful in spatial problems that involve the movement and conversion of a product. The dairy sector provides an excellent example of this.

For example, consider a cheese processor who buys raw milk (m) from two small dairy cooperatives and sells cheese (c) to two supermarkets. It takes approximately 10 pounds of raw milk to make one pound of cheese. Ignoring other factors of production, the cheese "production function" is:

$$c = 1/10m.$$

That is, every pound of raw milk produces 1/10 pound of cheese. Suppose that each of the two cooperatives sells raw milk in separate locations: cooperative 1 sells 200,000 pounds of raw milk per day, and cooperative 2 sells 400,000 pounds of raw milk per day to the cheese processor. The cheese processor has two cheese plants where the raw milk can be delivered and processed (assume both plants have unlimited plant capacity to handle the raw milk). Finally, supermarket 1 wants to buy 10,000 pounds of cheese per day, and supermarket 2 wants to buy 50,000 pounds of cheese per day from the processor. The following table summarizes the per unit transportation costs for raw milk and for cheese (cooperative to plant costs are expressed as dollars per pound of raw milk and plant to supermarket costs are expressed as dollars per pound of cheese).

From/To	Plant 1	Plant 2	Supermarket 1	Supermarket 2
Coop 1 (m)	$0.50	$0.60		
Coop 2 (m)	$0.40	$0.30		
Plant 1 (c)			$0.20	$0.30
Plant 2 (c)			$0.25	$0.20

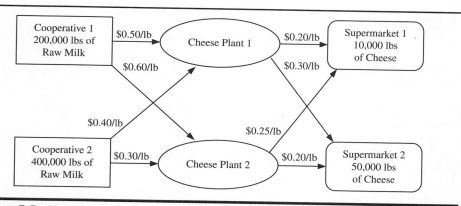

Figure 5.5   Network of transshipment model with product conversion.

The formulation of this problem is almost identical to the previous example. However, recognize that the units being shipped from cooperatives to the plant are not the same as those being shipped from the plant to the supermarkets (i.e., pounds of raw milk are not the same as pounds of cheese). Using the formula for the cheese production function, incorporate this conversion directly into the LP formulation, or convert all raw milk into cheese equivalents or vice versa.

Figure 5.5 gives a graphical overview of the network for this example. As before, the first set of constraints restricts each cooperative from selling more raw milk than it produces each day. Denoting each cooperative as c_1 and c_2, and each cheese plant as p_1 and p_2, this restriction is equivalent to:

$$c_1 p_1 + c_1 p_2 \leq 200{,}000, \text{ and}$$

$$c_2 p_1 + c_2 p_2 \leq 400{,}000.$$

Let s_1 and s_2 denote the two supermarkets. The transshipment constraints in this case require that the incoming raw milk be converted into cheese and then shipped out to the supermarkets. For the first plant, this requirement is equivalent to the following constraint:

$$c_1 p_1 + c_2 p_1 - 10 p_1 s_1 - 10 p_1 s_2 = 0.$$

This constraint implies that all milk going into plant 1 equals all cheese shipments going out of the plant, where the coefficient 10 gives the transformation from pounds of raw milk into pounds of cheese. Likewise, the equivalent constraint for plant 2 is:

$$c_1 p_2 + c_2 p_2 - 10 p_2 s_1 - 10 p_2 s_2 = 0.$$

Finally, the last set of constraints are the typical demand constraints, namely:

$$p_1 s_1 + p_2 s_1 \geq 10{,}000, \text{ and}$$

$$p_1 s_2 + p_2 s_2 \geq 50{,}000.$$

With an objective function of minimizing total transportation costs, the entire LP model for this problem is:

Min: $Z = 0.50 c_1 p_1 + 0.60 c_1 p_2 + 0.40 c_2 p_1 + 0.30 c_2 p_2 + 0.20 p_1 s_1$
$\qquad + 0.30 p_1 s_2 + 0.25 p_2 s_1 + 0.20 p_2 s_2 (0)$

s.t.:

$$c_1p_1 + c_1p_2 \leq 200,000 \tag{1}$$

$$c_2p_1 + c_2p_2 \leq 400,000 \tag{2}$$

$$c_1p_1 + c_2p_1 - 10p_1s_1 - 10p_1s_2 = 0 \tag{3}$$

$$c_1p_2 + c_2p_2 - 10p_2s_1 - 10p_2s_2 = 0 \tag{4}$$

$$p_1s_1 + p_2s_1 \geq 10,000 \tag{5}$$

$$p_1s_2 + p_2s_2 \geq 50,000 \tag{6}$$

$$c_1p_1, c_1p_2, c_2p_1, c_2p_2, p_1s_1, p_1s_2, p_2s_1, p_2s_2 \geq 0 \tag{7}$$

The Solver formulation of this problem is presented in Figure 5.6. The total minimum cost of the shipments in this problem is $233,000. Cooperative 1 should supply all 200,000 pounds of raw milk to cheese plant 1, while cooperative 2 should supply all of its 400,000 pounds of raw milk to cheese plant 2. Cheese plant 1 processes 20,000 pounds of cheese,

	A	B	C	D	E	F
1	Decision Variables		Plant 1	Plant 2	Supermarket 1	Supermarket 2
2		Coop 1	200,000	-		
3		Coop 2	-	400,000		
4		Plant 1			10,000	10,000
5		Plant 2			-	40,000
6						
7	Costs		Plant 1	Plant 2	Supermarket 1	Supermarket 2
8		Coop 1	0.5	0.6		
9		Coop 2	0.4	0.3		
10		Plant 1			0.2	0.3
11		Plant 2			0.25	0.2
12						
13	Objective					
14		Min Total Costs	$ 233,000			
15						
16	Constraints					
17	Production		Total		RHS	
18		Coop 1	200,000	≤	200,000	
19		Coop 2	400,000	≤	400,000	
20	Transhipment					
21		Plant 1	0	=	0	
22		Plant 2	0	=	0	
23	Demand					
24		Supermarket 1	10,000	≥	10,000	
25		Supermarket 2	50,000	≥	50,000	

Figure 5.6   Solver formulation of transshipment model with product conversion.

supplies 10,000 pounds to supermarket 1, and 10,000 pounds to supermarket 2. Cheese plant 2 processes 40,000 pounds of cheese, supplying all of it to supermarket 2.

The SPs are as follows:

$$SP_{sup1} = 0, SP_{sup2} = -0.21, SP_{tran1} = 0.5, SP_{tran2} = 0.51, SP_{dem1} = 5.2, SP_{dem2} = 5.3.$$

As an illustration of how these SPs are derived, consider the SP for supply 2, which equals -0.21. If cooperative 2 increased supply by one unit (one pound of raw milk), then one more unit would be shipped to cheese plant 2, which would increase transportation costs by $0.30. Transportation costs would also rise by $0.20/10, or $0.02 because an additional 1/10 pound of cheese is now shipped to supermarket 2 from cheese plant 2. Hence, total costs would increase by $0.32. With cheese plant 2 shipping an additional 1/10 pound of cheese to supermarket 2, cheese plant 1 will now ship 1/10 less pound of cheese to supermarket 2, which results in a cost savings of $0.30/10, or $0.03. Finally, since cheese plant 1 now needs one less pound of raw milk from cooperative 1, transportation will also decline by $0.50. In summary, given the $0.32 increase in costs and the $0.03 and $0.50 decrease in costs, SP for cooperative 2's SP is consequently -0.21. As extra practice, calculate the other SPs for this problem.

5.4   THE ASSIGNMENT MODEL

Another basic application of LP involves determining the most efficient way to assign tasks among people, machines, and other resources in order to carry out an assignment or job. This type of application is called an **assignment problem**. The assignment problem is a special case of the transportation problem where the decision maker wants to assign or match one "resource" to one "job." Hence, it can be thought of as a transportation model where each supply node has a fixed supply of 1 and each demand node has a fixed demand of 1.

To illustrate, suppose that a farmer has just hired three workers to carry out three jobs. Each of the workers can do any of the jobs; however, the productivity of each worker is different. The following table gives the number of hours required of each worker to complete each job:

Job	Worker		
	a	b	c
1	100	110	120
2	70	85	86
3	20	20	24

If the farmer's objective is to minimize the time involved in completing all three jobs, then the objective function is:

Min: $Z = 100a_1 + 70a_2 + 20a_3 + 110b_1 + 85b_2 + 20b_3 + 120c_1$
$\qquad + 86c_2 + 24c_3$ \hfill (0)

s.t.:

$$a_1 + a_2 + a_3 = 1 \qquad (1)$$

$$b_1 + b_2 + b_3 = 1 \qquad (2)$$

$$c_1 + c_2 + c_3 = 1 \qquad (3)$$

$$a_1 + b_1 + c_1 = 1 \qquad (4)$$

$$a_2 + b_2 + c_2 = 1 \qquad (5)$$

$$a_3 + b_3 + c_3 = 1 \qquad (6)$$

$$a_1, a_2, a_3, b_1, b_2, b_3, c_1, c_2, c_3 \geq 0 \qquad (7)$$

Here, a_i refers to worker a doing job i, b_i refers to worker b doing job i, and c_i refers to worker c doing job i.

The optimal solution to this problem is to have worker a do job 2, worker b do job 1, and worker c do job 3. Assigning workers in this way would "cost" the farmer 204 hours to complete all three jobs.

More generally, an assignment problem can be expressed as:

$$\text{Min: } Z = \sum_{i=1}^{n} \sum_{j=1}^{n} c_{ij} x_{ij} \qquad (0)$$

s.t.:

$$\sum_{j=1}^{n} x_{ij} = 1 \qquad i = 1, \ldots, n \qquad (1)$$

$$\sum_{i=1}^{n} x_{ij} = 1 \qquad i = 1, \ldots, n \qquad (2)$$

$$x_{ij} \geq 0 \qquad i = 1, \ldots, n; \, j = 1, \ldots, n \qquad (3)$$

An Example

The assignment problem need not have an objective function that is being minimized. Here is an example where total profits are maximized within an assignment problem. Consider an agricultural input company which sells farm inputs such as fertilizer, feed, herbicide, seed, and so on to farmers in the Upper Midwest. It has a team of nine salespeople who can cover nine territories in the Upper Midwest. The company needs to assign each salesperson to one and only one of the nine territories. The nine salespeople have different sales ability, and the president of the company estimates the following profitability for each salesperson in each territory:

Territory	Salespeople profits ($1,000 per month)								
	s_1	s_2	s_3	s_4	s_5	s_6	s_7	s_8	s_9
1	25	22	18	23	19	15	14	22	20
2	19	20	17	14	16	10	9	13	11
3	19	18	23	21	22	20	24	10	19
4	7	9	15	10	11	7	8	4	14
5	6	8	6	9	9	15	5	17	11
6	12	13	10	16	17	11	19	15	8
7	15	10	6	8	12	13	11	9	19
8	21	15	16	19	18	10	8	14	13
9	9	3	6	8	5	10	3	15	12

Assuming the company's sole objective is to maximize total profits from assigning the nine salespersons to the nine regions, the problem is:

$$\text{Max: } Z = \sum_{i=1}^{9} \sum_{j=1}^{9} c_{ij} s_{ij} \tag{0}$$

s.t.:

$$\sum_{j=1}^{9} s_{ij} = 1 \quad (i = 1, \dots, 9) \tag{1}$$

$$\sum_{i=1}^{9} s_{ij} \le 1 \quad (j = 1, \dots, 9) \tag{2}$$

$$s_{ij} \ge 0 \tag{3}$$

The first set of nine structural constraints in (1) restricts each territory to have only one salesperson. The second set of nine structural constraints in (2) restricts each salesperson to work in only one territory.

The optimal solution to this problem is: $s_{11}^* = s_{22}^* = s_{37}^* = s_{43}^* = s_{56}^* = s_{65}^* = s_{79}^* = s_{84}^* = s_{98}^* = 1$, $Z^* = 169$. That is, the firm can obtain a maximum profit of $169,000 per month by allocating salesperson 1 to territory 1, salesperson 2 to territory 2, salesperson 3 to territory 7, salesperson 4 to territory 3, salesperson 5 to territory 6, salesperson 6 to territory 5, salesperson 7 to territory 9, salesperson 8 to territory 4, and salesperson 9 to territory 8.

5.5   RESEARCH APPLICATION: U.S. DAIRY SECTOR SIMULATOR

There are many applications of LP transportation models that have been used to study optimal shipment patterns of milk in the U.S. dairy sector.[4] An excellent example is a large transshipment model developed by Pratt et al. (1996) called U.S. Dairy Sector Simulator (USDSS), which has been used to examine several real-world research problems.

The dairy industry is one of the most regulated industries in the United States, both in terms of safety and economic regulations. Most of the economic regulations date back to the Great Depression and the New Deal. One of the main economic programs impacting dairy farm prices is the Federal Milk Marketing Order Program (FMMOP). The federal government historically has made it a national goal to have all regions of the continental United States be self-sufficient in producing enough fluid milk to satisfy regional consumption. Traditionally, the Upper Midwest and Northeast have had an economic comparative advantage in producing milk relative to other parts of the nation such as the Southeast (more recently, some of the western states, such as California, have also become large producers of milk). To encourage all regions to be self-sufficient in milk produced for beverage purposes, the FMMOP has established regional milk markets that set minimum prices that fluid milk handlers must pay farmers. These minimum prices generally increase with distance from the Upper Midwest. Because milk is expensive to ship, the basic

---

[4]See, for example, the following studies: Babb et al. (1977); Beck and Goodin (1980); Boehm and Conner (1976); Buccola and Conner (1979); Francis (1992); Fuller et al. (1976); Jensen (1985); King and Logan (1964); Kloth and Blakely (1971); McLean et al. (1982); Novakovic et al. (1980); Pratt et al. (1986; 1996); and Thomas and DeHaven (1977).

idea is to provide market incentives via a higher price in milk deficit regions to encourage greater production. However, regional pricing of fluid or Class I milk has been highly controversial, with opposition coming mainly from the surplus milk regions, such as Wisconsin, who argue it is unfair because they receive a significantly lower milk price than dairy farmers in the Southeast.

The level of regional Class I milk prices has been debated in almost every piece of farm legislation in recent history. On more than one occasion, Congress has called upon the U.S. Department of Agriculture (USDA) to study whether the Class I price surface (i.e., regional variations in Class I prices throughout the U.S.) makes economic sense. One way to examine this question is through a transshipment model of the U.S. dairy industry. The USDSS model has been used to study this and other issues regarding regional milk production, distribution, location, and pricing issues.

USDSS is a transshipment model very similar to the milk example presented above, but with far greater detail and size. There are three markets in the USDSS network: (1) farm milk market, where raw milk is produced for further processing; (2) dairy product processing market, where raw farm milk is shipped, processed into five different dairy products, and shipped out of; and (3) dairy product consumption markets, where milk and dairy products are bought at supermarkets. In all three markets, products are expressed on a raw milk fat equivalent basis, and to handle milk component balancing, interplant transfers of intermediate dairy products are allowed. The model is a regional depiction of the 48 contiguous states.

The farm milk market is represented by 240 regional supply nodes. In reality, there are 100,000s of dairy farmers in the United States, and thus the 240 milk supply nodes are more of a representation of regional milk sheds than of individual dairy farms in the U.S. USDSS has 234 dairy product consumption markets throughout the United States with each market consuming five dairy products: (1) fluid milk, (2) soft products, (3) hard cheeses, (4) butter, and (5) dried, condensed, and evaporated products. Hence, there are 1,170 consumption nodes in total (5 × 234). The dairy processing market is represented by 507 regions where each of the five dairy products can be processed (2,535 nodes: 5 × 507). The model assumes that all raw milk is homogenous in its quality, fat content, and non-fat content.

Using actual data on regional milk production for the 240 farm milk nodes, dairy product demand for the 234 dairy consumption nodes, and assembly, processing, and distribution costs, USDSS provides optimal organization of milk, interplant, location, and distribution movements that minimizes the total costs for the U.S. dairy location network. It is assumed that raw farm milk can go to any regional processing node in the network, and consequently there are over 600,000 activities representing shipment flows of milk in the model.

The data demands of USDSS are quite enormous, and it would take many pages to describe here how Pratt et al. (1996) collected and estimated the model parameters. In general, milk supply for each of the 240 farm nodes in the network was calculated from USDA estimates of multicounty milk production for 1993, which is the base year for the model. Milk demand for the five dairy products for the 234 consumption nodes in the network was estimated on a county basis and then aggregated for each node. Various techniques were used to convert the final dairy product demand into a raw milk fat equivalent basis. Demand for most products is primarily based on prorating national demand to the population sizes of each consumption node. The transportation cost parameters, of which there are over 600,000, were estimated as a function of distance for each flow, and mapping software was used to compute the road distance for each flow. It was assumed that it cost $0.35 per hundred miles to move 100 pounds of raw milk. In 1993, there were 1,596 dairy processing plants in operation, and each was assigned to one of the 507 processing nodes in USDSS.

The output of USDSS is massive, and the following provides just some of the results. The optimal solution to the model provides several interesting pieces of information. First, it provides the optimal shipments from the raw milk supply nodes to the processing plants. Second, it provides which processing plants, by product, are active in the optimal solution. Third, it provides the optimal flow of processed dairy products from the plant to the consumption nodes.

Pratt et al. (1996) make several important conclusions from the solutions. First, the locations of fluid processing plants are generally closer to major population centers than other dairy processing plants. The majority of fluid milk processed by these plants is sold within the consumption area where the plant is located. This is due to the fact that packaged fluid milk products are much more expensive to ship than manufactured dairy products, and therefore, it is more efficient to locate fluid milk plants close to major consumption points.

Second, soft processing plants are generally located slightly farther away from major population centers than fluid milk plants, but nearer than other product plants. As with fluid milk, the majority of soft products processed is sold to supermarkets within the consumption area where the milk is processed.

Third, cheese, butter, and dry, condensed, and evaporated milk processing plants are primarily located near major farm milk supplies. While the farm milk supply is close to these processing plants, the distribution of processed cheese, butter, dry, condensed, and evaporated milk products can move a long way from where it is processed to multiple consumption nodes. This is due to the fact that transportation costs for these manufactured dairy products are much lower than for products like packaged fluid milk since manufactured products are less bulky.

The authors note two important observations regarding the optimal solution relative to the actual location of plants and milk movements. First, a comparison of the optimal solution to the actual network on milk and dairy product movements in the United States reveals numerous reasons for differences in results. However, based solely on efficiency or "market rationalization," the optimal solution can be thought of as a target for the industry to aspire to. Second, the solution is presented in terms of an annual situation, but in reality, milk movements occur on a daily basis 365 days per year. Because of seasonality in both supply and demand for milk and dairy products, some deviations between the optimal annual solution and the actual daily movements with the dairy industry should be expected.

Perhaps the most important piece of information from this model is the SPs on raw milk by location. That is, based on all the parameters of the model, what would an additional 100 pounds of raw milk delivered to a processor at each location be worth? The SPs for raw milk delivered to fluid processing plants can be used as a basis to evaluate the Class I price surface used by the FMMOP. Pratt et al. (1996) did this in the following manner. Since there are no production costs in the model, the SPs solely reflect the transportation component of the Class I price. To make these SPs more comparable to actual Class I prices, the authors added a constant, $1.20, to reflect the Class I differential in the largest fluid milk surplus market in the network, Minneapolis, Minnesota. The constant $1.20 is then added to all fluid milk SPs in the solution to represent what the Class I price surface would look like under optimal conditions.

The actual and optimal Class I price surface is shown in Figures 5.7 and 5.8. The optimal Class I price surface is similar to actual Class I prices. Prices increases with distance from the Upper Midwest states of Wisconsin and Minnesota to the southeastern U.S. in both cases. However, the optimal price structure has a greater degree of spatial specificity,

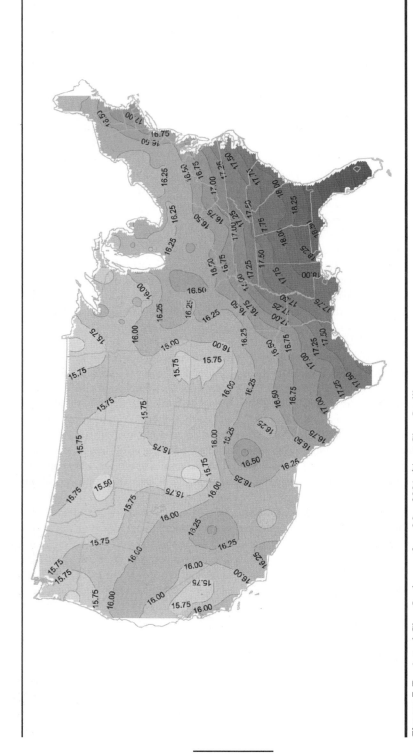

Figure 5.7 Actual Class I prices surface, May 2001, $/cwt Class I milk.

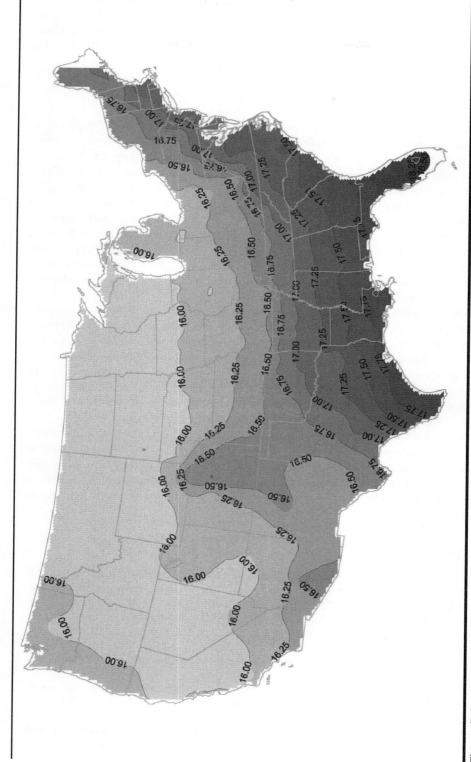

Figure 5.8 Optimal Class I prices surface, May 2001, $/cwt Class I milk.

especially in the Western states. In the late 1990s, the USDA used these results as a basis for revising the actual regional Class I prices in the U.S. and as a means for consolidation of federal milk marketing orders into fewer, but larger orders. The LP transshipment model proved to be an extremely useful tool for the policy debate and for justifying the resulting decisions reached by USDA.

SUMMARY

This chapter has dealt with three frequently used applications of LP: the transportation, transshipment, and assignment problems. Transportation problems involve determining how to move a product, or to schedule routes, in the most efficient way possible, given such constraints as available supply (plant capacity) from each production location, prevailing demand at each consumption destination, road systems, and so on. Generally, the objective is to minimize transportation costs, but other objectives may be modeled, such as maximization of profits net of transportation costs. Two examples of transportation models were highlighted for agricultural markets: an optimal ice cream distribution network, and optimal international trade flows under four different trade policies. Since agricultural commodities tend to be quite bulky and sometimes perishable, their transportation costs tend to be quite high relative to other products. Hence, determination of the most efficient distribution network is critical for this industry.

Transshipment models are transportation models with another set of nodes (intermediate transshipment nodes) included in the model. This makes it possible to extend the transportation model from plant to consumer networks to plant to wholesale to consumer networks, which is usually a more realistic depiction of actual distributional networks. An example of a warehouse transshipment model was provided, as well as an example of product transformation, such as raw farm milk processed into cheese. In addition, the chapter featured a research example of a large transshipment model for the U.S. dairy industry.

The last network model presented was the assignment problem. The assignment problem is a special case of transportation problems. These types of models are designed to minimize the cost of assigning resources to jobs on a one-to-one basis. Alternatively, they can be used to maximize revenue or profits associated with assigning different people to different jobs. An example of this was provided for an agribusiness that sells farm inputs in the Upper Midwest.

REFERENCES

Babb, E. M., Banker, D. E., Goldman, O., Martella, D. R., & Pratt, J. E. (1977). Economic model of federal milk marketing order policy simulator-model. *Agricultural Station Bulletin*, no. 158, Purdue Univ., West Lafayette, IN.

Beck, R. L., & Goodin J. D. (1980). Optimum number and location of manufacturing milk plants to minimize marketing costs. *Southern Journal of Agricultural Economics, 12*, 103–108.

Boehm, W. T., & Conner, M. C. (1976). Technically efficient milk assembly and hard product processing for the southeastern dairy industry. *Research Division Bulletin,* no. 122, VPI & SU, Blacksburg, VA.

Buccola, S. T., & Conner M. C. (1979). Potential efficiencies through coordination of milk assembly and milk manufacturing plant location in the northeastern United States. *Research Division Bulletin* no. 149, VPI & SU, Blacksburg, VA.

Francis, W. G. (1992). Economic behavior of a local dairy market under federal milk market order regulation. M.Sc. thesis, Department of Agricultural Economics, Cornell University, Ithaca, NY.

Fuller, S. W., Randolph, P., & Klingman, D. (1976). Optimizing sub-industry marketing organizations: A network analysis approach. *American Journal of Agricultural Economics, 58*, 425–436.

Jensen, D. L. (1985). Coloring and duality: Combinatorial augmentation methods. Unpublished doctoral dissertation, Department of Operations Research, Cornell University, Ithaca, NY.

King, G., & Logan, S. H. (1964) Optimum location, number and size of processing plants with raw product and final product shipments. *Journal of Farm Economics, 46*, 94–108.

Kloth, D. W., & Blakely, L. V. (1971). Optimum dairy plant location with economies of size and market share restrictions. *American Journal of Agricultural Economics, 53*, 461–466.

McLean, S., Kezis, A., Fitzpatrick, J., & Metzger, H. (1982). Transshippment model of the Maine milk industry. *Tech. Bul.*, no. 106, University of Maine, Orono.

Novakovic, A. M., Babb, E. M., Martella, D. R., & Pratt, J. E. (1980). An economic and mathematical description of the dairy policy simulator. A.E. Res. 80–21, Cornell University Agricultural Experiment Station, Ithaca, NY.

Pratt, J. E., Novakovic, A. M., Elterich, G. J., Hahn, D. E., Smith, B. J., & Criner, G. K. (1986). An analysis of the spatial organization of the northeast dairy industry. Search: Agriculture. Cornell University Agricultural Exp. Station, No. 32:84, Ithaca, NY.

Pratt, J. Novakovic, A., Stephenson, M., Bishop, P., & Erba E. (1996). U.S. Dairy Sector Simulator: A spatially disaggregated model of the U.S. Dairy industry. Agricultural Economics Staff Paper 96–06, Department of Agricultural Economics, Cornell University.

Thomas, W. A., & DeHaven, R. K. (1977). Optimum number, size, and location of fluid milk processing plants in South Carolina. *South Carolina Agricultural Experiment Station Bulletin*, no. 603, Clemson University, Clemson, SC.

EXERCISES

1. Due to poor weather, the shipping costs from Milwaukee have recently doubled. Solve the ice cream transportation problem presented in Section 5.1 and generate a sensitivity report using Solver.

2. Consider the transshipment example in Figure 5.5. Assume that the demand from a third supermarket is 20,000 lbs. Also assume that cooperative 1 now has 400,000 lbs of raw milk. Solve the problem using Solver. Compare the answer with that in the chapter. Conduct sensitivity analysis on the supply, demand, and transshipment constraints.

3. A company has three plants: a, b, and c, and there are two major distribution centers, d and e. In the current quarter, factories a, b, and c have the capacities 1,000, 1,500, and 1,200 units, respectively. The demands of distribution centers d and e are 2,300 and 1,400 items, respectively. The transportation is conducted by truck at the cost of 8 cents per item per kilometer. Design a transportation problem to minimize transportation costs. The distances are listed (in kilometers):

	d	e
a	80	215
b	100	108
c	102	68

4. A distillery produces two types of whiskey: a high end aged single malt Scotch and a more plebian blended Scotch whiskey. They can both be bottled in any combination of any of their three plants (p_1, p_2, and p_3) and need to be shipped to their five retail outlets, r_1 through r_5. The plants have capacities of 700, 400, and 650 cases, respectively. r_1 needs 100 cases of single malt and 150 cases of blended; r_2 needs 150 of each; r_3 needs 100 single malt and 175 blended; r_4 needs 150 single malt and 200 blended; and

r_5 needs 200 single malt and 250 blended. They also have two distribution centers, d_1 and d_2; however, only d_1 has the necessary environmental controls to handle the sensitive single malt. They have capacities of 550 and 650 cases, respectively. Given the following per case transport costs, find the shipping schedule to minimize costs.

	d_1	d_2	r_1	r_2	r_3	r_4	r_5
p_1	$4.00	$5.00	$10.00	$12.00	$13.00	$15.00	$14.00
p_2	$7.00	$3.00	$11.00	$15.00	$12.00	$13.00	$10.00
p_3	$5.00	$6.00	$15.00	$14.00	$13.00	$15.00	$11.00
d_1			$3.00	$4.00	$2.00	$6.00	$4.00
d_2			$2.00	$5.00	$3.00	$4.00	$7.00

5. A producer wholesaler has three factories, and the fruit and vegetables it ships are supplied to four different distribution centers. The table below gives unit shipping costs to each warehouse, along with factory capacities and warehouse demands.

Warehouse	1	2	3	4	Capacity
Factory 1	$0.40	$0.80	$0.30	$0.60	800
Factory 2	$1.60	$0.40	$1.20	$1.00	1,000
Factory 3	$1.20	$0.20	$0.80	$0.40	600
Demand	400	400	600	1,000	

Write the LP problem to minimize transportation costs.

6. A plant has four warehouses (a, b, c, and d) to ship the products to customers directly. The warehouses a, b, c, and d have 14, 16, 17, and 15 units in stock, respectively. The customer's demands are 18, 19, and 20 units for x, y, and z. The cost associated with each possible shipment is given below.

	Warehouse				
Customer	a	b	c	d	Demand
x	8	10	8	3	16
y	4	6	7	5	18
z	7	12	5	3	20
Stock	14	16	17	15	

Find the optimal transportation routes by minimizing transportation costs.

7. There are three warehouses in Detroit, Pittsburgh, and Buffalo with the following fixed supplies of watermelons: 250, 130, and 235, respectively. The business owner has the following fixed demand for watermelons: Boston (75), New York (230), Chicago (240), and Indianapolis (70). The unit cost of shipping the watermelons is:

From/To	Boston (BS)	New York (NY)	Chicago (CH)	Indianapolis (IN)
Detroit (DT)	15	20	16	21
Pittsburgh (PT)	25	13	5	11
Buffalo (BF)	15	15	7	17

Find the optimal transportation plan of this problem.

8. The following is a more realistic transportation problem for a flour processor.

Tom Miller, who owns a small, old-fashioned mill in Holyoke, Massachusetts, became aware of the great increase in demand for stone ground flour. More people were becoming health-conscious and learning that the heat caused by the steel-milling of wheat destroyed many of the natural B vitamins, whereas the slower and cooler stone milling process did not. Also, the freshness of the flour is a consideration since the vitamin content of the flour is reduced with age.

Our young entrepreneur, Tom, had access to considerable venture capital through his family. He saw how money could be made by buying up some of these old, water-powered, stone mills; refurbishing them; and training people to operate them.

After several years of hard work, Tom has acquired mills in these locations with the weekly output capabilities given in 100 lb. sacks of flour.

Holyoke, Massachusetts	150
Bingham, Maine	75
Carthage, New York	125
Woodstock, Vermont	300
Millerton, New York	250
Nashua, New Hampshire	100
Total weekly output capacity	1,000

Tom sells most of his flour to natural food distributors, a wholesale market. He has also entered the lucrative Boston retail market and has acquired a bagging facility and warehouse in Framingham. This serves as a collection point for weekly deliveries into Boston. He has a contract with a trucking company to make the run to his city customers (mostly natural food retail stores plus a few restaurants and small bakeries) with one dual-purpose truck with a capacity of 16 tons. He nets a profit of 5 cents per pound more than the wholesale on this retail flour even after the additional bagging and transportation costs.

He also has customers in the following locations with the following weekly demands (in 100 sacks).

Portland, Maine	40
Manchester, New Hampshire	60
Albany, New York	100
Rochester, New York	200
Waterbury, Connecticut	140
Framingham, Massachusetts	320
Total weekly demand	860

At the moment, there are 140 sacks per week excess production capacity. This doesn't cause any waste of flour because he doesn't grind any wheat he can't sell, but he knows he will make more money if he uses his mills to capacity.

Tom would like to break into the more lucrative New York City market. He will be able to get the same premium in New York that he now does in Boston. He is considering the same sort of arrangement with a trucking company to ship from his warehouse into the city. He has found a warehouse he can buy in Ossining, NY. To buy the warehouse and install a bagging facility would cost him $50,000. To supply the New York market, he will need another 320 sacks a week. He suspects that it would pay him well to abandon the Rochester market in favor of New York City. This would give him 200 sacks currently being sold in Rochester, and the remaining 120 sacks could be made up by his existing excess capacity.

He has asked for our help on two points:

1. What is the best shipment arrangement for the existing setup?
2. What profit, if any, is there in abandoning the Rochester and supplying New York instead?

His current transportation costs in dollars per 100 lb. bag is the following:

Demand Nodes	Weekly Demand	Supply Nodes					
		Holyoke	Bingham	Carthage	Woodstock	Millerton	Nashua
Weekly Output		50	75	125	300	250	100
Framingham	320	$0.15	$0.55	$0.65	$0.30	$0.30	$0.10
Portland	40		$0.25		$0.25		$0.20
Manchester	60	$0.20	$0.40		$0.15		$0.05
Albany	100	$0.15		$0.30	$0.25	$0.15	$0.30
Rochester	200	$0.60		$0.35	$0.70	$0.55	$0.80
Waterbury	140	$0.10		$0.50	$0.35	$0.10	$0.30
Ossining	320	$0.20		$0.55	$0.50	$0.15	$0.40

a. What is the objective function and constraint set for the existing situation? Write out the LP tableau.

b. What is the objective function and constraint set for the New York market situation? Write out the LP tableau.

9. Solve the LP problem in Exercise 8a using Solver.
   a. What are the weekly total transportation costs of all shipments?
   b. What is the SP of a sack of flour demanded in Rochester? What does this mean?
   c. Which mills have excess capacity?

10. Solve the LP problem in Exercise 8b using Solver.
   a. What changes have been made relative to the solution to Exercise 8a?
   b. The SP of shipping one more sack of flour to Framingham is 55 cents. Explain how this is derived and why it is that amount. Some of the SPs are very complicated. It may be necessary to alter the RHS of the constraint by one sack and then examine all the changes that happen. For example, the SP for Holyoke is 50 cents. That means that if one more sack of flour is produced in Holyoke, it will reduce the total cost by 50 cents. But none of the transportation arcs from Holyoke are 50 cents. So how is this arrived at? Alter the RHS of the Holyoke constraint (2) to 151. Alternatively, you could print out all these changes, but that is time consuming and unnecessary. Solve the problem. Now compare the solutions. Note the changes in the objective function values: BF = 14, WF = 206, HW = 106, and WW = 34. So this is what happens. That extra sack produced at Holyoke is shipped to Waterbury at a cost of 10 cents. That saves a sack being shipped from Woodstock to Waterbury, saving 35 cents. Woodstock now has an extra sack that gets shipped to Framingham at a cost of 30 cents, which saves a sack being

shipped from Bingham to Framingham, saving 55 cents. So altogether $55 - 30 + 35 - 10 = 50$ cents. (Don't forget to change the RHS of the Holyoke constraint back to 50 before you go on to the next SP.)

Now that you have had this one done for you, explain the SPs of Manchester and Waterbury.

11. Compare the two solutions in Exercises 9 and 10. How much flour is he selling in each case, and how much are his transportation costs?

12. Based on the solutions to Exercises 9 and 10, how much more per week is his income with the New York market than with Rochester? (Remember the 5 cents/lb. retail premium.)

13. Payments on a $50,000 loan at 12% for 5 years are $1,112.23 per month. Is it a good move for Tom to shut down the Rochester market and open one in New York based at Ossining? Explain your answer. (You do not need to know a price per pound for all the flour sold to discuss this issue intelligently. Simply knowing that the flour sold out of the Ossining warehouse goes for 5 cents more per pound than the wholesaled flour should be sufficient information.)

14. An organic apple processor produces apple juice in four locations in the United States with the following yearly capacities:

San Francisco	110,000
Chicago	50,000
San Antonio	40,000
New York	100,000
Total supply	300,000 gallons/year

It currently sells 300,000 gallons per year to the following regions of the United States:

East	150,000
South	25,000
West	75,000
North	50,000
Total demand	300,000 gallons/year

The company expects that the demand for the next year will increase by 30,000 gallons in the East, 5,000 gallons in the South, 10,000 gallons in the West, and 1,000 gallons in the North. Hence, total demand for the United States will increase from 300,000 to 346,000 gallons. In addition, the company has just negotiated a sale of 200,000 additional gallons to Japan. Thus, while the business currently can produce only 300,000 gallons, they face a demand of 546,000 gallons in the next year. The company has identified three options in meeting this new demand:

1. Purchase one new processing plant in Portland, Oregon, with a capacity of 300,000 gallons per year.

2. Purchase one new plant in Birmingham, Alabama, with a capacity of 150,000 gallons per year and another new plant in Syracuse, New York, with a capacity of 150,000 gallons.

3. Purchase a new plant in Seattle, Washington, with a capacity of 100,000 gallons, and expand their San Francisco plant from its current capacity of 110,000 to 310,000 gallons per year.

The total variable production cost (not including transportation costs) for each existing facility and the three expansion options are listed below:

Plant	Variable Production Costs ($/gallon)
San Francisco (110,000 capacity)	5.00
Chicago	4.50
San Antonio	4.00
New York	5.50
Portland	5.00
Birmingham*	5.76
Syracuse*	6.00
Seattle*	6.20
San Francisco (310,000 capacity)*	6.30

*Includes expansion costs.

The transportation cost per gallon from each supply to each demand node is:

Plant	East	South	West	North	Japan	Supply (1,000 gallons)
		(cents per gallon shipped)				
San Francisco	100	85	10	45	200	110
Chicago	50	50	55	5	300	50
San Antonio	75	5	80	75	400	40
New York	10	70	120	15	450	100
Portland	105	100	15	30	250	300
Birmingham	65	10	95	70	500	150
Syracuse	0	90	130	15	485	150
Seattle	110	110	25	25	220	100
San Francisco	100	85	10	45	200	310
New Demand	(1,000)	180	30	85	51	200

Assume that the firm's objective is to minimize the sum of total variable production costs and transportation costs. (Note that the transportation costs are in cents per gallon, while the variable production costs are in dollars per gallon.)

a. Formulate and solve an LP problem that involves the first option for expansion.

b. Formulate and solve an LP problem that involves the second option for expansion.

c. Formulate and solve an LP problem that involves the third option for expansion.

d. Based on the results of parts a, b, and c, which of the three options would you recommend?

15. Explain the shortcomings of the type of analysis in Exercise 14.

16. A lumber company produces premium mahogany hardwood for flooring in three plants and ships them to four sales areas. The three supply plants are located in Raleigh, Peoria, and Columbus. The four demand destinations are located in Atlanta, Buffalo, Chicago, and Denver. The fixed supply and demand in each market are expressed in units of tons. The unit transportation costs, fixed supply for each plant, and fixed demand for each sales area are summarized below:

| | Supply Regions | | | |
| | Raleigh (R) | Peoria (P) | Columbus (C) | Demand |
		($/ton)		(tons)
Atlanta (A)	200	500	400	13
Buffalo (B)	450	350	250	7
Chicago (C)	600	150	250	15
Denver (D)	900	500	600	10
Supply (tons)	25	15	10	

a.   Determine the least-cost method of distribution for the firm.

b.   Make a map of this problem and fill in the optimal shipments in the map.

c.   What is the optimal number of shipments that should be made from the plant at Raleigh?

d.   How many units should be shipped from Peoria to Buffalo?

e.   How much would costs be reduced in the optimal solution if additional lumber were produced at the Peoria plant?

f.   If demand in Atlanta were increased by three units, how much would the objective function value in the optimal solution increase?

g.   How much can the cost of shipping from Raleigh to Atlanta increase before a new optimal solution is required? How much could this cost decrease?

h.   How much must the cost of shipping from Columbus to Buffalo decrease before this shipping route could become part of the optimal transportation system?

i.   Which plant is the most efficiently located plant with respect to transportation costs?

j.   What is the meaning of the dual values on demand constraints in LP transportation models?

17. The following exercise is a larger example of a local dairy market. There are three dairy cooperatives that produce the following amounts of raw milk on an annual basis: cooperative 1 (c_1) produces 5 billion pounds of milk, cooperative 2 (c_2) produces 4.808 billion pounds of milk, and cooperative 3 (c_3) produces 12 billion pounds of milk. Each of the three cooperatives may sell any or all of their milk to two different cheese processors (p_1 and p_2) or to two different fluid milk bottlers (f_1 and f_2). Assume that the cheese and fluid milk production functions are:

cheese = 1/10 *raw milk*, where *cheese* is in pounds, and every pound of
raw milk produces 1/10 pound of cheese

fluid $= 1/8.63$ raw milk, where fluid is in pounds, and every pound
of raw milk produces EXR1/8.63 gallons of fluid milk

Assume that both cheese processors and both fluid bottlers have unlimited capacity to handle and process any amount of raw milk. There are three supermarket chain stores that demand the following amounts of cheese and fluid milk each year: supermarket 1 (s_1) needs 100 million pounds of cheese and 200 million gallons of milk, supermarket 2 (s_2) needs 400 million pounds of cheese and 800 million gallons of milk, and supermarket 3 (s_3) needs 300 million pounds of cheese and 600 million gallons of milk. The following table summarizes the per unit transportation costs for raw milk and for cheese. cwt $= 100$ pounds.

From/To	(unit)	p_1	p_2	f_1	f_2	s_1	s_2	s_3
c_1	cwt	$0.50	$0.60	$0.95	$0.75			
c_2	cwt	$0.40	$0.30	$0.99	$0.89			
c_3	cwt	$0.87	$0.99	$0.75	$0.99			
p_1	lb					$0.02	$0.01	$0.03
p_2	lb					$0.04	$0.06	$0.05
f_1	gal					$0.34	$0.32	$0.41
f_2	gal					$0.28	$0.29	$0.16

a. Write the LP tableau for this problem that minimizes total transportation costs of raw milk hauling from farm to the cheese-fluid plants and from cheese-fluid plants to the supermarkets.

b. Use Solver to find the optimal solution for this problem.

c. Summarize the optimal solution. What are the shipments from the cooperatives to the plants; from the plants to the supermarkets? Interpret the dual values (SPs) on the various nodes of the problem.

18. Recall Exercise 16 in which a company produces mahogany hardwood in three plants and ships them to four sales areas. The three supply plants are located in Raleigh, Peoria, and Columbus. The four demand destinations are located in Atlanta, Buffalo, Chicago, and Denver. The unit transportation costs, fixed supply for each plant, and fixed demand for each sales area are summarized below.

Demand Regions	Supply Regions			Demand (tons)
	Raleigh (R)	Peoria (P) ($/ton)	Columbus (C)	
Atlanta (A)	200	500	400	13
Buffalo (B)	450	350	250	7
Chicago (C)	600	150	250	15
Denver (D)	900	500	600	10
Supply (tons)	25	15	10	

This company is thinking about building two warehouses at two locations midway between its plants and the demand regions. These warehouses would serve as a location for aging all the wood. That is, all wood must now be shipped first to either warehouse 1 (w_1) or warehouse 2 (w_2). The four demand areas are then serviced from the two warehouses. Formulate this problem.

    a.   Compared to the answer to the previous question (Exercise 16), what are the new supply constraints for this problem for Raleigh, Peoria, and Columbus plants? (Define your activities.)

    b.   What are the new demand constraints for this problem for the Atlanta, Buffalo, Chicago, and Denver sales areas? (Define your activities.)

    c.   What are the transshipment constraints for the two warehouses for this problem? (Define your activities.)

19. Juice, Inc. produces and sells cranberry juice nationwide. They currently have five processing plants located around the country where cranberries are squashed and made into juice. These processing plants are located in the following five major cities: Los Angeles (LA), St. Paul (SP), Boston (B), Atlanta (A), and Dallas (D). Based on their plant technology, 1 ton of cranberries produces 75 gallons of cranberry juice.

    Crancoop Inc. is a large cranberry cooperative that sells bulk cranberries to Juice, Inc. They have four large transfer stations located in the North (NOR), South (SOU), East (EAS), and Western (WES) United States. These transfer stations are where all the cranberries from each region's farms are stored, and they represent the supply sources for each processing plant.

    After processing the raw cranberries into juice, Juice, Inc. ships the juice to four cold storage facilities in the nation, where supermarkets get their supply of cranberry juice. The cold storage facilities are located in California (CAL), Minnesota (MN), New York (NY), and Florida (FL).

    You have been hired as a consultant to determine the most efficient raw product–final product distribution network for Juice, Inc. You have been given the following unit transportation costs, supply capacities, and demand levels:

Cranberries: ($/ton of bulk cranberries)

From/To	LA	SP	B	A	D	Supply (tons)
NOR	500	75	80	650	700	900
SOU	450	455	666	150	100	100
EAS	1,000	225	50	100	950	500
WES	120	350	1,100	1,150	450	700

Cranberry Juice ($/gallon of cranberry juice)

From/To	CAL	MN	NY	FL
		($/gallon)		
LA	0.10	1.00	3.00	3.50
SP	1.00	0.05	1.25	2.35
B	2.50	0.75	0.35	1.45
A	3.00	1.10	0.80	0.55
D	0.90	1.05	2.00	0.95
Demand	60,000	20,000	40,000	30,000

a. Assuming the processing plants can handle unlimited amounts of cranberries, formulate an LP model that minimizes total distribution costs for transporting cranberries to the plants and cranberry juice to the cold storage facilities. Show either the LP tableau, or mathematical representation of the model.

b. Using Solver, solve the problem in part a.

c. Write a summary of the optimal solution.

d. Write a summary of your sensitivity analysis with respect to objective function coefficients.

20. A New York State wine maker owns two wineries, one in Niagara Falls (w_1) and one in the Finger Lakes (w_2). The wine maker also owns three grape farms that supply all grapes needed for his two wineries. Grape farm 1 is located in Tompkins County (g_1), grape farm 2 is located in Seneca County (g_2), and grape farm 3 is located in Niagara County (g_3). Define any shipment of grapes from grape farm i to winery j as $g_i w_j$. The unit transportation costs from each grape farm to each winery, as well as the annual supply of grapes are given below:

From/To	w_1	w_2	Total Supply
	($/ton of grapes)		(tons)
g_1	$400	$150	105
g_2	$250	$70	70
g_3	$25	$220	50

The wine maker has a contract to sell 25,000 bottles of wine to a distributor in Buffalo (d_1), 15,000 bottles to a distributor in New York City (d_2), and 70,000 bottles to a distributor in Albany (d_3). Define any shipment of wine from winery j to distributor k as $w_j d_k$ (*make j and k subscripts). The unit transportation costs and from each winery to each distributor are:

From/To	d_1	d_2	d_3
		($/bottle)	
w_1	$0.25	$1.10	$1.00
w_2	$0.70	$0.75	$0.65
Total Demand	25,000	15,000	70,000

One ton of grapes will make 500 bottles of wine. Assume that the wine maker's sole objective is to minimize total transportation costs and that he has infinite capacity at each winery.

a. Write this LP problem in general form using the notation outlined on the previous page, i.e., g_iw_j denotes grape shipments from grape farm i to winery j, w_jd_k denotes wine shipments from winery j to distributor k.

The following is the optimal solution for the problem.

Primal Problem Solution

Variable	Status	Value	Return/Unit	Value/Unit	Net Return
g_1w_1	Nonbasis	0.00	400.00	330.00	70.00
g_1w_2	Basis	100.00	150.00	150.00	0.00
g_2w_1	Basis	0.00	250.00	250.00	0.00
g_2w_2	Basis	70.00	70.00	70.00	0.00
g_3w_1	Basis	50.00	25.00	25.00	0.00
g_3w_2	Nonbasis	0.00	220.00	−155.00	375.00
w_1d_1	Basis	25,000.00	0.25	0.25	0.00
w_1d_2	Nonbasis	0.00	1.10	0.39	0.71
w_1d_3	Nonbasis	0.00	1.00	0.29	0.71
w_2d_1	Nonbasis	0.00	0.70	0.61	0.09
w_2d_2	Basis	15,000.00	0.75	0.75	0.00
w_2d_3	Basis	70,000.00	0.65	0.65	0.00

Dual Problem Solution

Constraint	Status	Dual Value	RHS Value	Usage	Slack
g_1sup	Nonbinding	0.00	105.00	100.00	5.00
g_2sup	Binding	−80.00	70.00	70.00	0.00
g_3sup	Binding	−305.00	50.00	50.00	0.00
w_1tran	Binding	0.66	0.00	0.00	0.00
w_2tran	Binding	0.30	0.00	0.00	0.00
d_1dem	Binding	0.91	25,000.00	25,000.00	0.00
d_2dem	Binding	1.05	15,000.00	15,000.00	0.00
d_3dem	Binding	0.95	70,000.00	70,000.00	0.00

Objective Row Ranges

Variable	Status	Value	Return/Unit	Minimum	Maximum
g_1w_1	Nonbasis	0.00	400.00	330.00	NONE
g_1w_2	Basis	100.00	150.00	70.00	220.00
g_2w_1	Basis	0.00	250.00	−55.00	295.00
g_2w_2	Basis	70.00	70.00	25.00	150.00
g_3w_1	Basis	50.00	25.00	NONE	330.00
g_3w_2	Nonbasis	0.00	220.00	−155.00	NONE
w_1d_1	Basis	25,000.00	0.25	−0.66	0.34
w_1d_2	Nonbasis	0.00	1.10	0.39	NONE
w_1d_3	Nonbasis	0.00	1.00	0.29	NONE
w_2d_1	Nonbasis	0.00	0.70	0.61	NONE
w_2d_2	Basis	15,000.00	0.75	−0.30	1.46
w_2d_3	Basis	70,000.00	0.65	−0.30	1.36

Right-Hand-Side Ranges

Constraint	Status	Dualvalue	RHS Value	Minimum	Maximum
$g_1 sup$	Nonbinding	0.00	105.00	100.00	NONE
$g_2 sup$	Binding	−80.00	70.00	65.00	170.00
$g_3 sup$	Binding	−305.00	50.00	45.00	50.00
$w_1 tran$	Binding	0.66	0.00	0.00	2,500.00
$w_2 tran$	Binding	0.30	0.00	−50,000.00	2,500.00
$d_1 dem$	Binding	0.91	25,000.00	25,000.00	27,500.00
$d_2 dem$	Binding	1.05	15,000.00	0.00	17,500.00
$d_3 dem$	Binding	0.95	70,000.00	20,000.00	72,500.00

b. Draw a network diagram of the optimal shipments. Include in the diagram the optimal quantities, unit transportation costs, and fixed supplies and demands.

c. The SP on the supply of grapes from grape farm 2 is −80. Explain how that number is derived.

d. Suppose that winery 1 has a capacity of handling 180 tons of grapes, and winery 2 has a capacity of 80 tons of grapes. Show how you would modify your model to account for these capacities.

e. From the wine maker's point of view, which distributor is in the most efficient location? Why?

f. How much would transportation costs change if additional grapes were grown on the first grape farm (g_1)? Why?

21. Solve the first example of a transshipment problem with warehouses presented in this chapter using Solver. Conduct sensitivity analysis on the supply, demand, and transshipment constraints.

22. Solve the transshipment with product conversion (fluid milk and cheese) example presented in this chapter using Solver. Conduct sensitivity analysis on the supply, demand, and transshipment constraints.

23. There are three dairy cooperatives that produce the following amounts of raw milk on an annual basis: Cooperative 1 (c_1) produces 8 billion pounds of milk, cooperative 2 (c_2) produces 10 billion pounds of milk, and cooperative 3 (c_3) produces 15 billion pounds of milk. Each of the three cooperatives may sell any or all of their milk to two different cheese processors (p_1 and p_2), or to two different fluid milk dealers (f_1 and f_2), or to two different butter manufacturers (b_1 and b_2). Assume that the cheese, fluid milk, and butter production functions are characterized by the following:

1 pound of cheese requires 10 pounds of raw milk.

1 gallon of milk requires 5 pounds of raw milk.

1 pound of butter requires 21 pounds of raw milk.

Assume that the cheese processors, fluid dealers, and butter manufacturers have unlimited capacity to handle and process any amount of raw milk. There are two supermarket chain stores that demand the following amounts of cheese, fluid milk, and butter each year: Supermarket 1 (s_1) needs 100 million pounds of cheese, 200 million gallons of milk, and 150 million pounds of butter. Supermarket 2 (s_2) needs 400 million pounds of cheese, 800 million gallons of milk, and 250 million pounds of butter. The following table summarizes the per unit transportation costs for raw milk and for cheese.

From/To	(unit)	p_1	p_2	f_1	f_2	b_1	b_2	s_1	s_2
c_1	lb	$0.05	$0.06	$0.11	$0.08	$0.15	$0.05		
c_2	lb	$0.04	$0.03	$0.09	$0.08	$0.04	$0.07		
c_3	lb	$0.07	$0.12	$0.08	$0.10	$0.16	$0.04		
p_1	lb							$0.02	$0.01
p_2	lb							$0.04	$0.06
f_1	gal							$0.34	$0.32
f_2	gal							$0.28	$0.29
b_1	lb							$0.03	$0.02
b_2	lb							$0.01	$0.03

a. Write the LP model for this problem that minimizes total transportation costs of hauling raw milk from the farm to the cheese, fluid, and butter plants and from the cheese, fluid, and butter plants to the supermarkets. Make sure to define all notation used in your model.

b. Show how you would modify the model in part a to account for the following capacity constraints: cheese plant 1 having a capacity of handling 3 billion pounds of milk, cheese plant 2 having a capacity of handling 5 billion pounds of milk, fluid plant 1 having a capacity of handling 4 billion pounds of milk, fluid plant 2 having a capacity of handling 3 billion pounds of milk, butter plant 1 having a capacity of handling 7 billion pounds of milk, and butter plant 2 having a capacity of handling 8 billion pounds of milk (for clarity, consider listing all of these separately instead of in paragraph form).

24. A large dairy farmer has nine employees to handle nine different jobs on his farm to complete a small construction project. Assume each employee takes the following amount of time to complete the nine jobs:

Hours to complete each job by employee(s)

Employee/Job	s_1	s_2	s_3	s_4	s_5	s_6	s_7	s_8	s_9
1	25	22	18	23	19	15	14	22	20
2	19	20	17	14	16	10	9	13	11
3	19	18	23	21	22	20	24	10	19
4	7	9	15	10	11	7	8	4	14
5	6	8	6	9	9	15	5	17	11
6	12	13	10	16	17	11	19	15	8
7	15	10	6	8	12	13	11	9	19
8	21	15	16	19	18	10	8	14	13
9	9	3	6	8	5	10	3	15	12

Solve this assignment problem using Solver to determine the one-to-one assignment of the nine workers to the nine jobs in a way that minimizes the total time to complete all tasks.

25. Modify Exercise 24 by assuming that rather than hours to complete each job, the parameters in the matrix are net revenues that the dairy farmer can make by assigning each worker to a specific project. Solve the new assignment exercise using Solver.

26. A dairy cooperative wants to improve employee morale and implements a survey to its milk truck drivers on the preferred routes that each would like to drive. Suppose there are five drivers and five routes, and that the drivers have indicated the following preferences for each route on a scale of 1–5 with 1 being their first choice and 5 being their last choice:

Driver	Route 1	Route 2	Route 3	Route 4	Route 5
1	1	1	2	3	1
2	4	5	4	2	2
3	5	2	5	1	3
4	3	3	3	5	4
5	2	4	2	4	5

Formulate this as an assignment problem that minimizes the sum of the numeric preferences of all drivers.

27. In a move to lower their carbon footprint and save money, a supermarket chain decides to re-evaluate its distribution of food from its warehouse to its six stores. It currently hires six independent truckers to transport food from the warehouse to each store with the following distances (in miles):

Truck	Store 1	Store 2	Store 3	Store 4	Store 5	Store 6
1	120	100	175	135	85	85
2	130	200	125	100	100	65
3	150	150	110	90	125	75
4	200	175	140	65	135	95
5	175	155	150	75	150	55
6	155	120	185	125	75	65

Solve this problem to minimize the total miles traveled by assigning each truck to each store.

28. Solve the salesperson assignment problem presented in this chapter using Solver. Conduct sensitivity analysis on the constraints.

6

Natural Resource and Environmental Economics Applications of Linear Programming

Decision making in the management of natural resources can be a complicated process as the planner needs to account for competing demands—the desire to protect ecosystems, the pressures of increased economic development, the demand for a sustainable and safe food supply, and concerns about preserving endangered species and scenic landscapes. In this chapter, several examples are presented that illustrate how linear programming (LP) can be used to improve decision making in a wide variety of contexts. The scope of applications of LP in the context of natural resource and environmental economics cannot be sufficiently covered in a single chapter. Instead, this chapter introduces some common applications and illustrates how these types of problems can be initially set up. More realistic and advanced models can then be developed from this foundation. The examples in this chapter include forest management, land use planning, wildlife management, agricultural production and irrigation decisions, optimal forest rotations, and the establishment of ecological corridors.

6.1 FOREST MANAGEMENT[1]

Consider a fictitious example of a forest inventory problem in the Pacific Northwest of the United States. By the mid-1950s, less than 15% of the Douglas fir inventory remained in the area compared to the 1800s. Johnsonville Timber Company wants to identify a harvest rotation plan to maximize its profit from its remaining stands of Douglas firs. After

---

[1]This example is based on a related problem outlined in Dykstra (1984). This problem set-up and solutions for this problem and the other problems discussed in this chapter are provided in the supplemental materials for Chapter 6 available at www.wiley.com/college/kaiser.

allowing the harvest of a section of the forest, Johnsonville would sell the logs, and the area would be replanted. Johnsonville needs to decide when to allow the harvest of each section of its forest to maximize its profits.

Johnsonville's forest tracks are divided into two age-classes: one is 12 hectares of 40-year-old trees and the other 24 hectares of 60-year-old trees. Douglas fir trees can live up to 200 years. Their growth function is nonlinear and can be estimated as $y = 2,113e^{0.36a}$, where y is the timber value per hectare and a is the age of the trees. However, since Johnsonville assumes that only timber from trees over 30 years old is merchantable and it cares only about making decisions for the next four decades, the forest managers of Johnsonville are able to estimate the timber value for the trees between the ages of 30 and 100 years using a linear approximation of this relationship of $y = -3,000 + 225a$. The per hectare timber values for each of the age classes is shown in the table below where the timber value is presented in dollars for the current time period.

Age Class	Timber Value ($/hectare)
30	3,750
40	6,000
50	8,250
60	10,500
70	12,750
80	15,000
90	17,250
100	19,500

Johnsonville wants to make its timber harvest schedule based on the assumption that harvesting would be made at the beginning of each decade. They want to maximize the timber value produced by its forest over a 40-year period. Thus, there are six types of management prescriptions that could be applied. All of them are listed in Table 6.1, where x denotes the choice to harvest trees and replant the parcel at the beginning of the decade.

Applying these six management prescriptions to the two forest age classes would generate different values of merchantable timber. These values are shown in Table 6.2.

Inspection of Table 6.2 shows the best strategy for Johnsonville is simply to not allow the harvest of the trees (management prescription 1) as after the four decades they will own 12 hectares of 80-year old timber valued at $15,000 per hectare and 24 hectares of 100-year old timber valued at $19,500. Therefore, Johnsonville's forests would be worth a total of $648,000.

When the local community learned of Johnsonville's forest management plan, they were concerned as they feared that their local economy would suffer if no timber harvesting

Table 6.1   Management Prescriptions

Management Prescription	Harvest in Planning Period (decade)			
	1	2	3	4
1 – No harvesting				
2 – Harvest in 1st and 4th decade	x			x
3 – Harvest in 1st decade only	x			
4 – Harvest in 2nd decade only		x		
5 – Harvest in 3rd decade only			x	
6 – Harvest in 4th decade only				x

Table 6.2   Timber Value by Management Prescription

Management Prescription	Value per Hectare of 40 Age Class	Value per Hectare of 60 Age Class
1 – No harvesting	$15,000	$19,500
2 – Harvest in 1st and 4th decade	6,000 + 3,750 = $ 9,750	10,500 + 3,750 = $14,250
3 – Harvest in 1st decade only	6,000 + 6,000 = $12,000	10,500 + 6,000 = $16,500
4 – Harvest in 2nd decade only	8,250 + 3,750 = $12,000	12,750 + 3,750 = $16,500
5 – Harvest in 3rd decade only	$10,500	$15,000
6 – Harvest in 4th decade only	$12,750	$17,250

occurred for at least 40 years. Therefore, local political leaders negotiated a deal with Johnsonville where they agreed to allow the harvest of at least five hectares of forest in the first decade, six hectares in the second decade, seven hectares in the third decade, and eight hectares in the fourth decade. To account for these additional constraints, Johnsonville's forest managers set up the following LP model where the variables x_{ij} represent the hectares of age class i (i=1 for the original 40 Age Class and i=2 for the original 60 Age Class) when assigned to management prescription j (j=1, ..., 6). Thus the problem is:

$$\text{Max: } Z = 15,000x_{11} + 9,750x_{12} + 12,000x_{13} + 12,000x_{14} + 10,500x_{15}$$
$$+ 12,750x_{16} + 19,500x_{21} + 14,250x_{22} + 16,500x_{23} + 16,500x_{24}$$
$$+ 15,000x_{25} + 17,250x_{26} \tag{6.1}$$

s.t.:

$$x_{11} + x_{12} + x_{13} + x_{14} + x_{15} + x_{16} = 12 \tag{6.2}$$

$$x_{21} + x_{22} + x_{23} + x_{24} + x_{25} + x_{26} = 24 \tag{6.3}$$

$$x_{12} + x_{13} + x_{22} + x_{23} \geq 5 \tag{6.4}$$

$$x_{14} + x_{24} \geq 6 \tag{6.5}$$

$$x_{15} + x_{25} \geq 7 \tag{6.6}$$

$$x_{12} + x_{16} + x_{22} + x_{26} \geq 8 \tag{6.7}$$

$$x_{ij} \geq 0 \tag{6.8}$$

Model Development

Figure 6.1 shows the Excel spreadsheet for this problem. The decision variables, x_{ij}, are listed from Cells B2 to C7, and the per hectare value for each of the six management prescriptions for the each of the forest stands are listed from Cell D2 to E7. Cells B15 to E20 represent the six harvest schedules for the management prescriptions shown in Table 6.1. Constraint 1—the total hectares of forest originally of the 40 Age Class—can be depicted by using the function "=SUM(B2:B7)" in Cell B8. Constraint 2 can be similarly developed in Cell C8. These constraints ensure that the hectares of forest assigned to the six management prescriptions are equal to the total available forest land. In Solver, these constraints can be written as "B8:C8 = B10:C10". Constraints 3 through 6 can be developed by first using the equation "=SUMPRODUCT(B15:B20,$B$2:$B$7) + SUMPRODUCT(B15:B20,$C$2:$C$7)" in Cell B21. Note that the use of the $ signs in the function allows for proper cell referencing, such that the equation in Cell B21 can be copied and pasted into Cells C21 to E21. These totals can then be used to ensure that the harvested hectares are greater

	A	B	C	D	E
1	Decision Variables (x_{ij})	40 Age Class (i=1)	60 Age Class (i=2)	Value for 40 Age Class	Value for 60 Age Class
2	j=1	0	15	$ 15,000	$ 19,500
3	j=2	0	5	$ 9,750	$ 14,250
4	j=3	0	0	$ 12,000	$ 16,500
5	j=4	2	4	$ 12,000	$ 16,500
6	j=5	7	0	$ 10,500	$ 15,000
7	j=6	3	0	$ 12,750	$ 17,250
8	Total	12	24		
9		=	=		
10	Forest Hectares	12	24		
11					
12	Maximize	$ 565,500			
13					
14	Management Prescription (j)	Decade 1	Decade 2	Decade 3	Decade 4
15	1				
16	2	1			1
17	3	1			
18	4		1		
19	5			1	
20	6				1
21	Total	5	6	7	8
22		≥	≥	≥	≥
23	Harvest Requirement	5	6	7	8

Figure 6.1   Solver spreadsheet for the forest management problem.

than or equal to the minimum harvesting requirement constraints as depicted in Cells B23 to E23. Finally, the objective of maximizing profit can be written in Cell B12 as "=SUMPRODUCT (B2:B7,D2:D7)+SUMPRODUCT(C2:C7,E2:E7)".

The results of this model are shown in Figure 6.1. To meet these new minimum harvest requirements, Johnsonville will harvest five and four hectares of trees from the 60 Age Class in the first two decades, respectively. Harvest two and seven hectares of trees from the 40 Age Class for the second and third decade, respectively. In the fourth decade, Johnsonville will harvest three hectares of trees from the 40 Age Class and five hectares of trees from the original 60 Age Class. The remaining 15 hectares from the 60 Age Class will be not be harvested during the next four decades (management prescription 1). The resulting profit for Johnsonville is $565,500, which is a loss of $82,500 (−14.6%) in comparison to the problem without the minimum harvest constraints.

This forest management problem is intended to provide the reader with a basic understanding about how this type of forestry problems can be modeled using LP. This model could be readily extended in a number of areas, such as having more frequent harvest choices, permitting different species to be planted after the harvest, accounting for variability in the growth rate depending upon the soil and location of the trees, including ending inventory constraints, and accounting for varying regulatory environments.

6.2   LAND USE PLANNING

Land use planning seeks to develop efficient and ethical ways to regulate and manage the allocation of land for a variety of competing uses. The most common application of LP to such decisions is a type of allocation (or assignment) problem in which a finite land resource must be divided among various uses. The example examined here is a modified version of a model developed by Dykstra (1984) that demonstrates the capabilities and limitations of LP when applied to land use planning.

This problem concerns the planning of land use for a small town that is confronted with challenges about how to both enhance its open space and parkland amenities and maintain a reasonable pace of economic development to enhance its residents' welfare, given an increasing population. Recent increases in the demand for residential development and municipal services along with a reduction in local open space have led to a strong interest in land use planning for the town's remaining undeveloped property. A special planning committee has been appointed to manage and coordinate the project as a series of general directives from the city council has been identified:

1. The prime purpose of the land use plan is to increase the property tax base as much as possible.

2. The property tax rate should not increase above its present level of 3%.

3. At least 30% of currently undeveloped fields should remain undeveloped (reserved land) for potential use by future generations.

4. In addition to the reserved areas, every acre assigned for development (defined as residential, commercial, or industrial) should be offset by at least one-third of an acre of open space land (defined as farm or park land).

5. A minimum of one-tenth of an acre of commercial land and one-sixteenth of an acre of industrial land should be developed for every acre of residential land to provide basic municipal and other services.

6. Due to concerns about excessive development, while seeking the objective of maximizing the property tax base, the city council expects that the total amount of land developed for commercial and industrial purposes will not provide more goods and services than demanded by the total increase in the local population.

7. At least two-fifths of an acre of forest land should be preserved for every acre of residential land developed.

8. Finally, the council wants this plan to account for the recreational needs of a growing population.

The planning committee includes not only government officials but also experts in demography, ecology, sociology, forestry, and economics. They have held a series of public meetings with current residents and have discussed the findings and their plans with the city council. To facilitate this process, an LP model is applied that integrates the available data and constraints into an acceptable plan for land allocation for the next decade.

Table 6.3 shows five undeveloped areas within the town and the number of acres within each area that are available for various kinds of uses. The total area of these five parcels is 9,240 acres, and the total area available for other land uses ranges from 3,850 acres for farming to 5,780 acres for industrial use. Note that the columns in this table do not need to sum to the total acreage for the undeveloped parcel because some areas are suitable for

Table 6.3   Acres Suitable for Each Proposed Land Use

Undeveloped Area (i)	Total Acreage (b$_i$)	Type of Land Use					
		Residential (j=1)	Commercial (j=2)	Industrial (j=3)	Forest (j=4)	Farm (j=5)	Park (j=6)
1	2,050	1,435	990	1,220	750	790	1,120
2	3,120	1,710	1,850	2,530	1,335	1,250	370
3	820	250	450	500	280	80	560
4	1,330	520	490	610	1,110	810	960
5	1,920	1,080	700	920	690	920	1,090
Total	**9,240**	**4,995**	**4,480**	**5,780**	**4,165**	**3,850**	**4,100**

Table 6.4   Expected Increase in the Property Tax Base and in the Cost of Municipal Services for Each Acre Allocated to Development (in $'000)

	Type of Land Use					
	Residential (j=1)	Commercial (j=2)	Industrial (j=3)	Forest (j=4)	Farm (j=5)	Park (j=6)
Increase in the Property Tax Base per Acre (c$_j$)	200	520	1,070	300	11.5	5
Cost of Municipal Services per Acre (d$_j$)	9	14	33.5	6.5	0.8	2

multiple types of development. Table 6.4 shows how much each kind of development would increase the city's property tax base and how much it would cost the city to provide municipal services.

Decision Variables and Constraints

In this model, the key decision is the allocation of land from undeveloped areas to one of six uses. The objective of the LP problem is to maximize the increase in the town's property tax base, subject to the policy restrictions and other limitations set forth. Therefore, the decision variables should measure the acreage from each proposed development area allocated to each land use type. These variables are defined as follows:

$$x_{ij} = \text{acres of area i allocated to use j} \quad i=1,\ldots,5; j=1,\ldots,7 \qquad (6.9)$$

where j = 1 for residential development, 2 for commercial development, 3 for industrial development, 4 for forests, 5 for farms, 6 for park, and 7 for reserved land.

To maximize the expected increase in tax base from all the new developed areas, the objective function is written as:

$$\text{Max: } Z = \sum_{j=1}^{6}\sum_{i=1}^{5} c_j x_{ij} \qquad (6.10)$$

where c_j are the increases in the property tax base per acre found in the first row of Table 6.4.

Constraint on the Property Tax Rate   The property tax rate is calculated from the amount of property tax required (i.e., the cost of new municipal services related to the development) minus the property tax base resulting from the development. The city council wants to ensure that the property tax rate will not be greater than the current 3%. This requires a constraint formulated as:

$$\sum_{j=1}^{6}\sum_{i=1}^{5}d_j x_{ij} - 0.03\sum_{j=1}^{6}\sum_{i=1}^{5}c_j x_{ij} \leq 0 \tag{6.11}$$

where d_j are values from the second row of Table 6.4, reflecting the estimated cost of municipal services required for each acre of new development.

Constraint on Excessive Development   Another important goal is to control for an "excessive" increase in commercial and industrial land. The planning commission suggests that the increase in acres of commercial and industrial land should not exceed the possible demand for the goods and services provided by those uses from the increased local population at the end of the tenth year.

The present population of the town is 32,753. Demographers estimate that the annual birth and death rates will be relatively stable for the next ten years, 0.8% and 0.5% respectively. Thus, the model assumes that the total population of the town will increase by 0.3% (0.8% − 0.5%) by the end of each year. Additionally, demographers estimate that there will be net immigration of 1,200 people every year. For the sake of simplicity, this model assumes that the immigration occurs at the end of each year. Based on this information, the town's total population by the end of the tenth year can be estimated using the following equation:

$$p_{i+1} = 1.003(p_i + 1200) \tag{6.12}$$

where p_i represents total population of the town at the beginning of the ith year. In this way, by the end of the tenth year, the estimated increase in the total population is 13,196. Experts estimate that, on average, every acre of new commercial land can service 11 new residents and every acre of new industrial land can support 16 new residents. Thus, the constraints for these types of land use can be set up as follows:

$$11\sum_{i=1}^{5}x_{i2} \leq 13,196, \quad \text{and} \tag{6.13}$$

$$16\sum_{i=1}^{5}x_{i3} \leq 13,196 \tag{6.14}$$

Minimum Development Rate   The plan should also consider the areas of residential, commercial, and industrial land necessary to assure the availability of adequate living, shopping, and other service facilities for new residents. Experts estimate that one acre of residential development can accommodate at most 22 people. In addition, at least one-tenth of an acre of commercial land and one-sixteenth of an acre of industrial land should be developed for every acre of residential land. These constraints can be written as follows:

$$22\sum_{i=1}^{5} x_{i1} \geq 13{,}196 \tag{6.15}$$

$$\frac{1}{10}\sum_{i=1}^{5} x_{i1} - \sum_{i=1}^{5} x_{i2} \leq 0, \text{ and} \tag{6.16}$$

$$\frac{1}{16}\sum_{i=1}^{5} x_{i1} - \sum_{i=1}^{5} x_{i3} \leq 0 \tag{6.17}$$

Open Land Requirement   Residents were concerned that the increase in development would diminish the scenic beauty of the town. Therefore, the council decided that the total acreage of farms and parks should be at least one-third as much as the amount of land allocated to residential, commercial, and industrial uses:

$$\sum_{i=1}^{5} x_{i5} + \sum_{i=1}^{5} x_{i6} - \frac{1}{3}\left(\sum_{i=1}^{5} x_{i1} + \sum_{i=1}^{5} x_{i2} + \sum_{i=1}^{5} x_{i3}\right) \geq 0 \tag{6.18}$$

Forest Land Requirement   Given the increasing population, the city council thought it was necessary to ensure sufficient forested land. To simplify this constraint, the model assumes that there should be at least two-fifths of an acre of forest for every acre of residential land:

$$\frac{2}{5}\sum_{i=1}^{5} x_{i1} - \sum_{i=1}^{5} x_{i4} \leq 0 \tag{6.19}$$

Recreation Requirement   The city council also wanted to increase the amount of park land. Therefore, they set the goal that by the end of the ten-year planning period, there should be at least one new acre of park land for recreation for every 50 new residents:

$$50\sum_{i=1}^{5} x_{i6} \geq 13{,}196 \tag{6.20}$$

Environmental Index   The ecologists involved with this project wanted to test an environmental measurement index (referred to as E-Index) that involves a 10-point grading system to evaluate the environmental quality provided by different land uses. Generally speaking, the more intensively a place is developed, the lower its score. Table 6.5 shows the scores for each of the different land uses considered.

Table 6.5   Environmental Benefit Score for Each Land Use

	Industrial ($j=3$)	Commercial ($j=2$)	Residential ($j=1$)	Farm ($j=5$)	Forest ($j=4$)	Park ($j=6$)	Reserved ($j=7$)
E-Index	1.5	2.5	3	4.5	6.5	8.5	10

To maintain a high-quality environment, the land planning commission proposed that the average E-Index for these five undeveloped areas should be no less than 6.0. The average E-Index is calculated from the weighted average of the seven different types of land by acreage. This restriction is written as:

$$\frac{1}{9,240}\left(3\sum_{i=1}^{5}x_{i1} + 2.5\sum_{i=1}^{5}x_{i2} + 1.5\sum_{i=1}^{5}x_{i3} + 6.5\sum_{i=1}^{5}x_{i4} + 4.5\sum_{i=1}^{5}x_{i5}\right.$$

$$\left. + 8.5\sum_{i=1}^{5}x_{i6} + 10\sum_{i=1}^{5}x_{i7}\right) \geq 6 \tag{6.21}$$

where 9,240 is the total acres of undeveloped land.

Development Feasibility Constraints   The first condition of feasibility is that all of the development, including land reserved, should not exceed the total amount of available undeveloped land:

$$\sum_{j=1}^{7}\sum_{i=1}^{5}x_{ij} \leq 9,240 \tag{6.22}$$

As directed by the city council, at least 30% of the undeveloped land should be reserved for future plans:

$$\sum_{j=1}^{6}\sum_{i=1}^{5}x_{ij} \leq (0.7)(9,240) \tag{6.23}$$

Additionally, constraints are required to ensure that for each of the five undeveloped areas, the sum of the acres of each type of land use does not exceed the upper bound (UB) of the suitable acres for that use:

$$x_{ij} \leq s_{ij} \quad i = 1, \ldots, 5; j = 1, \ldots, 7 \tag{6.24}$$

where s_{ij} is the number of acres in area i suitable for use j as given in Table 6.3.

Finally, the land allocated to all uses within each proposed development area cannot exceed the total available land for development:

$$\sum_{j=1}^{7}x_{ij} \leq b_{i} \qquad i = 1, \ldots, 5 \tag{6.25}$$

where b_i is the total number of acres in area i. From Table 6.3, $b_1 = 2,050$, $b_2 = 3,120$, $b_3 = 820$, $b_4 = 1,330$, and $b_5 = 1,920$.

Non-negativity Constraint   Since all the decision variables represent allocated acreage, all decision variables should be non-negative:

$$x_{ij} \geq 0 \qquad i = 1, \ldots, 5; j = 1, \ldots, 7 \tag{6.26}$$

Problem Set-Up and Finding a Solution

To solve this LP problem with Solver, four categories of information must be defined: the given information, the objective function, the decision variables, and the constraints.[2] The data listed in Tables 6.3, 6.4, and 6.5 are shown in the spreadsheet in Cells A2 through I10 of Figure 6.2. The formula for the objective function (Cell B13) is "=SUMPRODUCT(C8:H8,C22:H22)." As discussed in Chapter 3, when setting up this problem, it can be useful to enter a simple "guess" for the initial value. This helps to ensure that the formulas used in the model are functioning correctly and can help to identify any problems.

Cells C17 to I21 represent the decision variables (e.g., Cell D18 represents the acreage from undeveloped area 2 that will be allocated to commercial land use). Solver will recognize each cell in this area as an independent variable during the optimization process,

	A	B	C	D	E	F	G	H	I
1									
2	Undeveloped Area	Total	Residential	Commercial	Industrial	Forest	Farm	Park	Reserved
3	1	2,050	1,435	990	1,220	750	790	1,120	
4	2	3,120	1,710	1,850	2,530	1,335	1,250	370	
5	3	820	250	450	500	280	80	560	
6	4	1,330	520	490	610	1,110	810	960	
7	5	1,920	1,080	700	920	690	920	1,090	
8		C_i	200.0	520.0	1,070.0	300.0	11.5	5.0	
9		D_i	9.0	14.0	33.5	6.5	0.8	2.0	
10		E-Index	3.0	2.5	1.5	6.5	4.5	8.5	10.0
11									
12	**Objective**								
13	Max.	$ 2,525,344	Property Tax Base (1,000s)						
14									
15	**Decision Variables (acres)**								
16	Undeveloped Area	Total	Residential	Commercial	Industrial	Forest	Farm	Park	Reserved
17	1	2,050.0	-	-	-	-	-	263.9	1,786.1
18	2	3,120.0	509.8	124.4	-	889.1	610.7	-	985.9
19	3	820.0	90.0	450.0	-	280.0	-	-	-
20	4	1,330.0	-	220.0	-	1,110.0	-	-	-
21	5	1,920.0	-	405.3	824.8	690.0	-	-	-
22	Total	9,240.0	599.8	1,199.6	824.8	2,969.1	610.7	263.9	2,772.0
23									
24	**Constraints**								
25	Property Tax Rate	70,138	<=	75,760	Tax Rate:	2.78%			
26	Excessive Dev.	13,196	<=	13,196					
27		13,196	<=	13,196					
28	Min Dev. Rate	60	<=	1,200					
29		13,196	<=	13,196					
30		37	<=	825					
31	Open Land	875	<=	875					
32	Forest Land	240	<=	2,969					
33	Park land	13,196	<=	13,196					
34	E-Index	55,440	<=	58,047	Avg. E-index	6.28			
35	Feasibility	9,240	=	9,240					
36		6,468	<=	6,468					
37			$x_{ij} <= s_{ij}$						
38	1	2,050	<=	2,050					
39	2	3,120	<=	3,120					
40	3	820	<=	820					
41	4	1,330	<=	1,330					
42	5	1,920	<=	1,920					

Figure 6.2   Land use problem model.

---

[2]This problem set-up and solution are provided in the supplemental materials for Chapter 6 available at www.wiley.com/college/kaiser.

and by changing the values of these cells, will provide an optimal solution given the constraints and information provided in the problem.

The formulas that comprise the series of constraints are shown in the bottom part of Figure 6.2. For example, the first Excessive Development constraint (equation (6.13)) is in Cell B26. The formula is "=11*SUM(E17:E21)" and the associated right-hand-side (RHS) value for this constraint is in Cell D26.

Once all of the given information has been entered into the spreadsheet and the objective function and constraints have been properly set up, the problem is ready to be solved. Click the "Model" tab in Solver and select the option to maximize the property tax base (Cell B13). Continue by defining the decision variables and the constraints, and be especially careful to get the signs of the constraints correct. The completed model should look like Figure 6.3. Finally, the appropriate algorithm must be selected. In this case, the "Standard LP/Quadratic Engine" should be selected, and the "Assume Non-Negative" option should be set as "True."

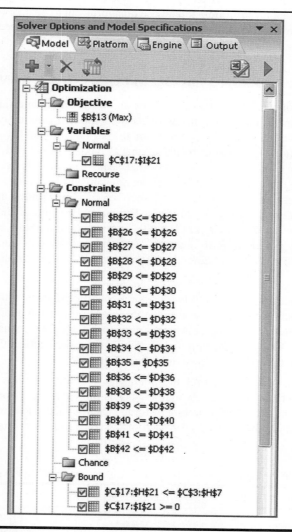

Figure 6.3 Land use model set-up.

Table 6.6 Optimal Allocation of Undeveloped Land

Undeveloped Area	Residential	Commercial	Industrial	Forest	Farm	Park	Reserved
1						263.9	1,786.1
2	509.8	124.4		889.1	610.7		985.9
3	90.0	450.0		280.0			
4		220.0		1,110.0			
5		405.3	824.8	690.0			
Total	599.8	1,199.6	824.8	2,969.1	610.7	263.9	2,772.0

The solution to this problem is shown in Table 6.6. Given this allocation plan, the total tax base will increase by $2,525,344, in thousands (Figure 6.2). The planning committee can also obtain some other important information from this model. For instance, the actual property tax rate will be 2.78% (Cell F25), which is lower than the city council's goal of 3.0%, and the E-Index for this optimal land use plan turns out to be 6.28 (Cell F34), which is higher than the recommended level of 6.0, meaning that the land use plan exceeds the expectations in this regard.

6.3 OPTIMAL STOCKING PROBLEM FOR A GAME RANCH

The management of wildlife in public reserves and on private game ranches often must account for the various competing needs of its inhabitants. In the case of private game ranches, owners seek to make a profit from the sustained viability of these natural resources. The following example builds upon LP models developed by Davis (1967) and Dykstra (1984), and uses a case study that seeks to assist Heart's Bluff Ranch in managing its three most popular wild game: deer, bison, and wild hogs.

Located in eastern Texas, Heart's Bluff Ranch offers game hunting and various recreational opportunities on nearly 5,000 acres. Based on 10 years of data on population trends for the game on the ranch, an LP model is developed to help Heart's Bluff better manage its game populations to increase customer satisfaction and, ultimately, to generate more profit. Several things are known about the ranch and its game farm business:

1. The ranch's land is divided into two primary conditions: areas with no special management and small cleared areas with higher forage production quantities. The forage production differs depending on whether and when the land is managed. Unmanaged lands result in forage production of only 5 kilograms per hectare. Cleared lands produce more but this effect declines over time (Table 6.7).

2. The ranch hires labor to patrol the land, check hunter permits, weigh harvested animals, and create clearings. Available labor is limited to 1,300 person-days during the busy hunting season. No hunting or forage clearing activities are done in the winter. The cost and labor required for each of the three main activities are provided in Table 6.8.

Table 6.7 Quantity of Forage for the Years after Treatment by Land Type (kg per hectare)

	1 Year	2 Years	3 Years	4 Years	5 Years
Unmanaged Land	5	5	5	5	5
Cleared Land	150	110	100	50	25

Table 6.8   Cost and Labor Requirements by Activity

	Cost ($)	Labor (person-days)
Game Clearing (1 hectare)	$50	3.00
Patrolling, Weighing, and Permit Checking (per animal harvested)	$30	2.67
Forage Harvest for Sale (kg)	$2	0.3

3. Historical data shows that about 78% of the wild hogs, 72% of the deer, and 80% of the bison can survive the winter if those animals successfully evade hunters during previous hunting season. All animals that survive winter are assumed to be available for harvest during the subsequent hunting season.

4. Fertility studies show that, on average, each surviving wild hog will lead to the birth of 0.4 live wild hogs the next year. Likewise, each surviving deer will lead to the birth of 0.3 live deer, and each surviving bison will lead to the birth of 0.4 live bison. For hunting purposes, it is assumed that hunters will not harvest the newborns because their smaller size makes them easy to identify as immature and, therefore, not permitted to be hunted. For each of the three animals, the newborns that survive the winter will reach sexual maturity by the following year.

5. Sufficient forage should be provided for each animal to survive, even though some will perish in winter. Forage requirements annually are estimated to be 200 kg per wild hog, 170 kg per deer, and 300 kg per bison.

6. Presently the ranch keeps 90 wild hogs, 150 deer, and 40 bison. To maintain viable populations and to ensure the scenic beauty of the ranch that can generate nonhunting revenue, the population of each species on the ranch has to be kept at or above a minimum number.[3] The wild hog population should be at least 20, the deer population should be at least 60, and the bison population should be at least 30.

7. Based on previous years, the ranch estimates that hunters can harvest up to 20 wild hogs, 30 deer, and 10 bison annually.

8. The price for harvesting a wild hog is $300, the price for a deer is $250, and the price for a bison is $350. Extra forage can be sold for $3 per kilogram.

Problem Formulation

The objective for this LP problem is to maximize profit over a five-year time horizon. The formulation of the problem is as follows:

$$\text{Max: } Z = \sum_{t=1}^{5} (c_f f_t + c_w w_{ht} + c_d d_{ht} + c_b b_{ht}) - \sum_{t=1}^{5} (50 l_{2t} + 30 w_{ht} + 30 d_{ht} + 30 b_{ht} + 2 f_t) \quad (6.27)$$

where the decision variables for each of the five years, t, are given as follows:

l_{1t} = hectares of unmanaged land

l_{2t} = hectares of land cleared for game

---

[3]Other than this constraint to ensure scenic beauty, this problem does not consider the nonhunting revenue generated by ranch guests.

f_t = kilograms of extra forage sold

w_{ht} = number of wild hogs harvested

d_{ht} = number of deer harvested

b_{ht} = number of bison harvested

w_{st} = number of wild hogs surviving

d_{st} = number of deer surviving

b_{st} = number of bison surviving

The revenue and selling prices in equation (6.27) are as follows:

c_f = revenue of forage sold per kilogram

c_w = selling price for a wild hog

c_d = selling price for a deer

c_b = selling price for a bison

Labor Constraint   The labor requirements for game clearings, hunting patrols, and forage harvests given in Table 6.8 are represented by:

$$3l_{2t} + 0.3f_t + 2.67\,(w_{ht} + d_{ht} + b_{ht}) \leq 1{,}300, \text{t} = 1, \ldots, 5 \qquad (6.28)$$

Population Constraints   The sum of the number of animals harvested and surviving in a particular year should be equal to the number of animals at the start of the year. For the first year, the constraint would be:

$$\text{Wild hogs: } w_{h1} + w_{s1} = 90 \qquad (6.29)$$

$$\text{Deer: } d_{h1} + d_{s1} = 150 \qquad (6.30)$$

$$\text{Bison: } b_{h1} + b_{s1} = 40 \qquad (6.31)$$

Additionally, the current number of animals (whether harvested in this year's hunting season or not) should be equal to the sum of the number of that species that survived the previous winter and the number of newborns that survived the winter and thus matured. For example, for wild hogs in year t, the constraint would be:

$$w_{ht} + w_{st} = (1.4)(0.78)\,w_{s,t-1} \quad t = 2, \ldots, 5 \qquad (6.32)$$

Similarly, the herd constraints for deer and bison would be:

$$d_{ht} + d_{st} - (1.3)(0.72)\,d_{s,t-1} = 0 \quad t = 2, \ldots, 5 \qquad (6.33)$$

$$b_{ht} + b_{st} - (1.4)(0.8)\,b_{s,t-1} = 0 \quad t = 2, \ldots, 5 \qquad (6.34)$$

Forage Production and Requirements   The three sources of forage are from (1) unmanaged lands, (2) land cleared in the current year, and (3) land cleared in the previous years. The forage production in year t must meet the demand of the animals that survived the hunting season. Extra forage can be sold.

The amount of forage produced from unmanaged land is given by $5l_{1t}$ (Table 6.7). Production declines with time on previously cleared lands. Forage produced from land cleared in the current year is given by $150l_{2t}$. For land cleared during the preceding year, the amount is $110l_{2,t-1}$, for land cleared two years prior it is $100l_{2,t-2}$, and so forth. Note that an assumption of this model is that all land in t = 0 is unmanaged land, so all land in t = 1 is either newly cleared or still unmanaged.

The quantity of forage provided in year t should be adequate to meet the herd's requirements. In winter, the total number of surviving wild hogs and wild hog infants would sum up to $1.4w_{st}$, which would require an amount of $200 \times 1.4w_{st}$ forage. All deer would need $170 \times 1.3d_{st}$ forage to survive winter, and bison would need $300 \times 1.4b_{st}$ forage. For year t = 5, the constraint could be constructed in the following form where the left-hand-side (LHS) presents the forage produced on the various types of land, and the RHS accounts for the forage needed for the three species and the amount of forage sold:

$$5l_{1t} + 150l_{2t} + 110l_{2,t-1} + 100l_{2,t-2} + 50l_{2,t-3} + 25l_{2,t-4} - (200)(1.4)w_{st}$$
$$- (170)(1.3)d_{st} - (300)(1.4)b_{st} - f_t = 0 \tag{6.35}$$

Genetic Diversity and Scenic View Sustainability As previously discussed, the managers of Heart's Bluff Ranch require that a minimum number of animals of each species be present at the ranch at any given time to ensure genetic diversity and scenic view sustainability:

$$0.78 \times 1.4w_{st} \geq 20 \quad t = 1, \ldots, 5 \tag{6.36}$$

$$0.72 \times 1.3d_{st} \geq 60 \quad t = 1, \ldots, 5 \tag{6.37}$$

$$0.8 \times 1.4b_{st} \geq 30 \quad t = 1, \ldots, 5 \tag{6.38}$$

Market Limitation Ranch managers estimated that hunters can harvest a maximum of 20 wild hogs, 30 deer and 10 bison annually:

$$w_{ht} \leq 20 \quad t = 1, \ldots, 5 \tag{6.39}$$

$$d_{ht} \leq 30 \quad t = 1, \ldots, 5 \tag{6.40}$$

$$b_{ht} \leq 10 \quad t = 1, \ldots, 5 \tag{6.41}$$

Land Availability Constraints are also needed to account for the limitations on the size of the ranch:

$$l_{11} + l_{21} = 2,000, \tag{6.42}$$

$$\sum_{j=1}^{t-1} l_{2j} + \sum_{i=1}^{2} l_{it} = 2,000 \quad t = 2, \ldots, 5 \tag{6.43}$$

For years 1 through 5, these constraints can be written as:

$$l_{11} + l_{21} \qquad\qquad\qquad = 2,000 \text{ (year 1)} \tag{6.44}$$

$$l_{12} + l_{21} + l_{22} \qquad\qquad = 2,000 \text{ (year 2)} \tag{6.45}$$

$$l_{13} + l_{21} + l_{22} + l_{23} \qquad = 2,000 \text{ (year 3)} \tag{6.46}$$

$$l_{14} + l_{21} + l_{22} + l_{23} + l_{24} \quad = 2,000 \text{ (year 4)} \tag{6.47}$$

$$l_{15} + l_{21} + l_{22} + l_{23} + l_{24} + l_{25} = 2,000 \text{ (year 5)} \tag{6.48}$$

Non-negativity Constraints   Finally, the decision variables should be non-negative.

Problem Set and Interpretation of the Solution

The model specification shows that the objective is to maximize profit (Cell B23) and that the decision variables are in Cells B14 to F22 (Figure 6.4). Figure 6.5 shows the references for the various constraints. The Standard LP/Quadratic Engine should be selected, and the "Assume Non-Negative" option should be set as "True".

The optimal solution will yield Heart's Bluff Ranch a total profit of $43,727, as can be seen in Table 6.9. The optimal solution would gradually reduce the animal populations to the minimum number required for genetic and scenic purposes. Consequently, after the fifth year, the ranch would need to permit limited hunting and focus its efforts only on managing the land to meet the needs of the animals and of selling forage.

	A	B	C	D	E	F	G
1		Population	Survival Rate	Selling Price	Growth Rate	Requirement	
2	Wild hogs	90	0.78	300	1.4	200	
3	Deer	150	0.72	250	1.3	170	
4	Bison	40	0.8	350	1.4	300	
5		Cost	Labor (person-days)	Selling Price			
6	Land clearing (per ha)	50	3				
7	Cost per animal harvested	30	2.67				
8	Forage harvest (per kg)	2	0.3	3			
9	Winter forage produced	Land Type 1	Land Type 2	Land Type 3	Land Type 4	Land Type 5	
10	Unmanaged land (kg/ha)	5	5	5	5	5	
11	Cleared land (kg/ha)	150	110	100	50	25	
12							
13	*Decision Variables*	Year 1	Year 2	Year 3	Year 4	Year 5	
14	1. Unmanaged land	1,661.2	1,638.6	1,638.6	1,572.6	1,507.8	
15	2. Cleared land	338.8	22.6	-	66.0	64.8	
16	3. Forage sold	411.0	3,602.1	4,121.1	3,434.7	3,571.9	
17	4. Wild hogs harvested	20.0	20.0	20.0	20.0	9.5	
18	5. Deer harvested	30.0	30.0	3.9	-	-	
19	6. Bison harvested	10.0	6.8	-	6.8	3.2	
20	7. Wild hogs surviving	70.0	56.4	41.7	25.5	18.3	
21	8. Deer surviving	120.0	82.3	73.2	68.5	64.1	
22	9. Bison surviving	30.0	26.8	30.0	26.8	26.8	
23	Profit (max)	37,339.8					
24	Herd Identity	Wild hogs(left = right)		Deer(left=right)		Bison(left=right)	
25	Year 1	90.0	90.0	150.0	150.0	40.0	40.0
26	Year 2	76.4	76.4	112.3	112.3	33.6	33.6
27	Year 3	61.6	61.6	77.1	77.1	30.0	30.0
28	Year 4	45.5	45.5	68.5	68.5	33.6	33.6
29	Year 5	27.8	27.8	64.1	64.1	30.0	30.0
30	Forest Production	Year 1	Year 2	Year 3	Year 4	Year 5	
31	Actual	59,131.0	48,848.1	44,557.5	36,959.6	34,116.8	
32	Requirment	59,131.0	48,848.1	44,557.5	36,959.6	34,116.8	Requirement
33	Land Used	2,000.0	2,000.0	2,000.0	2,000.0	2,000.0	2,000
34	Labor Used	1,300.0	1,300.0	1,300.0	1,300.0	1,300.0	1,300
35	*Scenic View sustainability*						
36	Wild hogs	76.4	61.6	45.5	27.8	20.0	20
37	Deer	112.3	77.1	68.5	64.1	60.0	60
38	Bison	33.6	30.0	33.6	30.0	30.0	30
39	*Max Hunting*						
40	Wild hogs	20.0	20.0	20.0	20.0	9.5	20
41	Deer	30.0	30.0	3.9	-	-	30
42	Bison	10.0	6.8	-	6.8	3.2	10

Figure 6.4   Game farm model set-up and solution.

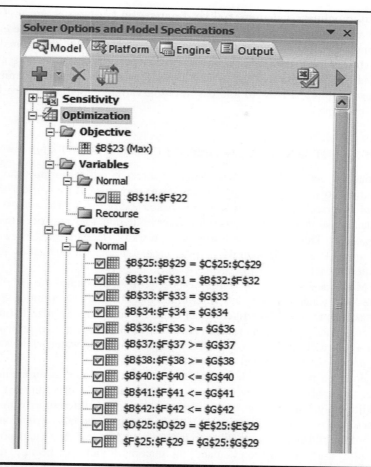

Figure 6.5   Game farm constraints.

Table 6.9   Game Ranch, Optimal Results

	Unmanaged Area (hectares)	Cleared Area (hectares)	Harvested			Survived			Forage Sold (kg)
			Wild Hogs	Deer	Bison	Wild Hogs	Deer	Bison	
Year 1	1661.2	338.8	20	30	10	70	120	30	411
Year 2	1638.6	22.6	20	30	6.8	56.4	82.3	26.8	3602.1
Year 3	1638.6	0	20	3.9	0	41.7	73.1	30	4121.1
Year 4	1572.6	66	20	0	6.8	25.5	68.5	26.8	3434.7
Year 5	1507.8	64.8	9.5	0	3.2	18.3	64.1	26.8	3571.9

Although this solution could make sense given that Heart's Bluff Ranch can also earn revenue from guests interested in nonhunting recreational activities who are drawn to the ranch as an attractive destination, this solution also arises from the model's assumptions. For example, the population growth functions were assumed to be linear.

Dealing with nonlinear functions will be discussed in Chapters 8 and 9. This model lacks stochastic components that could have been added into many of the constraints, such as winter survivability rates, the selling price of forage and the animals, and the growth of forage in a particular year. Finally, inspection of the results also shows the unrealistic assumption that the model allows for the harvesting of animals and the birth of partial animals. How to address decision variables that should be integers is discussed in Chapter 7.

6.4   EFFICIENT IRRIGATION AND CROPPING PATTERNS

Effective management of limited fresh water resources is critical to meet the needs of a growing human population and to support healthy ecosystems. As the world's largest water user, irrigation agriculture will increasingly be asked and/or required to use less water than it would ideally use. As farmers account for water limitation, they may need to adjust their cropping patterns. This example builds upon the work of Haouari and Azaiez (2001) and illustrates how LP can help identify optimal crop rotations and irrigation levels given various water restrictions.

Farmer Madeline Johnson owns a 100-hectare farm. At the end of the summer growing season, the local water authority has determined that she can use no more than 200,000 cubic meters (m^3) of water for irrigating her crops in the upcoming year. Farmer Johnson wants to develop a cropping plan for the entire year (winter and summer growing seasons) for her farm that will maximize her profits given the available water, various crop yields and water demands, and expected market conditions.

Farmer Johnson has several choices of cropping patterns: rice as an annual crop (a), wheat (w_1) and maize (w_2) as winter crops, and wheat (s_1) and cotton (s_2) as summer crops. Additionally, Farmer Johnson can choose one of three different irrigation levels for each of these crops: 100% irrigation (full), 80% irrigation, and 60% irrigation. When the crops are not fully irrigated, the yield from the crops will be less than the maximum yield. Data related to the full irrigation demand, the maximum yield per hectare, the expected profit per ton, and ratios of actual-to-maximum yield under the three different irrigation levels are shown for each of the crops in Tables 6.10 and 6.11.

Farmer Johnson understands that with crop rotations, the expected yield from a particular crop in the current growing season is dependent upon which crop was cultivated on that land the previous season. As shown in Table 6.12, discounting factors can express the impact that the cultivation of last season's crop will have on the yield of the current cultivated crop.

In the previous growing season, Farmer Johnson allocated 30 hectares for rice, 20 hectares for summer wheat, and 30 hectares for cotton. Additionally, she had 20 hectares of land enrolled in a conservation program; therefore, no crops were actively cultivated on those 20 hectares. This land will be available for cultivation in the subsequent growing season.

Table 6.10   Water Demand, Maximum Yield, and Profit by Crop

	Water Demand (m^3/hectare)	Maximal Yield (tons/hectare)	Profit ($/ton)
Rice (a)	1,200	7	180
Wheat (w_1)	800	6	150
Maize (w_2)	1,200	9	220
Wheat (s_1)	1,300	10	170
Cotton (s_2)	1,600	8	300

Table 6.11   Maximum and Actual Yield by Irrigation Percentage by Crop

	Maximum Yield by Irrigation Level			Actual Yield (tons/acre) by Irrigation Level		
	100%	80%	60%	100%	80%	60%
Rice (a)	100%	88%	72%	7.00	6.16	5.04
Wheat (w_1)	100%	85%	71%	6.00	5.10	4.26
Maize (w_2)	100%	56%	24%	9.00	5.04	2.16
Wheat (s_1)	100%	83%	73%	10.00	8.30	7.30
Cotton (s_2)	100%	69%	51%	8.00	5.52	4.08

Table 6.12   Discounting Factors on Yield with Respect to Crop Predecessors

	Crop from Last Year					
Crop for Current Year	None	Rice (a)	Wheat (s)	Maize (s)	Wheat (s)	Cotton (s)
Rice (a)	1.0	0.8	–	–	0.8	0.7
Wheat (w_1)	1.0	1.0	–	–	1.0	0.9
Maize (w_2)	0.9	1.0	–	–	0.9	1.0
Wheat (s_1)	1.0	–	0.8	0.9	–	–
Cotton (s_2)	1.0	–	1.0	0.6	–	–

Model Development

The objective of this problem is to select a crop pattern that maximizes Farmer Johnson's overall profit from her farm. Therefore, the decision variables will be related to the following three questions:

1. Which crops and what kinds of cropping patterns should be chosen for next year in order to get maximum profit?

2. For all of these selected crops, how much land should be allocated for each crop?

3. How much water should be allocated to each crop under the limitation of the water available?

Farmer Johnson's objective function can be expressed as:

$$\text{Max: } Z = \sum_{i=1}^{1}\sum_{j=0}^{3} a_{ij}x_{ij} + \sum_{i=1}^{2}\sum_{j=0}^{3} w_{ij}y_{ij} + \sum_{i=1}^{2}\sum_{j=0}^{2} s_{ij}z_{ij} \qquad (6.49)$$

Therefore, the profit from this year's crops, given the land use of the previous year, is derived using the following variables:

a_{ij} = profit per hectare for cultivating annual crop i (rice) on land that was planted last season with crop j

w_{ij} = profit per hectare for cultivating winter crop i (winter wheat or maize) on land that was planted last season with crop j

s_{ij} = profit per hectare for cultivating summer crop i (summer wheat or cotton) on land that was planted last season with crop j

x_{ij} = amount of land that was cultivated last season with crop j and is to be cultivated with crop i for all of next year

y_{ij} = amount of land that was cultivated last season with crop j and is to be cultivated with crop i for next winter

z_{ij} = amount of land that was cultivated last season with crop j and is to be cultivated with crop i for next summer

Land usage constraints   The land used in winter and summer should both be less than or equal to the 100 hectares of available land:

$$\sum_{i=1}^{1}\sum_{j=0}^{3} x_{ij} + \sum_{i=1}^{2}\sum_{j=0}^{3} y_{ij} \leq 100 \tag{6.50}$$

$$\sum_{i=1}^{1}\sum_{j=0}^{3} x_{ij} + \sum_{i=1}^{2}\sum_{j=0}^{2} z_{ij} \leq 100 \tag{6.51}$$

Other constraints need to be included to reflect the land availability for each type of crop given the previous land use. For instance, the following constraint accounts for the 20 hectares of land that is enrolled in a conservation program and is currently available for planting as either an annual or winter crop:

$$\sum_{i=1}^{1} x_{i0} + \sum_{i=1}^{2} y_{i0} + \sum_{i=1}^{2} z_{i0} \leq 20 \tag{6.52}$$

where x_{i0} is the land that was not cultivated with any crops last year and could be planted with crop i next year, and y_{i0} is the land that was not cultivated with any crops last year and could be planted with crop i next winter. Similarly, z_{i0} can be defined as the land that was not cultivated with any crops last year and would be planted with i next summer.

The other constraints on the cropping pattern are as follows:

$$x_{11} + \sum_{i=1}^{2} y_{i1} \leq 30 \tag{6.53}$$

$$\sum_{i=1}^{2} z_{id} - \sum_{j=1}^{2} y_{dj} \leq 0 \tag{6.54}$$

for every crop d planted in winter, and

$$\sum_{i=1}^{3}\sum_{j=1}^{1} x_{ij} + \sum_{i=1}^{3}\sum_{j=1}^{2} y_{ij} + \sum_{i=1}^{2}\sum_{j=1}^{2} z_{ij} \leq 100 \tag{6.55}$$

The first constraint reflects that last season, Farmer Johnson cultivated 30 hectares of rice, and, therefore, the amount of annual and winter crops that have rice as a precedent should be less than or equal to 30. Constraint (6.54) requires that the amount of land cultivated with a certain crop in summer should be less than or equal to the amount of land cropped with its predecessors in the winter season. Finally, constraint (6.55) is related to the amount of land that needs to be cultivated in the next winter season.

Irrigation Water Constraint   The irrigation needs for the various crops can be presented with the following constraint:

$$\sum_{i=1}^{3}\sum_{j=1}^{1} i_i x_{ij} + \sum_{i=1}^{3}\sum_{j=1}^{2} i_i y_{ij} + \sum_{i=1}^{2}\sum_{j=1}^{2} i_i z_{ij} \leq 200,000 \tag{6.56}$$

where i_i is the irrigation level for crop i. Irrigation water demand per unit area multiplied by the total areas cultivated by that crop represents the demand for irrigation water of that crop. This equation requires that the overall water demand should be less than or equal to the overall water supply.

Non-negativity Constraints   In this problem, all decision variables should be greater than or equal to zero.

Problem Set-up and Interpretation of the Solution

To be considered a linear model, the model should be developed to treat the same crop with different irrigation levels as different crops. Additionally, the land that is planted with a certain crop in winter should be planted again in the summer with a different crop.[4]

As shown in Figure 6.6, the first two tables in the first 16 rows, which are shaded in grey, contain the information provided in Table 6.10 through Table 6.12. Additionally, the cropping pattern from the previous season is shown in Cells B38 through G38. The spreadsheet also contains the given information regarding the size of the farm (100 hectares (ha)) and the maximum water supply (200,000 m³/ha).

The decision variables representing the cropping pattern are shown in Cells B21 through G35.[5] This matrix shows how the land that was cultivated last season will be cultivated in the upcoming year. For instance, a value of 30 in Cell C27 means that the 30 hectares that were cultivated with the annual crop of rice last year will now be planted with maize with a 100% irrigation level for next winter.

The annual and seasonal yields for each crop are totaled in Cells H21 to H35. The total water demand for each crop that arises from the selected cropping and irrigation plan is calculated in Cells J21 to J35. For example, the total water demand for maize under full irrigation is shown in Cell J27 and is calculated with the equation "=H27*B5", which multiplies the total area dedicated to fully irrigated maize (60 hectares) by its water demand (1,200 m³/ha). The total water demand for the cropping and irrigation plans is summed in Cell J36 (200,000 m³/ha).

Cells B41 through G55 calculate the profit for each crop and irrigation level combination, given each possible precedent crop. For instance, in the optimal solution, Cell C47 shows that Farmer Johnson will earn $59,400 profit from growing maize next year using a 100% irrigation level on 30 hectares of land that was cultivated with rice in the previous season. The formula used in Cell F47 is "=C27*C14*$H$5*$D$5", which is the product of the cropping pattern for 30 hectares of maize; the discount factor based on the predecessor crop (1.0), the final yield based on the 100% irrigation level selected (9 tons/ha), and the profit received per ton of maize ($220). The total profit for the cropping pattern and selected irrigation level is shown in Cell J49, which is the objective to be maximized in this problem.

With regard to model constraints, the previous year's summer cropping pattern is shown in Cells B38 through G38, and the inequalities and equalities in row 37 require that this year's winter cropping pattern be derived correctly. Note that the total land having the precedent of maize and winter wheat is restricted to being equal to zero since the cropping pattern is being determined after the summer. Cells H53 and H54 are the constraints on the total area (100 hectares) that can be cultivated in winter and summer, respectively. Note that the area of rice cultivation needs to be accounted for in both constraints as rice is an annual crop. Cell H55 is the constraint (6.52). Figure 6.7 shows the model specifications

---

[4]In this case, wheat cultivated in the summer is considered a different crop from wheat cultivated in the winter as the water needs and growing characteristics are different.

[5]To ensure that the model is linear, the same crop with different irrigation levels is treated as a different decision variable.

	Water Demand (m3/ha)	Max. Yield (tons/ha)	Profit ($/ton)	Maximum Yield by Irrigation Level			Actual Yield by Irrigation Percentage		
				100%	80%	60%	100%	80%	60%
Rice (a)	1,200	7	$ 180	100%	88%	72%	7.00	6.16	5.04
Wheat (w1)	800	6	$ 150	100%	85%	71%	6.00	5.10	4.26
Maize (w2)	1,200	9	$ 220	100%	56%	24%	9.00	5.04	2.16
Wheat (s1)	1,300	10	$ 170	100%	83%	73%	10.00	8.30	7.30
Cotton (s2)	1,600	8	$ 300	100%	69%	51%	8.00	5.52	4.08

Predecessors

	None	Rice (a)	Wheat (w1)	Maize (w2)	Wheat (s1)	Cotton (s2)
Rice (a)	1.0	0.8	-	-	0.8	0.7
Wheat (w1)	1.0	1.0	-	-	1.0	0.9
Maize (w2)	0.9	1.0	-	-	0.9	1.0
Wheat (s1)	1.0	-	0.8	0.9	-	-
Cotton (s2)	1.0	-	1.0	0.6	-	-

Decision Variables

Crop (Irrigation %)	None	Rice (a)	Wheat (w1)	Maize (w2)	Wheat (s1)	Cotton (s2)	Total (ha)	Total Water Demand
Rice (a) - 100%	-	-	-	-	-	-	-	-
Rice (a) - 80%	-	-	-	-	-	-	-	-
Rice (a) - 60%	-	-	-	-	-	-	-	-
Wheat (w1) - 100%	-	-	-	-	-	-	-	-
Wheat (w1) - 80%	-	-	-	-	-	-	-	-
Wheat (w1) - 60%	15.8	-	-	-	20.0	-	35.8	17,200
Maize (w2) - 100%	-	30.0	-	-	-	30.0	60.0	72,000
Maize (w2) - 80%	-	-	-	-	-	-	-	-
Maize (w2) - 60%	-	-	-	-	-	-	-	-
Wheat (s1) - 100%	-	-	-	-	-	-	-	-
Wheat (s1) - 80%	-	-	-	-	-	-	-	-
Wheat (s1) - 60%	-	-	-	60.0	-	-	60.0	46,800
Cotton (s2) - 100%	4.2	-	35.8	-	-	-	40.0	64,000
Cotton (s2) - 80%	-	-	-	-	-	-	-	-
Cotton (s2) - 60%	-	-	-	-	-	-	-	-
Total:	20.0	30.0	0.0	0.0	20.0	30.0		200,000
	<=	<=	=	=	<=	<=		<=
Previous Season:	20	30	0	0	20	30	Max water supply:	200,000

Profit	None	Rice (a)	Wheat (w1)	Maize (w2)	Wheat (s1)	Cotton (s2)
Rice (a) - 100%	-	-	-	-	-	-
Rice (a) - 80%	-	-	-	-	-	-
Rice (a) - 60%	-	-	-	-	-	-
Wheat (w1) - 100%	-	-	-	-	-	-
Wheat (w1) - 80%	-	-	-	-	-	-
Wheat (w1) - 60%	10,118	-	-	-	12,780	-
Maize (w2) - 100%	-	59,400	-	-	-	59,400
Maize (w2) - 80%	-	-	-	-	-	-
Maize (w2) - 60%	-	-	-	-	-	-
Wheat (s1) - 100%	-	-	-	-	-	-
Wheat (s1) - 80%	-	-	-	-	-	-
Wheat (s1) - 60%	-	-	-	67,014	-	-
Cotton (s2) - 100%	10,000	-	86,000	-	-	-
Cotton (s2) - 80%	-	-	-	-	-	-
Cotton (s2) - 60%	-	-	-	-	-	-

Total Profit: $304,712

Constraints Selected		Available
95.8	<=	100
100.0	<=	100
100.0	<=	100

Figure 6.6 Crop pattern and irrigation problem set-up and optimal solution.

for this problem including the objective function, the decision variables, and the constraints described above.

The optimal solution, Cells B21 to G35 of Figure 6.6, shows that during the next growing season, Farmer Johnson should plant 60 hectares of maize in winter, of which 30 hectares were cultivated with rice and 30 hectares were cultivated with cotton during the previous summer growing season. The maize will be fully irrigated and will achieve its maximum yield. Additionally, 35.8 hectares of wheat will be planted in winter with 60% irrigation. Of these 35.8 hectares, 15.8 will come from previously noncultivated areas, and 20 hectares will come from acres that were planted with wheat in the summer. In the summer, the 60 hectares of maize would be converted entirely to cultivating wheat with 60% irrigation. Note that both winter and summer wheat yield relatively high amounts when not

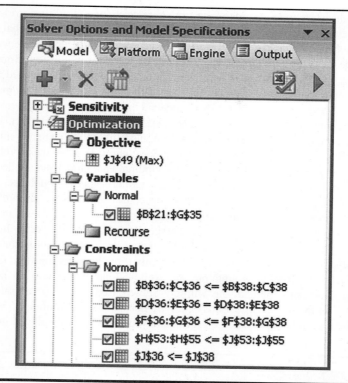

Figure 6.7 Crop pattern and irrigation model specifications.

receiving full irrigation, 71% and 73% of the maximum yield, respectively. The other 40 hectares will be planted with cotton with 100% irrigation. For cotton, 4.2 hectares of this land will come from previously uncropped land, and the other 35.8 hectares will come from land previously planted with wheat in the winter. Both maize and cotton experience large declines in yield when not receiving full irrigation. This entire cropping plan yields a profit of $304,712 for Farmer Johnson and exhausts the water allocation.

6.5 RESEARCH APPLICATION: OPTIMIZING GRIZZLY BEAR CORRIDOR DESIGN

Within the field of conservation biology, a branch of research has used mathematical programming to aid in selecting reserve sites. This research identifies areas to protect based on a variety of objectives, such as maximizing species richness, minimizing costs given a certain threshold of species protection (a type of covering problem), or optimization of certain biophysical attributes of an area. Economists have extended these models to include economic factors such as acquisition costs and the likelihood that the land would be developed in the absence of an acquisition program.

A recent focus of this research has been the spatial distribution of reserve sites. As habitat fragmentation has become recognized as a leading cause of species decline and extinction, several models have been proposed that emphasize the compactness and connectivity of protected land. One concept that has been advocated is a "wildlife corridor" or network of protected areas that connect existing reserves of protected land. There are several ecological advantages to wildlife corridors, including an increased area accessible to species that have large ranges, the ability to sustain greater genetic diversity, and the ability of species to escape natural disasters and respond to long-term climate change. Wildlife corridor programs have been established in many regions of the world, including the National Ecological Network in the Netherlands, the Siju-Rewak Corridor connecting elephant reserves near the India–Bangladesh border, and the Amapa Biodiversity Corridor in northern Brazil.

Most existing studies that focus on optimal corridor selection have formulated the problem as some variant of the traditional network problem and used a Least Cost Path model to connect reserves. A drawback of Least Cost Path models is that they can produce narrow pathways that ignore large areas of ecologically valuable land. Additionally, by only focusing on areas with the lowest cost, the models have an incentive to select areas with large amounts of unsuitable habitat if that habitat ends up being the lowest-cost option. To avoid these problems, Suter et al. (2008) proposed using a Budget Constrained Optimal Path model and applied this model to an optimal corridor design for grizzly bears in the northern Rocky Mountains.

Grizzly bears in particular are often targeted for conservation as an "umbrella species." An umbrella species is one that has fairly demanding ecological requirements and whose protection has beneficial spillover effects on other species in its ecosystem. In the case of grizzly bears, protecting bear habitat would, by extension, protect habitat for elk, moose, bison, and a host of other species. There are currently three large regions that are protected in the northern U.S. Rocky Mountains: the Yellowstone Ecosystem in northwestern Wyoming and southern Montana, the Salmon–Selway Ecosystem in central Idaho, and the Northern Continental Divide Ecosystem in northwestern Montana. The protected regions and the region being considered for conserved parcels are displayed in Figure 6.8.

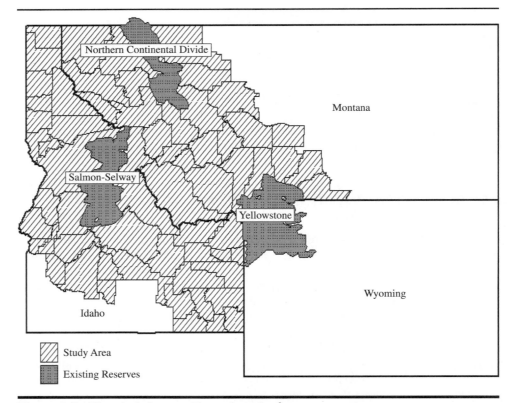

Figure 6.8   The grizzly bear corridor study region.[6]

---

[6]Map created by Jacob Fooks, University of Delaware.

The authors used an LP model to construct a corridor through Idaho and Montana to connect these three reserves. The corridor maximizes habitat suitability subject to a budget constraint.

The study focused on 64 counties in Idaho and western Montana. Habitat suitability was determined based on the Habitat Suitability Index (HSI) score developed by the Craighead Environmental Research Institute that ranks parcels on a scale from 2 to 4, with parcels scored as 4 being the most suitable. The cost to protect a parcel was calculated based on the amount of private land within the parcel multiplied by the average value of agricultural land within the county. The study considered varying size resolutions for the parcels under consideration for protection. Land was considered in six different grid resolutions ranging from 5 to 60 kilometers as well as on the whole-county level. There was also a variation for the 5-kilometer grid size that incorporated transaction costs.

To establish a connected corridor, the problem is set up as a connected sub-graph that is solved as a linear program as follows:

$$\text{Max: } Z = \Sigma \, u_i x_i \tag{6.57}$$

s.t.:

$$\Sigma \, c_i x_i \leq b \tag{6.58}$$

$$x_0 + y_{0t} = n \tag{6.59}$$

$$y_{0t} = \Sigma \, x_i \tag{6.60}$$

$$y_{ij} \leq n x_j, \text{ for all edges} \tag{6.61}$$

$$\Sigma y_{ij} = x_j + \Sigma y_{ji}, \text{ for all nodes} \tag{6.62}$$

$$x_t = 1, \text{ for terminal nodes t} \tag{6.63}$$

$$x_i = 0, 1, y_i \geq 0 \tag{6.64}$$

Each cell in the grid is represented by x_i, with x_0 being an artificial source cell that injects flow into the network. The cost for each cell is c_i, and the HSI score is represented by u_i. The x_i variable is binary choice, implying that the cell is either purchased (equals 1) or not purchased (equals 0). Binary constraints will be discussed in greater depth in Chapter 7. Constraint (6.58) is the budget constraint, and b is the total budget available. Constraints (6.59) through (6.62) establish the connected subgraph. This is somewhat similar to the networks seen in Chapter 5. Each cell, x_i, is a node in the network, and y_{ij} is an edge connecting nodes i and j. The total number of nodes in the network is n. Constraints (6.59) and (6.60) limit the total injected flow to the total number of nodes and ensure that all flow used by the network was injected by the source node. Constraint (6.61) causes the flow to decrease from node to node across the network so that each node retains one unit of flow. Constraint (6.62) causes the flow into a node to equal the flow out of a node plus the flow retained by the node so that flow is conserved across the network. Finally, Constraint (6.63) ensures that the path passes through each of the existing reserves or terminals.

Several different paths were calculated for different budget levels and parcel granularities. The least-cost corridors were determined for each budget level. Table 6.13 reports the minimum cost path for each resolution level. As would be expected, as the resolution increases, the area and cost decreases. The area goes from 9.6 to 1.4 million acres, and the total cost ranges from $1.9 billion to $11.8 million. However, the cost per acre also drastically decreases from $197.40 to $9.80, as a smaller parcel size offers greater possibilities for selecting parcels of entirely public (and thereby free) or inexpensive land. The total HSI also decreases at finer resolutions; however, the average HSI per acre increases.

Table 6.13   Minimum Cost Corridor Results

Parcel Size	Number of Parcels	Parcels Selected	Corridor Cost (thousand)	Total HIS (thousand)	Acres Preserved (thousand)	Percent Private	Cost per Acre ($)
County	64	5	1,904,355	7,038	9,649	27.2%	197.4
60km	118	11	1,657,740	7,188	8,234	27.1%	201.3
50km	167	12	1,329,090	5,902	6,777	30.7%	196.1
40km	239	16	891,052	5,807	5,409	13.6%	164.7
25km	570	23	449,430	3,743	3,408	12.5%	131.9
10km	3,296	120	99,341	3,679	4,096	1.9%	24.3
5km	12,788	265	10,865	2,147	1,637	0.5%	6.6
5km[†]	12,788	196	11,824	1,576	1,210	0.7%	9.8

[†]Includes a $5,000 transaction cost per parcel selected.

Table 6.14   Budget Constrained Maximum HIS Results

Parcel Size	Number of Parcels	Parcels Selected	Corridor Cost (million)	Total HIS (thousand)	Acres Preserved (thousand)	Percent Private	Cost per Acre
County	64	5	1,904	7,038	9,649	27.2%	197.3
60km	118	20	1,821	14,240	14,209	32.1%	128.2
50km	167	22	1,461	12,188	11,303	19.4%	129.3
40km	239	23	999	11,832	9,932	8.4%	100.6
25km	570	–	–	–	–	–	–
10km	3,296	–	–	–	–	–	–
5km	12,788	–	–	–	–	–	–

Table 6.15   50-kilometer Budget Variation Results

Budget (million)	Cost (million)	Total HIS (thousand)	Acres Preserved (thousand)	Percent Private	Cost per Acre ($)	HIS (per Acre)
–	1,329	5,902	6,777	30.7%	196.1	0.87
1,396	1,394	9.842	9,608	22.2%	145.1	1.02
1,462	1,461	12,188	11,303	19.4%	129.3	1.08
1,528	1,526	13,220	12,176	18.5%	125.3	1.09
1,595	1,594	14,145	12,874	15.5%	123.8	1.10
1,728	1,727	15,533	14,131	15.7%	122.2	1.10
1,861	1,857	16,777	15,119	15.2%	122.8	1.11
1,994	1,992	17,811	16,239	16.1%	122.7	1.10
2,658	2,658	22,151	20,105	16.2%	132.2	1.10
3,323	3,321	25,500	23,298	16.5%	142.5	1.09

As cell size becomes smaller, greater average habitat suitability scores are achievable at lower costs, but the number of cells also increases. The increasing cell number presents computational issues at too fine of a resolution for the cost-constrained HSI maximization problem. Because of this difficulty, this model was solved only from the county level to the 40-kilometer level. For these results, a maximum budget 10% higher than the minimum feasible corridor cost was used. These results are reported in Table 6.14.

Finally, the budget level considered for the 50-kilometer resolution ranged from the minimum cost corridor to 50% above minimum cost. These results are displayed in Table 6.15.

They show that increasing the budget above the minimum cost amount increased marginal habitat suitability benefits but at a decreasing rate, suggesting that the greatest benefits from optimization are when the budget is close to the minimum cost amount.

SUMMARY

This chapter introduced several natural resource and environmental problems where LP can be applied. These examples included a forest management example that demonstrated how forest harvest rotations could be determined. A land use allocation model was developed where city planners sought to increase the local tax base by using previously undeveloped lands while also setting goals to prevent excessive development and ensure the protection of open space including forests, parks, and farms. In another example, LP was applied to wildlife and forage management in a wild game park. The results suggest a plan that manages hunting to bring the population down to a minimum threshold necessary for scenic and genetic purposes, while maximizing profits. The fourth example showed how, in the face of mandated water restrictions, a farmer could use LP to determine optimal cropping patterns and irrigation practices to maximize profits. Finally, a research example was presented where LP was used in a type of covering program to derive an optimal habitat corridor for grizzly bears in the northern Rocky Mountains.

These examples demonstrated how LP can address a diverse set of circumstances and yield information to decision makers that can enhance the achievement of the natural resource and environmental objectives. In many cases, natural resource and environmental problems involve aspects that linear models, at their best, can be reasonably good approximations for the real world. In later chapters more advanced techniques will be introduced that build upon this understanding of linear models to account for more complex relationships.

REFERENCES

Davis, L. S. (1967). Dynamic programming for deer management planning. *Journal of Wildlife Management, 31,* 667–679.

Dykstra, D. P. (1984). *Mathematical programming for natural resource management.* New York, NY: McGraw-Hill.

Haouari M., & Azaiez, M. N. (2001). Theory and methodology: Optimal cropping patterns under water deficits. *European Journal of Operational Research, 130,* 133–146.

Suter, J., Conrad, J., Gomes C., van Hoeve W. J., & Subharwal, A. (2008). Optimal corridor design for grizzly bear in the U.S. Northern Rockies. *Agricultural Economics Association Annual Meetings.* July 2008, Orlando, Florida.

EXERCISES

1. Consider the land use planning problem presented earlier in this chapter. The city council is seeking to balance the conflict between development and environmental protection. Notice in this problem that the minimum value of the E-index of 6 is likely to influence the final tax base increase. Keeping the other information constant, plot the relationship between the actual tax base increase and the minimum E-index value. In particular, evaluate the influence of the minimum E-index value in the range from 5.8 to 7.2 with intervals of 0.2.

2. Reconsider the land use planning problem presented earlier in this chapter. Suppose the environmental issues are the primary concern for the public, such that the city

council determines that the primary purpose of the land use plan is to maximize the E-index described, as long as the increase in the property tax base is at least $2.2 billion. Meanwhile, the resource economist on this planning committee notices that, according to the problem set-up, farmland can neither make a significant contribution to the tax base nor can it help increase the E-index. Consequently, the model may suggest a dramatic decrease in farmland. This observation draws considerable public attention, such that the city council decides to add a new requirement for farmland development. The new requirement is that farmland acreages should be at least one half of residential land use. Given that all the other constraints remain the same, formulate and solve this exercise in Solver.

3. A county is planning to develop 750 acres of available land in an environment-friendly way. Available land use types include parks, farmland, wildlife reserves, residential areas, and forests. The county expects to develop at least 90 acres of parks and 110 acres of wildlife reserve areas. The county wants to ensure a diversity of land uses so it would like to restrict wildlife reserve areas to a maximum of 350 areas. Additionally, the county wants to make sure that at least 50 acres of land is allocated each to farmland and forest, and 70 acres of land is allocated to residential areas. Formulate and solve this exercise to maximize the environmental benefit for the county. The yearly environment benefits for each type of land use per acre are scored as follows:

Types	Environment Benefit
Park	108
Farmland	75
Wildlife Reserve	150
Residential Area	53
Forest	116

4. Suppose that in Exercise 3, the county also needs to consider if it has enough budget for the land use project. It is estimated that the available budget is $4 million for the next year. The yearly costs needed for each type of land use per acre are estimated as:

Types	Cost ($1,000)
Park	9.6
Farmland	8.1
Wildlife Reserve	6.6
Residential Area	2.9
Forest	11.2

In this case, derive and graph the input demand function given the budget.

5. Suppose in Exercise 4, the official estimate is that the county can provide at most $3.5 million for the land use project in the coming year. Otherwise, the county has to apply for a loan from a bank with a yearly interest rate at 2%. In order to obtain the highest environmental benefit score, how much of a loan should the county apply for, and how much interest will the county pay during the upcoming year?

6. A group of experts is considering building different biogas power generation projects at five possible locations in Newport County. The experts would ideally like to build a small biogas power plant at each location to convert agricultural material and

residues into usable biomethane, which can then be used to generate electricity. However, the total budget for these biogas projects is only $8 million which is not enough to build all five power plants. Construction costs and future revenue of the biogas power plant project at each location are summarized in the following table (in thousands). Given the budget, decide which locations should be selected to build power plants so that net revenue (revenue minus construction cost) can be maximized.

Locations	1	2	3	4	5
Construction Cost	80	120	160	240	270
Revenue	40	48	64	84	85

7. Consider the previous biogas power plant exercise. Due to more sophisticated considerations, the experts decide that only one power plant can be built on either location 1 or location 2. Also, power plants should be built on both locations 1 and 5, otherwise neither location should be considered to locate a power plant. Given the $8 million budget, decide which locations should be selected to build the power plants so that net revenue can be maximized.

8. Reconsider the wild game management example discussed in this chapter. If the manager's objective is to have the maximum number of animals harvested over a five-year period, how can you modify the model to achieve this new objective?

9. Reconsider the wild game management example discussed in this chapter. What if the available budget is only $20,000 each year to do jobs such as land clearing, foraging, and animal harvesting? The labor force, on the other hand, is for practical considerations unlimited in this area as a large number of volunteers are interested in helping. Given this situation, what would you recommend to the managers so that they can maximize profit? How much labor will be required each year?

10. Again consider the wild game management example discussed in this chapter. Note that the time value of money is ignored in this example. Assume a discount rate of 10% and make adjustments to the model and calculate the maximum discounted value of profit at the end of five years.

11. The U.S. Forest Service needs an allocation plan for its forest firefighters in Florida. It currently hires 200 experienced firefighters at the beginning of the year. The workload required in the forest varies from season to season as the firefighters not only work directly on combating existing fires, but also in a variety of fire prevention activities. The estimated work requirement is as follows: 31,000 hours in the spring, 32,000 hours in the summer, 40,000 hours in the autumn, and 38,000 hours in the winter. Approximately 10% of the firefighters leave at the end of each season, and new trainees are hired. For each new trainee recruited, it takes 90 hours of experienced firefighters' time to conduct training. An experience firefighter is expected to finish 170 hours of work during every season. A trainee usually does 100 hour of work due to lack of experience. After one season, a trainee is considered an experienced firefighter. However, about 20% of trainees leave after the first season. An experience firefighter costs $6,500 per season. A trainee costs $3,500 per season. Formulate an LP model to decide how many trainees the U.S. Forest Service should hire at the beginning of every season to minimize total costs.

12. Consider Exercise 11, if the manager wants at least 200 forest firefighters to stay at the end of year, what should the hiring plan be for the U.S. Forest Service?

13. The State of Virginia is developing five forest protection sites. Each project has its own required starting year and duration. The table below provides the basic information about these projects.

	Year 1	Year 2	Year 3	Year 4	Year 5	Cost ($ million)	Ecosystem Benefits ($ million)
Project 1	Start		End			4.5	0.050
Project 2		Start			End	8.0	0.070
Project 3	Start				End	14.0	0.135
Project 4			Start	End		2.0	0.030
Budget ($ million)	2	4	2	6	7		

The Ecosystem Benefits score evaluates each project's yearly contribution and calculates this contribution in dollar terms. Projects 1 and 4 must be finished within the duration period. For the other projects, they can be partially finished within budget limitation. These protection sites can start functioning and realize partial annual benefits even when unfinished. For example, if 10% of project 1 is completed in year 1 and the remaining 90% is completed in year 3, then the total benefits (measured in millions of dollars) that can be calculated as 0.1×0.05 (year 2) $+ 0.1 \times 0.05$ (year 3) $+ (0.1 + 0.9) \times 0.05$ (year 4) $+ (0.1 + 0.9) \times 0.05$ (year 5) $= 0.11$ million dollars. Help the project coordinator decide an optimal schedule for the projects that will maximize the total benefits over five years.

14. Reconsider Exercise 13, what would be the best schedule if a constraint was added such that at least 25% of each project was finished by the end of the five-year time period?

15. A fishery in Maine has about 120 tons of fish, including 30 tons of salmon, 50 tons of tuna, and 40 tons of sardines. Every year, fishermen capture a certain amount of fish and sell them to the market. Assume that in 2009, the market price for salmon was $6.00 per kg, the market price for tuna was $5.00 per kg, and the market price for sardines was $5.20 per kg. To maintain an ecological equilibrium, the fishery manager would like to keep the amount of tuna less than twice the amount of salmon. Suppose the overall average reproduction rate of these three fish is 12% per year. The fishery manager would like to keep the expected amount of fish in the fishery after five years to still be 120 tons. Additionally, the fishery manager would like to have at least 12 tons of salmon, 25 tons of tuna, and 16 tons of sardines in the fishery at the end of the five years. Form an LP model to help the manager find the optimal solution.

16. Paul is a farmer in Georgia and he is planning to grow rice, wheat, and cotton on his farm. He has 100 hectares of land and 120,000 m³ of irrigation water to use. Farmer Paul is planning to plant 30 hectares of rice, 40 hectares of wheat, and 30 hectares of cotton this year. The first table contains the fully irrigated water demand, the maximum yield per hectare, and the expected profit per ton for these three crops. The ratios of actual-to-maximum yield for each of these three crops under different irrigation levels are shown in the second table. Help Farmer Paul find an optimal cropping pattern so that he can maximize his profit for this year.

Water Demand, Maximum Yield, and Profit by Crop

	Water Demand (m³/ha)	Maximal Yield (tons/ha)	Profit ($/ton)
Rice	1,300	9	190
Wheat	1,200	12	180
Cotton	1,500	8	240

Maximum and Actual Yield by Irrigation Percentage by Crop

	Maximum Yield by Irrigation Level			Actual Yield by Irrigation Level (tons/acre)		
	100%	80%	60%	100%	80%	60%
Rice	100%	92%	78%	9.00	8.28	7.02
Wheat	100%	88%	80%	12.00	10.56	9.60
Cotton	100%	72%	69%	8.00	5.76	5.52

17. Farmer Paul has planted 30 hectares of rice, 40 hectares of wheat, and 30 hectares of cotton this year as in Exercise 16. The current year's planting period ends in the summer, and Farmer Paul is planning for his cropping pattern for next year (starting from this summer). For Paul's farm, rice is an annual crop, and wheat, maize and cotton are seasonal crops. Assume wheat can be planted in both winter and summer while maize can only be planted in winter and cotton can only be planted in summer. The discounting factors with respect to the predecessor crops are shown in the table below. In this case, a total of 180,000 m³ of irrigation water is available. The subsequent table shows the fully irrigated water demand, the maximum yield per hectare and the expected profit per ton for the five crops. The final table contains the ratios of actual-to-maximum yield for each of the five crops under different irrigation levels. Help Farmer Paul to find the optimal cropping pattern for next year. In this exercise, consider winter wheat and summer wheat as different crops. Note that it is possible that not all land will be used in each season.

Discounting Factors with Respect to Predecessor Crops

	Predecessors				
	Rice (a)	Wheat (w_1)	Maize (w_2)	Wheat (s_1)	Cotton (s_2)
Rice (a)	1	0	0	0.9	0.8
Wheat (w_1)	0.8	0	0	1	0.9
Maize (w_2)	0.9	0	0	0.8	0.9
Wheat (s_1)	0	1	0.8	0	0
Cotton (s_2)	0	0.8	0.7	0	0

Water Demand, Maximum Yield, and Profit by Crop

	Water Demand (m³/ha)	Max. Yield (tons/ha)	Profit ($/ton)
Rice	1,200	8	$180
Wheat (w_1)	1,000	8	$180
Maize	1,200	9	$220
Wheat (s_1)	1,400	10	$170
Cotton	1,600	7	$280

Maximum and Actual Yield by Irrigation Percentage by Crop

	Maximum Yield by Irrigation Level			Actual Yield by Irrigation Percentage (tons/hectare)		
	100%	80%	60%	100%	80%	60%
Rice	100%	90%	76%	1	0.9	0.76
Wheat (w_1)	100%	85%	72%	0.8	0.68	0.576
Maize	100%	72%	50%	1	0.72	0.5
Wheat (s_1)	100%	83%	73%	1	0.83	0.73
Cotton	100%	69%	56%	1	0.69	0.56

18. Farmer Paul's family can provide 1,600 hours of labor per year. The labor requirement per hectares for the five different crops are as follows: 10 hours for rice, 7 hours for winter wheat, 9 hours for winter maize, 9 hours for summer wheat, and 12 hours for summer cotton. If all the other conditions are the same as in Exercise 17, help Farmer Paul determine his optimal cropping pattern for next year. What is the optimal profit of this problem? Compare the solution with Exercise 17 and describe the similarities and differences that you find.

19. Compare the optimal solution for Exercises 17 and 18. Does the labor availability become a constraint for the optimal cropping pattern? Conduct a sensitivity analysis of the total labor available and explain the results. If Farmer Paul can hire at most 300 hours of labor from his neighbor at a cost of $12.00 per hour, what is the new optimal cropping pattern?

20. The Gila River is a tributary of the Colorado River that runs through the city of Phoenix from the south. The river is a major source of the city's water, and the government has a plan to expand the city's development in the area along the river. A total of 100 acres of land is projected to be needed for residential, business, and recreational use. According to the plan, at least 20 acres of land should be designed for residential development, 30 acres will be used for industrial development, and a recreational park will be built on at least 10 acres. The initial investment cost for residential land is $8 million for the first 20 acres of land and $300,000 for every extra acre of land thereafter. The initial investment costs for industrial land and recreational land are $20 million and $12 million, respectively, while the costs for additional land are $500,000 and $400,000 per acre, respectively. An acre of residential land can yield a profit of $50,000 per year, and the expected profits for industrial and recreational land are $120,000 and $150,000, respectively. On average, every acre of residential land will use 20 m³ of water per month, and every acre of industrial land and recreation land consumes 40 m³ and 25 m³ of water per month, respectively. The budget is $80 million, and the regulation of water use from the Gila River is 40,000 m³. Set up an LP model to find the annual optimal profit for this land development project.

21. Fertile Landscapes Associates (FLA) is a company that produces fertilizers for crop use. They have just developed a new general fertilizer named Fertilizer A. Before widely introducing Fertilizer A to the market, they are planning to do some research on the efficiency and sales price of Fertilizer A. By selecting a sample of 50 farms, FLA tested the efficiency of Fertilizer A under different cropping patterns and irrigation levels. A 120-acre farm in Alabama was selected for the experiment. On this farm,

rice can be planted on a yearly basis. Cabbage and maize can be planted in the winter, and wheat and peanuts can be planted in the summer. The average amount of irrigation water that can be allocated to this farm is 220,000 m³. Water demand, profit, and maximum yield for each crop are shown in the first table below. The second table provides information about the maximum yield by irrigation level in percentage. The third table shows the demand of Fertilizer A for each crop under different irrigation levels and the effects on the maximum yield after Fertilizer A is used. This farm is now planted with 50 acres of rice, 30 acres of wheat and 40 acres of peanuts. The feasible cropping pattern for this farm is shown in the last table. The initial selling price for Fertilizer A is set to be $12.00/kg. Evidence shows that a farm with 120 acres can use up to 1,600 kg of fertilizer. Compare the maximum profit of the farm when Fertilizer A is used and not used. Give a range for the selling price in which the use of Fertilizer A is still profitable for this farm.

Water Demand, Maximum Yield, and Profit by Crop

	Water Demand (m³/acre)	Maximal Yield (tons/acre)	Profit ($/ton)
Rice (a)	1,200	12	190
Cabbage (w_1)	1,100	10	180
Maize (w_2)	1,200	11	190
Wheat (s_1)	1,300	10	180
Peanuts (s_2)	1,800	8	260

Maximum Yield by Irrigation Percentage by Crop

	Maximum Yield by Irrigation Level		
	100%	80%	60%
Rice (a)	100%	90%	82%
Cabbage (w_1)	100%	82%	72%
Maize (w_2)	100%	78%	56%
Wheat (s_1)	100%	83%	75%
Peanuts (s_2)	100%	72%	56%

Demand and Effects of Fertilizer A for each crop in different irrigation

	Fertilizer Demand (kg)			Utilization of Fertilizer A		
	100%	80%	60%	100%	80%	60%
Rice (a)	60	50	42	120%	120%	115%
Cabbage (w_1)	70	60	52	120%	110%	106%
Maize (w_2)	65	57	50	118%	112%	102%
Wheat (s_1)	80	68	55	115%	110%	105%
Peanuts (s_2)	70	56	45	130%	120%	105%

Feasible Cropping Pattern

| | Predecessors | | | | |
	None	Rice	Cabbage	Maize	Wheat	Peanuts
Rice (a)	0	1	0	0	0.9	0.8
Cabbage (w_1)	0	0.8	0	0	1	0.9
Maize (w_2)	0	0.9	0	0	0.8	0.8
Wheat (s_1)	1	0	1	0.9	0	0
Peanuts (s_2)	1	0	0.7	0.8	0	0

22. Global climate change models frequently suggest that some areas will experience changes in the levels of expected rainfall. Assume that for this Alabama farm, these climate models estimate that less rainfall will occur such that the water demand for each of the crops is going to increase by 50% as more irrigation water is needed. Assuming that farm can still obtain 220,000 m³ of irrigation water what is the maximum profit from the farm when Fertilizer A is used and not used? Provide the selling price for the use of Fertilizer A given this situation.

23. Imagine that you are the Executive Director of a non-profit conservation group named the Pangaea Conservancy, which has a budget of $10 million to purchase land from private landowners so that these lands can be permanently protected. The six parcels available for purchase are shown in light gray in the figure below, labeled A–F. Pangaea already has two protected national parks, which are shown in this figure as dark gray areas. Ecologists and conservation professionals from the Pangaea Conservancy have evaluated each of the available parcels and assigned each with a parcel-specific ecological benefit score, as shown in the table below, where a higher score indicates a higher ecological benefit. As the Executive Director, which parcels would you recommend that the Pangaea Conservancy acquire? Why?

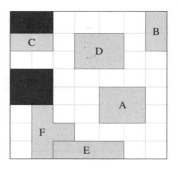

Parcel ID	Ecological Score	Cost ($m)
A	2	$6
B	5	$3
C	4	$2
D	6	$6
E	2	$3
F	9	$4

24. Most conservation organizations and government agencies in the United States and throughout the world use what is called "Benefit Targeting" (also referred to as "Rank-Based Models") to select which parcels to acquire for conservation (Messer and Allen 2010). With Benefit Targeting, the organization prioritizes the parcels based solely on the parcels' benefits—in this case the Ecological Score—and then acquires the highest ranked parcel first, the second highest parcel second, and so forth, until the budget is exhausted. Assume that the Pangaea Conservancy uses the Benefit Targeting approach to solving Exercise 23. Which parcels would it select? Comment on whether these parcels are similar to or different than the selections you recommended in Exercise 23. In your comparison of the selected parcels, evaluate a number of criteria including the total ecological score achieved, the total cost, the average values of the selected parcels, and the spatial location of the parcels.

25. Environmental economists have raised concerns about the use of Benefit Targeting, as this method does not take into account the costs of the selected parcels except when determining whether there are sufficient funds. As an alternative, economists often have recommended that the selection be done based on benefit–cost ratios, where the parcel with the highest ratios should be acquired first, the parcel with the second highest ratio should be acquired second, and so forth, until the budget is exhausted. This technique is frequently referred to as Cost Effectiveness Analysis or Benefit-Cost Targeting. A parcel's benefit-cost ratio is calculated by simply dividing its benefit score by its costs. For example, Parcel A would be assigned the value of 0.33, as its ecological score of 2 is divided by its cost of $6 million. (Note, to facilitate interpretation, the ratio is often multiplied by a large number. As long as the same large number is used for each parcel, then this multiplication does not change the overall results). Given the same information that you used in Exercises 23 and 24, which parcels would the Pangaea Conservancy select if it used Cost Effectiveness Analysis? Comment on whether these parcels are similar or different than the selections you recommended in Exercise 23 and the parcels selected by Benefit Targeting in Exercise 24. In your comparison of the selected parcels, evaluate a number of criteria including the total ecological scores, the total cost, the average values of the selected parcels, and parcels' spatial location. Given the results from Exercises 23, 24, and 25, what method of selection would you suggest that the Pangaea Conservancy use? Why?

26. The Pangaea Conservancy is considering protecting another area. This area already has four protected areas, shown below in dark gray. In this area, the Pangaea Conservancy has budgeted $25 million to purchase land from private landowners so that it can permanently protect these areas. The 12 parcels available for purchase by the Pangaea Conservancy are lettered from A-L below and are shown in light gray. The Pangaea Conservancy has used a new and improved benefit assessment technique which calculates two benefit scores as shown below.[7] For both of these measures, the higher the score signifies the higher the quality. Assuming that the Pangaea Conservancy considers the ecological score and the scenic value to be of equal importance, which parcels would you recommend that it acquire if it wants to use Benefit Targeting? Describe the selected parcels.

---

[7] Data for this example is from Messer, 2006.

Parcel ID	Ecological Score	Scenic Value	Total Benefits	Cost ($m)
A	100	126	226	$4.50
B	100	143	243	$5.00
C	130	130	260	$6.00
D	150	60	210	$5.00
E	80	185	265	$6.50
F	140	140	280	$7.00
G	110	95	205	$7.00
H	50	60	110	$6.50
I	60	25	85	$9.00
J	15	10	25	$4.50
K	150	150	300	$12.00
L	75	75	150	$8.00

27. Given the information provided with Exercise 26, identify the parcels that the Pangaea Conservancy would select if it used Cost Effectiveness Analysis. Discuss these results in comparison to the results of Exercise 26.

28. Given the information provided in Exercise 26, identify the parcels that the Pangaea Conservancy would select if it used binary interger programming. The binary variables should be either 0 (not selected) or 1 (selected), and can be multiplied by the original environmental benefit scores to calculate the overall benefits of the selected parcels. For example, if Parcel A is selected, then by multiplying the Total Benefits score of 226 by 1, the entire amount can be added into the aggregate Total Benefits calculated for the selected parcels. If Parcel A is not selected, then by multiplying the Total Benefits score by 0, makes the resulting value zero. Discuss these results in comparison to the results of Exercises 26 and 27.

29. The Board of Directors for the Pangaea Conservancy are concerned that the aggregate ecological scores are lower in the analysis than desired. They would like to see that the selected parcels achieve a minimum value of 500 for the Ecological Score. Which method—Benefit Targeting, Cost Effective Analysis or Binary Integer Programming— is best able to solve this problem? Using your preferred technique, identify a solution that addresses this concern while continuing to maximize the weighted total of the ecological score and scenic values given a budget of $25 million. Discuss the advantages and disadvantages of adding this type of minimum value threshold.

Part 2

RELAXING THE ASSUMPTIONS OF LINEAR PROGRAMMING

7

Integer and Binary Programming

While linear programming (LP) is one of the more widely used problem-solving approaches in quantitative methods, there are cases where the assumptions of LP may be too restrictive and unrealistic to apply to the problem at hand. For instance, the assumption of divisibility may not be appropriate for many capital budgeting problems, where the decision may involve determining investment decisions on a variety of projects. In this case, the level of many of the decision variables are required to take on only integer values, which violates the assumption of divisibility. Likewise, the assumptions of additivity and proportionality may not be reasonable for some applications. If, for example, there is economies of scale in production, then LP may not be an accurate way to model the industry's production technology. Furthermore, the assumption of certainty of all parameters is usually unrealistic for crop farmers. These farmers face uncertainty in virtually all the parameters of their decision problem, including prices, yields, availability of field time, and others. This section of the book examines mathematical programming models that relax these assumptions.

So far in this book, we have concentrated on models assuming that the decision variables are perfectly divisible. However, in reality, decisions are often constrained to be integers. For example, when deciding to harvest a tree, generally a forester must cut down the entire tree, not just a fraction of the tree. The allocation of seafood harvest quotas to different fishing operations cannot be fractional amounts. Obviously, a quota system cannot allow the catching of 107.54 lobsters, 33.33 crabs, and 255.79 tuna. In most cases, integer activity values are the most common units for decision making, and examples are abundant throughout the areas of agricultural, natural resource, and environmental economics, such that most decision problems are characterized by activities that should be integer values.

One way to achieve integer values is to formulate the problem as an LP model and then round off the optimal decision variables. This may cause two problems, however. First, rounding creates uncertainty that this will give the optimal solution. Second, the rounded-off integer solution may actually be infeasible, even though the LP solution is feasible. Infeasibility in maximization problems may occur if activities are rounded up; in minimization problems it may occur if activities are rounded down. To deal with such potential problems, **integer programming (IP)** has been developed.

Scientists began working on IP algorithms in the early 1960s. Thus, initially IP applications were relatively slow due to the large computational requirements that even modest IP problems require. However, with the dramatic increase in computer power, most IP problems can be solved within reasonable time allowances. Despite these improvements, it should be stressed that IP should only be used whenever the assumption of perfect divisibility is clearly inappropriate as requiring integer solutions adds another constraint that limits the feasible space and potentially reduces the value of the objective function.

Integer programming is basically the same as LP, with the exception that some or all variables are restricted to be integers. Integer programming can be all or mixed; **all-integer programming** means all decision variables are constrained to be integers, while **mixed-integer programming** means that at least one decision variable is constrained to be an integer and at least one activity is divisible.

In this section, the basic concepts underlying IP are presented. Specifically, the most efficient IP solution procedure to date, known as the branch-and-bound method, is examined. This is followed by a discussion of **binary linear programming** and several important applications of IP.

7.1   BACKGROUND ON INTEGER PROGRAMMING

Consider the following LP problem where H-Bolt, a manufacturer of automobile batteries, is deciding how many batteries they should make in the next week. Small batteries are used in gas-electric hybrid engines for cars and generate \$4,000 profit each, whereas larger batteries are designed for plug-in electric cars and generate 50% more profit—\$6,000 each. The smaller batteries used in gas-electric hybrid cars are frequently referred to as "nickel metal hydride" batteries due to their reliance on nickel, while the larger batteries designed for plug-in electric batteries are referred to as "lithium ion" batteries due to their reliance on lithium.

In this example, each small battery is manufactured with 0.5 pounds of nickel, 2 pounds of lithium, and 97.5 pounds of other metals. In contrast, each large battery requires 20 pounds of lithium and 136.5 pounds of other metals. In any week, H-Bolt can obtain 2 pounds of nickel, 70 pounds of lithium, and 682.5 pounds of other metals. If we assume that no more of the metals can be purchased during that week and there are no limits on the number of hybrid batteries that H-Bolt can sell this week, how many small and large hybrid batteries should H-Bolt make this week to maximize its profit?

$$\text{Max: } Z = 4x_1 \quad + 6x_2 \tag{0}$$

s.t.:

$$0.5x_1 \qquad \leq 2 \qquad \text{(Nickel Constraint)} \tag{1}$$

$$2x_1 + \quad 20x_2 \leq 70 \qquad \text{(Lithium Constraint)} \tag{2}$$

$$97.5x_1 + 136.5x_2 \leq 682.5 \qquad \text{(Other Metals)} \tag{3}$$

$$x_1, \qquad x_2 \geq 0 \quad \text{(Non-negativity Constraint)} \tag{4}$$

The optimal solution to this LP problem, which is shown graphically in Figure 7.1, is:

$$x_1^* = 2.44, \ x_2^* = 3.26, \ Z^* = 29.30 \ \text{(in thousands)}.$$

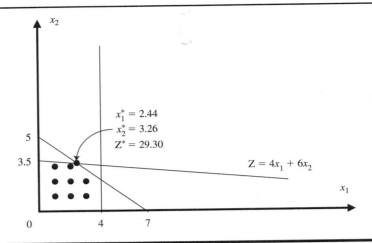

Figure 7.1   Graphical solution to H-Bolt example.

Now suppose that the decision variables are restricted to be integers. A naive way of handling this would be to round off the optimal LP solution activities to their nearest integers. If you round off x_1^* and x_2^* in this case, the solution is:

$$x_1^* = 2, x_2^* = 3, Z^* = 26.$$

The first question is: Is this solution feasible? To see, substitute $x_1^* = 2$ and $x_2^* = 3$ into the three constraints. It should be obvious that this solution will be feasible since both activities were rounded down rather than up.

The next question is: Is this an optimal integer solution? The answer to this question is no. The optimal IP solution is:

$$x_1^* = 4, x_2^* = 2, Z^* = 28.$$

The value of the optimal solution for a maximization IP model will always be less than or equal to the value of the optimal solution of its non-integer LP counterpart. For minimization problems, the reverse is true. This is due to the fact that IP adds an additional constraint to LP.

The feasible region to an integer linear program is found in two steps:

1. Graph all constraint lines and find the border where all constraints are satisfied. This is exactly the same as before.

2. The feasibility region corresponds to all integer values lying within the feasible region determined in step 1.

Because IP violates the standard LP assumption of perfect divisibility, the simplex method could not be applied to IP to obtain an optimal solution. However, as will be shown shortly, the simplex method plays an integral part in obtaining the optimal IP solutions.

7.2   THE BRANCH-AND-BOUND SOLUTION PROCEDURE

The **branch-and-bound method** uses the simplex procedure along with an iterative process which resembles a decision tree to solve an IP problem (see Figure 7.2). To illustrate this procedure, consider an IP problem where all of the activities must be integers.

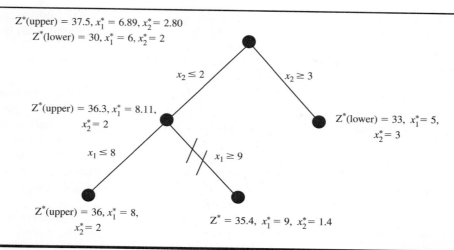

Z^*(upper) = 37.5, $x_1^* = 6.89$, $x_2^* = 2.80$

Z^*(lower) = 30, $x_1^* = 6$, $x_2^* = 2$

$x_2 \leq 2$

$x_2 \geq 3$

Z^*(upper) = 36.3, $x_1^* = 8.11$, $x_2^* = 2$

Z^*(lower) = 33, $x_1^* = 5$, $x_2^* = 3$

$x_1 \leq 8$

$x_1 \geq 9$

Z^*(upper) = 36, $x_1^* = 8$, $x_2^* = 2$

$Z^* = 35.4$, $x_1^* = 9$, $x_2^* = 1.4$

Figure 7.2   Decision tree for alternative energy example.

For instance, imagine that the government is going to provide subsidies for renewable energy sources. The government has a budget of $6,666,000, and it is considering providing subsidies to two different energy production sources: solar power panels (x_1) and wind power turbines (x_2). The government has determined that due to the higher efficiencies involved with wind power, the public benefits of wind power ($6 million) are twice as high as the benefits of solar power ($3 million).

In order to get private investments in the various energy sources, they will need to get $900,000 for each wind power plant and $600,000 for each solar power plant. The government has also agreed to provide $560,000 to help build the smart grid transmission lines necessary to reach these new energy sources, which cost 10 times more for wind power ($160,000) than solar power ($16,000). Solar power plants can be built close to the existing energy infrastructure, but wind power plants need to be placed in locations that receive high winds, which are often located in remote areas far removed from the existing infrastructure. Finally, due to the concerns that environmentalists raised about wind power's potential impacts on migratory birds, they have forbidden any more than four new wind power plants to be built. If the government agency seeks to maximize public benefits given its objective function, subsidy budget, transmission, and total construction constraint, the LP problem is as follows:

Max: $Z = 3x_1 + 6x_2$ (0)

s.t.:

$600x_1 + 900x_2 \leq 6{,}666$ (1)

$16x_1 + 160x_2 \leq 560$ (2)

$x_2 \leq 4$ (3)

$x_1, \quad x_2$ are integers   (4)

$x_1, \quad x_2 \geq 0$ (5)

To start off the branch-and-bound procedure, the LP equivalent of the problem must be solved. This step is sometimes called **initialization**.

The LP solution to this problem is $Z^* = 37.5$ (in millions), $x_1^* = 6.89$, $x_2^* = 2.80$. Note that both activities in this solution are non-integers; hence the solution violates the integer constraint specified above.

From a property stated earlier in this section, it is known that the optimal IP solution can never exceed the optimal LP solution. Hence, 37.5 becomes the **upper bound** (UB) for this problem as it is the optimal LP objective function value.

A **lower bound** (LB) for the optimal IP should be determined by rounding down all optimal LP activities to the nearest integer values, and by substituting this value into the objective function.[1] In this case, the LB is 30 since, $3(6) + 6(2) = 30$. Although we have not yet determined the optimal IP solution, we have established a lower and an upper bound for what the optimal IP objective function value can be. That is, now the optimal IP solution is known to have an objective function value between 30 and 37.5. This information is useful because if this range is relatively small, then this process could be stopped at this point, and the LB solution could be used as a "nearly optimal solution." While that solution would not be optimal, it may be acceptable, especially considering the amount of work that is necessary in finding the true optimal IP solution.

If the process is stopped here, then it would be possible to compute a "maximum percentage error" (MPE), which gives the maximum possible error (in objective function values) that is possible if the LB is used instead of the true IP optimal solution. The maximum percentage error formula is equal to:

$$\text{MPE} = \frac{\text{UB} - \text{LB}}{\text{UB}} \times 100.$$

In this case, the maximum percentage error is 20% because:

$$\text{MPE} = \frac{37.5 - 30}{37.5} \times 100 = 20\%.$$

This formula implies that if the current all-integer solution found by rounding down the initial LP solution is used, then the value of that objective function would be at most 20% lower than the value of the true optimal IP solution.[2] If this is not acceptable, then proceed to the next phase of the branch-and-bound procedure.

The next phase is called **branching**. To branch, use the following rule:

Using the initial LP solution, choose the activity value that is the furthest away from being an integer value to branch on.

In this case, $x_2^* = 2.8$ is selected. For the optimal IP solution, x_2 will either be less than or equal to 2, or ≥ 3 since it is restricted to being an integer. This suggests that branch 1 should be based on the initial problem modified by including the structural constraint $x_2 \leq 2$, and branch 2 should be based on the initial problem modified by including the structural constraint $x_2 \geq 3$.

Branch 1 is the following LP:

Max: $Z = 3x_1 + 6x_2$ (1.0)

s.t.:

$$600x_1 + 900x_2 \leq 6,666 \tag{1.1}$$

$$16x_1 + 160x_2 \leq 560 \tag{1.2}$$

$$x_2 \leq 4 \tag{1.3}$$

---

[1]In the case of minimization problems, generally one should round up all activities instead of rounding down.
[2]The maximum percentage error formula for minimization problems is the same as it is for maximization problems. However, in this case, the UB is not as "good" as the LB since smaller objective function values are preferred to larger values.

$$x_2 \leq 2 \tag{1.4}$$

$$x_1, \quad x_2 \geq 0 \tag{1.5}$$

The optimal solution for branch 1 is: $Z^* = 36.3$, $x_1^* = 8.11$, $x_2^* = 2$.
   Branch 2 is the following LP:

Max: $Z = 3x_1 + 6x_2$ $\tag{2.0}$

s.t.:

$$600x_1 + 900x_2 \leq 6,666 \tag{2.1}$$

$$16x_1 + 160x_2 \leq 560 \tag{2.2}$$

$$x_2 \leq 4 \tag{2.3}$$

$$x_2 \geq 3 \tag{2.4}$$

$$x_1, \quad x_2 \geq 0 \tag{2.5}$$

The optimal solution for branch 2 is: $Z^* = 33$, $x_1^* = 5$, $x_2^* = 3$.
   Notice that x_2 is an integer value for both branch solutions. The value of the objective function for the first branch is a UB for all solutions that include the structural constraint $x_2 \leq 2$ since this is an LP solution. Also, the value of the objective function for the second branch is a UB for all solutions that include the structural constraint $x_2 \geq 3$ since it is an LP solution.
   The next step is to compute a new UB using the following rule:
   The new UB will always be the highest objective function value of the LP problem for the current branches.
   In this case, the UB will be the larger of branch 1 and branch 2 objective function values. Since the first branch has a higher objective function value, its value (36.3) becomes the new UB. What about the LB?
   The LB value will always be the highest objective function value for the most recent all-integer solution.
   Recall that the original LB was 30. However, note that the second branch solution is also an all-integer solution that has a higher objective function value of 33. Consequently, the new LB becomes 33. The new maximum percentage error can now be calculated:

$$\text{MPE} = \frac{36.3 - 33}{36.3} \times 100 = 9.1\%.$$

After one iteration the maximum percentage error has been reduced from 20% to 9.1%. If the new maximum percentage error is acceptable, the process could be stopped at the current LP solution of $x_1^* = 5$, $x_2^* = 3$, $Z^* = 33$.
   The stopping rule for when the branch-and-bound method reaches the optimal IP solution is as follows:
   The optimal IP solution has been found whenever the iteration results in the UB being equal to the LB, in which case the maximum percentage error is zero.[3]

---

[3]For complicated problems that may require a long time to solve, you may consider establishing cut-off rules other than having the maximum percentage error of zero. However, in general, it is best to initially set the Solver Tolerance to zero and to allow some small level of error.

In this case, since the UB is larger than the LB, the process needs to continue in order to find the true optimal IP solution. Since the first branch has a higher objective function value, branch from this problem. Because x_2 is an integer in the first branch solution, the next two branches will be based on adding the constraints for x_1 which is not currently an integer. As the current value of x_1 is 8.11, the third branch will be identical to the first branch, except that we add the structural constraint $x_1 \leq 8$. Similarly, let the fourth branch be the same as the first, except that we add the constraint $x_1 \geq 9$. These two additional branches and their solutions are shown below.

Branch 3 is:

Max: $Z = 3x_1 + 6x_2$ (3.0)

s.t.:

$$600x_1 + 900x_2 \leq 6{,}666 \tag{3.1}$$

$$16x_1 + 160x_2 \leq 560 \tag{3.2}$$

$$x_2 \leq 4 \tag{3.3}$$

$$x_2 \leq 2 \tag{3.4}$$

$$x_1 \qquad \leq 8 \tag{3.5}$$

$$x_1, \quad x_2 \geq 0 \tag{3.6}$$

The optimal solution for the branch 3 is: $Z^* = 36$, $x_1^* = 8$, $x_2^* = 2$.

Branch 4 is:

Max: $Z = 3x_1 + 6x_2$ (4.0)

s.t.:

$$600x_1 + 900x_2 \leq 6{,}666 \tag{4.1}$$

$$16x_1 + 160x_2 \leq 560 \tag{4.2}$$

$$x_2 \leq 4 \tag{4.3}$$

$$x_2 \leq 2 \tag{4.4}$$

$$x_1 \qquad \geq 9 \tag{4.5}$$

$$x_1, \quad x_2 \geq 0 \tag{4.6}$$

The optimal solution for branch 4 is: $Z^* = 35.4$, $x_1^* = 9$, $x_2^* = 1.4$.

Clearly, the branch 3 solution is superior to the branch 4 solution because it has a higher objective function value. Therefore, the branch 3 solution becomes the new UB. In addition, branch 3's objective function value of 36 becomes the new LB since it is the most recent all-integer solution. Is branch 3 the optimal solution? The answer is yes because the UB and LB solutions are equal.

To generalize, the following steps illustrate the logic behind the branch-and-bound method.

1. Set up the problem as an LP without integer restrictions to initialize the problem and to compute the original UB.

2. If the optimal solution is all integers, then stop because the optimal IP solution is equal to the optimal LP solution.

3. If the optimal solution is not all integers, then round down (for a maximization problem) all non-integer values, and compute the LB. The maximum percentage error can then be computed, that is:

$$\text{MPE} = \frac{\text{UB} - \text{LB}}{\text{UB}} \times 100.$$

4. Stopping Rule: If the UB is equal to the LB, then the solution is optimal with the value equal to the lower and upper bound value. Stop when this condition is achieved.

5. If the UB is greater than the LB, then branch on the LP with the highest objective function value. If this is immediately following the initial LP, then branch on the non-integer variable that is the furthest away from being an integer. Create two branches for the variable, one with $x_i \leq xi_i$ and the other with $x_i \geq xi_i + 1$, where xi_i is the integer value of x_i.

6. Solve each branch as an LP problem without the integer restriction.

7. Recompute the upper and LBs. The UB is the LP solution with the largest objective function value for which there are no branches. The LB is the most recent (largest valued) all-integer solution.

8. Go to step 2 and continue the process until you have satisfied the stopping rule.

7.3   MIXED-INTEGER PROGRAMS

The branch-and-bound method is also applicable to mixed-IP. The procedures are essentially identical, with the exception that the LB values are calculated using only integer values for the variables constrained to be integers. To illustrate, suppose that only x_2 in the previous example is constrained to be an integer. That is:

Max: $Z = 3x_1 + \quad 6x_2$ (0)

s.t.:

$$600x_1 + 900x_2 \leq 6{,}666 \tag{1}$$

$$16x_1 + 160x_2 \leq 566 \tag{2}$$

$$x_2 \leq 4 \tag{3}$$

$$x_2 \text{ is an integer} \tag{4}$$

$$x_1, \quad x_2 \geq 0 \tag{5}$$

Using the branch-and-bound method, once again start off with the initialization phase, where the problem is solved as an LP. The LP solution is: $Z^* = 37.5$, $x_1^* = 6.89$, $x_2^* = 2.80$.

The LB is computed next. However, this time substitute $x_1 = 6.89$ and $x_2 = 2$ into the objective function to get the LB since x_1 can be a real number. Therefore, the LB in the case of mixed-IP will always be greater than or equal to the LB for its all IP counterpart. In this case, the LB is 32.67 (recall that the equivalent LB for the all-IP was 30). Using the UB and LB values, the maximum percentage error for this problem is 12.9%.

Therefore, branch only on the x_2 variable for this problem. The first and second branches are identical to the case before.

Branch 1 is the following LP:

$$\text{Max: } Z = 3x_1 + 6x_2 \tag{1.0}$$

s.t.:

$$600x_1 + 900x_2 \leq 6{,}666 \tag{1.1}$$

$$16x_1 + 160x_2 \leq 560 \tag{1.2}$$

$$x_2 \leq 4 \tag{1.3}$$

$$x_2 \leq 2 \tag{1.4}$$

$$x_1, \quad x_2 \geq 0 \tag{1.5}$$

The optimal solution to this problem is: $Z^* = 36.3$, $x_1^* = 8.11$, $x_2^* = 2$.
    Branch 2 is the following LP:

$$\text{Max: } Z = 3x_1 + 6x_2 \tag{2.0}$$

s.t.:

$$600x_1 + 900x_2 \leq 6{,}666 \tag{2.1}$$

$$16x_1 + 160x_2 \leq 560 \tag{2.2}$$

$$x_2 \leq 4 \tag{2.3}$$

$$x_2 \geq 3 \tag{2.4}$$

$$x_1, \quad x_2 \geq 0 \tag{2.5}$$

The optimal solution to this problem is: $Z^* = 33$, $x_1^* = 5$, $x_2^* = 3$.

    The new UB is given by the value of the branch 1 objective function since it is larger than the branch 2. The new LB is also given by the branch 1 objective function because it represents the highest value solution that restricts x_2 to be an integer. Since the UB is equal to the LB, branch 1 represents the optimal mixed-IP solution. Comparing the solution to this problem to the solution to the all-integer problem reveals that the mixed-integer solution yields a higher objective function value than the all-integer solution. Again, this is due to the fact that the mixed-integer problem is less constrained than the all-integer problem.

    It should be clear that mixed-IP problems are easier to solve than all-IP problems. Consequently, only the activities that are absolutely restricted to being integers should be constrained. All other activities should not be constrained to be integers.

7.4   SOLVER'S INTEGER AND BINARY PROGRAMMING OPTIONS

Using the branch-and-bound procedure, Solver can handle both all-IP problems and mixed-IP problems. To invoke IP, additional constraints need to be added in the Solver dialogue box. To activate IP in Solver, first highlight the decision variable(s) that should be constrained to an integer value and then click the "Constraints" folder. Then use the drop-down menu to select "int" to signify an integer constraint. As shown in Figure 7.3, this can be accessed in the Solver dialogue box where we have previously adjusted the inequality constraints.

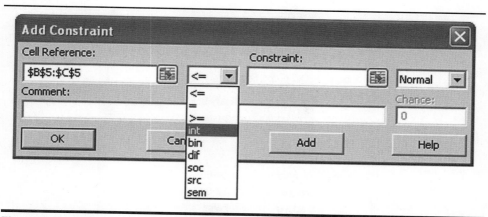

Figure 7.3 Setting integer constraints in Solver.

Solver can also handle a further restricted class of IP problems called **binary integer programming**. This means that all integer variables are restricted to be either 0 or 1. Solver also uses the branch-and-bound method to solve binary IP problems. To constrain a decision variable to be binary select the "bin" (for binary) option in the constraint section of the Solver dialogue box and then continue to solve as normal (see Figure 7.3).

A key option with integer and binary programming is the Integer Tolerance parameter, which sets the stopping rule for Solver. In earlier versions of Solver, the default Integer Tolerance level was 5%, which means that Solver would stop if it found a solution with an objective function value that was within 5% of the optimal (nonconstrained) solution. Given the improvements in computer speed and the capability of Solver, the default setting is now 0% (Figure 7.4). We recommend you keep this setting at 0, since, if a solution exists, this solution will be guaranteed to be optimal. If you are dealing with more complicated problems that are more difficult to solve quickly, then the Integer Tolerance can be increased, which should decrease the amount of time needed to find a solution.

In the following sections, several applications of IP are demonstrated and are designed to help you become more familiar with setting integer and binary constraints.

7.5 CAPITAL BUDGETING—A CASE OF WATER CONSERVATION

Capital budgeting involves the allocation of a finite amount of capital to alternative projects. Capital can mean money or it can mean human-made resources, such as machinery. In business applications, capital is often defined as a type of money or financial measure, such as cash, stocks, bonds, and savings. Economists generally define capital more broadly. For example, McConnell, Brue, and Flynn (2010) defines capital as "all manufactured aids used in producing consumer goods and services. Included are all tools, machinery, equipment, factory, storage, transportation, and distribution facilities." (p. 10.) Hence, the uses of capital budgeting may include monetary investments among alternative projects, or the allocation of human-made aids to production or project alternatives.

The use of IP in capital budgeting is particularly appropriate as investments in projects tend to be either all or nothing, rather than continuous. It is often unrealistic to assume

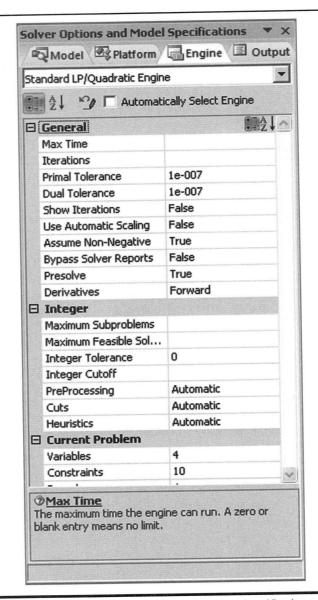

Figure 7.4   Integer tolerance in the Solver Options and Model Specification menu.

that fractional amounts of projects like constructing buildings, power plants, or highways can be done.

Consider, for example, the following application. Suppose that a metropolitan water district in the arid southwest is evaluating how much to invest in several projects, which will yield different expected water conservation benefits over a five-year period. A summary of the expected water savings, capital requirements (cost of investment), and available capital for each project is given below. The water district is considering investment in three water savings projects: a public outreach campaign to encourage voluntary water conservation (*cons*), a plan to identify and replace existing pipes that have leaks (*pipes*), and a one-time program to offer subsidies for homeowners buying low flow toilets (*lflow*). Additionally, the water district is required to fund the provision of new service to a recently built housing project (*home*).

Project	Water Savings (acre-feet)	Cost of Investment				
		Year 1	Year 2	Year 3	Year 4	Year 5
Conservation Campaign (*cons*)	500,000	100,000	110,000	75,000	60,000	60,000
Replace Bad Pipes (*pipes*)	600,000	100,000	100,000	100,000	100,000	100,000
Low Flow Toilets (*lflow*)	200,000	110,000	0	0	0	0
New Housing Project (*home*)	−50,000	15,000	15,000	15,000	15,000	15,000
Available Capital		350,000	220,000	210,000	210,000	110,000

The linear (non-integer) programming problem is:

Max: $Z =$ $\quad 500cons + 600pipes + 200lflow - 50home$ $\qquad\qquad$ (0)

s.t.:

(Year 1 Cap) $\quad 100cons + 100pipes + 110lflow + 15home \le 350$ $\qquad$ (1)

(Year 2 Cap) $\quad 110cons + 100pipes \qquad\quad + 15home \le 220$ $\qquad$ (2)

(Year 3 Cap) $\quad 75cons + 100pipes \qquad\quad + 15home \le 210$ $\qquad$ (3)

(Year 4 Cap) $\quad 60cons + 100pipes \qquad\quad + 15home \le 210$ $\qquad$ (4)

(Year 5 Cap) $\quad 60cons + 100pipes \qquad\quad + 15home \le 110$ $\qquad$ (5)

(Max *cons*) $\quad cons \qquad\qquad\qquad\qquad\qquad\qquad \le 1$ $\qquad$ (6)

(Max *pipes*) $\qquad\qquad pipes \qquad\qquad\qquad\qquad \le 1$ $\qquad$ (7)

(Max *lflow*) $\qquad\qquad\qquad lflow \qquad\qquad \le 1$ $\qquad$ (8)

(Min *home*) $\qquad\qquad\qquad\qquad\qquad home = 1$ $\qquad$ (9)

$\qquad cons, \qquad pipes, \qquad lflow, \qquad home \ge 0$ $\qquad$ (10)

The objective function here is to maximize the expected water savings (measured in acre-feet) among four capital projects over the next five years. The first five constraints restrict the investment requirements for all projects from exceeding a specific annual budget for each year. The next three constraints insure that an investment is not made in more than one of each of the projects; for instance, investments cannot take place in two public outreach efforts promoting voluntary water conservation. Finally, the last structural constraint requires that the service to the new housing development, which actually entails a loss of water, is conducted. Presumably this project is required by law, and the water district must provide this service. As shown in Figure 7.5, the optimal solution to this problem is:

$$cons^* = 1, pipes^* = 0.35, lflow^* = 1, home^* = 1, Z^* = 860.$$

That is, the company should adopt the public outreach conservation program and the subsidy program for low flow toilets, as well as providing service to the new housing development. However, the solution says to invest only 35% in replacing leaky pipes.

	A	B	C	D	E	F	G	H	I
1			cons	pipes	lflow	home			
2	Decision Variables						Max Savings		
3	Objective Function		500	600	200	-50	0		
4	Constraints						Total		RHS
5		Year 1	100	100	110	15	0	≤	350
6		Year 2	110	100		15	0	≤	220
7		Year 3	75	100		15	0	≤	210
8		Year 4	60	100		15	0	≤	210
9		Year 5	60	100		15	0	≤	110
10		Max cons	1				0	≤	1
11		Max pipes		1			0	≤	1
12		Max lflow			1		0	≤	1
13		Max home				1	0	=	1

Figure 7.5 Solution for water conservation problem.

While replacing 35% of the leaking pipes may seem like a realistic outcome, it may not make sense if a large part of the cost involves identifying which pipes are leaking in the first place and acquiring the equipment to replace them. In this case, investing just 35% of the needed budget could yield no water savings as no pipes are actually replaced.

The non-integer solution to this problem is unrealistic for two reasons. First, we are not using IP, and consequently all solution activities are completely divisible. Second, we are implicitly assuming that project revenue (capital) cannot be carried over between years in this five-year period. If the model includes the ability to transfer capital that is not used in the present year for use in succeeding years, then there might be enough money to invest in all four projects. In fact, with the optimal solution shown above (with divisibility but no transfers), the first four constraints are slack yielding unspent funds of $90,000, $60,000, $85,000, and $100,000, respectively, for the first four years. Therefore, before the IP counterpart is introduced, this problem will be modified by introducing **transfer activities** and then solved as an LP problem.

Transfer Activities and Capital Budgeting

To introduce transfer activities into the LP, let:

$$c_i = \text{Unused capital in year i with i} = 1, \dots, 5.$$

Assuming that these new activities do not influence the value of the objective function, these objective function coefficients should be assigned values of zero. So the new objective function is:

$$\text{Max: } Z = 500cons + 600pipes + 200lflow - 50home$$
$$+ 0c_1 + 0c_2 + 0c_3 + 0c_4 + 0c_5.$$

In the former model, any unused capital in a given year was discarded: that is, any slack was not available for future years. Under the new formulation, it is assumed that if there is slack in the first year, then it will become available for use in the next year. Hence, the first constraint becomes:

$$\text{(Year 1 Cap)} \quad 100cons + 100pipes + 110lflow + 15home + 1c_1 = 350.$$

Rearranging this constraint by solving for c_1 gives the definition of c_1, that is:

$$c_1 = 350 - 100cons - 100pipes - 110lflow - 15home.$$

Since $c_1 \geq 0$, capital savings in Year 1 will always be non-negative. If $c_1 > 0$, then the amount of c_1 becomes available for use in Year 2. To model this, we write the second constraint as:

$$\text{(Year 2 Cap)} \quad 110cons + 100pipes + 15home - c_1 + c_2 = 220.$$

Rearranging terms, this constraint is equivalent to:

$$\text{(Year 2 Cap)} \quad 110cons + 100pipes + 15home = 220 + c_1 - c_2.$$

That is, the amount of capital available for Year 2 is equal to the original amount (220) plus the amount carried over from Year 1 (c_1) minus the amount that will be carried over to Year 3 (c_2). The third, fourth, and fifth constraints are:

$$\text{(Year 3 Cap)} \quad 75cons + 100pipes + 15home - c_2 + c_3 = 210,$$

$$\text{(Year 4 Cap)} \quad 60cons + 100pipes + 15home - c_3 + c_4 = 210,$$

$$\text{(Year 5 Cap)} \quad 60cons + 100pipes + 15home - c_4 + c_5 = 110.$$

These three constraints have the same interpretation as constraint (2). The new model is:

$$\text{Max: } Z = \quad 500cons + 600pipes + 200lflow - 50home + 0c_1 + 0c_2 + 0c_3 + 0c_4 + 0c_5 \tag{0}$$

s.t.:

(Year 1 Cap)	$100cons + 100pipes + 110lflow + 15home +$	c_1	$= 350$	(1)
(Year 2 Cap)	$110cons + 100pipes +$	$15home -$	$c_1 + c_2 = 220$	(2)
(Year 3 Cap)	$75cons + 100pipes +$	$15home -$	$c_2 + c_3 = 210$	(3)
(Year 4 Cap)	$60cons + 100pipes +$	$15home -$	$c_3 + c_4 = 210$	(4)
(Year 5 Cap)	$60cons + 100pipes +$	$15home -$	$c_4 + c_5 = 110$	(5)
(Max cons)	$cons$		≤ 1	(6)
(Max pipes)	$pipes$		≤ 1	(7)
(Max lflow)	$lflow$		≤ 1	(8)
(Min home)	$home$		$= 1$	(9)
	$cons, \quad pipes, \quad lflow, \quad home, \quad c_1, c_2, c_3, c_4, c_5 \geq 0$		(10)

The optimal solution to this problem is:

$$cons^* = 1, pipes^* = 1, lflow^* = 1, home^* = 1, c_1^* = 25,$$

$$c_2^* = 20, c_3^* = 40, c_4^* = 75, c_5^* = 10, Z^* = 1{,}250.$$

		cons	pipes	lflow	home	c_1	c_2	c_3	c_4	c_5			
2	Decision Variables	1	1	1	1	25	20	40	75	10	Max Savings		
3	Objective Function	500	600	200	-50	90	60	1	1	1	$ 4,825		
4	Constraints										Total		RHS
5	Year 1	100	100	110	15	1					350	=	350
6	Year 2	110	100		15	-1	1				220	=	220
7	Year 3	75	100		15		-1	1			210	=	210
8	Year 4	60	100		15			-1	1		210	=	210
9	Year 5	60	100		15				-1	1	110	=	110
10	Max cons	1									1	≤	1
11	Max pipes		1								1	≤	1
12	Max lflow			1							1	≤	1
13	Max home				1						1	=	1

Figure 7.6 Revised problem formulation for water conservation problem, transfer and capital budgeting.

The company can now adopt all four projects and increase its water savings by 390,000 acre-feet. Hence, in this example the need to use IP has been eliminated by allowing for the possibility of unused capital in year t to transfer to the next year $t + 1$. The use of transfer activities, however, may not always result in an all-integer solution.

Integer Programming Formulation of Capital Budgeting

To formulate the original problem (without transfer activities) as an IP problem, simply eliminate constraints (6) through (8) and specify that all activities must be binary integers (0 or 1). Figure 7.6 shows the revised formulation of this problem. The solution to this problem is:

$$cons^* = 1, pipes^* = 1, lflow^* = 1, home^* = 1, Z^* = 4,825.$$

Multiple Choice and Mutually Exclusive Constraints

Suppose that the public outreach campaign to encourage voluntary water conservation (cons) actually has three possible projects that represent that amount and type of advertising that will be purchased as part of this effort. To help determine which of these three options is best for the metropolitan water district given its other choices, additional binary constraints can be added to the problem:

$$cons_i = 1 \text{ if original leaky project is adopted for } i = 1,2,3; 0 \text{ otherwise.}$$

Also, assume that only one of these three projects can be selected. To reflect this, the following multiple choice constraint is added:

$$cons_1 + cons_2 + cons_3 = 1.$$

Since $cons_1$, $cons_2$, and $cons_3$ can either be 0 or 1, this constraint will cause only one project to be selected. For example, if $cons_2 = 1$, then:

$$cons_1 + 1 + cons_3 = 1.$$

Solve for $cons_1$ yields:

$$cons_1 = 1 - 1 - cons_3, \text{ or}$$

$$cons_1 = 0 - cons_3.$$

The only possible value for $cons_1$ and $cons_3$ is zero, if the binary requirement and non-negativity are to hold. Note also that this constraint requires that one public outreach effort must be adopted. If you wanted to constrain the problem so that at most one public outreach project is adopted (which includes the possibility of no public outreach projects in the solution), then the constraint is:

$$cons_1 + cons_2 + cons_3 \leq 1.$$

This is called a mutually exclusive constraint. Alternatively, constraints can be used to require that at least one project be selected.

$$cons_1 + cons_2 + cons_3 \geq 1.$$

Binary constraints can also be used in a number of other ways: for instance if project 3 can only be done if project 2 is also done, the constraint is $cons_3 \leq cons_2$.

7.6   DISTRIBUTION SYSTEM DESIGN

In the transportation model introduced in Chapter 5, we derived the optimal shipment flows given fixed locations and amounts of supply and demand. Integer programming can also be used to find optimal plant locations.

Consider a planning problem in which m potential plant locations with plant capacities s_i and n retail outlets with demand d_j have been identified. Recall that in this problem, the objective is to select the amount to ship from site i to j, x_{ij}, with transportation costs c_{ij}. Now the problem is to minimize transportation and plant costs and find optimal plant locations.

Let $y_i = 1$ if a plant is constructed on site i; 0 otherwise,

$f_i =$ fixed cost of constructing plant; with capacity s_i.

We need a constraint that specifies that nothing can be shipped from site i if a plant is not constructed. This is accomplished by the following constraint:

$$\sum_{j=1}^{n} x_{ij} \leq s_i y_i \qquad i = 1, 2, \ldots, m, \text{ or}$$

$$\sum_{j=1}^{n} x_{ij} - s_i y_i \leq 0 \qquad i = 1, 2, \ldots, m.$$

If site i is not selected, then $y_i = 0$, and the sum of all shipments from i must be zero. If site i is selected, then $y_i = 1$ and the sum of all shipments from i cannot exceed the capacity of i (s_i). Additionally, another term must be included in the objective function to represent the fixed cost of plant construction, that is:

$$\sum_{i=1}^{m} f_i y_i$$

The complete model is:

$$\text{Min: } Z = \sum_{i=1}^{m}\sum_{j=1}^{n} c_{ij}x_{ij} + \sum_{i=1}^{m} f_i y_i \tag{0}$$

s.t.:

$$\sum_{j=1}^{n} x_{ij} - \quad s_i y_i \le 0 \quad i = 1, 2, \ldots, m \tag{1}$$

$$\sum_{i=1}^{m} x_{ij} \quad\quad \ge d_j \quad j = 1, 2, \ldots, n \tag{2}$$

$$x_{ij} \quad\quad \ge 0 \quad \text{for all } i \text{ and } j \tag{3}$$

$$y_i = 0, 1 \quad i = 1, 2, \ldots, m \tag{4}$$

A Distribution System Design Example

A person is thinking about going into the organic yogurt distribution business. The person has solicited business from four supermarkets located in different geographical areas: Locations A, B, C, and D. There are six possible plant sites to rent for organic yogurt factories: Locations 1, 2, 3, 4, 5, and 6. The weekly costs of renting out the factories are: \$100 (Factory 1), \$1,000 (Factory 2), \$2,000 (Factory 3), \$400 (Factory 4), \$1,300 (Factory 5), and \$1,000 (Factory 6). The weekly output of organic yogurt from the factories is: 500 (Factory 1), 700 (Factory 2), 1,000 (Factory 3), 1,000 (Factory 4), 750 (Factory 5), and 700 (Factory 6). The demand from each location is 500 cases of yogurt (Location A), 400 cases of yogurt (Location B), 600 cases of yogurt (Location C), and 800 cases of yogurt (Location D). The transportation costs (per case of yogurt) from each potential factory to each demand location are:

Supply Location	Demand Location			
	A	B	C	D
	(cents/case of yogurt)			
1	20	40	30	25
2	10	15	45	20
3	20	10	50	20
4	100	95	60	50
5	60	60	15	15
6	120	130	10	40

Assuming the objective is to minimize weekly factory and transportation costs, the IP problem can be set up as follows:

$$\text{Min: } Z = 100s_1 + 1000s_2 + 2000s_3 + 400s_4 + 1300s_5 + 1000s_6 + 0.2x_{1A}$$
$$+ 0.4x_{1B} + 0.3x_{1C} + 0.25x_{1D} + 0.1x_{2A} + 0.15x_{2B} + 0.45x_{2C}$$
$$+ 0.2x_{2D} + 0.2x_{3A} + 0.1x_{3B} + 0.5x_{3C} + 0.2x_{3D} + x_{4A} + 0.95x_{4B}$$
$$+ 0.6x_{4C} + 0.5x_{4D} + 0.6x_{5A} + 0.6x_{5B} + 0.15x_{5C} + 0.15x_{5D}$$
$$+ 1.2x_{6A} + 1.3x_{6B} + 0.1x_{6C} + 0.4x_{6D} \tag{0}$$

s.t.:

$$-500s_1 + x_{1A} + x_{1B} + x_{1C} + x_{1D} \le 0 \tag{1}$$

$$-700s_2 + x_{2A} + x_{2B} + x_{2D} + x_{2C} \le 0 \tag{2}$$

$$-1000s_3 + x_{3A} + x_{3B} + x_{3C} + x_{3D} \leq 0 \tag{3}$$

$$-1000s_4 + x_{4A} + x_{4B} + x_{4C} + x_{4D} \leq 0 \tag{4}$$

$$-750s_5 + x_{5A} + x_{5B} + x_{5C} + x_{5D} \leq 0 \tag{5}$$

$$-700s_6 + x_{6A} + x_{6B} + x_{6C} + x_{6D} \leq 0 \tag{6}$$

$$x_{1A} + x_{2A} + x_{3A} + x_{4A} + x_{5A} + x_{6A} \geq 500 \tag{7}$$

$$x_{1B} + x_{2B} + x_{3B} + x_{4B} + x_{5B} + x_{6B} \geq 400 \tag{8}$$

$$x_{1C} + x_{3C} + x_{4C} + x_{5C} + x_{6C} + x_{2C} \geq 600 \tag{9}$$

$$x_{1D} + x_{2D} + x_{3D} + x_{4D} + x_{5D} + x_{6D} \geq 800 \tag{10}$$

$$s_1, s_2, s_3, s_4, s_5, s_6 = 0, 1 \tag{11}$$

$$\text{Non-negativity} \tag{12}$$

In the model, s_i denotes supply location i (i=1, … , 6), and x_{ij} denotes shipment from supply location i to demand location j.

The optimal IP solution for this problem is presented in Figure 7.7. The solution indicates that factories 1, 2, 4, and 6 should be rented for yogurt production. These four sites

	A	B	C	D	E	F	G	H	I
1					**Factories**				
2	**Decision Variables**		s_1	s_2	s_3	s_4	s_5	s_6	**Total**
3		Factory Used	1	1	0	1	0	1	
4		Cases Shipped to A	200	300	0	0	0	0	500
5		Cases Shipped to B	0	400	0	0	0	0	400
6		Cases Shipped to A	0	0	0	0	0	600	600
7		Cases Shipped to A	300	0	0	400	0	100	800
8		Total	500	700	0	400	0	700	
9									
10	**Objective Function Coefficients**								
11		Rental Cost Per Week	100	1,000	2,000	400	1,300	1,000	
12		Shipping Cost to A	0.20	0.10	0.20	1.00	0.60	1.20	
13		Shipping Cost to B	0.40	0.15	0.10	0.95	0.60	1.30	**Minimize**
14		Shipping Cost to C	0.30	0.45	0.50	0.60	0.15	0.10	**Total Costs**
15		Shipping Cost to D	0.25	0.20	0.20	0.50	0.15	0.40	$ 3,005
16									
17	**Constraints**		Used	Capacity	Available	Shipped	Total		RHS
18	s_1 Output		1	-500	-500	500	0	≤	0
19	s_2 Output		1	-700	-700	700	0	≤	0
20	s_3 Output		0	-1,000	0	0	0	≤	0
21	s_4 Output		1	-1,000	-1,000	400	-600	≤	0
22	s_5 Output		0	-750	0	0	0	≤	0
23	s_6 Output		1	-700	-700	700	0	≤	0
24	A Demand					500	500	≥	500
25	B Demand					400	400	≥	400
26	C Demand					600	600	≥	600
27	D Demand					800	800	≥	800

Figure 7.7 Solution to the distribution system design example.

are optimal given their location and rental costs. In addition, the solution shows the seven routes that are optimal to minimize costs given the demands from the different locations.

7.7   SENSITIVITY ANALYSIS IN INTEGER PROGRAMMING

Sensitivity analysis for IP is generally more critical than for LP, as a very small change in one coefficient can lead to large changes in the value of the optimal solution, as well as changes in the optimal activity values. The primary reason for these potential large changes is the discontinuous nature of IP problems.

Consequently, effort should be dedicated to conducting sensitivity analysis on all key parameters in IP models. This analysis shows that if the model is very sensitive (i.e., tiny parameter changes result in substantial changes in the optimal solution), then the validity of the model's results could be questioned. This potential problem is particularly acute when the accuracy of the parameters in the model is insufficient.

Unfortunately, due to the discontinuous nature of these problems, Solver does not provide a sensitivity report as it does with standard LP problems. Therefore, to conduct sensitivity analysis, the **Optimization Parameter tool** offered by Solver should be used. This automates the **Parameter Analysis process**, which can be used to perform sensitivity analysis on IP models. This approach can also be used to extend sensitivity analysis for linear programs beyond the reports discussed in Chapter 3 and automate Parametric Programming in applications like the input supply and output demand derivation discussed in Chapter 4.

Solver can quickly vary a parameter over a large range of possible values and report the optimization results for each value of the parameter. A cell that contains a parameter in the Excel spreadsheet can be replaced with a **PSI Optimization Parameter** formula. This formula defines a set of values over which that parameter may be varied. This set of values may be defined by the minimum and maximum values of an interval, or explicitly by referring to a range of cells containing different parameter values. The syntax for this function is "=PSIOPTPARAM(MinVal, MaxVal)" for the interval definition and "=PSIOPTPARAM(RangeRef)" for the range definition, where **MinVal** and **MaxVal** define the bounds of the interval and **RangeRef** is a reference to a range of values.

Once a cell is defined as a PSI Optimization Parameter, click on the Reports button, and bring up the Optimization menu where the Answer, Sensitivity, and Limits Reports are usually found. From this menu, clicking on the Parameter Analysis button will bring up the **Multiple Optimizations Report** window, which should look similar to Figure 7.8. The top-left pane offers a list of the decision variables and the objective function value, any of which may be reported in the Parameter Analysis output. To track the value of an individual cell, click on its reference in the list and then click on the ">" button between the top-left and top-right panes to move it to the top-right pane of the window. To add all the values at once to the top-right pane, click on the ">>" button.

The bottom-left pane of this window lists the defined PSI Optimization Parameter cells. Click on the reference cell that should be tracked, and then click on the ">" button between the bottom-left and bottom-right panes to move this reference cell to the bottom-right pane of this window. The **Major Axis Points** field in the bottom-left of this window specifies how many values in the interval defined for the PSI Optimization Parameter will be used in the sensitivity analysis. For example, if the interval is from 0 to 10, then 11 major axis points are chosen, and in the sensitivity analysis the parameter will be varied over 11 values: 0, 1, 2, ..., 10. To generate a new Analysis Report click OK. Solver will

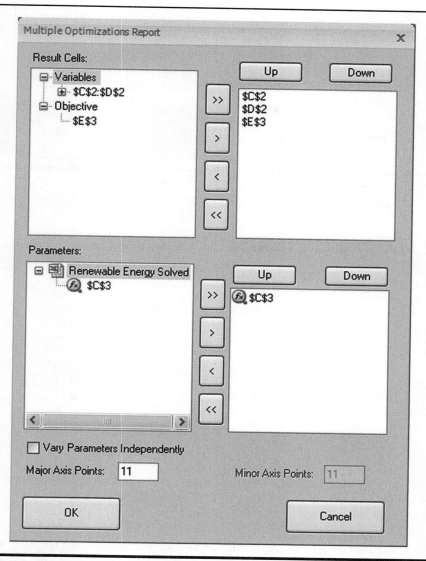

Figure 7.8   The Multiple Optimizations Report window.

then produce a report that lists all of the parameter values as well as the values of the tracked variables for the corresponding optimal solution.

To illustrate how to do these reports, let us again consider the example presented earlier in this chapter regarding alternative energy investments. In this situation, we found that the third branch of the problem yielded the optimal solution of $Z^* = 36$, $x_1^* = 8$, $x_2^* = 2$. Inspection of the result shows that none of the constraints are binding in the optimal solution (Figure 7.9). A reasonable sensitivity analysis in this case is to test how much the optimal solution would change by varying the constraints by $\pm10\%$. As shown in Table 7.1, the results of the sensitivity analysis show that the only constraint that affects the optimal solution with a change of

◢	A	B	C	D	E	F	G
1			x_1	x_2			
2	**Decision Variables**		8	2	**Max Benefits**		
3	**Objective Function Coefficients**		3	6	36		
4	**Constraints**				**Total**		**RHS**
5		Plant Cost	600	900	6,600	≤	6,666
6		Transmission Line Cost	16	160	448	≤	560
7		Max Wind Plants	0	1	2	≤	4

Figure 7.9 Optimal solution to alternative energy investments problem.

Table 7.1 Sensitivity Analysis for the Alternative Energy Investments Problem

Constraint 1	Z	Constraint 2	Z	Constraint 3	Z
5,999.4	33	504	36	3.6	36
6,132.72	33	515.2	36	3.68	36
6,266.04	33	526.4	36	3.76	36
6,399.36	33	537.6	36	3.84	36
6,532.68	33	548.8	36	3.92	36
6,666	36	560	36	4	36
6,799.32	36	571.2	36	4.08	36
6,932.64	36	582.4	36	4.16	36
7,065.96	36	593.6	36	4.24	36
7,199.28	36	604.8	36	4.32	36
7,332.6	39	616	36	4.4	36

plus or minus 10% is Constraint 1. When the value for this constraint is changed in this 10% range, the optimal solution varies from 33 to 39. Thus, the sensitivity analysis suggests that of the three constraints, Constraint 1 is the most binding.

7.8   RESEARCH APPLICATION: OPTIMIZING AGRICULTURAL LAND PROTECTION IN DELAWARE[4]

Between 1990 and 2005, the population of the State of Delaware grew at a rate nearly 28% faster than the rest of the United States. Most of this population growth was accommodated by converting agricultural land to residential use. Due to the large population growth and relatively small area, the American Farmland Trust designated the Mid-Atlantic coastal plain, including all of Delaware to be "endangered" (American Farmland Trust, 1997).

The Delaware Agricultural Lands Preservation Foundation (DALPF) was formed in the early 1990s with the goals of preserving agricultural open space and supporting the agricultural economy. The program traditionally receives more offers from landowners willing to sell conservation easements on their agricultural lands than the DALPF program can afford to acquire. According to the Delaware Department of Agriculture, as of 2009,

---

[4]This example is based on Messer and Allen (2010).

DALPF had acquired easements on over 500 farms (of a total state population of 2,300 farms) consisting of nearly 91,000 acres (17.5% of the state's 520,000 acres) for a total cost of approximately $150 million.

To determine which lands it will acquire, DALPF selects parcels using a sealed-bid auction mechanism and purchases conservation easements on the selected properties. The selection is determined by the highest percentage discount submitted by the landowner relative to the parcel's appraised market value. For example, if the easement is appraised at $1 million and a landowner offers a 40% discount, DALPF would pay the landowner $600,000 for the easement.

This selection approach by DALPF is a variant of a "greedy agent" algorithm as it seeks to acquire lands with the greatest discount until a constraint is met—in this case, the annual program budget. This selection approach—hereto referred to as the DALPF Algorithm—can be compared to a grocery shopper who buys a food item only because it is marked down in price more than any other item. However, this selection process has potential problems since no assurances exist that the item purchased, due to its high percentage discount, will have the taste and/or nutrition attributes that the shopper likes. Likewise, problems can arise if the foods with the most deeply discounted prices are also the most expensive (for instance, caviar or truffles) so that they are relatively more expensive, even with the large discount, than other high-quality foods with a smaller percentage reduction in price.

An alternative selection mechanism that is commonly used by conservation foundations is the Benefit Targeting (BT) Algorithm. Like the DALPF Algorithm, BT is a greedy agent algorithm except that instead of selecting the parcels being offered at the highest discount, BT selects, in an iterative process, parcels that have the highest levels of agricultural value until the budget is exhausted. Despite its widespread use in the conservation community, BT can lead to inefficient results from both an economic and conservation perspective (see, e.g., Underhill (1994); Rodrigues et al. (2000); Rodrigues and Gaston (2002); Azzaino et al. (2002); Messer (2006)). The source of the problem is that a parcel's price is only explicitly factored into the decision process to determine whether there is enough money still available.

As shown in Messer and Allen (2010), substantial efficiency gains can be achieved using binary LP. Their analysis used cost data from 509 willing sellers who had submitted offers to sell conservation easements to DALPF in its first decade of existence. Benefit information regarding agricultural suitability and the quality of the green infrastructure (Figure 7.10) was derived from geographic information systems (GIS) data. Relative weightings were determined through an Analytical Hierarchy Process involving 23 stakeholders representing 18 private conservation partners and local, state, and federal government agencies. The budget level was set at $93 million, approximately the total amount spent by DALPF in its first decade.

As seen in Table 7.2, the DALPF and BT greedy agent algorithms and binary LP are used to select acquisitions from the set of 509 parcels given a $93 million budget. The DALPF Algorithm protected 65,683.4 acres with an aggregate agricultural suitability score of 4,460,437 and an aggregate ecological services score of 1,736,429.[5] The aggregate results from the BT analysis were consistent with those of the DALPF Algorithm in terms of the number of acres protected (71.5 acres fewer), and the aggregate agricultural suitability scores were 1.3% higher. The most significant difference was that

---

[5] Unlike number of acres, aggregate scores for LESA and Core GI are not necessarily intuitive to interpret since they have been scaled by parcel size. However, the numbers are cardinal.

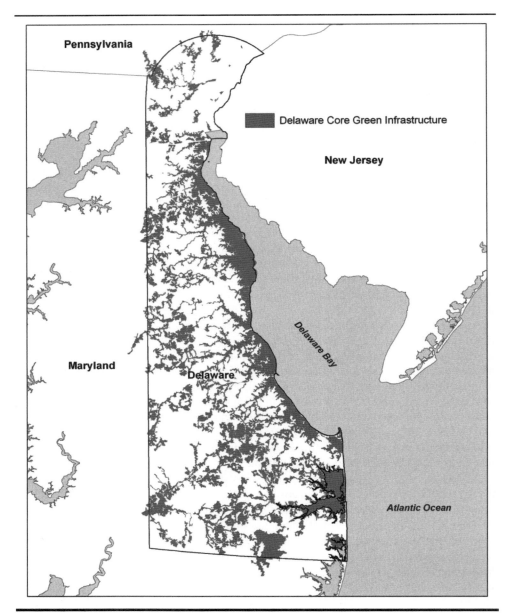

Figure 7.10   Map of Core Green Infrastructure areas in the state of Delaware.

BT Algorithm produced those results by selecting easements on 38.6% fewer farms, as the average selected farm was 277 acres for BT compared to 170.2 acres for DALPF.

Binary LP produced more conservation benefits than either the DALPF or BT Algorithms as it protected 447 farms (15.8% percent more than DALPF and nearly double the number protected by BT) with the same $93 million budget. Binary LP also protected 20.5% more acres (13,446) and yielded higher levels of aggregate agricultural values—as measured by Land Evaluation and Site Assessment (LESA) scores—and higher ecological values—as measured by Core Green Infrastructure (Core GI) scores. Relative to the

Table 7.2 Benefit Results

Selection Algorithm	Number of Farms Protected	Total Cost ($)	Acres	Aggregate Values	
				Agricultural Suitability	Ecological Services
DALPF	386	92,986,682	65,683.4	4,640,437	1,736,429
Benefit Targeting	237	92,997,985	65,611.9	4,701,728	1,831,548
Binary LP	447	92,999,225	79,129.5	5,597,928	2,067,438

aggregate scores obtained by DALPF, these scores were 20.6% and 19.1% higher, respectively. Similarly, in comparison to BT, binary LP protected 20.6% more acres and produced aggregate LESA and Core GI values that were 19.1% and 12.9% higher. More importantly, these gains in conservation benefits did not occur by purchasing smaller farms—in fact, the size of the average farm protected by binary LP was 7 acres (4.0%) larger than the one protected by the DALPF Algorithm.

Calculations of the relative cost effectiveness, which measured the amount of the additional funds that would be needed for the DALPF and BT Algorithms to achieve an equivalent number of acres (79,129.5) as binary LP with a $93.0 million budget, suggest that the DALPF and BT Algorithms would have required an additional $20.7 million and $19.8 million, respectively.

7.9 RESEARCH APPLICATION: FARMLAND CONSERVATION WITH A SIMULTANEOUS MULTIPLE-KNAPSACK MODEL

This research application introduces a multiple-knapsack binary IP model and shows how in situations where there are multiple conservation programs, higher levels of benefits can be achieved by developing a model that simultaneously considers all potential funding sources and selects an optimal set of parcels for each program. The application of this approach is farmland preservation in Baltimore County, Maryland. Baltimore County has one of the 10 largest agricultural protection programs in the United States (Sokolow, 2006). According to staff estimates, they will have protected over 30,000 acres of farmland by 2006.

Baltimore County has gained recognition in the conservation community by being the first conservation program to incorporate the concept of benefit-cost ratio targeting in its selection of which agricultural land to preserve from a pool of willing sellers. Benefit-cost ratio targeting has been a selection approach advocated by economists as a way of getting improved aggregate benefit results that can approach the optimal results described above with binary IP (Messer 2006).

According to Wally Lippincott, Baltimore County Land Preservation Administrator, "After trying for years to balance price with farm quality using rank based methods, we switched to optimization through benefit–cost ratio targeting. In the first three years, Baltimore County has been able to protect an additional 680 acres for the same amount of funds that would otherwise have been spent. This also translates into a savings of approximately $5.4 million."

Baltimore County staff seeks to preserve agricultural lands by receiving funds from a variety of sources including the Maryland Agricultural Land Preservation Foundation (MALPF), the State's Rural Legacy Program, and direct financing through the County's own budget process. Interestingly, while these programs tend to share the definition of how to measure the benefits of agricultural land, they differ with respect to how much they are willing to pay for it. In the case of MALPF, the program conducts independent appraisals of the properties easement value (i.e., the non-agricultural value of the property that the landowner could receive if she sold the land for residential or commercial development). MALPF also asks landowners to provide a discount on this appraised amount when they want to sell their easement to the program in exchange for increasing the likelihood of being accepted. In contrast, for funds coming from Baltimore County's budget, the staff and an advisory board have developed a formula that determines the maximum amount that the County will pay for the land based on the size of the parcel, the type of soils on the property, and other factors. Thus, in any given year, the cost of conservation for a particular parcel may differ depending on the program funding source. Owners are aware of this discrepancy and therefore decide whether to submit their offer to sell to one or both of the programs.

Traditionally, the selection method used by the staff and advisory board of Baltimore County proceeds in a sequential manner where first selections are made for the MALPF program. For the second phase of the selection process, the County selects among the parcels that are not selected in the initial MALPF phase (and are still interested in being considered for County funds) and the parcels that applied only for county funds. Similarly, the remaining parcels are considered for a third phase, which represents the Rural Legacy Program, and, in some cases, a fourth phase is added which includes end-of-year discretionary funding from Baltimore County. This sequential approach fails to take advantage of the disparity between the programs. Additionally, by making the selections sequentially, Baltimore County cannot take full advantage of the remainder of program budgets, which can frequently be large, given the costs of conservation traditionally range in the hundreds of thousands of dollars per project.

The process of selection of parcels for conservation can be viewed as a classic 0-1 knapsack problem. The mechanism behind a knapsack model is to pick some of the available items to achieve maximum total utility, while the total weight of the chosen items must not exceed the stated limit of the knapsack. This research considers the separate conservation programs as separate knapsacks and then evaluates the overall benefits that can be achieved by making the selection for each knapsack sequentially or simutanously. The budget of each of the programs in the simultaneous case is taken as the weight limit of the knapsack, and the conservation benefits are maximized using binary IP.

The model specification for the sequential knapsack method is as follows. The decision variables of the model are defined as $x_{ij} = \{0,1\}$ where 0 denotes parcel i is not selected and 1 denotes parcel i is selected in program j. The objective function seeks to maximize the conservation benefit for conservation program j.

$$\text{Max: } V_j(x) = \sum_i^I x_{ij} v_i \tag{1}$$

s.t.:

$$\sum_{j=1}^{J} x_{ij} \leq 1 \tag{2}$$

$$\sum_{i=1}^{I} c_{ij} x_{ij} \leq b_j \tag{3}$$

where $i = 1, 2, \ldots, I$ denotes an index for land parcels, j denotes an index for conservation programs and also the order of program participation. In this simplified example, $j = 1$ denotes the state MALPF program and $j = 2, \ldots, J$ denotes Baltimore County programs. Additionally, v_i denotes the conservation value for parcel i, b_j denotes budget for program j, c_{ij} denotes cost of parcel i appraised by program j, and v is the aggregate conservation value, such that:

$$v = \sum_j v_j \qquad (4)$$

In contrast, the simultaneous multiple-knapsack model can be expressed as the following:

$$\text{Max: } V(x) = \sum_i^I \sum_j^J x_{ij} v_j \qquad (5)$$

s.t.:

$$\text{For all } i \quad \sum_{j=1}^J x_{ij} \leq 1 \qquad (6)$$

$$\text{For all } j \quad \sum_{i=1}^I c_{ij} x_{ij} \leq b_j \qquad (7)$$

Note that for both models, a parcel may not be able to participate in all programs. If parcel i fails to participate in program j, then there is no x_{ij} in decision variables. Equations 2 and 6 imply that once a parcel is purchased, this parcel is not available for further consideration, and a parcel need not be selected in any program.

This research used data from the applicants to Baltimore County for acquisition with MALPF and/or County funding in 2008 and 2009. As shown in Table 7.3, the simultaneous multiple-knapsack model yields significant improvements in the overall conservation outcome. For instance, in 2008, the simultaneous model would have protected an additional 9.6% more conservation benefit and 7.2% more acres compared to the

Table 7.3   Results of 2008 and 2009 Portfolios

	Aggregate Conservation Benefit	Acres	Parcels Selected	Amount Spent ($)
2008 Data				
Simultaneous Model	134,649	2,016	29	10,728,994
Sequential Model	122,879	1,880	29	10,725,157
Improvement	+11,770	+136		
	+9.6%	+7.2%		
2009 Data				
Simultaneous Model	46,929	594	11	3,596,608
Sequential Model	43,744	568	10	3,560,051
Improvement	+3,815	+26		
	+7.3%	+4.6%		

achievements of the sequential multiple knapsack model. In 2009, 7.3% more conservation benefit and 4.6% more acres would be protected. Thus in summary, by using the simultaneous multiple-knapsack approach, Baltimore County could use the same financial resources and protect an additional 162 acres of high quality agricultural land valued at approximately $1.1 million. Recall these benefits would be in addition to the nearly 700 acres and more than $5 million Baltimore County had already received in benefits by moving away from the benefit targeting and selecting a more optimal approach.

SUMMARY

Real-world problems often involve integer solutions: for instance, hybrid battery producers cannot build half a battery, and many investment decisions require a commitment to either do it or not do it (a form of a binary solution). The theory behind integer and binary programming advanced a great deal with the introduction of the branch-and-bound algorithm and other IP methods not discussed here, such as Gomory cuts and implicit enumeration. However, applications lagged behind due primarily to the absence of computing power sufficient to drive large IP models. This situation has since changed a great deal. While larger IP models can still be complicated and time consuming, advanced computers have made it possible to solve more realistic integer problems without sacrificing much with regard to speed.

This chapter introduced integer and binary programming and discussed how a variety of problems could be set up to involve all-integer, mixed-integer, and binary solutions. A simple example was provided that illustrated the procedures involved with the branch-and-bound method and how it uses an iterative process to move from an initial set-up to an optimal, feasible integer solution by identifying and comparing upper and lower bounds.

A variety of problems were presented that illustrated the use of integer and binary programming. These problems and research examples were related to hybrid battery choice, water conservation investments, distributive system designs, and agricultural preservation. This chapter also discussed how models can be constructed to include multiple choice or mutually exclusive constraints, which can be used in the development of more sophisticated models.

Finally, this chapter discussed how to solve integer and binary programming problems using Solver and what model specifications, such as "Integer Tolerance," are important for finding optimal solutions. Additionally, this chapter discussed how to conduct sensitivity analysis in the context of IP.

REFERENCES

American Farmland Trust. (1997). *Farming on the Edge*. Washington, D.C: American Farmland Trust.

Amundsen, O. M., Messer, K. D., & Allen, W. L. (2009). The next big step in strategic land conservation: Conservation optimization. Eastern Lands Resource Council White Paper.

Azzaino, Z., Conrad, J. M., & Ferraro, P. J. (2002). Optimizing the riparian buffer: Harold Brook in the Skaneateles Lake Watershed, New York. *Land Economics, 78*, 501–514.

McConnell, C. R., Brue, S. L., & Flynn, S. M. (2010). *Microeconomics: Principles, Problems, and Policies*, 18th ed. Boston, MA: McGraw-Hill Irwin.

Messer K. D. (2006). The conservation benefits of cost-effective land acquisition: A case study in Maryland. *Journal of Environmental Management, 79*, 305–315.

Messer, K. D., & Allen W. (2010). Applying optimization and the analytic hierarchy process to enhance agricultural preservation strategies in the state of Delaware. *Agricultural and Resource Economics Review, 39*(3), 442–456.

Rodrigues, A. S. L., Cerdeir J. O., & Gaston, K. J. (2000). Flexibility, efficiency, and accountability: Adapting reserve selection algorithms to more complex conservation problems. *Ecography, 23,* 565–574.

Rodrigues, A. S. L., & Gaston K. J. (2002). Optimization in reserve selection procedures—Why Not? *Biological Conservation, 107,* 123–129.

Sokolow, A. D. (2006). A national view of agricultural easement programs: Profiles and maps— report 2. American Farmland Trust and Agricultural Issues Center, DeKalb, Illinois.

Sokolow, A. D., & Zurbrugg, A. (2003). A national view of agricultural easement programs: Profiles and maps–report 1. American Farmland Trust and Agricultural Issues Center, DeKalb, Illinois.

Underhill, L. G. (1994) Optimal and suboptimal reserve selection algorithms. *Biological Conservation, 70,* 85–87.

EXERCISES

1. At every harvest season, the wine estate has to manage the transportation of grapes from its three vineyards (denoted as v_1, v_2, and v_3) to its two wineries (denoted as w_1 and w_2). It is estimated that this year the total output from the three vineyards is 300 crates, 250 crates, and 350 crates, and stock capacity of the two wineries is 300 crates and 250 crates. Handling cost of shipments from each vineyard to each winery is listed in the following table.

	v_1 ($/crate)	v_2 ($/crate)	v_3 ($/crate)	Demand (crate)
w_1	10	7.5	4.5	300
w_2	5.5	7	6	250
Supply (crate)	200	250	350	

   Formulate the problem as an IP problem to minimize handling cost, and then solve it in Solver.

2. Suppose in the previous wine estate case, in order to simplify the shipping pattern, the owner of the wine estate would like that either all crates from one vineyard be fully shipped to one winery or not at all, given the condition that the stock capacity of each winery should be fully used. In the case that the output quantity from the vineyard is greater than the stock capacity of winery, $3 of overstock fee will occur for each crate over the stock capacity. Formulate the exercise as a binary programming problem and solve it in Solver.

3. Solve the following problem using the branch-and-bound method.

   Max: $Z = 20x_1 + 30x_2$ (0)

   s.t.:

   $$x_1 + 5x_2 \leq 20 \tag{1}$$

   $$2x_1 + x_2 \leq 115 \tag{2}$$

   $$x_2 \leq 3.5 \tag{3}$$

   $$x_1, \quad x_2 \geq 0 \text{ and integer} \tag{4}$$

4. Solve the following problem using the branch-and-bound method.

$$\text{Max: } Z = 35x_1 + 40x_2 \tag{0}$$

s.t.:

$$8x_1 + 12x_2 \leq 5300 \tag{1}$$

$$4.5x_1 + x_2 \leq 600 \tag{2}$$

$$x_1, \quad x_2 \geq 0 \text{ and integer} \tag{3}$$

5. Solve the following problem using the branch-and-bound method.

$$\text{Max: } Z = 6x_1 + 3x_2 \tag{0}$$

s.t.:

$$3.42x_1 + 4.15x_2 \leq 14 \tag{1}$$

$$3x_1 + x_2 \leq 70 \tag{2}$$

$$1.4x_1 \leq 25 \tag{3}$$

$$x_1, \quad x_2 \geq 0 \text{ and integer} \tag{4}$$

6. Solve the following problem using the branch-and-bound method.

$$\text{Max: } Z = 4x_1 + 6x_2 \tag{0}$$

s.t.:

$$97.5x_1 + 136.5x_2 \leq 682.5 \tag{1}$$

$$2x_1 + 20x_2 \leq 70 \tag{2}$$

$$0.5x_1 \leq 2 \tag{3}$$

$$x_1, \quad x_2 \geq 0 \text{ and integer} \tag{4}$$

7. Solve the following problem using the branch-and-bound method.

$$\text{Max: } Z = 60x_1 + 48x_2 \tag{0}$$

s.t.:

$$3x_1 + 6.5x_2 \leq 33.5 \tag{1}$$

$$4x_1 + 2.5x_2 \leq 27.5 \tag{2}$$

$$x_1, \quad x_2 \geq 0 \text{ and integer} \tag{3}$$

8. A farmer wishes to invest \$12,000 in one of the four crops A, B, C or D in the farm so as to maximize its profit. The fixed cost of each crop and their revenues are summarized in the table on the next page. According to the market situation, the farmer decides to include the following principles: if C is planted, then D has to be planted; if A is planted, then B cannot be planted.

Project	A	B	C	D
Price ($)	6000	5000	5000	4000
Value ($)	8000	7000	8000	6000

Formulate an integer program to select the best crop.

9. An international trade company is considering opening warehouses in four possible states adjacent to the water: California, Texas, Louisiana, and South Carolina. The company's business is mainly importing latex around the world to supplement the U.S. latex market. The four exporters are from Thailand, Indonesia, Mexico, and Vietnam. Suppose each warehouse has to supply the U.S. market with at least 60 tons of latex per week. The Thai exporter can supply 80 tons per week, the Indonesia exporter can supply 70 tons per week, the Mexican exporter can supply 60 tons per week, and the Vietnamese exporter can supply 40 tons per week. The shipping costs per ton are shown below (in thousands of dollars).

From/To	California	Texas	Louisiana	South Carolina
Thailand	20	40	42	45
Indonesia	26	18	23	48
Mexico	18	15	18	35
Vietnam	24	50	45	50

Model the exercise as an IP problem to help decide the best shipping pattern that will minimize the total cost of importing latex.

10. Reconsider the previous international latex trade example. Suppose the weekly fixed cost of keeping each warehouse open is $7,000 for California, $4,000 for Texas, $3,000 for Louisiana, and $3,500 for South Carolina. Moreover, only three sites will be selected from the four possible locations. In order to minimize the total operation cost, rebuild the exercise as a mixed-integer problem and solve it in Solver.

11. Using the example in Exercise 10, instead of selecting three sites out of four possible locations, add the following conditions:

   a. If the California warehouse is opened, the Texas warehouse must be opened.
   b. At least two warehouses should be opened.
   c. Either the Texas or Louisiana warehouse must be opened, but not both.

   What are your additional constraints?

12. First Farm is an agricultural processer in Miami who supply mango smoothies to restaurants at Miami beaches. They can import mangos from the Philippines or from Mexico. Mangos from the Philippines cost $3.25 per case, and mangos from Mexico cost $4.05 per case. They have a budget of $200 per month to invest in mangos. From a survey, they found that in restaurants, customers prefer smoothies made with Mexican mangos, which they have been able to sell for $3.99 each. Smoothies made with mangos from the Philippines sell for $3.25 since customers don't seem to like

these as much. Making a smoothie from Mexican mangos takes about 1 minute while from Filipino mangos, it takes 3 minutes. Employees make $12 per hour, so Mexican mango smoothies cost the company $0.20 each and Filipino smoothies cost the company $0.60 each in labor. There is a total of $480 per week available for labor cost, and since the company uses Mexico mangos already, they estimate that 0.5 cases of Mexican mangos are always available. Employees only make whole smoothies, so no partial smoothies are made. Using the branch-and-bound method maximize the restaurant's profit from smoothies.

13. Consider the following integer LP problem to select a combination of projects from projects x_1 through x_6:

$$\text{Max: } Z = 4x_1 + 8x_2 + 6x_3 + 3x_4 + 4x_5 + 7x_6$$

s.t.:
$$500x_1 + 700x_2 + 550x_3 + 400x_4 + 450x_5 + 750x_6 \leq 2200$$
$$10x_1 + 7x_2 + 9x_3 + 9x_4 + 8x_5 + 5x_6 \leq 35$$
$$x_1, \quad x_2, \quad x_3, \quad x_4, \quad x_5, \quad x_6 = 0, 1$$

a. What is the optimal solution for this problem as stated?
b. Formulate constraints and find the optimal solutions for the following conditions:
   i. Exactly two projects out of x_2, x_3, x_4, and x_5 must be selected.
   ii. Project x_1 may be selected if and only if project x_6 is selected.
   iii. If project x_2 is selected, projects x_4 and x_5 must both be selected.
   iv. If projects x_1 and x_2 are both selected, x_6 must be selected.

14. David has three different kinds of crops, A, B, and C, that he could plant on his farm. The total budget he has is $1,000. The cost and expected annual revenue per acre for each crop is:

Crop	Fixed Cost ($)	Annual Revenue ($)
A	140	20.00
B	100	10.00
C	25	3.50

There are labor costs that apply to these three crops:

Labor cost for A = $0 if no acres of A are cropped, or

$100 if one or more acres of A are cropped

Labor cost for B = $0 if no acres of B are cropped, or

$80 if one or more acres of B are cropped

Labor cost for C = $0 if no acres of C are cropped, or

$25 if one or more acres of C are cropped

If David has enough land available to plant crops, what is his optimal cropping pattern to maximize the annual profit (expected annual revenue minus labor cost) of the farm.

15. A fabric store sells three products each week: cotton, wool, and hemp. The unit selling prices for the three goods are $3.75 per kilogram of wool, $2.00 per kilogram of cotton, and $4.25 per kilogram of hemp. The store owner faces two constraints. First, the owner only has 500 square feet for stocking these three products. Assume that each kilogram of wool requires 1 square foot of shelf space, each kilogram of cotton requires 0.20 square feet, and each kilogram of hemp requires 0.9 square feet. Second, the store cannot spend more than $5,000 each week to stock these three products. Assume that the costs for the three products (before the sin tax) are $2.00 per kilogram of wool, $1.50 per kilogram of cotton, and $1.50 per kilogram of hemp.

The state has just passed a new sin tax on cotton and hemp to discourage stores from selling these two products and to help finance a growing state budget deficit. Specifically, the state applies the following weekly fixed tax on each store that sells cotton or hemp:

Weekly tax on cotton = $0 if store does not sell any kilograms of cotton over the week, or $1500 if store sells 1 or more kilograms of cotton over the week

Weekly tax on hemp = $0 if store does not sell any kilograms of hemp over the week, or $500 if store sells 1 or more kilograms of hemp over the week

Formulate and solve an IP problem that chooses the weekly amount of cotton, wool, and hemp to sell that maximizes the store's net profit (gross revenue minus variable costs minus the fixed sin taxes) subject to the square footage and budget constraints. Clearly define all variables and constraints in your answer.

16. With a monthly budget of $350, Arial wants to shop for the following shopping items from a grocery store, which has a nutrition value of at least 2,300 RDA, while maximizing the taste index points. Use IP and formulate the best grocery items Arial should buy based on the table below. Also, due to storage constraints, Arial cannot buy more than seven of each item.

Items	Soda	Chicken	Banana	Juice	Pastry	Eggs	Milk
Price	14	9	15	10	17	7	9
RDA	50	80	100	110	70	90	105
Taste	15	10	8	12	17	8	7

17. Paul is a farmer in Maryland. He is planning to plant 30 hectares of rice, 40 hectares of wheat, and 30 hectares of cotton in his farm this year. He has 100 hectares of land and 125,000m³ of irrigation water to use. The first table below contains the fully irrigated water demand, the maximum yield per hectare, and the expected profit per ton for the three crops. The ratios of actual-to-maximum yield for each of these three crops under different irrigation levels are shown in the second table. Assume that Farmer Paul can select only one irrigation level for each crop. Help Farmer Paul find an optimal cropping pattern so that he can maximize his profit for this year.

	Water Demand (m³/ha)	Maximal Yield (tons/ha)	Profit ($/ton)
Rice	1,300	9	190
Wheat	1,200	12	180
Cotton	1,500	8	240

	Maximum Yield by Irrigation Level			Actual Yield (tons/acre) by Irrigation Level		
	100%	80%	60%	100%	80%	60%
Rice	100%	92%	78%	9	8.28	7.02
Wheat	100%	88%	80%	12	10.56	9.6
Cotton	100%	72%	69%	8	5.76	5.52

18. Recall our farmer from Chapter 1, Exercise 1. She was deciding how many acres of corn (with a gross profit of $40 per acre, tractor time requirement of 1 hour per acre, and a maximum suitable acreage of 400) and soybeans (with a gross profit of $45 per acre and tractor time requirement of 1.5 hours per acre) to plant subject to a 600 acre farm and 750 hours of tractor time set of constraints. Since then, the farmer has decided that she wants to set aside 200 acres (specifically the 200 that had been unsuitable for corn production) to participate in a government subsidy program and now only has 400 acres available. However, her neighbor has decided he may want to give up farming, so he is offering to rent her a 300-acre lot, which is entirely suitable for both corn and soybeans for the year at a cost of $5,500. Additionally, he has contracted for 250 hours of tractor time, which he will sell to her for $2,600. Formulate and solve a mixed binary LP to maximize her profit under these new conditions.

19. As the leader of a wildlife exploration venture, you must determine the best selection of four out of eight possible sites. The sites are labeled as $s_1, s_2, \ldots, s_8$ and the expected associated benefits quantified and given in the table below:

Site	s_1	s_2	s_3	s_4	s_5	s_6	s_7	s_8
Benefit	3	4	6	4	2.5	7	2	4.5

If site s_2 is explored, then site s_3 must also be explored. Furthermore, regional restrictions are such that exploring sites s_1 and s_7 will prevent you from exploring site s_8. Exploring sites s_3 or s_4 will prevent you from exploring site s_5.

20. The U.S. Forest Service needs to set up sites for district rangers. The forest is made up of a number of districts, as illustrated in the following figure.

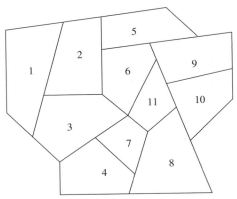

A district ranger can be placed in any district and is able to handle the job of protecting the forest resources for future generations and to protect visitors for both its district and any adjacent districts. The objective is to minimize the number of district rangers hired.

Use an IP model to solve this problem.

21. Love Apple is a company that produces and sells apples in Delaware. In 2008, they had an overall harvest of 1,400 tons of apples. The company has several routes to sell these apples. They can export these apples to foreign countries, sell them to supermarkets, transport them to some companies in the food industry for reprocess, or directly sell them in the local market. If some apples cannot be sold or handled by these routes, the company should dispose all the remaining apples under federal regulation. For each of these selling routes, certain operations are needed for preparation. The apples selling and distribution process for the company is shown in the following graph.

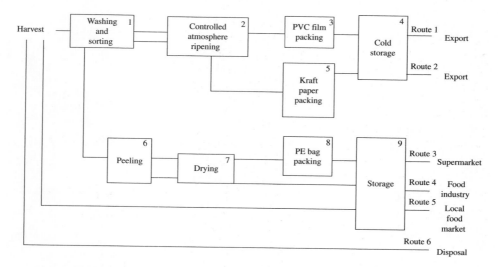

For each operation, the fixed charge for the equipment (a_n, $/year), unit fixed cost for the capacity of operation (b_n, $/t), unit processing cost for the apples (d_n, $/t) and the capacity of the operation (LB: L_n; UB: U_n) are shown in the table below. The annual fixed cost C_n for operation n is defined as a fixed charge function reflecting economies of scale: $C_n = a_n y_n + b_n f_n$.

n	a_n ($/year)	b_n ($/t)	d_n ($/t)	L_n (t/yr)	U_n (t/year)
1	1,300	40	42	60	1,400
2	7,200	70	117	300	1,400
3	4,200	43	120	300	1,400
4	7,000	105	80	150	1,400
5	2,000	95	88	150	1,400
6	600	30	32	100	1,400
7	1,200	20	75	150	1,400
8	3,600	60	126	300	1,400
9	700	26	18	100	1,400

The following table contains the final selling price (S_m), the distribution cost (I_m) and the capacity (LB: L_m; UB: U_m) for each route.

m	I_m ($/t)	S_m ($/t)	L_m (t/year)	U_m (t/year)
1	70	1,200	100	500
2	80	1,100	200	400
3	80	1,000	25	120
4	90	700	70	150
5	30	400	100	200
6	0	0	0	—

The supply cost is fixed in $200/t. We do not need to consider transportation cost in this process. The labor has been accounted into the cost in operations. As the whole process is in a short time period, inflation is not considered. Try to help the manager of Love Apple find the optimal route selections so that the company can maximize profits.

8

Optimization of Nonlinear Functions

In many real-world problems, the assumption of linear relationships is not realistic, and we must therefore turn to nonlinear models to depict the problem at hand. For example, natural resource problems often need to account for the growth rate of the resource in question, and these growth rates are often nonlinear. In this chapter, we introduce the general solution procedure for solving nonlinear problems. This chapter explains the conditions that are both necessary and sufficient for determining optimal solutions for any nonlinear function for both unconstrained and constrained problems.

   A basic understanding of calculus is required in order to understand these procedures. Hence, this chapter begins with a very elementary review of several concepts of differential calculus, followed by a discussion of the procedures for determining optimal solutions for unconstrained nonlinear functions. Students proficient in introductory differential calculus may choose to skip Sections 8.1 and 8.2 and begin with Section 8.3. Section 8.3 is followed by a section on procedures for solving constrained nonlinear problems with emphasis on how to use Solver to find the solution to such problems. Solver is capable of handling both linear and nonlinear constrained optimization problems. The use of nonlinear models is then illustrated with a fisheries example. The chapter concludes with two research applications of nonlinear programming (NLP). The first application relates to agricultural marketing and illustrates an NLP model that optimally allocates advertising across various media types to maximize net revenue subject to several structural constraints. The second example is from the field of environmental economics, which presents an NLP model for water pollution abatement in India.

8.1 SLOPES OF FUNCTIONS

Recall that the slope of a linear function $y = a + bx$ is equal to:

$$b = (\Delta y)/(\Delta x),$$

where Δ means "change in."

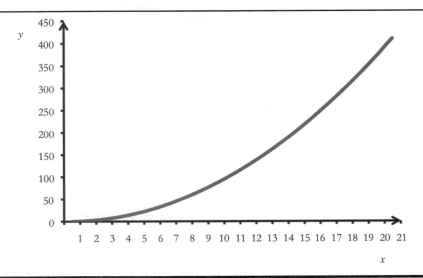

Figure 8.1   Graph of function $y = x^2 + 0.5x$.

Furthermore, the slope (b) is the same at every point along any linear function. What about the slope for a nonlinear function? Consider the following nonlinear function, which is also presented graphically in Figure 8.1:

$$y = x^2 + 0.5x \tag{8.1}$$

Let x_1 and x_2 represent two different points within the domain of the function y, and y_1 and y_2 be their respective values determined by function (8.1). The changes in x and y are defined as:

$$\Delta x = x_2 - x_1 \tag{8.2}$$

$$\Delta y = y_2 - y_1$$

Now consider calculating the slope of this function as was done above for specific values of Δx and Δy. First, let $x_1 = 1$ and $x_2 = 5$. Then the slope is equal to 6.5 since $y_1 = 1.5$ and $y_2 = 27.5$, that is:

$$b = \Delta y/\Delta x = (27.5 - 1.5)/(5-1) = 6.5.$$

On the other hand, if the slope of this function is computed for a smaller change in x, then the answer will be different. Consider $x_1 = 1$ and $x_2 = 2$. Then $y_1 = 1.5$ and $y_2 = 5$. In this case,

$$b = 3.5.$$

Finally, let $x_1 = 1$ and $x_2 = 1.25$. Then $y_1 = 1.5$ and $y_2 = 2.2$ and

$$b = 2.8.$$

Note that the slope of this nonlinear function is approximated by the slope of a straight line passing between two points (x_1, y_1) and (x_2, y_2). In this example as the change in x becomes smaller, the slope of the line becomes flatter (b becomes smaller) until the two points on the curve are close enough such that this line just touches the curve but does not cut through it. In other words, the slope of a nonlinear function at any point on this function is given by the slope of a line that is **tangent** to the curve. Since this is defined for very small changes in x, let Δx be a very small number. First consider a specific case where the slope is calculated for the following point along the function $(x_1, y_1) = (1, 1.5)$. Then define x_2 using Δ as being equal to $x_1 + \Delta x$, that is:

$$x_2 = x_1 + \Delta x, \text{ or}$$

$$x_2 = 1 + \Delta x, \text{ since } x_1 = 1 \text{ in this case} \tag{8.3}$$

Substituting this into y, calculate the value for y_2:

$$y_2 = x_2^2 + 0.5x_2, \text{ or}$$

$$y_2 = (1 + \Delta x)^2 + 0.5(1 + \Delta x), \text{ or}$$

$$y_2 = (\Delta x)^2 + 2.5\Delta x + 1.5 \tag{8.4}$$

Recall that $y_1 = 1.5$. Now calculate Δy. $\Delta y = y_2 - y_1$, by substituting $y_1 = 1.5$ and using (8.4), Δy is:

$$\Delta y = (\Delta x)^2 + 2.5\Delta x + 1.5 - 1.5, \text{ or}$$

$$\Delta y = (\Delta x)^2 + 2.5\Delta x \tag{8.5}$$

So the slope is:

$$b = \Delta y/\Delta x = ((\Delta x)^2 + 2.5\Delta x)/\Delta x, \text{ or}$$

$$b = \Delta x + 2.5 \tag{8.6}$$

As Δx becomes smaller, the slope of this line approaches the slope of the tangent line. This is equivalent to what is called taking the **limit** of $\Delta y/\Delta x$ as the value of Δx approaches (but does not equal) zero. This is stated mathematically as:

$$\text{Limit } \{\Delta x + 2.5\} = 0 + 2.5 = 2.5 \tag{8.7}$$
$$\Delta x \to 0$$

The statement in (8.7) means that as Δx approaches zero, the slope of this nonlinear function at the point $(x_1, y_1) = (1, 1.5)$ approaches 2.5.

Unlike linear functions, the slope of a nonlinear function will vary by the point along the curve. More generally, let

$$x_1 = x \tag{8.8}$$

$$x_2 = x + \Delta x$$

Then, the values of y_1 and y_2 are found by plugging in x_1 and x_2 into function (8.1).

$$y_1 = x^2 + 0.5x, \text{ and} \tag{8.9}$$

$$y_2 = (x + \Delta x)^2 + 0.5(x + \Delta x), \text{ or}$$

$$y_2 = x^2 + (\Delta x)^2 + 2x\Delta x + 0.5x + 0.5\Delta x$$

Using (8.9), the general expression for Δy is:

$$\Delta y = y_2 - y_1, \text{ or} \tag{8.10}$$

$$\Delta y = x^2 + (\Delta x)^2 + 2x\Delta x + 0.5x + 0.5\Delta x - (x^2 + 0.5x), \text{ or}$$

$$\Delta y = (\Delta x)^2 + 2x\Delta x + 0.5\Delta x$$

Now, the slope becomes:

$$b = ((\Delta x)^2 + 2x\Delta x + 0.5\Delta x)/\Delta x, \text{ or} \tag{8.11}$$

$$b = \Delta x + 2x + 0.5$$

Taking the limit of (8.11) yields:

$$\text{Limit } \{\Delta x + 2x + 0.5\} = 0 + 2x + 0.5 \tag{8.12}$$
$$\Delta x \to 0$$

The statement in (8.12) means that as Δx approaches zero, the slope of this nonlinear function approaches $2x + 0.5$. So, for $x = 1$, $b = 2.5$. For $x = 2$, $b = 4.5$. For $x = 3$, $b = 6.5$, and so on. Taking the limit of $\Delta y/\Delta x$ as Δx approaches zero gives the slope of a function at any point along the function. This is the definition of the **derivative** of a function y with respect to x (denoted as dy/dx).

8.2   SHORTCUT FORMULAS FOR DERIVATIVES

The **first derivative** for a nonlinear function of the form $y = cx^n$ (where c is a constant) is the following:

$$dy/dx = ncx^{(n-1)} \tag{8.13}$$

Consider the following functions and their first derivatives:

$$y = x^2,$$

$$dy/dx = 2x^{(2-1)} = 2x.$$

$$y = 5x^9,$$

$$dy/dx = (9)(5)x^{(9-1)} = 45x^8.$$

$$y = 55 + x^{-1},$$

$$dy/dx = -1x^{(-1-1)} = -x^{-2}.$$

$$y = 0.5x^{0.5},$$

$$dy/dx = (0.5)(0.5)x^{(0.5-1)} = 0.25x^{-0.5}.$$

The **second derivative** is defined as the derivative of the first derivative. While the first derivative gives the slope of the function at any particular point along the function, the second derivative measures the rate of change in the slope of the function. That is, the second derivative is a measure of the curvature of the function. The second derivative is calculated exactly the same way as the first derivative, except now you are taking the derivative of the derivative instead of the original function. Let d^2y/dx^2 denote the second derivative and consider the following functions, their first derivatives, and their second derivatives:

$$y = x^2,$$

$$dy/dx = 2x^{(2-1)} = 2x,$$

$$d^2y/dx^2 = 2.$$

$$y = 5x^9,$$

$$dy/dx = (9)(5)x^{(9-1)} = 45x^8,$$

$$d^2y/dx^2 = 8(45)x^{(8-1)} = 360x^7.$$

$$y = 55 + x^{-1},$$

$$dy/dx = -1x^{(-1-1)} = -x^{-2},$$

$$d^2y/dx^2 = (-2)(-1)x^{(-2-1)} = 2x^{-3}.$$

$$y = 0.5x^{0.5},$$

$$dy/dx = (0.5)(0.5)x^{(0.5-1)} = 0.25x^{-0.5},$$

$$d^2y/dx^2 = (-0.5)(0.25)x^{(-0.5-1)} = -0.125x^{-1.5}.$$

8.3   UNCONSTRAINED OPTIMIZATION

There are two types of optima: local and global. A **local optimum** (i.e., maximum or minimum) is defined as the highest (lowest) function value $y(x_1)$ for all x values in the neighborhood (in proximity) to x_1. A **global optimum** (maximum or minimum) is defined as the highest (lowest) function value $y(x_1)$ for all x in the domain of y.

Both local and global optima have the same **first-order conditions** (FOCs), namely that in order for some point x_1 to be an optimum, the slope of $y = f(x)$ at x_1 must be zero. Hence, the FOCs consist of deriving the derivative of a function and setting it equal to zero. Consider, for example, the following function, which is depicted graphically in Figure 8.2:

$$y = f(x) = 10 + x - x^2 \tag{8.14}$$

The derivative of this function is:

$$dy/dx = 1 - 2x \tag{8.15}$$

The FOCs for optimization require taking the derivative of y with respect to x, setting it equal to zero, and solving for x. Setting (8.15) equal to zero yields the following value for x:

$$x^* = 0.5 \tag{8.16}$$

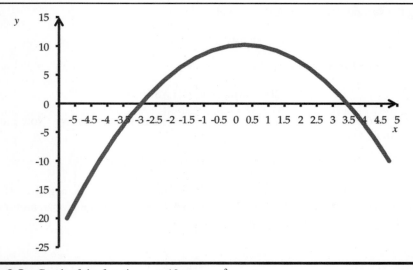

Figure 8.2   Graph of the function $y = 10 + x - x^2$.

The FOC tells us that at $x = 0.5$, we have a point that satisfies a **necessary condition** for it to be optimal (this is called a **critical** or **stationary point**). This is a necessary but not sufficient condition.

In order to distinguish whether this point is a maximum, minimum, or neither, we need to check the **second-order sufficient conditions** (SOCs). The SOCs use the second derivative of the function, evaluated with the optimal value of x, and examine whether it is positive, negative, or zero. The second derivative of this function is found by taking the derivative of (8.15) with respect to x (i.e., the derivative of the first derivative).

$$d^2y/dx^2 = -2 \qquad (8.17)$$

In the case of a single variable function, for $y = f(x)$ to be a maximum, $d^2y/dx^2 < 0$. For $y = f(x)$ to be a minimum, $d^2y/dx^2 > 0$. For $y = f(x)$ to be neither a maximum nor a minimum, $d^2y/dx^2 = 0$. In this example, $x^* = 0.5$ is a maximum for this function since $d^2y/dx^2 = -2 < 0$. Plot this function and examine the rationale for these three rules.

So far, we have not distinguished between local and global optima. In order to ascertain whether the critical point is a local or global optimum, find all the critical points in the domain of x ($dy/dx = 0$), plug them into the function to get y, and compare y. In this example, there is only one critical point ($x^* = 0.5$), and hence we know that this is a global optimum. Furthermore, we know that this point is a global maximum since the second derivative evaluated at ($x^* = 0.5$) is negative. However, there are many nonlinear functions that have multiple critical points. The equation:

$$y = f(x) = 55 + 2x - 10x^2 + x^3 \qquad (8.18)$$

is such an example. Differentiating (8.18) yields:

$$dy/dx = 2 - 20x + 3x^2 \qquad (8.19)$$

Setting (8.19) equal to zero and using the quadratic formula[1] to solve for x^*, we get:

$$x^* = 6.5651 \text{ or } 0.1015 \qquad (8.20)$$

Taking the second derivative of (8.18) yields:

$$d^2y/dx^2 = -20 + 6x \qquad (8.21)$$

Plug $x^* = 6.5651$ into (8.21) yields:

$$-20 + 6(6.5651) = 19.391 > 0 \qquad (8.22)$$

Hence, $x^* = 6.565$ is a minimum since its second derivative is positive. Now plug $x^* = 0.1015$ into (8.21):

$$-20 + 6(0.1015) = -19.391 < 0 \qquad (8.23)$$

Hence, $x^* = 0.1015$ is a maximum since its second derivative is negative. However, plotting this cubic function, illustrates that $x^* = 6.5651$ is a local, not global, minimum and $x^* = 0.1015$ is a local, not global, maximum.

---

[1] For a quadratic function $y = ax^2 + bx + c$ (where a, b, and c are constant parameters), the quadratic formula states that the solution to $ax^2 + bx + c = 0$ is $x^* = [-b + (b^2 - 4ac)^{0.5}]/2a$ and $[-b - (b^2 - 4ac)^{0.5}]/2a$.

8.4 MULTIVARIATE FUNCTIONS

So far the discussion has focused on single variable functions. These results may be extended to multivariate functions. Consider the following:

$$y = f(x, z) = x^2 + z^2 + xz \qquad (8.24)$$

Now y partially depends upon X and partially depends on z. Hence, we introduce the notion of a **partial derivative**. A partial derivative gives the slope of a function y with respect to one of the exogenous variables (in this case x or z). The rule for taking a partial derivative of y with respect to x or z is the same as before, except that you treat the other exogenous variables as constants. In cases where the exogenous variables (x and z) are not multiplicative (e.g., $x + z$ or $x^2 + z^6$), the partial derivative of y with respect to x (which we will denote as $\partial y/\partial x$) is found by taking the derivative of all terms in the function that contain x, ignoring the z terms. The partial derivative of y with respect to z ($\partial y/\partial z$) is found by taking the derivative of all terms in the function that contain z, ignoring the x terms. Some examples of nonmultiplicative functions will help clarify this:

$$y = f(x, z) = 2x^2 + 3z^2,$$

$$\partial y/\partial x = 4x,$$

$$\partial y/\partial z = 6z.$$

$$y = f(x, z) = 3x^6 + 5x^3 + 3x^2 + z^8 + 7z,$$

$$\partial y/\partial x = 18x^5 + 15x^2 + 6x,$$

$$\partial y/\partial z = 8z^7 + 7.$$

$$y = f(x, z, w, t, v) = x^2 + z^3 - 7w + 2t^6 - v^3,$$

$$\partial y/\partial x = 2x,$$

$$\partial y/\partial z = 3z^2,$$

$$\partial y/\partial w = -7,$$

$$\partial y/\partial t = 12t^5,$$

$$\partial y/\partial v = -3v^2.$$

In cases where the exogenous variables (x and z) are multiplicative (e.g., xz or x^2z^6), $\partial y/\partial x$ is found by taking the derivative of all terms in the function that contain x, treating the z term as a constant. The partial derivative of y with respect to z ($\partial y/\partial z$) is found by taking the derivative of all terms in the function that contain z, treating the x term as a constant. Some examples of multiplicative functions will help clarify this:

$$y = f(x, z) = x^3z^7,$$

$$\partial y/\partial x = 3x^2z^7,$$

$$\partial y/\partial z = x^3(7)z^6 = 7x^3z^6.$$

$$y = f(x, z) = x^2z^3 + xz + x + 7z,$$

$$\partial y/\partial x = 2xz^3 + z + 1,$$

$$\partial y/\partial z = 3x^2z^2 + x + 7.$$

$$y = f(x, z, t) = xzt + x^2,$$

$$\partial y/\partial x = zt + 2x,$$

$$\partial y/\partial z = xt,$$

$$\partial y/\partial t = xz.$$

The FOCs for multivariate functions are the same as before, except that now you must set each partial derivative to zero and solve for the critical points. The SOCs are a little more complicated, so only the case for a two-variable function is given.

Consider the function $y = f(x, z)$. Define $f_{xx} = \partial^2 y/\partial x^2$, $f_{zz} = \partial^2 y/\partial z^2$, and $f_{xz} = \partial^2 y/\partial x \partial z$ (where f_{xz} means the partial derivative is first taken with respect to x and then with respect to z in that order, and $f_{xz} = f_{zx}$ if the derivatives exist). For $y = f(x, z)$ to be a maximum, the SOCs are the following:

$$f_{xx} < 0, f_{zz} < 0, \text{ and } f_{xx} f_{zz} - f_{xz}^2 > 0.$$

For the same function to be a minimum, the SOCs are:

$$f_{xx} > 0, f_{zz} > 0, \text{ and } f_{xx} f_{zz} - f_{xz}^2 > 0.$$

Note that the last term needs to be positive for both maximum and minimum. This condition is equivalent to assuring that the function is concave (maximum SOCs) or convex (minimum SOCs). Since several matrix algebra concepts not covered in this class are necessary to show the logic behind these SOCs, it will not be covered here.

8.5 CONSTRAINED OPTIMIZATION WITH EQUALITY CONSTRAINTS[2]

There are several steps that are necessary to solve a nonlinear constrained optimization problem with equality constraints. Consider the following general problem to illustrate these basic steps:

Max or Min: $Z = f(x_1, \dots, x_n)$,

s.t:

$$g_i(x_1, \dots, x_n) = b_i \ (i = 1, \dots, m, m < n).$$

Notice that non-negativity on all variables is **not** imposed in this problem.

There are several steps to solving a constrained optimization problem, which start with forming the **LaGrange function** for the problem. The LaGrange function (L) is a mathematical technique used for solving constrained optimization problems, and is essentially the problem re-expressed in functional form, and includes two components. The first component is simply the objective function for the problem. The second component is the sum of each structural constraint times its respective **LaGrange multiplier** (λ_i). The optimal solution for the LaGrange multipliers is the shadow price (SP) for each structural constraint.

Step 1: Form the **LaGrange function** for the problem. The LaGrange function for this problem is:

$$L(x_1, \dots, x_n, \lambda_1, \dots, \lambda_m) = f(x_1, \dots, x_n) + \sum_{i=1}^{m} \lambda_i(b_i - g_i(x_1, \dots, x_n)).$$

---

[2]For a more detailed presentation of optimization theory, see Sundaram (1996).

Note that the variables of the LaGrange function include the original activities in the model and the LaGrange multipliers.

Step 2: Take the first-order necessary conditions for the problem. This is done by partially differentiating L with respect to all decision variables (x_i) and all LaGrange multipliers (λ_i) and setting the resulting expressions equal to zero. In this example, the FOCs are:

$$L_{x1} = f_{x1} - \sum_{i=1}^{m} \lambda_i g_{ix1} = 0 \qquad (1)$$

$$\vdots \qquad\qquad \vdots$$

$$L_{xn} = f_{xn} - \sum_{i=1}^{m} \lambda_i g_{ixn} = 0 \qquad (n)$$

$$L_{\lambda 1} = b_1 - g_1(x_1, \ldots, x_n) = 0 \qquad (n+1)$$

$$\vdots \qquad\qquad \vdots$$

$$L_{\lambda m} = b_m - g_m(x_1, \ldots, x_n) = 0 \qquad (n+m)$$

where:

L_{xi} = partial derivative of L with respect to x_i

f_{xi} = partial derivative of the objective function with respect to x_i

g_{ixi} = partial derivative of the ith constraint with respect to x_i

$L_{\lambda 1}$ = partial derivative of L with respect to λ_1

Note that the partial derivative of L with respect to λ_i is simply the ith constraint of the original problem.

Step 3: Find all values for x_i^* and λ_i^* that satisfy equations (1) through (n+m).

Step 4: Check the SOCs to determine whether the solution (x_i^*, λ_i^*) is a maximum, a minimum, or neither. For a two-variable, one-constraint problem, the sufficient conditions are found by substituting (x_i^*, λ_i^*) into the following expression:

$$S = -(g_{x2})^2 L_{x1x1} - (g_{x1})^2 L_{x2x2} + 2g_{x1}g_{x2}L_{x1x2},$$

where:

g_{xi} = partial derivative of the constraint (g) with respect to x_i (i = 1, 2)

L_{xixi} = second partial derivative of L with respect to x_i, and

L_{xixj} = cross partial derivative of L_{xi} with respect to x_j (for i not equal j)

If S is positive, then (x_i^*, λ_i^*) is a maximum; if S is negative, then (x_i^*, λ_i^*) is a minimum; and if S is zero, then (x_i^*, λ_i^*) is neither a maximum nor a minimum. More complicated matrix algebra is necessary to show the SOCs for problems involving more than two variables. Consult any basic textbook in calculus or mathematical economics to learn the SOCs for problems involving more than two variables.

Example 1

Consider the following constrained maximization problem with one equality constraint:

Max: $Z = 2x - 1y$,

s.t.:

$$-x^2 + 2y = 100.$$

The LaGrange function for this problem is:

$$L(x, y, \lambda) = 2x - 1y + \lambda(100 + x^2 - 2y).$$

The FOCs are:

$$L_x = 2 + 2\lambda x = 0 \tag{8.25}$$

$$L_y = -1 - 2\lambda = 0 \tag{8.26}$$

$$L_\lambda = 100 + x^2 - 2y = 0 \tag{8.27}$$

To solve (8.25) through (8.27) for (x^*, y^*, λ^*), first solve (8.26) for λ^*. From (8.26),

$$\lambda^* = -0.5 \tag{8.28}$$

Next, substitute $\lambda^* = -0.5$ into (8.25) to solve for x^*. This yields

$$2 + 2(-0.5)x = 0, \text{ or}$$

$$x^* = 2 \tag{8.29}$$

Next, substitute $x^* = 2$ into (8.27) and solve for y^*:

$$100 + (2)^2 - 2y = 0, \text{ or}$$

$$y^* = 52 \tag{8.30}$$

Finally, check to see whether the solution $(x^*, y^*, \lambda^*) = (2, 52, -0.5)$ is indeed a maximum. To do this, evaluate the following expression:

$$S = -(g_y)^2 L_{xx} - (g_x)^2 L_{yy} + 2g_x g_y L_{xy}, \text{ or}$$

$$S = -(-2)^2(2\lambda) - (2x)^2 (0) + 2(2x)(-2)(0), \text{ or}$$

$$S = -8\lambda = 4.$$

Since $S > 0$, this solution is a maximum. The optimal value of the objective function is:

$$Z^* = 2(2) - 1(52) = -48 \tag{8.31}$$

Example 2

To illustrate a constrained minimization problem with one equality constraint, consider the following example:

Min: $Z = x^2 + 4xy + 4y^2 - x - y$,

s.t.:

$$x + y = 200.$$

The LaGrange function for this problem is:

$$L(x, y, \lambda) = x^2 + 4xy + 4y^2 - x - y + \lambda(200 - x - y).$$

The FOCs are:

$$L_x = 2x + 4y - 1 - \lambda = 0 \tag{8.32}$$

$$L_y = 4x + 8y - 1 - \lambda = 0 \tag{8.33}$$

$$L_\lambda = 200 - x - y = 0 \tag{8.34}$$

To solve (8.32) through (8.34) for (x^*, y^*, λ^*), first solve (8.32) and (8.33) for λ to get:

$$\lambda = 2x + 4y - 1 \tag{8.35}$$

$$\lambda = 4x + 8y - 1 \tag{8.36}$$

Next, equate (8.35) and (8.36) and solve for either x or y. Solving for x yields:

$$2x + 4y - 1 = 4x + 8y - 1, \text{ or}$$

$$-2x = 4y, \text{ or}$$

$$x = -2y \tag{8.37}$$

Substitute $x = -2y$ into (8.34) and solve for y^*

$$200 - (-2y) - y = 0, \text{ or}$$

$$y^* = -200 \tag{8.38}$$

Plug $y^* = -200$ into (8.37) to get x^*

$$x^* = -2(-200) = 400 \tag{8.39}$$

Next, substitute (8.38) and (8.39) into either (8.35) or (8.36) to get λ^*. Substituting (8.38) and (8.39) into (8.35) yields:

$$\lambda^* = 2(400) + 4(-200) - 1 = -1 \tag{8.40}$$

Finally, check to see whether the solution $(x^*, y^*, \lambda^*) = (400, -200, -1)$ is indeed a minimum:

$$S = -(g_y)^2 L_{xx} - (g_x)^2 L_{yy} + 2g_x g_y L_{xy}, \text{ or}$$

$$S = -(-1)^2 (2) - (-1)^2 (8) + 2(-1)(-1)(4), \text{ or}$$

$$S = -2 - 8 + 8 = -2.$$

Since S is negative, this solution is a minimum. The optimal value of the objective function is:

$$Z^* = (400)^2 + 4(400)(-200) + 4(-200)^2 - 400 - (-200) = -200 \tag{8.41}$$

8.6   KUHN–TUCKER CONDITIONS AND CONSTRAINED OPTIMIZATION WITH INEQUALITY CONSTRAINTS

Consider the following problem:

Max: $Z = f(x)$,

s.t.:

$$x \geq 0.$$

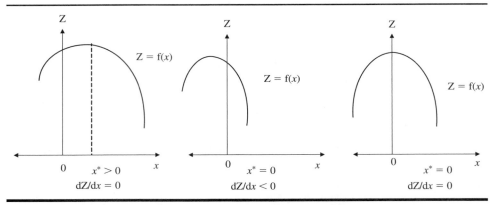

Figure 8.3   Three possible solutions for max: $Z = f(x)$ s.t.: $x \geq 0$.

Unlike the previous problems, this problem restricts x to be a non-negative value. This restriction introduces a special problem that makes the previously described first-order necessary conditions no longer appropriate. This can be seen by examining Figure 8.3, which shows the three possible solutions for this maximization problem: (1) $x^* > 0$ and $dZ/dx = 0$, (2) $x^* = 0$ and $dZ/dx < 0$, and (3) $x^* = 0$ and $dZ/dx = 0$. The graph on the left-hand-side (LHS) of this figure shows the case where the optimal solution occurs for a positive level of x^*. In this case, the slope or derivative of Z with respect to x is zero at x^*, which is positive. This solution is similar to the previous cases and is sometimes called an **interior solution** because the solution occurs in the interior of quadrant 1 rather than on one of the axes. The middle graph illustrates the case where the maximum value of Z actually occurs for $x < 0$. However, since x is restricted to be zero or positive, the optimal constrained solution occurs at the origin ($x^* = 0$). Notice that at $x^* = 0$, the derivative of Z with respect to x is negative. Finally, the graph on the right-hand-side (RHS) of this figure shows the situation where the maximum value of the function occurring at $x^* = 0$.

Kuhn and Tucker used these three possible solutions to restate the first-order necessary conditions for constrained optimization problems with inequality constraints. The **Kuhn–Tucker FOCs** simply require that a maximum will occur where either $dZ/dx = 0$ if x is strictly positive, or where $dZ/dx \leq 0$ if x is equal to zero. Restate the Kuhn–Tucker conditions for this problem using the following mathematical restrictions:

$$dZ/dx \leq 0, \; x \geq 0, \text{ and } (dZ/dx)\, x = 0.$$

These three conditions simply state that (1) if $x > 0$, then $dZ/dx = 0$; and (2) if $x = 0$, then $dZ/dx \leq 0$. The same type of logic applies to minimization problems, but $dZ/dx \geq 0$. For example, consider the following minimization problem:

Min: $Z = f(x)$

s.t.: $x \geq 0$

The three possible solutions for this problem are illustrated in Figure 8.4:

$$x^* > 0 \text{ and } dZ/dx = 0 \tag{1}$$

$$x^* = 0 \text{ and } dZ/dx > 0, \text{ and} \tag{2}$$

$$x^* = 0 \text{ and } dZ/dx = 0 \tag{3}$$

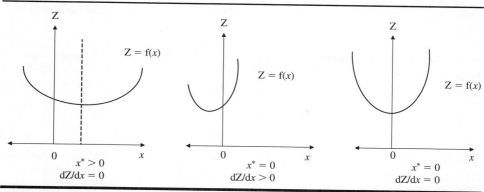

Figure 8.4   Three possible solutions for min: $Z = f(x)$ s.t.: $x \geq 0$.

The graph on the LHS of this figure shows the interior solution where the optimal solution occurs for a positive level of x^*. In this case, the slope or derivative of Z with respect to x is zero at x^*, which is positive. The middle graph illustrates the case where the minimum value of Z actually occurs for $x < 0$. However, since x is restricted to be zero or positive, the optimal constrained solution occurs at the origin ($x^* = 0$). Notice that at $x^* = 0$, the derivative of Z with respect to x is positive. Finally, the graph on the RHS of this figure shows the situation where the minimum value of the function occurs at $x^* = 0$ and the derivative of Z with respect to x is also zero. The Kuhn–Tucker FOCs are:

$$dZ/dx \geq 0,\ x \geq 0,\ \text{and } (dZ/dx)\, x = 0.$$

Consider the following NLP problem:

Max: $Z = xy$ (0)

s.t.:

$$x + y \leq 10 \tag{1}$$

$$x + 2y \leq 18 \tag{2}$$

$$x,\quad y \geq 0 \tag{3}$$

The LaGrange function for this problem is:

$$L(x,\ y,\ \lambda_1,\ \lambda_2) = xy + \lambda_1(10 - x - y) + \lambda_2(18 - x - 2y).$$

The Kuhn–Tucker FOCs are:

$$L_x = y - \lambda_1 - \lambda_2 \leq 0,\ x \geq 0,\ (y - \lambda_1 - \lambda_2)x = 0 \tag{8.42}$$

$$L_y = x - \lambda_1 - 2\lambda_2 \leq 0,\ y \geq 0,\ (x - \lambda_1 - 2\lambda_2)y = 0 \tag{8.43}$$

$$L_{\lambda 1} = 10 - x - y \geq 0,\ \lambda_1 \geq 0,\ (10 - x - y)\lambda_1 = 0 \tag{8.44}$$

$$L_{\lambda 2} = 18 - x - 2y \geq 0,\ \lambda_2 \geq 0,\ (18 - x - 2y)\lambda_2 = 0 \tag{8.45}$$

The last part of constraints (8.44) and (8.45) is the product of the LaGrange multiplier (λi) and constraint i (i = 1, 2). These conditions are sometimes called the **complementary slackness** conditions. These conditions imply that the LaGrange multiplier must equal zero if the constraint is not binding or the constraint has to be binding if the LaGrange multiplier is not zero.

Note that three possible solutions ($x^* = y^* = 0$, $x^* > 0$, $y^* = 0$, and $x^* = 0$, $y^* > 0$) can be ruled out immediately since each would result in $Z^* = 0$ because the objective function is the product of x and y. This implies that the optimal solution will be $x^* > 0$ and $y^* > 0$. Because both x^* and y^* will be positive, conditions (8.42) and (8.43) may be rewritten as:

$$L_x = y - \lambda_1 - \lambda_2 = 0, x > 0, (y - \lambda_1 - \lambda_2)x = 0 \qquad (8.46)$$

$$L_y = x - \lambda_1 - 2\lambda_2 = 0, y > 0, (x - \lambda_1 - 2\lambda_2)y = 0 \qquad (8.47)$$

That is, in order for x^* and y^* to be positive, L_x and L_y must be equal to zero. Therefore, there are three possible solutions to this problem defined by the possibilities for the two structural constraints:

1. $x + y = 10$ and $x + 2y = 18$,

2. $x + y < 10$ and $x + 2y = 18$,

3. $x + y = 10$ and $x + 2y < 18$.

Each one of the three cases must be solved for (x^*, y^*, λ_1^*, λ_2^*). Then the values must be substituted into the Kuhn–Tucker conditions to check if any condition is violated. If the solution for a case violates one of these conditions, then the case is ruled out because it represents an infeasible solution.

Case 1: Both Constraints are Binding

In this case, because the two constraints are assumed to be binding, simply solve the two equations for x and y using the simultaneous equation method:

$$x + y = 10 \qquad (8.48)$$

$$x + 2y = 18 \qquad (8.49)$$

Solve (8.48) for y:

$$y = 10 - x \qquad (8.50)$$

Plug (8.50) into (8.49) solve for x^*:

$$x + 2(10 - x) = 18, \text{ or}$$

$$x + 20 - 2x = 18, \text{ or}$$

$$x^* = 2 \qquad (8.51)$$

Next, plug $x^* = 2$ into (8.50) and solve for y^*:

$$y^* = 8 \qquad (8.52)$$

To obtain solutions for λ_1, λ_2, substitute $x^* = 2$ and $y^* = 8$ into the first two FOCs and solve for λ_1, λ_2:

$$y - \lambda_1 - \lambda_2 = 0, \text{ or } 8 - \lambda_1 - \lambda_2 = 0 \qquad (8.53)$$

$$x - \lambda_1 - 2\lambda_2 = 0, \text{ or } 2 - \lambda_1 - 2\lambda_2 = 0 \qquad (8.54)$$

Now solve (8.53) for λ_1 to get:

$$\lambda_1 = 8 - \lambda_2 \qquad (8.55)$$

Then substitute (8.55) into (8.54) and solve for λ_2^*:

$$2 - 8 + \lambda_2 - 2\lambda_2 = 0, \text{ or}$$

$$\lambda_2^* = -6 \tag{8.56}$$

Finally, substitute $\lambda_2^* = -6$ into (8.55) to get λ_1^*:

$$\lambda_1^* = 14 \tag{8.57}$$

For case 1, the solution is

$$(x^* = 2, y^* = 8, \lambda_1^* = 14, \lambda_2^* = -6).$$

Is this solution feasible? To see, check if it satisfies all of the Kuhn–Tucker conditions. Obviously this solution is not feasible since $\lambda_2^* = -6$ violates the non-negativity condition listed in FOC (8.45).

Case 2: Constraint 1 is Not Binding and Constraint 2 is Binding

In this case, the two constraints can be written as:

$$x + y < 10 \tag{8.58}$$

$$x + 2y = 18 \tag{8.59}$$

Because constraint (8.58) is not binding, λ_1^* must equal zero because of the third Kuhn–Tucker condition. From (8.46) we have:

$$y - \lambda_1 - \lambda_2 = 0.$$

Since $\lambda_1 = 0$, we get

$$y = \lambda_2 \tag{8.60}$$

From (8.47) we have

$$x - \lambda_1 - 2\lambda_2 = 0.$$

Noting that $\lambda_1 = 0$ and $\lambda_2 = y$:

$$x = 2\lambda_2 \tag{8.61}$$

Equations (8.60) and (8.61) imply that

$$x = 2y \tag{8.62}$$

Substituting (8.62) into constraint (8.59), that is, $x + 2y = 18$ yields

$$2y + 2y = 18, \text{ or} \tag{8.63}$$
$$y^* = 4.5$$

Substitute $y^* = 4.5$ into (8.62),

$$x^* = 2(4.5) = 9 \tag{8.64}$$

Substitute $y^* = 4.5$ into (8.60) to get λ_2^*:

$$\lambda_2^* = y = 4.5 \tag{8.65}$$

Hence, the solution for case 2 is:

$$(x^* = 9, y^* = 4.5, \lambda_1^* = 0, \lambda_2^* = 4.5).$$

Is this a feasible solution? To see, again check if the solution satisfies all of the Kuhn–Tucker conditions. This solution is not feasible because Kuhn–Tucker condition (8.44) is violated.

Case 3: Constraint 1 is Binding and Constraint 2 is Not Binding

The two structural constraints for this case may be written as:

$$x + y = 10 \tag{8.66}$$

$$x + 2y < 18 \tag{8.67}$$

Because constraint (8.67) is not binding, λ_2^* must equal zero in order to satisfy Kuhn–Tucker condition (8.45). From (8.46) we have:

$$y - \lambda_1 - \lambda_2 = 0.$$

Since $\lambda_2 = 0$, we get:

$$y = \lambda_1 \tag{8.68}$$

From (8.47) we have:

$$x - \lambda_1 - 2\lambda_2 = 0.$$

Noting that $\lambda_2 = 0$ and $\lambda_1 = y$ implies:

$$x = \lambda_1 = y \tag{8.69}$$

Using information from (8.69), solve (8.66) for either x or y. Solving for x yields:

$$x + x = 10, \text{ or}$$

$$2x = 10, \text{ or}$$

$$x^* = 5 \tag{8.70}$$

Substitute $x^* = 5$ into (8.69) to get y^* and λ_1^*:

$$y^* = 5 \text{ and } \lambda_1^* = 5 \tag{8.71}$$

Hence, the solution for case 3 is:

$$(x^* = 5, y^* = 5, \lambda_1^* = 5, \lambda_2^* = 0).$$

Is this a feasible solution? This solution is feasible because none of the Kuhn–Tucker conditions are violated. To determine whether the solution is a maximum, it would be necessary to check the SOCs. These conditions will not be derived here due to the matrix algebra required to learn them. This solution, however, is indeed a maximum.

There are several useful observations to consider regarding optimization problems with inequality constraints and the Kuhn–Tucker conditions. These include:

1. For each constraint, there will be one LaGrange multiplier and complementary slackness condition.

2. If there is only one constraint, then there are two possibilities: the constraint is either binding or not binding.

3. If there are n constraints, then there are 2^n possibilities in terms of binding and not binding constraints. In the example above, since there were two constraints, there were four possibilities ($2^2 = 4$).

The logic in solving these problems with the Kuhn–Tucker conditions is fairly complicated. You should work through the above problem several times to see the logic more clearly.

Solving Nonlinear Programming Problems

Three categories of methods have been used to computationally solve NLP models: (1) **separable linear programming**, (2) special-purpose algorithms for quadratic programming (QP), and (3) general NLP algorithms. Separable LP models use the simplex method to solve linearized versions of the nonlinear problem. This approximation is done by using piecemeal linear functions as proxies for the nonlinear objective function and/or nonlinear constraint equations. For the second category, **Wolfe's method** can be used to solve a special type of NLP model known as QP, where the objective function is a quadratic form, but all constraints are linear. This method uses a modified version of the simplex method because with the exception of the complementary slackness conditions, all other Kuhn–Tucker conditions are linear equations. Finally, there are other methods to solve more general NLP models such as gradient, heuristic, and interior point methods. Some of these will be discussed in the next section and Chapter 9.

8.7 SOLVING CONSTRAINED OPTIMIZATION PROBLEMS WITH SOLVER

Nonlinear programming problems can be set up and solved with Solver in the same manner as LPs. However, several different methods exist for solving nonlinear problems. Therefore, several different engines in Solver need to be used to analyze and solve a problem. For instance, quadratic programs can be solved using the LP/Quadratic (referred to as LP/QP) Engine exactly as in past chapters. For more complex nonlinear problems, a different engine will have to be selected. Traditionally, Solver offered one nonlinear engine that used the GRG, or Generalized Reduced Gradient algorithm. For a maximization problem, the GRG algorithm starts from some initial point and calculates the direction of greatest increase in the objective function at that point. The algorithm moves a given distance in that direction and repeats until it reaches a point at which the objective function can no longer be increased, at which time the algorithm declares the solution to the maximization problem. This GRG is a fast and flexible engine, but it has a couple of significant weaknesses. Because the stopping criterion is met whenever the algorithm gets to a "flat spot," there is no way to know whether the solution found is a local or global optimum. In fact, the solution may not even be an optimum, but instead simply a flat area or "saddle point." Consequently, this engine is very sensitive to the initial solution entered into the spreadsheet, and it would be worthwhile to try several different initial solutions, especially if the topology of the feasible region is not well understood. Also, sometimes re-running the engine from the solution given by a prior run will offer an improvement to the initial solution.

Another difference between analyzing linear and nonlinear programming in Solver is in the Sensitivity Analysis. The Sensitivity and Limits reports are generated by the user the exact same way; however, the Sensitivity reports will offer different information depending on the engine used.[3] The primary difference in the sensitivity reports arises in the ranges of optimality, which are not easily calculated for nonlinear programs. Therefore, the Allowable Increases and Decreases are not reported. The reports do provide the LaGrange multiplier, which is exactly the same as the SPs offered on the LP Sensitivity report.

This section presents two examples of nonlinear problems that can use the LP/QP Engine and one example that requires the GRG Engine. The recent version of Solver offers several more advanced engines with global solution techniques for certain types of nonlinear problems, which will be covered in the next chapter.

---

[3]Since the default engine in Solver is frequently the GRG Nonlinear Engine, this can result in fewer informative sensitivity reports than would be provided if the LP/QP Engine was used.

A Constrained Quadratic Maximization Example[4]

Consider a problem where the objective is to minimize a quadratic function subject to a set of linear constraints. For example,

$$\text{Min: } Z = x^2 + y^2 + xy \tag{0}$$

s.t.:

$$2.5x + y \geq 100 \tag{1}$$

$$x \qquad \geq 70 \tag{2}$$

$$y \geq 45 \tag{3}$$

$$x, \quad y \geq 0 \tag{4}$$

To solve this problem using Solver, three pieces of information need to be put into a spreadsheet: Decision Variables, Objective Function, and Constraints. Figure 8.5 illustrates the Excel spreadsheet used for data input to solve the above example.

The formula for the objective function is entered using the cells that define the decision variables. Recall that in order to enter a formula in Excel, first input an equation sign in the cell, then the formula itself in terms of the cells that define the decision variables, and then press the "Enter" key to finish the input. In this example, Cell B5 is the location for inputting the objective function $Z = x^2 + y^2 + xy$. The function formula can be entered as

$$\text{"} = (B2\text{^}2) + (C2\text{^}2) + B2*C2\text{"}$$

where ^ is used to indicate an exponent and * represents the multiplication operator. The value appearing in Cell B5 will represent the result of the objective function for the values of x and y in Cells B2 and C2.

Now enter the formulae that define all constraints for the problem. In this step, it is always good practice to first input the technological coefficients matrix (i.e. the a_{ij} coefficients) and RHS values into the spreadsheet. Based on what has been inputted in this step, enter the

	A	B	C	D	E	F
1		*x*	*y*			
2	**Decision Variables**	70	45			
3						
4	**Objective**					
5	Min Total Costs	$10,075				
6						
7	**Constraints**	x	y	**Total**		**RHS**
8		2.5	1	220	≥	100
9		1		70	≥	70
10			1	45	≥	45

Figure 8.5 Excel spreadsheet for quadratic example.

---

[4]The problems, solutions, and corresponding sensitivity analysis for the following three examples are shown in the Chapter 8 supplemental materials available at www.wiley.com/college/kaiser.

formulae for each of the constraints. For instance, the first constraint, $2.5x + y \geqslant 100$, can be inputted as "=SUMPRODUCT(B2:C2,B8:C8)" into Cell D8. Note that the non-negativity constraints of the problem do not need to be entered. Instead, the "Assume Non-negative" option can be selected from the Engine tab in Solver.

Add the objective, variables, and constraints to the Solver model the same way as was done for an LP in Chapter 3. The completed model should look like Figure 8.6. On the Engine tab set the Assume Non-negative option to True, and, since the objective function is quadratic, select the LP/QP Engine and solve the problem.

In this example, the optimal value of the objective function is 10,075, which is accomplished by producing 70 units of x and 45 units of y.

A Two-Product Firm Example

A large produce cooperative produces and sells two products—apples and oranges (denoted as x and y in this example)—and wishes to determine simultaneously how

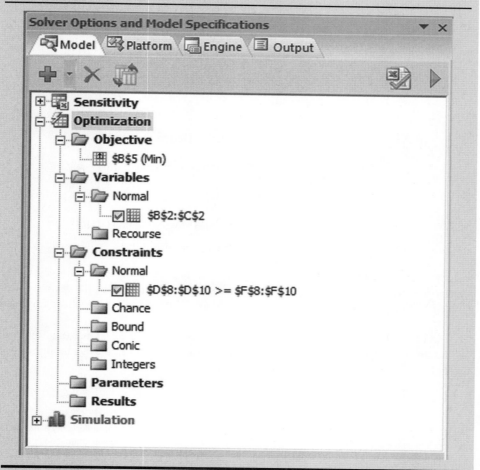

Figure 8.6   Completed model for quadratic example.

much of x and y to produce and sell and how much the price should be for x and y.[5] The firm's economist has estimated the following demand functions for x and y:

$$x = 100 - 10p_x + 0.3p_y,$$
$$y = 200 - 25p_y + 0.2p_x,$$

Note that (1) the demand functions are linear and (2) x and y are substitute products since p_y appears with a positive coefficient in the demand for x, and p_x appears with a positive coefficient in the demand for y. The firm's variable costs are \$0.50 per unit for producing x and \$1.20 per unit for producing y. Finally, the firm faces the following resource requirements and endowments in producing both products:

Technical Coefficients

Resource	x	y	Resource Endowment (hours)
Labor (L)	1.5	2.5	450
Machine 1 (M1)	1.4	1.0	150
Machine 2 (M2)	1.0	1.0	140

This problem can be formulated as the following nonlinear program:

Max: $Z = (p_x - 0.50)x + (p_y - 1.20)y$

s.t.:

$$10p_x + x - 0.3p_y = 100 \text{ (Define Demand for } x)$$
$$-0.20p_x + 25p_y + y = 200 \text{ (Define Demand for } y)$$
$$1.5x + 2.5y \leq 450 \text{ (Labor Constraint)}$$
$$1.4x + 1.0y \leq 150 \text{ (Machine 1 Constraint)}$$
$$1.0x + 1.0y \leq 140 \text{ (Machine 2 Constraint)}$$
$$p_x, \quad x, \quad p_y, \quad y \geq 0 \text{ (Non-negativity)}$$

The objective function here is to maximize total net revenue, which is equal to the price of each product minus the variable cost of each product times the quantity produced. Figure 8.7 illustrates this problem entered into Excel spreadsheet form for Solver.

The optimal solution is to sell 47.29 units of x at a price of \$5.41 per unit, and sell 83.8 units of y at a price of \$4.69 per unit. This strategy would result in a total profit of \$524.85.

[5]Notice that in this nonlinear example, price is no longer a constant parameter, but rather a decision variable that depends on the amount that is supplied. In other words, both quantity and price are endogenous variables, which is different from all the LP models illustrated so far in the book. This example also implies that the cooperative has some market power (i.e., is not a "price-taker") because it can control the price by setting supply. See the later chapter on price-endogenous mathematical programming for a more detailed discussion of these models.

	A	B	C	D	E	F	G	H
1		x	y	p_x	p_y			
2	**Decision Variables**	47.3	83.8	5.4	4.7			
3								
4	**Objective**							
5	Max Total Net Revenue	525						
6								
7	**Constraints**	x	y	p_x	p_y	**Total**		**RHS**
8	Demand for x	1.0	0.0	10.0	-0.3	100	=	100
9	Demand for y	0.0	1.0	-0.2	25.0	200	=	200
10	Labor Constraint	1.5	2.5			280	≤	450
11	Machine 1	1.4	1.0			150	≤	150
12	Machine 2	1.0	1.0			131	≤	140

Figure 8.7   The Solver model for two-product firm problem.

A Utility Maximization Example[6]

The prior examples of NLP had a quadratic functional form. Solver can also handle other nonlinear functional forms.

Consider, as an example, the following basic economics problem of utility maximization. Suppose that a consumer's utility (or welfare) depends upon the consumption level of two goods, x and y. Furthermore, assume that the person's utility can be measured by the following nonlinear utility function:

$$U(x,y) = x^{1/3} \, y^{1/3}.$$

The consumer has up to $500 that she can spend on goods x and y over the next week, and assume the price of x is $2.00 and the price of y is $1.00. The problem is to maximize utility subject to the budget constraint that one cannot spend more than $500 on these two goods. This constrained nonlinear maximization problem can be entered into Excel as the following model, shown in Figure 8.8.

This time, trying to solve using the Standard LP/QP Engine will produce the error message, "The linearity conditions required by this Solver engine are not satisfied" in the bottom of the Solver window. Instead, in the drop-down menu at the top of the Engine tab, select "Standard GRG Nonlinear Engine." Now solve the problem as before. If the initial values for x and y in the spreadsheet were both zero or both negative, Solver may converge to the solution $x = 0$, $y = 0$. This is because $(0, 0)$ is a critical point of the function, which means that it is "flat" at this point or its gradient is the 0 vector, so the Standard GRG Engine cannot find a direction to improve it. Instead select some positive initial values for x and y, and Solver will converge

---

[6]This problem, solution, and corresponding sensitivity analysis are shown in the Chapter 8 supplemental materials available at www.wiley.com/college/kaiser.

	A	B	C	D	E	F
1		*x*	*y*			
2	**Decision Variables**	125	250			
3						
4	**Objective**					
5	Max Utility	31				
6						
7	**Constraint**	*x*	*y*	**Total**		**RHS**
8	Budget	2	1	500	≤	500

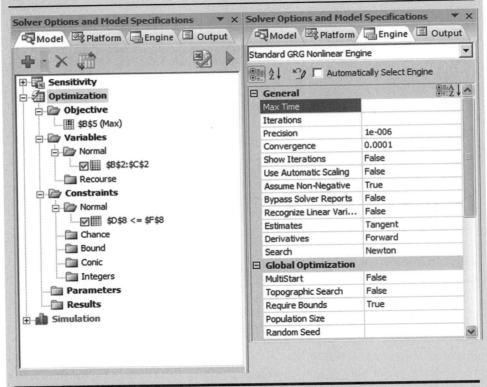

Figure 8.8 The Solver model for utility maximization problem.

Figure 8.9 The model and engine set-up for utility maximization problem.

to the optimal solution. The complete model and engine options are displayed in Figure 8.9.

In this case, the consumer will maximize utility by purchasing 125 units of x and 250 units of y. Looking at the Sensitivity Report, the SPs, or LaGrange multipliers, can be identified for each constraint. For instance, the SP for the budget constraint is \$0.042 and should be interpreted as the marginal utility of money.

8.8 FISHERY MANAGEMENT USING NONLINEAR PROGRAMMING

In Chapter 6, a linear function was used to describe the growth rate of Douglas fir trees. While linear approximations may be helpful for harvest decisions about a resource with relatively consistent growth rates, in most cases, growth rates of renewable resources are nonlinear. This example considers the optimal management of a fishery. In this **fishery stocking** problem, managers are asked to develop an optimal harvest plan that maximizes the present value of profits. This type of analysis may be helpful to prevent overfishing due to the well-known problem of the "tragedy of the commons." In this case, individuals pursue individual profits in an open-access fishery such that the fishery is depleted and the overall profits from the fishery are diminished. By knowing the harvest amounts that can lead to optimal profits, managers can establish policies such as individual transferrable quotas, harvest season length, and fishery equipment regulations to achieve a more sustainable and socially optimal result.

To solve this problem, several functions related to the fishery and harvest effort need to be established.[7] First, a general function describing the population of the renewable fish stock is as follows:

$$x_{t+1} - x_t = F(x_t) - y_t \tag{8.69}$$

where x_t denotes the stock of the renewable resource at the beginning of year t, $F(x_t)$ is the net growth in year t, and y_t is the amount of fish harvested in year t. Thus, the LHS in the equation is the change of stock from year t to year t + 1, and the RHS is the difference between the net increase in the fish stock and the harvest amount. If the amount harvested each year equals to the amount of net growth, a steady-state will be achieved where the size of the fish stock does not change:

$$y_t = F(x_t) \tag{8.70}$$

The natural growth of the fish stock can be described using a logistic growth function:

$$F(x_t) = rx_t \left(1 - \frac{x_t}{k}\right) \tag{8.71}$$

where $r > 0$ is the intrinsic growth rate, such that a larger number indicates stronger reproductive ability of fish, and $k > 0$ is the environmental carrying capacity that limits the size of the fish stock that the ecological environment can sustain. Figure 8.10 shows the nonlinear change of F(x) according to x. At the points $x = 0$ and $x = k$, the growth function denotes two steady-states, absent of fishing, where the size of fish stock will not increase

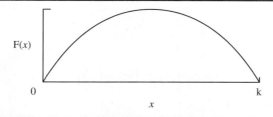

Figure 8.10   Logistic growth function.

[7]For an excellent review of the modeling of renewable resources, see Conrad (1999).

through time. When $x = 0$, there are no fish to reproduce and when $x = k$, the fish stock reaches the environmental carrying capacity such that the rising death rate offsets the net growth rate.

With the harvesting of fish, equations (8.69) and (8.71) can be combined to determine the fish stock size x_{t+1} at the beginning of year t+1.

$$
\begin{aligned}
x_{t+1} &= x_t + F(x_t) - y_t \\
&= x_t + rx_t\left(1 - \frac{x_t}{k}\right) - y_t
\end{aligned}
\tag{8.72}
$$

Let $\pi_t = \pi\,(x_t, y_t)$ denote the net profits in year t from harvesting y_t from a fish stock of size x_t. π_t is directly related to y_t in that generally harvesting more fish generates more revenue. Though the market for fish is limited and a flooding of the market with fish would generally decrease the market price, this example assumes that the fishery is sufficiently small that harvesting in this fishery does not have a significant influence on the overall market price for the fish. If the fish stock size, x_t is small, the fish population could be more scattered, and the cost of fishing would thus rise. The following function can be used to denote the net profit where a > 0 is the market price for fish (in bulk), y_t is the harvest level in year t, and c > 0 is the cost parameter reflecting the cost of effort for fishing. Thus, net profits in year t can be written as:

$$
\pi_t\left(x_t, y_t\right) = ay_t - c\frac{y_t}{x_t}
\tag{8.73}
$$

Next, assume that δ is the **discount rate** that captures the effect of the change in the value of money over time. For example, in traditional banking situations, the value of $100 after 10 years when $\delta = 0.05$ can be calculated as $100(1 + \delta)^{10}$. Therefore, after 10 years, the original $100 is now worth $162.89. In this fishery example, the discount rate is used to determine the **present value** of a future payment. For example, the value today of a $100 payment made in 10 years would be calculated by multiplying the future payment by the function $\left(\dfrac{1}{1+\delta}\right)^t$. Thus, the present value of a payment made in 10 years with a 0.05 discount rate is only $61.39. Therefore, the present value of a payment of π_t made in year t can be expressed as $\pi_t\left(\dfrac{1}{1+\delta}\right)^t$. Converting equation (8.73) into present value terms yields:

$$
\begin{aligned}
P_t &= \left(\frac{1}{1+\delta}\right)^t \pi_t\left(x_t, y_t\right) \\
&= \left(\frac{1}{1+\delta}\right)^t \left(ay_t - c\frac{y_t}{x_t}\right)
\end{aligned}
\tag{8.74}
$$

In this example, the fishery manager's objective is to maximize the present value of profits, P_t. Assume that after t $= T$ years, the fishery will be well managed and continue to operate in a steady-state condition, $y_t = F(x_t)$, and that the annual harvest amount will equal the net growth of the fish stock. In other words, the size of fish stock, x_t, will remain unchanged, and the net profit, $\bar{\pi}_t$, made every year until T will also remain unchanged, such that by using equations (8.70), (8.71), and (8.73), π_t can be expressed as:

$$\bar{\pi}_t = \pi_t\left(x_t, y_t\right)$$
$$= \pi_t\left(x_t, F\left(x_t\right)\right)$$
$$= aF(x_t) - c\frac{F(x_t)}{x_t} \tag{8.75}$$
$$= arx_t\left(1 - \frac{x_t}{k}\right) - \frac{crx_t\left(1 - \dfrac{x_t}{k}\right)}{x_t}$$

Building upon equations (8.74) and (8.75), all profits made after year T can be defined as

$$\bar{P}_T = \sum_{t=T+1}^{\infty}\left(\frac{1}{1+\delta}\right)^{t-T-1}\bar{\pi}_t = \sum_{t=0}^{\infty}\left(\frac{1}{1+\delta}\right)^t \bar{\pi}_t$$
$$= \left[1 + \left(\frac{1}{1+\delta}\right)^1 + \left(\frac{1}{1+\delta}\right)^2 + \cdots\right]\bar{\pi}_t$$
$$= \left(\frac{1+\delta}{\delta}\right)\bar{\pi}_t \tag{8.76}$$
$$= \left(\frac{1+\delta}{\delta}\right)\left[arx_t\left(1 - \frac{x_t}{k}\right) - \frac{crx_t\left(1 - \dfrac{x_t}{k}\right)}{x_t}\right]$$

As a result, the present value of $\bar{P}_T$ at the beginning when t = 0 is

$$\left(\frac{1}{1+\delta}\right)^T \bar{P}_T = \left(\frac{1}{1+\delta}\right)^T\left[arx_t\left(1 - \frac{x_t}{k}\right) - \frac{crx_t\left(1 - \dfrac{x_t}{k}\right)}{x_t}\right]\left(\frac{1+\delta}{\delta}\right)$$
$$= \left[arx_t\left(1 - \frac{x_t}{k}\right) - \frac{crx_t\left(1 - \dfrac{x_t}{k}\right)}{x_t}\right]\left(\frac{1}{\delta(1+\delta)^{T-1}}\right) \tag{8.77}$$

Model Set-Up

This fishery management problem can then be solved by developing a model as shown in Figure 8.11. In this spreadsheet, Cells C2 to C6 contain the given parameters that do not change over time. In this example, assume the market price for fish in bulk is $10,000, the cost of each unit of fishing effort is $1,000, the environmental carrying capacity is 1,000 fish (in thousands), and the discount rate is 0.05.

The initial stock level of fish, x_0, is given in Cell D10 as 200 (in thousands). The function for x_1 follows equation (8.72) such that Cell D11 can be written as =D10+$C$4*

	A	B	C	D	E
1	**Parameters**				
2	price for fish	a = $	10,000		
3	cost of effort	c = $	1,000		
4	growth rate	r =	0.50		
5	environmental carrying	K =	1,000		
6	discount rate	δ =	0.05		
8					
9		**Year (t)**	**Harvest amount, y_t (in 1000s)**	**Fish stock, x_t (in 1000s)**	**Present value of profits, p_t**
10		0	0.00	200.00	$ -
11		1	0.00	280.00	$ -
12		2	48.61	380.80	$ 440,783
13		3	123.77	450.09	$ 1,068,940
14		4	123.77	450.07	$ 1,018,004
15		5	123.73	450.06	$ 969,248
16		6	123.76	450.08	$ 923,280
17		7	123.76	450.08	$ 879,357
18		8	123.75	450.07	$ 837,369
19	**T**	9	126.27	450.08	$ 813,770
20		10			447.56 $ 15,175,624
21					
22					**Total $ 22,126,375**

Figure 8.11   Initial solution to nonlinear fishery problem.

D10*(1–D10/$C$5)–C10. Note that by using proper cell referencing notation, where "$" signs are used to indicate the absolute location of the given information, the fish stock for years 2 through 9 can be calculated by simply dragging down the equation in Cell D11 to Cells D12 through D20, so that the function in Cell D20 is =D19+$C$4* D19*(1–D19/$C$5)–C19.

The present value profits, P_t, can be calculated in Cell E10 by using the function = ((1/(1+$C$6))^B10)*($C$2*C10–$C$3*C10/D10) as depicted in equation (8.74). Likewise, this equation can be dragged down to capture the present value of profits in years 1 through 9.

Cell E20 captures the stream of present value profits that occur when the fishery and its annual harvest are in a steady-state as depicted in equation (8.77). This function captures the present value profit for all years after year T and can be written as:

=((1/(1+$C$6))^B20)*($C$2*$C$4*D20*(1–D20/$C$5)–$C$3*$C$4*D20
    *(1–D20/$C$5)/D20)/$C$6.

Finally, the objective that the fishery manager wants to maximize, net present profit, is expressed as the function =SUM (E10:E20) in Cell E22.

In Solver, the decision variables are defined as the harvest amounts, y_t, as shown in Cells C10 to C19. Constraints should also be set to ensure that the harvest amounts and the fish stock are non-negative. Finally, given the nonlinear nature of the problem, the Standard GRG Nonlinear Engine should be used with the starting values set at zero.

The solution to this problem is shown in Figure 8.11. To maximize present value profits no harvesting should occur in the fishery during the first two years. This will allow the fish stock to grow. Harvest levels then increase until they reach their highest level of 126.25 (in thousands) by year 9. The resulting present value profit for this management plan is $22,126,375.

As with most problems of this type, numerous real-world technological and political constraints can be incorporated into the model to evaluate how these constraints impact the management decisions and overall profit. In this case, one could imagine that the fishing community may be reluctant to accept a closure of the fishery for two years, even if it would lead to higher long-term profits. Therefore constraints can be set that raise the minimum harvest amount from 0 to 70. In this case, the optimal harvest amount would remain at the minimum permitted amount until the eighth year when fish stocks would have recovered sufficiently to permit more fishing. Interestingly, in this fictitious example, despite these higher levels of initial harvesting, the present value profit is just 4% less ($21,232,927).

8.9   RESEARCH APPLICATION: OPTIMAL ADVERTISING

Dairy farmers in the United States collectively contribute over $250 million per year to be used to advertise and promote milk products. Much of this money is used to advertise fluid milk and dairy products. The popular "Got Milk?" advertisements are but one example of these marketing activities. The national dairy promotion program finds its roots in the Dairy and Tobacco Adjustment Act (1983), which authorized the current assessment of 15 cents per hundred pounds of milk marketed (which is usually about 1% of the price received by farmers) in the continental United States. Because of the significant investment over time on these advertisements, there have been a lot of studies conducted on whether farmers are getting the biggest bang for their buck. In this section, we review one such study that used NLP to answer this question.

Pritchett, Liu, and Kaiser (1998) conducted a study that evaluated whether fluid milk advertising was being optimally allocated across the four major types of media: television, radio, print, and outdoors. The authors combined an econometric model of demand and supply with a dynamic, NLP model to determine the optimal allocation of milk advertising across media outlets. The econometric model, which will not be covered here, was used to derive the key parameters for the NLP model. These include the advertising elasticities of demand, own price elasticities of demand and supply, and other elasticities.[8] This was the first study to determine optimal mixes of advertising across type of media over time.

The model consists of a dynamic objective function (i.e., a quarterly model that allocates the advertising budget across types of media) and a set of constraints on supply and demand. The objective function for the NLP problem is to maximize:

$$H = \sum_{t=1}^{T-1} \rho^t \left\{ p_t^b \, s_t - \sum_{i=1}^{4} a_{i,t} \right\} + \rho^T \, V(d_T, s_T),$$

where $\rho = (1 + r)^{-1}$, r is the interest rate, p_t^b is the milk price received by farmers in time period t, s_t is the milk supply in period t, $a_{i,t}$ is advertising expenditures in time t spent on media type i (i = television, radio, print, and outdoor), and $V(d_T, s_T)$ is a salvage term

---

[8]Recall from economic principles that elasticity measures the percentage change in demand or supply given a small (e.g., 1%) change in one of its determinants. For example, the television advertising elasticity of demand measures the percentage change in demand given a 1% change in television advertising.

including terminal cash flow values and the terminal value of the decision variables d_T and s_T.[9] The objective is to maximize the discounted net revenue stream for a selected period of time (t = 1,2, ... , T) by choosing advertising expenditures for the ith media outlet {$a_{i,}$: t = 1, ... , T} so as to drive the decision variables, fluid milk sales {d_t: t = 1, ... , T} and farm milk supply {s_t: t = 1, ... , T}, to the optimal path. This objective is maximized subject to a set of dynamic (quarterly) constraints.

The first constraint in the model defines farm milk supply in time period t + 1, and is given by:

$$s_{t+1} = f(p_t^b, s_t, w_{t+1}).$$

That is, milk supply in period t+1 is a function of the price farmers receive for their milk in period t, milk supply in period t, and other determinants of milk supply such as technology, input prices, etc (w_{t+1}). An explicit milk supply function was estimated econometrically, and its result is used in the optimization model. Note that the authors used a "naïve price expectations" assumption that milk supply in t + 1 depends upon the price in the previous period, t. It is called "naïve" since it assumes that price observed in the most recent past period is what farmers collectively believe the price will be in the next period for making supply decisions.

The second constraint defines the milk price that farmers receive:

$$p_t^b = \delta_t \, d_t/s_t + p_t,$$

where δ_t is a premium that farmers receive for milk sold to fluid milk processors (i.e., Class I premium that was discussed in the transportation chapter) in time period t, d_t is the demand for milk by fluid milk processors in period t, s_t is total farm milk supply available for fluid and nonfluid demand in period t, and p_t is the base price for milk for nonfluid use (this is sometimes called the Class IV price) in period t. Note that this equation is nonlinear.

The next constraint limits fluid milk demand in period t to be less than or equal to total farm milk supply in period t, that is:

$$d_t \leq s_t.$$

Finally, the last constraint in the model limits the sum of advertising across all four media types in period t to be less than or equal to its total advertising budget for that period:

$$\sum_{i=1}^{4} a_{i,t} \leq \bar{a}_t.$$

To discount the stream of net revenue over time, the interest rate, r, is defined as 25% of the effective annual rate index. The authors used 6.155% as the annual rate index, since it was the average rate on six-month Treasury Bills between 1985 and 1995. The optimization problem is solved for the period beginning in the first quarter of 1984 through the final quarter of 1993. The authors used two estimated fluid milk demand models with the NLP model (Model A and Model B) in their analysis. The only difference between the two models is the way each addressed the carry-over effect advertising has on demand. One

---

[9]In any dynamic model like this, you need a "terminal value" for the decision variables because at the end of the optimization time period, some or all of these variables still have value and must be added to the value of the objective function.

model restricted the advertising carry-over effect according to a predetermined pattern (Model A), while the other placed no restriction on the carry-over effect (Model B).

The results of the optimization suggest there are important benefits to redistributing media expenditures. Using Model A, the optimal media mix increases the discounted profit in (1) by $950 million ($95 million per year) over the simulation period. While this represents only a small portion of the revenue stream in (1), this additional profit can be obtained virtually cost-free. A similar result was obtained with Model B, but profits increase by only $427 million ($43 million per year). Increased profits are due to an increase in fluid milk demand resulting from the reallocation of advertising dollars across various types of media.

What does the optimal media allocation look like? In absolute terms, it would be expected that the most effective media outlet has the largest share of expenditures. The econometric results indicate that television has the largest advertising elasticity. As a result, television has the largest share of expenditures relative to print, radio, and outdoor advertising expenditures. However, due to diminishing marginal returns, overspending on television is possible, which would reduce profitability. To improve profitability, funds should then be diverted from television to the media outlet with the greatest marginal benefit. The optimality principle dictates that the marginal benefit must be equal for each media outlet.

The result from both optimization models supports this principle. From 1984 to 1993, the actual average allocation of advertising dollars was to invest 88% on television, 5.2% on print, 4.4% on radio, and 2.4% on outdoor. Obviously, the actual plan heavily favored television as the dominant media type for advertising. The two optimal solutions significantly reduce television's share of the budget. The optimization results from Model A, for instance, reduce television to 70%, with outdoor increasing to 15%, print to 9%, and radio to 6%. The optimization results from Model B reduce television to 58%, increases outdoor to 29%, print to 8%, and radio to 5%. Regardless of optimization model, the qualitative results suggest that a reallocation away from television to other media types was in order during the 1984 to 1993 simulation period. These results suggest that print, radio, and outdoor advertising are more cost effective at the margin than was envisioned by the advertising agency running the dairy farmers' marketing campaign.

While the policy direction is that television expenditures should have been reduced in favor of other media outlets, several caveats apply. First, this analysis evaluates the overall performance of advertising campaigns over a period of 10 years. Hence, the effectiveness of specific campaigns within subperiods of the analysis are not measured. It is entirely possible that television advertising expenditures were optimal for a specific campaign, while they were overused for other campaigns on average. Further research on specific campaigns over the study period might be useful in resolving this issue. Second, the study considers the optimal mix of media expenditures for the period 1984–1993, and the results may not necessarily be applicable to the future. This would be particularly true if a more effective television campaign is developed. In fact, the "Got Milk?" campaign, which was not part of the time period in this study, may be a case in point. Third, it is possible that the dairy promotion unit receives price discounts for high-volume media purchases. In that event, shifting funds from television to other media outlets might compromise these discounts. Obviously, the validity of this issue can be best assessed by the program managers themselves. Finally, the model assumes a national milk marketing order. In truth, there are regional differences in both advertising responses and utilization percentages of dairy products. Further analysis is needed to address these issues.

8.10   RESEARCH APPLICATION: WATER POLLUTION ABATEMENT POLICIES

Goldar and Pandey (2001) applied a nonlinear mathematical programming model to examine alternative government policies aimed at reducing water pollution in India. A major problem in India has been firms' practice of using groundwater to dilute effluent streams in order to satisfy government environmental standards. Firms do this because the environmental regulations define limits on the maximum concentration of pollution, but not on the volume of wastewater pollution. This has given polluters an incentive to dilute effluent levels by mixing groundwater with wastewater until the concentration levels are just under those mandated by the law. As a result, the volume of wastewater discharged by factories is not regulated and is significantly higher than it would be in the absence of the regulations.

Water is underpriced in India, which induces this behavior. Without government intervention, water pricing does not incorporate the true social costs of pollution, and hence polluting entities have no economic incentive to reduce effluents. This study focused on water pollution caused by the distilleries industry in India, and addressed the following issue: how should the government price water to distilleries to bring down the level of pollution to adequate levels prescribed by existing legislation?

Two general policy scenarios were examined. The first was the actual command and control (CAC) instruments used by the government. The inclusion of this scenario in the optimization model was to determine whether current policies with cheap groundwater induce firms to dilute effluent with groundwater. The second scenario was a replacement of the CAC instruments with pollution taxes based on imputed pollution load. Under this scenario, both the tax rate and price of groundwater have an important impact on the extent of dilution used. The pollutant chosen for the study was biological oxygen demand (BOD), which is a measure of the amount of oxygen required to oxidize various compounds present in water.

A NLP model was used to investigate several alternative water pricing strategies. The objective function is a pollution abatement cost function for treating wastewater emitted from the distilleries. Water pollution abatement requires reducing influent concentrations to target acceptable levels of effluent concentrations. Influent is the wastewater resulting from production before being treated, while effluent is the residual emitted after treatment. The nonlinear cost function was estimated econometrically with cross-sectional data on 45 distilleries. The authors used a Cobb–Douglas (exponential) functional form in estimating the model, that is:

$$c = A \, q_I^{\beta 1} \, q_E^{\beta 2} \, q_I^{\beta 3} \, p_L^{\beta 4} \, p_K^{\beta 5} \, p_E^{\beta 6} \, p_M^{\beta 7},$$

where:

c = pollution abatement cost function

A = constant term

q_I = quantity of influent

q_E = pollution level of effluent

q_I = pollution level of influent

p_L = price of labor

p_K = price of capital

p_E = price of energy

p_M = price of materials

βi = parameters of cost function to be estimated, β4 + β5 + β6 + β7 = 1

Using the cross-sectional data, the authors estimated this cost function. The resulting cost function was used as one of the main parts of the objective function in the NLP model.

The NLP model is:

Min: Z = ct + cgw + tbrdn   (0)

s.t.:

$$- \ln ct + \text{CONSTANT} + 0.943 \ln volw + 0.923 \ln prtbod - 0.0998 \ln pstbod = 0 \quad (1)$$

$$- cgw + (prgw)(gwp) = 0 \quad (2)$$

$$- tbrdn + \text{TR}[(fbod - 30)volw] = 0 \quad (3)$$

$$- volw + 15out + 10out = 0 \quad (4)$$

$$- fbod + (pstbod)(volw)/(volw + gwp) = 0 \quad (5)$$

$$prtbod - pstbod \geq 0 \quad (6)$$

$$pstbod \geq 30 \quad (7)$$

Non-negativity   (8)

where:

ct = cost of treatment, based on estimated cost function

cgw = cost of groundwater extraction

tbrdn = tax burden to distillery

volw = volume of water treated

pstbod = post-treatment BOD concentration level

gwp = volume of groundwater extracted

fbod = BOD concentration level in final water discharge

ln = natural logarithmic operator

prtbod = pretreatment BOD, which is the index of pollutants in wastewater, and is set to 46,000 mg/liter

CONSTANT = price variables in estimated cost function set at sample means

prgw = price of groundwater, which is assigned different values under various scenarios

TR = tax rate, which is assigned different values under various scenarios

out = annual output of alcohol from distillery, set at 10,000 KL

The objective function consists of three costs: treatment, groundwater extraction, and pollution tax burden. Constraint (1) defines the treatment cost for the firm, which is based on the econometrically estimated cost function. Constraint (2) defines the cost of groundwater extraction as the product of volume of groundwater extracted times the price of groundwater, which is set to different levels in various scenarios. Constraint (3) defines the cost of total pollution taxes levied on the distillery for BOD concentration levels larger than 30. If the BOD concentration levels are 30 or less, then the total taxes are zero. Constraint (5) sets the total volume of water treated, which depends upon wastewater spent on wash generation per KL of alcohol produced, and wastewater process water generation per KL of alcohol produced. Constraint (5) defines the level of BOD concentration levels in the final water discharged. Constraints (6) and (7) require the post-treated BOD concentration levels to be lower than the pretreatment BOD levels of 30 or higher.

Under the first set of scenarios, the model was run assuming that the distillery is required to reduce BOD concentrations of its final discharge to 30 mg/liter. It can do so by treatment of wastewater in the effluent treatment plant (ETP) and/or by dilution with clean groundwater. The extent of each practice depends on the relative cost of groundwater versus the ETP. The authors found that current prices for groundwater are so cheap that the optimal solution for the distillery is to use groundwater rather than the ETP. As the price of groundwater is increased, the optimal solution relies less and less on groundwater for dilution and more on the ETP. The authors simulated nine scenarios increasing the groundwater price from 0.25 rupees per KL (current price) to 3.50 rupees. At 3.5 rupees, no groundwater was used to dilute the wastewater. A main policy conclusion is that more realistic pricing of groundwater to reflect its true cost is one solution to this pollution abatement problem.

The second set of scenarios involved taxing pollution. Under this scheme, there is an incentive to use groundwater to dilute wastewater after treatment in order to lessen the tax burden. The authors found that high tax rates coupled with low groundwater prices provide a strong incentive for firms to use groundwater to dilute the wastewater. However, as the price of groundwater increases, this incentive goes down.

Based on these results, the authors have three recommendations in terms of curbing the use of clean groundwater to dilute wastewater from distilleries. First, raise the price of groundwater to its economic value, taking into account its many diverse uses such as drinking water, irrigation, and so on. The price should be different for different regions of the country. Second, set a pollution tax at a level such that distilleries and other industries have little incentive to dilute their effluent. The level of the tax should also be set regionally in accordance to the price of groundwater. Third, set effluent standards taking into account regional water quality and the absorptive capacity of the environment in each region.

SUMMARY

This chapter presented an overview of NLP. The chapter began with a basic review of calculus concepts required for unconstrained and constrained nonlinear optimization problems. We introduced the general solution procedure for solving nonlinear problems using differential calculus. This included a discussion of the first-order necessary conditions and SOCs for determining a solution to a maximum and minimum problem. The fundamental necessary and sufficient conditions for determining optimal solutions for any nonlinear function for both unconstrained and constrained problems were presented. As was discussed in the introduction, a basic understanding of calculus is necessary to understand these procedures.

Next, the Kuhn–Tucker conditions and procedures for constrained nonlinear problems were examined. Constrained optimization problems are a little more complicated to solve because of the possibility of corner solutions. We demonstrated how to use the Kuhn–Tucker conditions to solve such problems.

Also, an overview of how to solve these problems using Solver was presented. Several simple examples were presented to illustrate how to use Solver for NLP. It is important to note that while these were small problems, Solver is capable of handling much larger and more realistic NLP problems with exactly the same logic. Hence, it is a nice tool for applied research, though it is important to understand the basic problem to avoid finding local instead of global solutions.

The chapter concluded with two research applications of NLP in the agricultural marketing and environmental economics literature. In the first application, NLP was used to determine the optimal allocation of advertising by media type over time. In the second application, an NLP model of water pollution abatement in India was presented.

In the next chapter, greater detail is provided on the issues and concepts central to understanding global approaches to nonlinear optimization.

REFERENCES

Conrad, Jon M. (1999). *Resource Economics*. Cambridge: Cambridge University Press.

Goldar, B., & Pandey, R. (2001) Water pricing and abatement of industrial water pollution: Study of distilleries in India. *Environmental Economics and Policy Studies, 4*, 95–113.

Pritchett, James G., Liu, Donald J., & Kaiser, Harry M. (1998). Optimal choice of generic milk advertising expenditures by media type. *Journal of Agricultural and Resource Economics, 23*, 155–169.

Sundaram, Rangarajan K. (1996). *A first course in optimization theory*. Cambridge: Cambridge University Press.

EXERCISES

1. For each of the following functions, find the critical point and determine whether the critical point is a maximum, minimum, or neither.

   a. $y(x) = x^2 - 15x + 7$.
   b. $y(w,z) = w^2 - z^2wz$.
   c. $y(a,b) = a^2 - 5ab + 100$.

2. Determine the minimum of the following function:

$$y(x) = x^3 - 10x^2 + 100.$$

3. Determine the maximum of the following function:

$$y(x) = 200x - 10x^2 - 100.$$

4. A firm faces the following inverse demand curve:

$$p = 500 - 0.3q,$$

where q is the monthly production, and p is price, measured in dollars per unit.

The firm also has a total cost (TC) function of:

$$TC = 6{,}000 + 20q.$$

Assuming the firm maximizes profits, answer the following:

a.   Compute the total and marginal revenue for the firm.

b.   Assuming the firm operates as a monopolist, calculate the following: price, quantity, and profit. (*Hint*: the firm maximizes profit by equating marginal revenue and marginal costs). Graph and show the equilibrium price and quantity.

c.   Assuming perfect competition, what are the price, quantity, and profit? Show on the graph from above.

5.   You have been assigned the task of helping the Midland Milk Producers Marketing Co-operative Association (MMPMCA), the sole marketing agency for milk produced in the Island State of Midland. In the past, the producers have been marketing their milk as a homogeneous commodity to all processors of milk. However, you have studied the market extensively and realize that the market can actually be segmented into two separate units: (1) the market for fluid milk (milk for drinking) and (2) the market for processing milk (for manufacturing of cheese, etc.). In fact, you have done some preliminary analysis and generated demand curves for the two separate markets for the MMPMCA's milk output.

Fluid milk market:

$$\text{Inverse demand curve:}\quad p_{\text{fluid}} = 10 - 2q_{\text{fluid}}.$$
$$\text{Marginal revenue curve:}\quad MR_{\text{fluid}} = 10 - 4q_{\text{fluid}}.$$

Processing milk:

$$\text{Inverse demand curve:}\quad p_{\text{processing}} = 5 - 0.5q_{\text{processing}}.$$
$$\text{Marginal revenue curve:}\quad MR_{\text{processing}} = 5 - 1q_{\text{processing}}.$$

Cost of production:

Assume that individual firms have the same cost function as follows:

$$TC = 10 + 3q, \text{ where}$$
$$q = q_{\text{fluid}} + q_{\text{processing}}.$$

a.   Determine the marginal cost of milk production.

b.   What is the desired allocation of milk production between the two markets, and what is the price they should charge for fluid and processing milk in the respective markets, assuming that the MMPMCA wishes to maximize profits for the farmers they represent? In other words, given that they can practice price discrimination in the market, how much should they sell into each of the fluid milk and processing milk markets, and at what price?

c.   Given that the price of milk without price discrimination is $4.50, how have prices changed for fluid milk and processing milk buyers?

d. How have total revenues changed if the actual total production of milk remains unchanged?

e. Calculate the own-price elasticities of demand for fluid and processing milk at the equilibrium values for p and q.

6. Assume that the market demand, marginal revenue, and total costs faced by a food firm are:

$$p = 1000 - 10q \quad \text{(demand)},$$
$$\text{TR} = 1000q - 10q^2 \quad \text{(total revenue), and}$$
$$\text{TC} = 300q \quad \text{(total cost)},$$

where p is the price, and q is the total quantity (in thousands).

a. Assuming that the firm acts as a monopolist, calculate the optimal price, quantity, and profit (i.e., TR – TC) for the organization. After using algebra to find the solution, present your results graphically.

b. The Department of Justice claims that the monopoly solution in part a is socially inefficient, and passes a law that requires the firm to set its output and price levels as if it were a perfect competitor. Assuming the firm follows the new law, what would the price, quantity, and profit now be?

7. Solve the following consumer utility maximization problem.

Max: $U(q_1, q_2) = q_1 q_2$

s.t.:

$$p_1 q_1 + p_2 q_2 = y,$$

where q_1 and q_2 are goods, p_1 and p_2 are the prices of the goods, and y is income.

8. A person's utility function (U) is given by the following expression:

$$U(x, y) = (x + 2)(y + 1),$$

where x and y are products the consumer may purchase. Suppose that the price of x is $2.00 and the price of y is $5.00, and the consumer wants to spend exactly $51.00 on both products. The consumer's objective is to maximize her utility function subject to the stated budget constraint.

a. Write the LaGrange function for this problem.

b. Derive the optimal purchases for x and y. Show your work.

c. Verify whether the SOC for a maximum is satisfied. Show your work.

9. To reduce pollution, two devices can be implemented (a and b). Assume the total cost function for these two devices is:

$$Z(a, b) = 2a^2 + 4b^2 + 2ab - 6a - 8b + 25.$$

Solve this function for the cost minimizing values of a and b.

10. A supermarket chain has devised the following formula for maximizing its sales:

$$S = -30x^2 + 5a^2 - ax + 100x + 1{,}000,$$

where s = sales, x = square feet of the store in units of 50,000 square feet, and a = annual advertising in units of 100,000. Suppose $a = 10$. Find the square footage for the store that maximizes sales.

11. Solve the following constrained optimization problem with equality constraints (make sure to write out the first- and second-order conditions):

$$\text{Max: } Z = -x^2 - 2y^2 + 8x + 12y - 34 \tag{0}$$

s.t.:

$$-2x - 4y = -8 \tag{1}$$

12. Solve the following constrained optimization problem with equality constraints (make sure to write out the first- and second-order conditions):

$$\text{Min: } Z = x^2 + y^2 - 12x - 10y - 61 \tag{0}$$

s.t.:

$$20x + 30y = 60 \tag{1}$$

13. A firm manufactures a product using two inputs (x and y) and has the following cost function:

$$\text{Cost} = f(x,y) = 35x^2 + 2xy + 10y^2 - 200x - 100y.$$

The firm may use any combination of the inputs x and y so long as $x + y = 100$. Find the amount of x and y that should be used to minimize cost while satisfying the restriction that $x + y = 100$. In addition, what is the minimum cost and SP for the constraint?

14. Write the LaGrange function and take the Kuhn–Tucker FOCs for the following problem:

$$\text{Max: } Z = x^{0.2} \, y^{0.4} \tag{0}$$

s.t.:

$$x + y \leq 100 \tag{1}$$

$$y \leq 75 \tag{2}$$

15. Consider the following constrained optimization problem with inequality constraints:

$$\text{Max: } Z = xy \tag{0}$$

s.t.:

$$x + y \leq 100 \tag{1}$$

$$x, \quad y \geq 0 \tag{2}$$

    a.   Give the Kuhn–Tucker FOCs conditions for this problem.

    b.   Solve this problem by hand.

16. Consider the following problem:

Max: $Z = xy$ (0)

s.t.:

$$x + y \leq 200 \tag{1}$$

$$x, \quad y \geq 0 \tag{2}$$

    a. Write the Kuhn–Tucker FOCs for this problem.
    b. Solve this problem for x^*, y^*, and λ_1^*.
    c. Verify whether the SOC for a maximum is satisfied.

17. Consider the following maximization problem:

Max: $Z = -a^2 - 10b^2 - 6ab + 15a + 40b$ (0)

s.t.:

$$a + \quad b \leq 4 \tag{1}$$

$$a, \quad b \geq 0 \tag{2}$$

    a. Write the Kuhn–Tucker FOCs for this problem.
    b. Solve this problem for a^*, b^*, and λ_1^*.
    c. Verify whether the SOC for a maximum is satisfied.

18. Write the Kuhn–Tucker conditions for the following problem:

Max: $Z = 2x - y$ (0)

s.t.:

$$x^2 + y^2 \leq 2 \tag{1}$$

$$x, \quad y \geq 0 \tag{2}$$

19. Consider the following NLP problem:

Min: $Z = (x - 2)^2 + (y - 3)^2$ (0)

s.t.:

$$x \quad + y \geq 8 \tag{1}$$

$$x, \quad y \geq 0 \tag{2}$$

    a. Write the Kuhn-Tucker FOCs for this problem.
    b. Solve this problem by hand.
    c. Solve this problem using Solver.

20. An economist has derived the following formula for a profit function of an agribusiness firm as follows:

$$\pi = -a^2 - 2b^2 + 10a + 15b + 1{,}000,$$

where a and b are inputs of production. The firm also faces the following production constraint:

$$2a + 2b \leq 6{,}000.$$

Solve this problem by hand. Then verify your answer using Solver.

21. Solve the following problem using Solver:

Max: $Z = -10a^2 + b^2 - 5c^2 + 100a - 25b + 50c - 100$ (0)

s.t.:

$$
\begin{aligned}
a + b &\leq 60 && (1)\\
a - b + c &\leq 15 && (2)\\
c &\leq 7 && (3)\\
a, b, c &\geq 0 && (4)
\end{aligned}
$$

9

Global Approaches to Nonlinear Optimization

An astoundingly large number of problems not only in economics but also in mathematics, engineering, and the sciences can be formulated as optimization problems. This realization is not as useful as it might seem at first because most optimization problems are intractably difficult to solve with the current technology and techniques available to researchers. So far, most of this book has focused on the useful and very well understood special case of linear problems. In Chapter 8, we introduced nonlinear optimization problems, discussed some basic mathematical techniques, including using calculus to find optima for both unconstrained and constrained functions, and briefly discussed the **Generalized Reduced Gradient (GRG) Algorithm**. We also saw some of the difficulties with the GRG Algorithm, specifically its tendency to converge to nonglobal solutions and saddle points.

Dealing with nonlinear problems can be very challenging in reality, and much of the current research in the field is devoted to developing new approaches to different classes of nonlinear programs. Some difficult problems require the development of altogether new algorithms specific to their particular attributes. Solver offers several different engines built on relatively recently developed algorithms that can be applied successfully to fairly general categories of nonlinear problems and that will usually find global optimal, or at least near-optimal, solutions.

In this chapter, we will first go into greater detail about some of the issues and concepts central to understanding global optimization techniques. We will then venture beyond the standard linear programming (LP) and GRG Solver Engines that we have used in prior chapters and explore some of the other tools that Solver offers for special types of problems. First, we will look at the SOCP (**Second-Order Conic Problem**) Barrier Engine, which can be used for a broad range of convex problems, allowing a substantial relaxation of the linear constraint requirement. Then we will discuss Solver's **Evolutionary Engine**. This engine uses a variation on a popular **metaheuristic algorithm** to search for the feasible region semi-randomly for good solutions; given sufficient time, it will usually find a very good, though not necessarily optimal, solution. Finally, we will introduce the **Interval Global Engine**, which uses an algorithm related to the branch-and-bound algorithm to find global solutions to any relatively small-constrained optimization problem, regardless of its structure.

The algorithms discussed in this chapter are simplified versions of the ones used by Solver. They are described to provide greater understanding, comfort, and competence with the processes by which the solutions are obtained than would be possible with just a "black box" understanding of the techniques. The details of some of these algorithms can be rather complex. For practical purposes, understanding the inner workings of the algorithms detailed here is less important than understanding the appropriateness, strengths, and limitations of each one.

9.1 DEVELOPMENT OF NONLINEAR PROBLEMS

In analyzing a problem, there are a variety of decisions and tradeoffs involved in formulating, designing, solving, and presenting the results. The deeper the understanding an analyst has of both the problem and the solution techniques available, the more effective the analyst will be in finding a solution that meets the needs of the problem and in communicating that solution. A basic understanding of a few issues that come up in optimization theory can be valuable in approaching and solving more complex problems. These issues should be kept in mind when considering the techniques detailed later in this chapter.

Problem Formulation

Often for complex problems, a well-formulated model can make obtaining a solution much easier. Sometimes functions that are not linear can easily be made linear through algebraic manipulation or techniques like the use of binary variables and weighting to represent logical constraints.

Say, for instance, that there is a constraint-limiting option a out of activities a, b, and c to be no more than 20% of the total amount. The natural formulation of this would be:

$$a/(a + b + c) \leq 0.2,$$

which is clearly nonlinear. However, with some algebraic manipulation this becomes:

$$a \leq (0.2a + 0.2b + 0.2c), \text{ or}$$

$$0.8a - 0.2b - 0.2c \leq 0,$$

which is now linear.

Another possibility is a situation where there is a fixed cost that must be paid before any number of units of a variable activity may be performed. This could be modeled with a logical condition like =IF(Activity1 > 0, 1, 0). However this would not only be nonlinear, but would also be nondifferentiable, so the GRG Engine would not be applicable. Instead, this situation can be modeled with a binary variable associated with the fixed cost in the objective function and a constraint linking it to the variable activity.

If the variable activity is represented by the variable x, and the binary variable is b, such a constraint would look something like $x - mb \leq 0$, where m is an upper bound (UB) to x. In minimizing, the algorithm will try to push b to zero since there is a cost associated with it; however if there is a nonzero value associated with x, b will have to be set to 1, or the constraint will be violated. This can then be solved as a linear MIP.

Other nonlinear functions can be closely approximated by linear functions on the particular domain of the problem. For instance, often for a small interval with midpoint x_1, the first-order **Taylor polynomial approximation**, defined by:

$$f(x) \approx f(x_1) + [df/dx(x_1)](x - x_1),$$

can give a reasonable linear estimate of a smooth function of one variable on that interval. Furthermore, second-order Taylor Polynomials, defined as:

$$f(x) \approx f(x_1) + [df/dx(x_1)](x - x_1) + \{[df^2/d^2x(x_1)]/2\}(x - x_1)^2,$$

can provide a convex approximation. For many nonlinear algorithms it is important, or at least useful, to have definite bounds on the decision variables.

Finally, for many algorithms, especially nonlinear ones, the starting values used (which would be the values in the decision cells when using Solver) play a big role in how the algorithm performs. Generally, all initial values should be feasible and ideally in the same general neighborhood as the optimal solution.

Convexity

In terms of difficulty in solving problems, the distinction has classically been between linear and nonlinear problems. As technologies and techniques advance however, the distinction is increasingly shifting towards **convex** versus **nonconvex** problems. As we discussed in Chapter 1, a convex set is essentially one for which a line segment connecting any two points in the set will be totally contained within the set. A function is convex if the area above the function is convex.

A moment's consideration should reveal that linear problems are always convex, as are many quadratic objective functions and quadratically constrained problems. Many other types of problems are convex, can be transformed, or are at least approximated as convex problems. Commonly used techniques can be helpful in making problems convex, but a full exploration of these techniques requires a fairly high degree of mathematical sophistication. Therefore, readers interested in pursuing this topic further should consult one of the excellent texts available on **Convex Optimization**.[1]

As more research is done in the field and more stable algorithm implementations are developed, convex optimization is becoming increasingly accessible to applied analysts. The SOCP Barrier Engine that we discuss in the following section can be used to solve most types of convex problems.

Deterministic versus Nondeterministic Algorithms

One key difference between different types of algorithms is whether they are **deterministic**, meaning that given the same initial input, they will follow the same path. In general, deterministic algorithms tend to have better (or at least more rigidly) defined decision and stopping criteria and are very predictable.

The evolutionary algorithm presented below is **nondeterministic**, as are other popular "heuristic" algorithms. The decision path for the Evolutionary Solver commonly involves some random elements and relies upon arbitrary time limits or convergence criteria to determine when to stop. One implication is that you can let the algorithm run for an arbitrarily long period of time, and could potentially continue to see increases in the objective function. Therefore, you can never be certain that the solution is entirely optimal; however, the longer you run the algorithm, the more certain you can be that it is at least a good solution.

Because of their randomness, nondeterministic algorithms tend to be much more adept with problems that have irregular feasible regions.

---

[1]See, for example, Berkovitz (2002) or Boyd and Vandenberghe (2004).

Algorithm Efficiency

In theory, any problem could be solved to a certain level of precision using an exhaustive algorithm. Simply start at the lowest feasible values and try every possible combination of feasible solutions to a certain level of decimal precision. This process is impractical for even small problems, so ideally we want our algorithms to be faster or more efficient than an exhaustive search.

The relative speed or efficiency of algorithms essentially depends upon the number of calculations required to solve the problem. Often, the time required to solve a problem grows exponentially, in which case the time required to solve reasonably large or complex problems can be astronomical. Interior point algorithms, like that used by the SOCP Engine detailed below, are usually fairly fast.

The simplex method is a bit of a special case. Depending on the composition of the feasible region, it is always possible that the simplex method will evaluate every extreme point before it finds the optimal solution. When this scenario occurs, the simplex method can be quite slow. In reality this seldom happens, and the simplex method is usually very efficient. The Global Interval method, also described below, is usually very slow, so it is really only useful for small problems.

A slow algorithm could take days, weeks, or possibly even longer to solve a moderately sized problem. For instance, in the last chapter we offered an example with 30 decision variables and about 20 constraints (which is relatively small when compared to common real-world problems). Using the LP/Quadratic Engine, it took several seconds to solve on a fairly new personal computer. Solving it using the Interval Global Engine on the same computer took 5 hours and 17 minutes. This was unnecessary since the problem is linear; however, it illustrates the disparity in efficiency between different algorithms and emphasizes that it is always preferable to state or approximate models in forms that lend themselves to fairly efficient methods of computation.

9.2   SECOND-ORDER CONIC PROBLEM BARRIER SOLVER

Linear programming and quadratic problems are both subcategories of **convex problems**. They can both be further generalized into the category of SOCPs, a mathematical categorization that is essentially equivalent to convex problems with quadratic objective functions and constraints. They are referred to as conic because the decision variables when taken as an n-dimensional vector can be thought of as constrained within a closed n-dimensional cone. To approach these types of problems, SOCP Barrier Solver uses a **Barrier** or **Interior Point** algorithm.

In contrast to the simplex algorithm, which traverses the boundary of the feasible region examining extreme points for optimality, interior point algorithms start at a feasible point and then move through the interior of the feasible region to converge on the optimal solution. Barrier algorithms work by assigning a penalty to the objective function with a **Barrier function** that grows as the function approaches a constraint.

As a result, the boundaries to the feasible region will essentially push the optimal solution to the barrier-adjusted problem towards the center of the feasible region. Then, by iteratively decreasing the weight given to the barrier function, the progressive solutions will trace an increasing path towards the most extreme point. This process will cause the best solution found to approach, but never quite reach, the optimal solution. We can continue to decrease the weighting on the barrier function until the successive iterations demonstrate a desired level of convergence. Another way this can be visualized is that the barrier function represents a distortion to the objective function. We are simultaneously optimizing the distorted objective function while having it converge to the true objective function.

There are many variations on barrier algorithms. For illustrative purposes, consider the following problem:

$$\text{Max: } Z = -e^x + (25x)^{0.5} + x \tag{0}$$

s.t.:

$$x^2 + x \le 3 \tag{1}$$

$$x \ge 0 \tag{2}$$

This problem has both a quadratic objective function and quadratic constraints, which means that the LP/Quadratic Engine cannot be used as in Chapter 8. Instead, consider the barrier function:

$$\Phi(x) = 1/x + 1/(3 - x^2).$$

Applying a weighting factor, α, to the barrier function and using it to penalize the original function $f(x)$ gives the new function:

$$f_1(x, \alpha) = f(x) - \alpha^* \Phi(x).$$

Maximizing this function gives the unconstrained maximization problem:

$$\text{Max: } Z = -e^x + (25x)^{0.5} + x - \alpha[1/x + 1/(3 - x^2)]$$

Since $1/x$ approaches infinity as x approaches 0, and $1/(3 - x^2)$ approaches infinity as x^2 approaches 3, the closer the function is to the original constraints, the more it will be penalized. Then, by using decreasing values of α such as 1, 0.5, 0.1, 0.05, and so on, $f_1(x, \alpha)$ can be forced arbitrarily close to $f(x)$. Note in particular that $f_1(x, 0) = f(x)$ except at roots of the constraints.

Starting with $\alpha = 1$ and an initial value of $x = 0.75$, optimizing using the GRG method provides an optimal solution of $x^* = 0.9438$, $Z^* = 1.1395$. From that solution as a starting point, decreasing α to 0.5 and then optimizing again reveals a new optimal solution of $x^* = 0.9847$, $Z^* = 2.2834$. Continuing, the first several steps are displayed below.

This will eventually converge to the solution $x^* = 1.1911$, $Z^* = 3.3573$. The original function is displayed in Figure 9.1a, and a series of $f_1(x, \alpha)$ for $\alpha = 1, 0.5, 0.1, 0.05$ is displayed in Figure 9.1b, along with the interior path followed by the algorithm. Notice that the series $f_1(x, \alpha)$ has vertical asymptotes at $x = 0$ and $x = 1.3027$ which represent the two constraints. Notice also that in this case the maximum was not at a boundary; however, even if it were, it would still converge to the proper solution.

α	x	Z
	0.75	1.0372
1	0.9438	1.1395
0.5	0.9847	2.2834
0.1	1.0789	3.1062
0.05	1.1109	3.2224
0.01	1.1605	3.3264
0.005	1.1605	3.3264
0.001	1.1726	3.3412

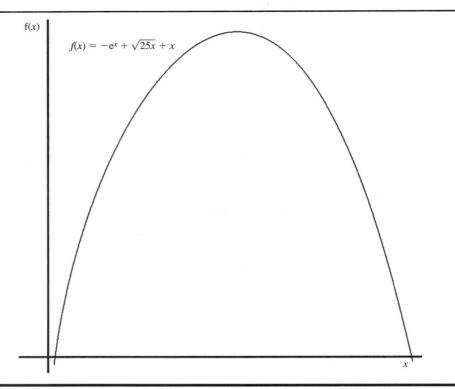

$f(x) = -e^x + \sqrt{25x} + x$

Figure 9.1a   The nonlinear function f(x).

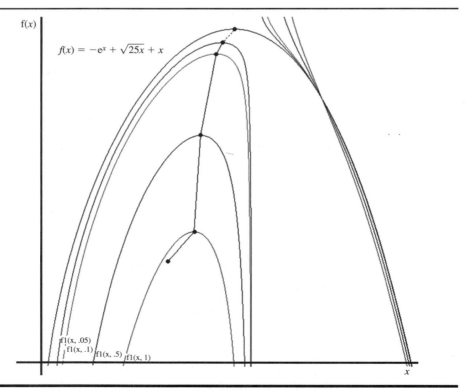

$f(x) = -e^x + \sqrt{25x} + x$

f1(x, .05)
f1(x, .1)
f1(x, .5)
f1(x, 1)

Figure 9.1b   Approximations of f(x) for progressive values of α.

Using Second-Order Conic Problem Barrier Solver

The SOCP Barrier Engine can solve problems with up to 200 variables, 8,192 (or 2^{13}) general constraints, 400 bounds on the variables, and 200 integer constraints. The engine settings are shown in Figure 9.2. The following settings are of interest when using Solver's SOCP Barrier Engine:

- **Gap Tolerance**: Gap tolerance is the primary convergence criteria used by the SOCP Barrier Solver. The algorithm implemented works with both primal and dual formulations to the problem simultaneously. When both the primal and dual solutions get within the amount specified, the problem is considered to be solved. The gap tolerance must be between 0 and 1. A larger gap tolerance will lead to a quicker but less precise solution.

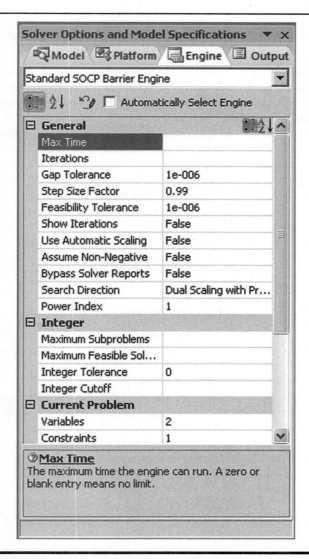

Figure 9.2   SOCP Barrier Engine settings.

- **Step Size Factor**: Step size factor determines the rate at which α decreases, or the rate at which the modified objective function is allowed to approach constraints. The step size factor must be between 0 and 0.99.

- **Feasibility Tolerance**: Because the algorithm is working with both the primal and dual problems, there is a possibility that a solution may be feasible for one, but not the other. This determines how close a solution must be to being feasible for both problems in order to be acceptable. The feasibility tolerance must be between 0 and 1.

- **Search Direction**: Search direction specifies the method used to determine the direction in which the algorithm will search for an improved solution for each iteration, each with the option of a prediction-correction term.

- **Power Index**: Power index must be greater than zero. It can be used to specify the behavior of search direction methods.

9.3   EVOLUTIONARY SOLVER

Evolutionary algorithms are one of a variety of metaheuristic algorithms, which are a class of generalizable search algorithms. Evolutionary algorithms are nondeterministic approaches which use a type of semistochastic method, often modeled after natural phenomena, to sample a search space in order to find a "best" solution. These algorithms work well with nonlinear functions and do not have the same issues with local maxima and discontinuous feasible regions that gradient algorithms do. That said, as the feasible region becomes more irregular, the likelihood of finding the true optimal solution in a reasonable amount of time decreases. Unlike deterministic methods, evolutionary algorithms use arbitrary stopping criteria, like maximum time or the number of iterations since the last improvement, which means there is no way of knowing for sure that the best known solution is truly optimal.

Another major drawback of such algorithms is that they usually take much longer than deterministic algorithms to converge on a solution, especially as the number of decision variables increases. Since the size of the solution space grows exponentially as new decision variables are added, problems with hundreds to thousands of decision variables may take a prohibitively long amount of time to converge on a reasonably optimal solution.

Description of Evolutionary Algorithms

Evolutionary algorithms are based on the theory of evolution and use a filtering technique analogous to natural selection to iteratively improve solutions over a series of several generations. First, a population of some number of possible feasible solutions is generated and evaluated based on a **fitness criterion**. Then a proportion of the solutions that best satisfy the fitness criterion is kept to become **parent solutions**. These are interacted, and a new generation of **offspring solutions** is generated using a **cross-over method**, which combines attributes of different parent solutions along with a random element to produce **mutations**. The length of time required to achieve a reasonable solution can vary and is a function of several factors including the level of certainty desired, the amount of randomness introduced, the amount of time between identifying new solutions, and the speed of the computer and processor.

The evolutionary algorithm implemented by Solver adds an additional step of conducting a local search after a new optimum is found to see if it can be further improved upon. By default it uses a gradient search, but other methods can be selected with the Local Search option, which is useful in dealing with functions that are not particularly smooth. In general, evolutionary algorithms tend to have problems with more than a handful of inequality constraints. Solver, however works around this in situations like integer and equality constraints by using various algorithms to make infeasible offspring solutions feasible.

Example of an Evolutionary Algorithm

First let us look at how the Evolutionary Solver would go about solving a simple problem. Consider an analyst who is seeking to understand the relationship between vehicle weight and fuel efficiency (measured in miles per gallon in city driving conditions). The analyst is particularly interested in looking at SUVs since those vehicles pose the greatest challenge to companies as they seek to meet the 2009 revisions to the Corporate Average Fleet Economy (CAFE) standards, which call for each carmaker's fleet to average 42 miles per gallon by the 2016 model year.[2]

However, the x-y Scatter Plot of the data (Figure 9.3) suggests that the relationship between weight and fuel efficiency is nonlinear (at least in this simple model). The analyst's supervisor suggests that a power curve ($y = \alpha x^{\beta}$) be estimated and parameters α and β found to minimize the sum of squared errors. This problem can be solved using the Evolutionary Solver.

For an illustration of this process, let us look at what happens with a population of six, and let us assume that the analyst knows something about the data and thus can limit the search for possible answers for alpha between 0 and 500,000 and the possible answers for beta between 0 and −1.

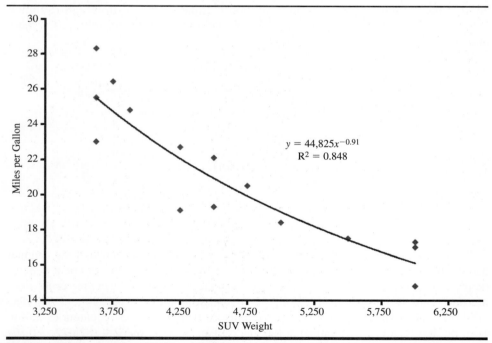

Figure 9.3   Scatter plot of SUV weight versus fuel efficiency.

---

[2]A sample of the data publically available from the U.S. Environmental Protection Agency (www.epa.gov/otaq/tcldata.htm) suggests that the analyst is correct in assuming that as the weight of an SUV without four-wheel drive increases, its fuel efficiency decreases (Table 9.1).

Table 9.1   Sample of SUV Fuel Efficiency

Manufacturer	Car Line	Weight	MPG
Chrysler	Grand Cherokee	4,500	19.3
Chrysler	PT Cruiser	3,625	23
Ford	Escape	3,625	28.3
Ford	Expedition	6,000	17
GM	Envoy	5,000	18.4
GM	Tahoe	5,500	17.5
Honda	Element	3,875	24.8
Honda	Pilot	4,750	20.5
Hyundai	Tucson	3,625	25.5
Mitsubishi	Endeavor	4,250	19.1
Nissan	Murano	4,250	22.7
Land Rover	Range Rover	6,000	14.8
Mazda	Mazda 5	3,750	26.4
Toyota	Highlander	4,500	22.1
Toyota	Sequoia	6,000	17.3

The initial six possible solutions, presented in Step 1 below, are generated at random and are referred to as the **parent solutions**. The **fitness criterion** in this case is the sum of squared errors, which we want to minimize. Not surprising given their random generation, these solutions have a wide variation of fitness ranging from 288 (Solution 1-1) to over 16 billion (Solution 1-3). The average fitness of the six solutions is more than 3.1 billion.

Step 1: Random Numbers

Solution	Alpha	Beta	Fitness
1-1	62,349	−0.98	358
1-2	96,039	−0.91	9,384
1-3	51,353	−0.05	17,023,858,049
1-4	82,672	−0.41	102,302,852
1-5	36,738	−0.87	208
1-6	93,738	−0.26	1,653,924,926

The next six **offspring solutions** *evolve* by keeping one of the **genes** of the original parent solution. For instance in Step 2, the first two solutions retain the beta gene of their parent solutions (1-1 and 1-2), but exchange their alpha value using the **cross-over method**. Thus, Solution 2-1 is (96,039, −0.98) and Solution 2-2 is (62,349, −0.91). The third offspring solution (77,779, −0.05) retains the beta gene from its parent (Solution 1-3), but has its alpha determined at random. Offspring solutions 2-4 (82,672, −0.87) and 2-5 (36,738, −0.41) retain the alpha value from the parent solutions and exchange their beta values. Finally, the sixth solution has its beta value generated randomly (93,738, −0.10).

While the average fitness of these offspring solutions is actually worse than the original parent solutions (approximately 10.6 billion), some of the individual offspring solutions have better fits than their parents.

Step 2: Cross-Over and Random

Solution	Alpha	Beta	Fitness
2-1	**96,039**	−0.98	329
2-2	**62,349**	−0.91	1,171
2-3	**77,779**	−0.05	39,069,356,586
2-4	82,672	**−0.87**	17,757
2-5	36,738	**−0.41**	19,792,233
2-6	93,738	**−0.10**	24,455,270,249

Step 3 selects the six solutions with the best fit and makes them the new parent solutions and discards the other solutions from consideration. This new population of parent solutions has a dramatically better average fit—down to 4,868.

Step 3: Best Six

Solution	Alpha	Beta	Fitness
2-4	82,672	−0.87	17,757
1-1	62,349	−0.98	358
2-1	96,039	−0.98	329
1-2	96,039	−0.91	9,384
1-5	36,738	−0.87	208
2-2	62,349	−0.91	1,171

As with Step 2, the cross-over method is applied to this new population of solutions where the beta genes are held constant and the alpha values are exchanged for the first two new offspring solutions (62,349, −0.87) and (82,672, −0.98), respectively. The fourth and fifth solutions retain the same alpha gene and exchange their beta values (96,039, −0.87) and (36,738, −0.91), respectively. The third solution retains its beta gene and has a randomly selected alpha value (41,411, −0.98) while the sixth solution has its random gene entering in the beta gene while it retains its parent's alpha value (36,738, −0.85).

Step 4: Cross-Over and Random

Solution	Alpha	Beta	Fitness
3-1	**62,349**	−0.87	6,440
3-2	**82,672**	−0.98	46
3-3	**41,411**	−0.98	1,617
3-4	96,039	**−0.87**	28,276
3-5	36,738	**−0.91**	237
3-6	36,738	**−0.85**	973

Step 5 is similar to Step 3 as it selects the best six solutions to date, as the new population of parent solutions. Note that since the six best solutions were already selected in Step 3, the previously discarded solutions no longer need to be considered. Therefore, the best six solutions will come from either the parents' solutions identified in Step 3 or the new offspring solutions developed in Step 4.

In this case, three of the new offspring solutions were among the six best-fitting solutions and thus form a new population of parent solutions from which further evolutions of potential solutions could occur. In this case, the lowest identified fitness is from Solution 3-2 (fitness of 46), and the average fitness of the new parent population is down to 358.

Step 5: Best Six

Solution	Alpha	Beta	Fitness
3-2	82,672	−0.98	46
1-5	36,738	−0.87	208
3-5	36,738	−0.91	237
2-1	96,039	−0.98	329
1-1	62,349	−0.98	358
3-6	36,738	−0.85	973

This example illustrates how the cross-over method used in the Evolutionary Solver uses an iterative process to find better and better solutions for the problems. In this case, the fitness criteria of minimizing the sum of squared errors improved each time the best six solutions were selected.

For evolutionary algorithms to be effective they require a somewhat larger population mutated over many iterations, so this is necessarily implemented on a computer. Solver contains the Evolutionary Solver Engine, which is based on this algorithm, with a few alterations to make it more efficient under some conditions. To solve this problem using the Evolutionary Solver (Figure 9.4), one needs to first set the objective function as the minimization of the sum of squared errors between the actual mpg values (D6:D20) and the predicted values (E6:E20). The predicted values are calculated based on the values of Cells B2 and C2, so, for instance, E6 should contain the formula "=$B$2*(C6^$C$2)". We can then calculate the difference as shown in the column F6:F20, and use the "=SUMSQ" formula in F22 to get our fitness criteria. A constraint can be put on alpha (B2) that represents its lower bound (LB) (0) and its UB (500,000). The bound constraint for beta is −1 and 0 for the lower and upper bounds, respectively. The final model is shown in Figure 9.5.

The dialogue box in Figure 9.6 shows the selection of the "Standard Evolutionary Engine" and a number of important user-defined options are available, which are particularly important since the Evolutionary Solver does not guarantee an optimal solution. Therefore, the user must determine when enough searching is sufficient to find a *reasonable* solution.

Note that the solution derived from Evolutionary Solver (55,917, −0.94) after less than 30 seconds has a better fit (35.2) than the trend line estimated by Excel (44,825, −0.912) which had a fit of 35.5.

Using Evolutionary Solver

The Evolutionary Engine can solve problems with up to 200 variables, 100 general constraints, 400 bounds on the variables, and 200 integer constraints. The engine settings are shown in Figure 9.6. The following settings are of interest when using Solver's Evolutionary Engine:

- **Tolerance**: As the Evolutionary Solver identifies possible solutions, it will evaluate any new solution against the previous best. Tolerance establishes the bound by which

▲	A	B	C	D	E	F
1		**Alpha**	**Beta**			
2	**Parameters**	55,917.16	-0.94			
3						
4						
5	**Manufacturer**	**Car Line**	**Weight**	**MPG**	**Predicted**	**Difference**
6	Chrysler	Grand Cherokee	4500	19.3	21.0	(1.7)
7	Chrysler	PT Cruiser	3625	23	25.7	(2.7)
8	Ford	Escape	3625	28.3	25.7	2.6
9	Ford	Expedition	6000	17	16.0	1.0
10	GM	Envoy	5000	18.4	19.0	(0.6)
11	GM	Tahoe	5500	17.5	17.4	0.1
12	Honda	Element	3875	24.8	24.1	0.7
13	Honda	Pilot	4750	20.5	19.9	0.6
14	Hyundai	Tucson	3625	25.5	25.7	(0.2)
15	Mitsubishi	Endeavor	4250	19.1	22.1	(3.0)
16	Nissan	Murano	4250	22.7	22.1	0.6
17	Land Rover	Range Rover	6000	14.8	16.0	(1.2)
18	Mazda	Mazda 5	3750	26.4	24.9	1.5
19	Toyota	Highlander	4500	22.1	21.0	1.1
20	Toyota	Sequoia	6000	17.3	16.0	1.3
21						
22				**Sum of Squared Errors:**		35.2

Figure 9.4   Evolutionary problem set-up.

a new solution is considered an *improvement* over the previous one. It is best to keep it at its default value of 0, unless the problem is taking considerable time to solve.

- **Max Time without Improvement**: This parameter sets the time (in seconds) that Evolutionary Solver will continue to search for a new *best* solution before stopping the search. As described above, Evolutionary Solver will use the definition of an improvement based on the tolerance set for the program. The default value is 30 seconds, which in most cases may provide a good initial estimate. However, to help ensure that the *best possible* solution is identified, this constraint should be relaxed considerably in order to help avoid merely identifying local maxima. For important problems, an analyst should set the time very high after an initial good answer is identified and let the computer seek improvements for hours or overnight.

- **Population**: In general, the more solutions considered, the more likely that a better answer will be identified. Therefore, having a large population has its advantages; however, since this Evolutionary Solver uses an iterative process, the number of solutions considered is mostly determined by the tolerance permitted by the user and the maximum time allowed without improvements in the objective function. By default, it uses 10 times the number of decision variables with a maximum population size of 100.

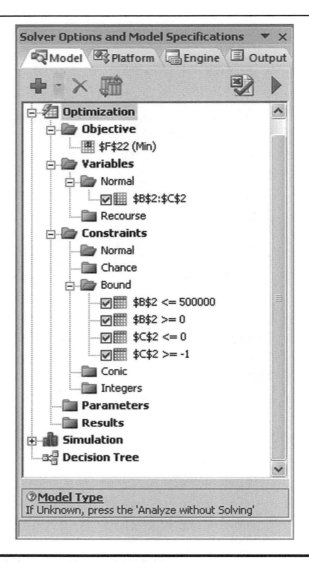

Figure 9.5   Evolutionary problem model definition.

- **Mutation Rate**: The more mutations, the more likely that Evolutionary Solver will find a global instead of a local solution. However, a high mutation rate also means that a large number of possible solutions will be introduced for consideration that will result in poor fitness for many candidate solutions.

- **Random Seed**: For generating mutations, Solver uses an algorithm that is initialized with some value and generates a series of random numbers. If a unique value is specified for the random seed, it will generate the same series for every run of the algorithm, so the results will be exactly reproducible. If a value is not specified, it uses the value of the system clock to seed the algorithm.

- **Require Bounds**: Whenever possible, set bounds on the decision variables as it will limit the range of random values selected for random genes. The tighter the bounds

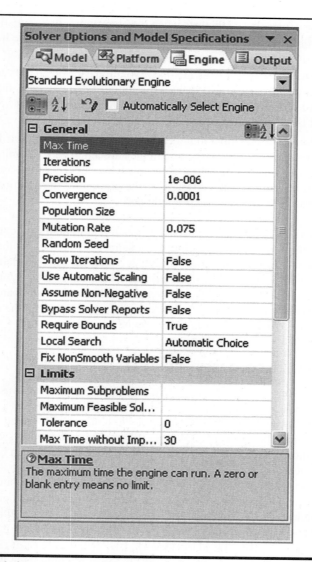

Figure 9.6   Evolutionary engine settings.

are, the more rapidly Evolutionary Solver will converge; however, be careful not to exclude potential solutions by making the bounds too restrictive. If the variables cannot be bounded, the required bounds option can be set to "False." However, this will substantially decrease the likelihood of achieving the true global optimum.

- **Iterations**: Max Time and Iterations both place limits on how long the algorithm can run before stopping.

- **Precision**: Instead of dropping infeasible offspring, Evolutionary Solver uses a variety of techniques to "repair" the solution, depending on the nature of its infeasibility. Precision dictates how close to the constraint the repaired solution must be. It must have a value between 10^{-4} and 10^{-9}. For most problems, the default value should usually be acceptable.

- **Convergence**: Convergence dictates when the algorithm will be satisfied with a solution. If the objective function changes less than this amount for five iterations, it will stop on that value.

- **Local Search**: Determines what method Evolutionary Solver uses to find a better local optimum once a new best solution is found. For relatively smooth functions, the Gradient Local option should work well. If the function is not very smooth, the Randomized Local option will probably be more effective.

- **Limits Options**: Dictates how long before the algorithm gives up if it is not able to satisfy the convergence criteria. By default they are left at zero, meaning that they will not be used. They should be left at zero unless the algorithm is having issues converging. If they are used and activated by the algorithm, the solution provided will likely not be as good as it otherwise might be.

The Evolutionary Solver offers a special analysis report to help gauge the effectiveness of the most recent optimization. The **population report** is found under the Reports menu where the sensitivity, answer, and limits reports are located. The population report gives statistics for the entire population of solutions considered by the algorithm including mean, standard deviation, minimum, and maximum values. If the standard deviations are small and minima and maxima are fairly consistent across trials, then the answer is likely optimal. If not, the engine's settings may need to be adjusted.

9.4   INTERVAL GLOBAL SOLVER

Interval Global Solver uses a technique called Interval Analysis that was first developed during the early 1960s. Instead of considering discrete numbers (also referred to as scalars), Interval Analysis considers a subset of the real numbers called an interval. For example, in one dimension, the interval $x = [1, 10]$ represents all numbers with $1 \leq x \leq 10$. Arithmetic operations can be extended to intervals; hence, functions can be generalized to operate on intervals.

Thus, an operation applied to two intervals results in a new interval, such that when the scalar equivalent to that operation is performed on any values from the original intervals, the result will be in the new interval. For instance, if we let $x = [1, 10]$ and $y = [4, 11]$, then we know that $x + y = [1, 10] + [4, 11] = [1 + 4, 10 + 11] = [5, 21]$. Additionally, we know that $x - y = [1, 10] - [4, 11] = [1 - 11, 10 - 4] = [-10, 6]$. It can be easily verified that choosing numbers from the two intervals and combining them with the standard arithmetic operations will yield a number in the respective resultant interval. Interval operations are similar to their scalar counterparts and have many similar properties, though there are important differences. The distributive property, for example, does not hold as strongly for intervals as it does with scalars.

Evaluating an **interval function**, $f(x)$ will map to an interval y, which will result in a box for every (x, y) instead of a point. A set of x's that covers the domain of a function, when evaluated with the corresponding interval function, will result in a set of boxes, which approximate and completely contain the original function. **Interval branch-and-bound** algorithms use an iterative process similar to that introduced in the prior chapter to consider increasingly fine covers of the domain.

If x represents the entire feasible range for a function f, considering the interval equivalent function $f(x) = [y_1, y_2]$, then y_1 and y_2 are initial lower and upper bounds for the optimal objective function solution. x is then branched into two subintervals x_1 and x_2 with $f(x_1) = [y_3, y_4]$ and $f(x_2) = [y_5, y_6]$. The greater of y_3 and y_5 is the new LB, and the greater of y_4 and y_6 is the new UB. If $y_5 < y_4$, x_2 can be dropped from consideration, and if $y_3 < y_6$, x_1 can be dropped. The remaining intervals are branched, and the process is repeated until the upper and lower bounds converge within an acceptable level of error. An example of such a process is very computationally intensive and requires a deeper treatment of interval algebra, so it will not be covered here.

The algorithm that Solver uses in its Interval Global Solver Engine is based on a classic version of a method known as the **Moore–Skelboe** algorithm. The algorithm was originally published in 1974 and has been the basis for many modified approaches since. There has been substantial research on different branching methods and other acceleration techniques. Interval branch-and-bound algorithms are exciting because they can find the optimal solution with certainty to within a given level of error for absolutely any problem. Nonetheless, they also tend to be extremely slow, especially for problems with more than a few variables. As was mentioned above, a problem with just a couple of dozen variables and constraints can take several hours or more to solve. Since solution time can grow exponentially as more variables are added, it becomes impossible to solve large problems within any practical time frame.

Using Interval Global Solver

The Interval Global Engine can solve problems with up to 200 variables, 100 general constraints, 400 bounds on the variables, and 200 integer constraints. The engine settings are shown in Figure 9.7. The following settings are of interest when using Solver's Interval Global Engine:

- **Accuracy**: Accuracy must be between 0 and 1. It represents how small an interval must become before it is considered a solution. The smaller this number is, the more precise your solution will be, and the longer it will take to converge.

- **Resolution**: Resolution must be between 0 and 100. It is a percentage difference used in comparing two different intervals. If the intervals' relative difference is less than the resolution, the two solutions are considered identical. If the relative difference is greater than the resolution, then a new solution is treated as a unique value. The smaller this value is, the slower and more precise the algorithm will be.

- **Max Time without Improvement**: This represents the amount of time in seconds that Solver will run without finding new optimal solutions.

- **Absolute versus Relative Stop**: If this is set to "True," Solver will use a scaled relative comparison between the difference between the UB and the objective function and the Accuracy; if it is set to "False," it will use an absolute difference comparison.

- **Assume Stationary**: This will speed up Solver considerably if set to "True"; however, Solver will not consider values of the decision variables at their bounds.

- **Method**: Method presents two different techniques for solving the problem. The Classic Interval method uses interval bounding and interval gradients, and can employ the Second-Order method to accelerate it. The Linear Enclosure method is an alternative method that approximates a series of linear boundaries that enclose the problem. If Linear Enclosure is used, the LP options can be used to break this approximation into a series of LPs, and approximate solutions and bounds using the simplex method can be used.

- **Second-Order**: The Second-Order option can only be applied when the Classic Interval method is being used. It can increase the speed with which the algorithm is able to shrink or discard intervals, usually leading to a faster solution time.

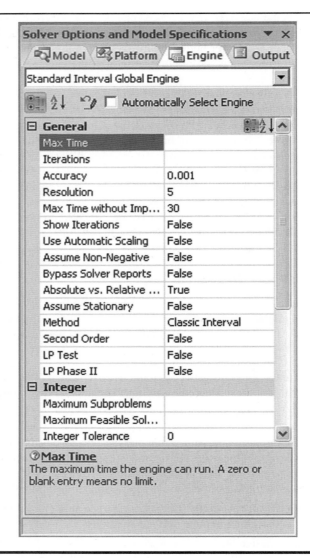

Figure 9.7   Interval global engine settings.

- **LP Test**: When the Linear Enclose method is being used and LP Test is set to "True," Phase I of the simplex method will be applied to the problem defined by linear enclosure on the current box.
- **LP Phase II:** When the Linear Enclose method is being used and LP Phase II is set to "True," LP Test will be run, and the simplex method will be used to solve and possibly update the bound on the globally optimal solution.

9.5   A FORESTRY EXAMPLE USING NONLINEAR EXCEL FUNCTIONS

Excel offers a variety of functions that can be useful in building models, but they are necessarily nonlinear. For example, suppose that a lumber company has eight teams of loggers that they want to deploy among 25 different lots for selective harvesting. Each lot has a different size and density score. Each team has a known productivity rate, and the amount that each team harvests is a function of the lot's density multiplied by the team's productivity. Additionally, each lot has a maximum amount of timber that is allowed for harvest set by the federal agency that manages the land. The problem is to match each team with a lot to maximize the amount of lumber produced. This type of problem is known as a **Combinatorial Optimization problem**, because the objective is to find the best combination from a discrete set of possibilities. These types of problems tend to be very computationally intensive and have their own set of specialized techniques and algorithms; however, this problem can be modeled in Excel using some specialized logical and lookup functions. Using this approach, a reasonably good solution can be obtained.

This model is displayed in Figure 9.8. The decision cells are the Assignment and Amount Harvested columns. The Potential Harvest column calculates the maximum that the team could potentially harvest from a lot based on its productivity multiplied by the lot density corresponding to the value in the Assignment column. In Excel, this can be done using the **VLOOKUP function**. The first cell, Cell L2, contains the function "=VLOOKUP(I2,A2:B26,2)*H2" that means look-up the row that has I2 in the first cell from the range A2:B26 and return the value in the second column. In this case, I2 is 15, so the corresponding density is 400. This value is then multiplied by the value in Cell H2, or 1.30, for a total potential harvest of 520.

The Total Harvested column uses the **SUMIF function** to calculate the total amount harvested from each lot. Cell C2, for example, contains the formula "=SUMIF(I2:I9, "="&A2, J2:J9)." This function compares every cell in the range I2:I9 to see if they are equal to the value in Cell A2. Note that the use of the "&" symbol in the equation refers the value in Cell A2 as text (also known as a "string"), which is required for this SUMIF function. Those cells that are equal to this value are added together. These totals are constrained to be less than the Maximum Harvest amount.[3]

The combination of logical, lookup, and arithmetic functions lead to a very irregular, noncontinuous feasible region. Since the problem is nonlinear, nonconvex, and nonsmooth, the LP, SOCP, and Interval Global engines will not work. The GRG Engine will run, but will likely not be able to offer much improvement beyond whatever values the user initially supplies. Therefore, this type of problem is best suited for Solver's Evolutionary Engine. Because the solution space for this problem is so irregular, it might be helpful to increase the mutation rate, and to loosen some of the time and convergence constraints to give the algorithm more time to search for a good solution. For this example, several trial runs with different settings produced a greatest maximum value of 3,035, though better solutions may be possible.

Excel has several other functions, such as IF, COUNTIF, HLOOKUP, and INDEX, which can be used in similar situations. These are useful tools and can sometimes make modeling easier and clearer. As shown with this example, these functions are nonlinear, thereby making it difficult for Solver to find a good solution. Therefore, at times, it may be easier to find creative ways to reformulate the problem to make it linear, or at least convex or continuous.

---

[3]It might be tempting to set the Amount Harvested to be equal to the lesser of the potential and maximum harvest instead of having it as a decision cell and using Maximum Harvest as a constraint. However, this approach would require a constraint that all Assignment cell values be unique. This additional constraint would make the problem much more computationally difficult in the search for a feasible solution.

	A	B	C	D	E	F	G	H	I	J	K	L
1	Lot	Density	Total Harvested		Maximum Harvest		Team	Productivity	Assignment	Amount Harvested		Potential Harvest
2	1	380	0	≤	400		1	1.30	15	459.7	≤	520
3	2	440	444	≤	490		2	1.26	6	469.6	≤	617
4	3	120	0	≤	210		3	1.01	2	444.4	≤	444
5	4	350	0	≤	390		4	0.97	13	440.0	≤	446
6	5	290	0	≤	350		5	0.89	21	391.6	≤	392
7	6	490	470	≤	470		6	0.77	24	384.6	≤	385
8	7	320	0	≤	350		7	0.69	10	304.8	≤	305
9	8	230	0	≤	280		8	0.57	11	140.6	≤	217
10	9	220	0	≤	290				Max Harvest	3,035.2		
11	10	440	305	≤	470							
12	11	380	141	≤	390							
13	12	300	0	≤	320							
14	13	460	440	≤	440							
15	14	300	0	≤	330							
16	15	400	460	≤	460							
17	16	420	0	≤	450							
18	17	100	0	≤	180							
19	18	300	0	≤	340							
20	19	160	0	≤	250							
21	20	130	0	≤	200							
22	21	440	392	≤	440							
23	22	120	0	≤	210							
24	23	300	0	≤	340							
25	24	500	385	≤	460							
26	25	150	0	≤	230							

Figure 9.8 Model of forestry problem.

9.6   RESEARCH APPLICATION: CROP FARMING IN NORTHEAST AUSTRALIA

Generally, the problems that apply nonlinear optimization methods are complex. Evolutionary and genetic algorithms have been applied extensively to problems in agriculture and natural resources, especially in models that involve several interconnected decision systems, like large agricultural operations or problems that have an intensive spatial component and are formulated in a GIS environment. Interior point methods have been used in problems that involve game theory or equilibrium systems, such as supply and demand models. In this chapter an example of each type is considered.

The first example comes from a study by deVoil et al. (2006), which considers a crop rotational model similar in concept to those discussed in Chapters 4 and 6. Specifically, the authors look at dryland crop farming in northeastern Australia. Farmers in this region face highly uncertain seasonal rainfall conditions. There are several types of crops commonly planted, each with different water requirements, which may be planted across two seasons. Planting too intensively will deplete soil moisture and make crop yields more dependent on the uncertain seasonal rainfall. Because of this uncertainty, a mix of crops is usually planted to hedge against unfavorable weather conditions. Recently there has also been increased attention to preventing nutrient depletion and soil erosion, which not only offers an external environmental benefit, but can also affect the long-term profitability of the operation.

One interesting facet to this study is that instead of a single objective value, this model searches for a continuous set of optimal solutions. The multiple objectives of maximizing gross return, minimizing soil erosion, and minimizing financial risk are all considered. If the entire space of combinations of all three attributes is considered with constraints to the cropping systems available, there will be an outer shell of solutions that are all **Pareto-optimal solutions,** meaning that from this optimal point, an increase in any one objective will lead to a decrease in another. The model is used to determine the set of Pareto-optimal solutions, so that the tradeoffs involved between different goals can be examined.

The decisions involve land use during the summer and winter over two years and the amount of moisture in the soil before the different crops are planted. The four crops considered in the model are cotton and sorghum, which can be planted during the summer, and wheat and chickpea, which can be planted during the winter. All have different costs, market prices, and possible levels of soil moisture thresholds. Historical rainfall data were used, and crops were planted if the rain received over a three-day period during certain planting windows was sufficient such that the total amount of water in the soil exceeded the minimum threshold for that crop.

This model was optimized several times using an Evolutionary Algorithm with different replacement rates, along with several variations on population size, mutation rate, and cross-over rate. The authors report that the final model required several minutes to evaluate each generation, so a network of computers was used simultaneously to evaluate populations in parallel so that a single optimization could be completed in a matter of hours. Sixty generations were used for each run, and it was found that populations that were smaller or had lower mutation rates were more likely to converge to a Pareto-optimal solution.

The results of the model emphasized the tradeoff between gross margin and soil erosion, and the tradeoff between gross margin and financial risk. In general, planting sorghum instead of cotton during the summer increased profit, but also increased risk. Also, planting wheat as a winter crop substantially decreased erosion, but represented a considerable loss of profit versus chickpeas. The authors offer these results along with several tables and graphical representation of the frontiers obtained to assist in decision making for these crop planting systems.

9.7   RESEARCH APPLICATION: AN ANALYSIS OF ENERGY MARKET DEREGULATION

One common model that lends itself to interior point algorithm solutions is a **Mathematical Program with Equilibrium Constraints (MPEC)**. This type of model is used by Hobbs et al. (2000) in analyzing the effects of deregulation of energy markets. The composition of energy markets can be quite complex. One reason for this complexity is due to geographical distribution issues: the market tends to be very regionalized with producers having a large degree of market power within their local markets. Because of this, even in the absence of regulation, competition tends to be fairly low, and prices can be frequently maintained above marginal costs. This study considers the short-run effects of deregulation on prices and market power, and the interaction between the two.

The model used by Hobbs et al. (2000) is based on a pricing system where producers offer bids to an Independent Service Operator (ISO), a type of organization established by the Federal Energy Regulatory Commission, which is in charge of setting prices and coordinating the market. Producers place linear bid functions for a megawatt-per-hour rate in the form of a linear supply curve with a set slope. It is assumed that firms have the power to shift their supply up or down. The ISO determines pricing and distribution based on an **Optimal Power Flow function**.

This study considers two cases. The first is a single-firm case, in which the model considers the bid of one "leader" firm based on how the other firms are expected to follow. The second is the multiple-firm case where the actions of each firm are considered as attempting to maximize profit based on both the other firms and the market as a whole. In this case, equilibrium should be reached in which no firm can increase profits by changing its strategy.

The problem is stated as considering the ISO's behavior setting electricity prices to maximize consumer welfare based on the firm's bids, constrained by production capacities, transmission network constraints, and constraints dealing with the physical properties of electrical circuits. Then, based on the ISO's pricing behavior, the producer will make their supply bid to maximize their profit. The model covers a network with 30 nodes, 41 arcs, 12 loops, 6 suppliers, and 21 consumers.

The case of a single firm is considered first using an interior point algorithm. For this case, the solution obtained represents the equilibrium to a **Stackleberg competition**. Then, a multiple-firm case in which several firms use strategic pricing behavior is solved with a similar but somewhat more complex interior point algorithm than the one used in the single-firm case. In this case the equilibrium prices are lower in some cases and higher in others than the Stackleberg equilibrium. The primary conclusion from this research is that the network asymmetry inherent in electrical power networks will cause firms to operate differently from the way they would under pure competition, suggesting that deregulation analysis will require different assumptions about firm behavior from those usually employed.

SUMMARY

In this chapter further topics and techniques useful in nonlinear optimization were presented. Some specific issues important to consider in building and solving linear problems, including aspects of problem formulation, categorization of different types of problems, and some properties useful in categorizing algorithms were presented and discussed.

Next, three Solver engines useful in solving nonlinear problems were introduced, including descriptions of the concepts behind the underlying algorithms and some of the significant options they use. The first engine, the SOCP Barrier Engine, uses an interior point, or barrier, algorithm. It is able to find the optimal solution to a very broad class of convex problems, known as "Second-Order Conic" problems. This includes LPs and problems with concave quadratic objective functions and constraints. The second engine, the Evolutionary Engine, uses a metaheuristic algorithm to semirandomly search the feasible region for a very good, but not necessarily optimal solution. How good the solution ultimately is depends on the length of time the algorithm is allowed to run, along with other user-defined options. The third engine was the Interval Global Engine. It uses a variation of the branch-and-bound method applied to successively finer intervals until it converges on the optimum. It is capable of finding the absolute optimal solution to any problem, but is prohibitively slow for many situations.

Finally, research examples were presented. The first one used the Evolutionary Algorithm to solve a cropping problem given the highly uncertain rainfall patterns in northeastern Australia. The second example used the SOCP Barrier Engine to evaluate market power in a deregulated electricity market.

REFERENCES

Berkovitz, L. D. (2002). *Convexity Optimization in* R^n. New York, NY: John Wiley & Sons.

Boyd, S. P., & Vandenberghe, L. (2004). *Convex Optimization*. New York, NY: Cambridge University Press.

deVoil, P., Rossing, W. A. H., & Hammer, G. L. (2006). Exploring profit—Sustainability trade-offs in cropping systems using evolutionary algorithms. *Environmental Modelling & Software, 21*, 1368–1374.

Hobbs, B. F., Metzler, C. B., & Pang, J. S. (2000). Strategic gaming analysis for electric PowerSystems: An MPEC approach. *IEEE Transactions on Power Systems, 15*, 638–645.

EXERCISES

1. Maximize the following equation:

$$y = (x - 2)(x + 4) + x(x - 3)(x - 1) \quad \text{for } 5 \le x \le 8.$$

2. Maximize the following equation:

$$y = x^6 + 2x^5 - x^4 - 3x^3 + 6x^2 \quad \text{for } 1 \le x \le 10.$$

3. Minimize the following:

$$y = (x_1 - 6)^2 + (x_2 - 8)^2 \tag{0}$$

s.t.:

$$x_1 \qquad\qquad \le 7 \tag{1}$$

$$x_2 \le 5 \tag{2}$$

$$x_1 + \quad 2x_2 \le 12 \tag{3}$$

$$x_1 + \quad x_2 \le 9 \tag{4}$$

$$x_1, \quad x_2 \ge 0 \tag{5}$$

4. Minimize the following:

$$y = 20{,}000 - 440x_1 - 300x_2 + 20x_1^2 + 12x_2^2 + x_1 x_2 \tag{0}$$

s.t.:

$$x_1 + \quad x_2 = 100 \tag{1}$$

$$x_1, \quad x_2 \ge 0 \tag{2}$$

5. Find the global minimal point of the following nonlinear function:
$$f(x) = x^2 - 2x + 1.$$

6. Find the minimal point of the following nonlinear function:
$$f(x) = x^3 - 2x^2 + x + 1,$$

where $-10 \le x \le 10$.

7. Find the global minimal point of the following nonlinear function:
$$f(x) = x^3 - x^2 + x + 1.$$

8. Consider the following problem:

Max: $Z = x^3 - 5x^2 + 2x - 3y^2 + 4xy$ (0)

s.t.:
$$x^2 - 3 \quad \le 13 \qquad (1)$$
$$x \ge 0 \qquad (2)$$

 a. Solve using GRG, starting from $x = 0$, $y = 0$, as well as from $x = 2$, $y = 0$.
 b. Solve using the Interval Global Engine.

9. Solve the following problem (by hand and using Solver):

Max: $Z = x + x^2 + x^3 + \cdots\cdots + x^{20}$ (0)

s.t.:
$$0 \le x \le 1 \qquad (1)$$

10. Consider the function:
$$f(x, y) = x! \,/\, y! \,(x - y)!$$

Find the maximum of $f(x, y)$ with the constraints:
$$x^2 + y^2 \le 100,$$
$$x, y \ge 0.$$

11. Consider the function:

$y = (x - 2)(x - 4)(x - 6)(x - 8)(x - 10)$ (0)

s.t.:
$$2 \le x \le 10 \qquad (1)$$

Find the local or global maxima.

12.

Max: $Z = 20{,}000 - 440x_1 + 20x_1^2 + 12x_2$ (0)

s.t.:
$$x_1 + \qquad\qquad x_2 = 7 \qquad (1)$$
$$x_1 \le 7, \qquad\quad x_2 \le 5 \qquad (2)$$

13. Maximize the following equation:

$Z = x(x - 2)(x - 4)(x - 6)(x - 1)$ (0)

s.t.:
$$0 \le x \le 6 \qquad (1)$$

14. Maximize the following equation:

$$Z = 0.01x_1^2 + x_2^2 - 20 \tag{0}$$

s.t.:

$$10x_1 - x_2 \geq 10 \tag{1}$$

$$2 \leq x_1 \leq 50 \tag{2}$$

$$0 \leq x_2 \leq 50 \tag{3}$$

15. John is considering investing in an annuity. He expects to invest anywhere between 5% and 20% of his monthly income. John currently makes $72,000 per year and expects an average raise of 4% in his salary until he retires; he doesn't know exactly when he will retire but expects that it will be anywhere from 15 to 20 years from today. The annuity's interest rates (depending on the time period) are listed below:

Time Length of Annuity	Expected Effective Annual Interest Rate (%)
15	4.25
16	4.00
17	3.95
18	3.75
19	3.60
20	3.50

Find the optimal annuity value for John, that is, for how many years John should work and how much he should invest every month.

16. Cola Company has introduced a new Raspberry Cola in the market, which customers initially find attractive; however, once there is too much cola the price reduces, so we have to develop a model that gives us the maximum profit with a single price model. Raspberry Cola has four customers from North, East, West, and South. Develop the model using a unit price of $0.50; the respective market size of each customer is 10, 25, 18, and 12 in thousands.

17. Consider the following problem:

$$\text{Max: } Z = \ln(x) + (36x)^{0.5} + x^3 \tag{0}$$

s.t.:

$$x^2 - 4x \leq 35 \tag{1}$$

$$x \geq 0 \tag{2}$$

To solve, use the following equation:

$$\text{Max: } Z = \ln(x) + 6(x)^{0.5} + x^3 - \alpha(1/x + 1/(35 - x^2 + 4x)).$$

Find a maximum value of Z with initial value of $x = 1$ and α ranging from 0 to 1. Describe the distribution of Z.

18. What are some benefits of the following engines?
  a. Standard LP
  b. GRG

c. SOCP Barrier Engine
d. Solver's Evolutionary Engine
e. Interval Global Engine

19. What are some drawbacks of the following engines?

a. Standard LP
b. GRG
c. SOCP Barrier Engine
d. Solver's Evolutionary Engine
e. Interval Global Engine

20. Determine the optimal solution for a dairy firm which produces two kinds of ice cream. One is priced ($/box) at $P_1 = 220 - 0.4x_1$, the other is $P_2 = 180 - 0.2x_2$. There are 800 hours available in the production department and 500 hours available in the inspection department. The firm maximizes its profit:

Max: $(P_1 - 60)x_1 + (P_2 - 45)x_2$ (0)

s.t.:

$$2x_1 + 3x_2 \leq 800 \qquad (1)$$
$$2x_1 + x_2 \leq 500 \qquad (2)$$

21. Paul is considering opening a book club in New York City. The monthly cost includes a facility fee, rent, and wages. Also, he needs to invest $30,000 initially on a total of 5,000 books and $10,000 on the furniture. To decide how much he should charge on every book rented, he did an online survey with a random size of 500 people. After receiving the responses, he categorized them into three groups based on their income level. The final result is as follows:

Number of books read (per month)	Price Willing to Pay (marginal value of books)		
	Low	Medium	High
1	2.00	2.50	3.35
2	1.50	2.01	2.56
3	1.20	1.52	2.45
4	1.02	1.33	2.01
5	0.92	1.03	1.34
6	0.89	0.95	1.03
7	0.80	0.76	1.02
8	0.63	0.60	0.98
9	0.42	0.26	0.87
10	0.25	0.22	0.79

Assume that books depreciate over one year and the furniture depreciates over five years.

a. What should the rent for each book be?
b. Given the fixed cost for each month in the table below, should Paul open the book club? Why?

Fixed Cost ($)		
Rent	Facility	Wage
1,000	200	512

10

Risk Programming Models

Risk and uncertainty are constant problems for agricultural decision makers. In fact, there is no other sector in the economy that faces such extreme volatility. Farmers and agribusinesses face risk in many dimensions, including price, production, and finance. For some commodities, it is not uncommon for price to fluctuate by over 50% between months, a phenomenon demonstrated by milk prices in the past decade. Within crop agriculture, production faces the most extreme volatility due to its dependence on many uncontrollable factors such as weather conditions, plant disease, and pests. Crop yields can easily fluctuate over 50% for a given farm from year to year. Like other businesses, farmers rely on bank credit to finance their production. Therefore, financial credit risk in the form of interest rate variability adds yet another source of risk to the agricultural sector.

Agricultural decision analysis has emerged as an important topic in the research literature within the last several decades. Considerable evidence suggests that farmers adjust their farm plans according to their risk posture. The impacts of risk and uncertainty on investment, marketing, production, and resource allocation decisions in agriculture have been studied extensively. One conclusion that has resulted from these studies is that profit-maximizing models, which ignore risk preferences by farmers, have failed to give accurate normative or positive economic results when applied to many farming situations.[1] Such models tend to overstate optimal output levels, exaggerate specialization in cropping patterns, give biased estimates of supply elasticities, overvalue resources, and incorrectly predict technology choices on the part of producers (Hazell, 1985). Thus, in order to properly study most farm-level decision-making problems, the decision environment must be formulated in such a way that risk and uncertainty are critical components in the model.

This chapter is devoted to the topic of mathematical programming models that incorporate risk and uncertainty into the decision environment. There are several risk programming models that have been extensively used in food and agricultural applications, including quadratic risk programming (i.e., mean-variance analysis), minimization of total absolute deviations (MOTAD), target MOTAD, chance-constrained programming, and

---

[1]"Normative" analyses refer to studies that focus on explaining how decision makers ought to behave in order to be consistent with their goals and plans. "Positive" analyses are concerned with describing or predicting how decision makers actually behave.

discrete sequential programming.[2] Each of these techniques is discussed in this chapter. Since these models are based on basic statistics and decision theory, the chapter begins with some definitions of basic statistical moments and an overview of the theory of farm decision analysis under risk and uncertainty. We then examine the more popular risk programming techniques used in applied economics. The chapter concludes with a summary of three empirical applications of risk programming. The first involves applying quadratic risk programming to identify optimal production and marketing plans for a representative cotton–grain farm in Texas. The second application develops a discrete stochastic sequential programming (DSSP) model for a representative Minnesota corn-soybean farm. The last application links a DSSP model for a representative Minnesota corn–soybean farm with a climate and agronomic model to simulate farm adaptation strategies in response to several climate change scenarios.

10.1 EXPECTED VALUE, VARIANCE, AND COVARIANCE

Let x and y be two random variables. Let the possible random events be $x_1, x_2, \ldots, x_n$ with probabilities $p(x_1), p(x_2), \ldots, p(x_n)$ for x; and $y_1, y_2, \ldots, y_n$ with probabilities $p(y_1), p(y_2), \ldots, p(y_n)$ for y. Then the **expected values** for x and y are defined as:

$$\text{Expected value of } x = E(x) = \sum_{i=1}^{n} x_i\, p(x_i), \text{ and}$$

$$\text{Expected value of } y = E(y) = \sum_{i=1}^{n} y_i\, p(y_i).$$

If we assume equally likely probabilities (i.e., $p(x_1) = p(x_2) = \cdots = p(x_n)$), then the expected values are the simple **means** or **averages** of the observations:

$$E(x) = \sum_{i=1}^{n} x_i/n, \text{ and}$$

$$E(y) = \sum_{i=1}^{n} y_i/n.$$

A measure of the variability of these random variables is given by the **variance**. The variances of $E(x)$ and $E(y)$ are defined as:

$$V(x) = E(x - E(x))^2 = \sum_{i=1}^{n} (x_i - E(x_i))^2 p(x_i), \text{ and}$$

$$V(y) = E(y - E(y))^2 = \sum_{i=1}^{n} (y_i - E(y))^2 p(y_i).$$

The estimated variances for x and y from a sample of the population are:

$$S(x) = \sum_{i=1}^{n} (x_i - \bar{x})^2/(n-1),$$

$$S(y) = \sum_{i=1}^{n} (y_i - \bar{y})^2/(n-1),$$

where $\bar{x}$ and $\bar{y}$ are sample means.

---

[2]There are additional mathematical programming models that incorporate risk. These include mean-gini analysis, focus loss models, Wicks and Guise models, and others. The interested reader can refer to Boisvert and McCarl (1990) for an excellent and thorough overview of mathematical risk programming models.

If we assume that x and y are independent, that is, they do not influence one another, then the covariance of x and y is zero. The **covariance** is a measure of how x will vary with y, and vice versa. For example, if the covariance is positive, then high values of x will tend to be associated with high values of y. The covariance for dependent random variables x and y is:

$$C(x,y) = E[(x - E(x)) (y - E(y))].$$

The estimated covariance for x and y from a sample of the population of n and m observations are:

$$C(x,y) = \sum_{i=1}^{n} \sum_{j=1}^{m} (x_i - \bar{x})(y_j - \bar{y})/(n + m - 2).$$

10.2 AGRICULTURAL DECISION ANALYSIS UNDER RISK AND UNCERTAINTY

Decision making involves selecting a course of action from a set of potential actions that offer different outcomes for some intended purpose. A decision problem exists when possible consequences are perceived as important, but there is uncertainty regarding the best course of action. When the decision maker is uncertain of the future consequence of a current decision, the decision maker is said to face a **risky** choice.

Since agricultural decisions occur over time with current decisions dependent upon uncertain future events, the decision-making process should be based on a probabilistic decision model. **Bayesian decision theory** is one way to represent sequential decision making under uncertainty.[3] The Bayesian decision problem under risk can be divided into several components. The first element of the decision problem is the identification of the set of acts or **actions** available to the decision maker for which a choice or plan must be made. For example, a set of actions (denoted as all a_j belonging to a, j = 1, ... , n) that could be identified for a producer includes all potential risk-reducing strategies to manage production and marketing variability. Decision analysis improves as the list of potential acts identified approaches the set of actual opportunities and when each act is clearly defined.

The next component of the decision problem is the identification of the possible **events** or **states of nature** (denoted as all s_i belong to s, i = 1, ... , m), which are exogenous random variables beyond the decision maker's control. Several state variables critical to farm planning include crop yields and the number of field days available, which depend primarily on weather conditions, and harvest and post-harvest prices, which depend primarily on market conditions. Specification of the set of possible events, similar to identifying the acts, improves the decision process when the set is comprehensive and clearly defined.

The third major part of decision analysis is quantifying the decision maker's subjective beliefs about the probabilities associated with the occurrence of the states. This step involves specifying the agent's perceptions of the **probability distribution** (denoted as all $p(s_i)$ belonging to p, i = 1, ... , m). The next element of the problem is identifying the **consequences** (c_{ij} belonging to C, i = 1, ... , m; j = 1, ... , k) of the decision maker's action, a_j, when state of nature s_i occurs. The consequences are frequently defined in terms of payoffs, such as net income.

Finally, the last major component of the decision process involves the selection of evaluative or **choice criterion**, which provides the basis for selecting the course of action, given the other dimensions of the problem. The choice criterion should be closely tied with the objectives of the decision maker. Studies concerned with the role of risk and uncertainty in the decision process usually specify maximization of expected utility as the major choice criterion.

---

[3]The discussion on the components of the Bayesian decision problem is based on an excellent presentation in Anderson et al. (1977).

The Expected Utility Hypothesis

The **expected utility hypothesis** (EUH), which was first developed by Bernoulli in 1738[4] and later refined by von Neumann and Morgenstern in the 1940s, provides the basis for many evaluative criteria used to help farmers make decisions under risk. Simply put, the EUH states that an agent will prefer one risky action, a_1, over another risky action a_2, if the former has a greater expected utility associated with it. A utility function $[u(a_j)]$ is an analytical method of expressing an agent's preferences, which translates the outcome of the choice of some action, a_j, to a real number index of its desirability, $u(a_j)$. When used with a decision rule such as maximizing utility, the utility function serves as the basis for determining optimal decision strategies for the decision maker. In order for a well-defined single-dimensional utility function to exist, the decision maker's risk preferences must satisfy three axioms, which are sufficient conditions in deducing the EUH. These conditions are represented by the following three axioms.

1. **Ordering and Transitivity.** Ordering implies that for any two risky prospects,[5] a_1 and a_2 belonging to a, the agent either prefers a_1 to a_2, a_2 to a_1, or is indifferent between them. Transitivity is a logical extension of ordering for situations involving more than two risky prospects. Transitivity implies that for any three risky prospects a_1, a_2, and a_3 belonging to a, if the decision maker prefers a_1 to a_2 (or is indifferent between them) and prefers a_2 to a_3 (or is indifferent between them), then the agent will prefer a_1 to a_3 (or be indifferent between them).

2. **Independence.** Independence implies that for any three risky prospects a_1, a_2, and a_3 in a, if the agent prefers a_1 to a_2, then the agent will prefer a lottery composed of a_1 and a_3 as its possible outcomes to a lottery involving a_2 and a_3 as possible outcomes when the probabilities of a_1 and a_2 are equal.

3. **Continuity.** Continuity implies that for any three risky prospects a_1, a_2, and a_3 in a, if a_1 is preferred to a_2, and a_2 is preferred to a_3, then there exists a probability p, other than zero or one, which will make the decision maker indifferent between a_2 for certain and a lottery composed of a_1, with probability p, and a_3, with probability $(1 - p)$.

The EUH or Bernoulli's principle can now be formally stated, having defined these axioms. The EUH states that if a decision maker's preferences are consistent with the axioms of ordering and transitivity, continuity, and independence, then a utility function defined on risky prospects exists that assigns a single real number utility value for each prospect and has the following properties:

1. If a_1 is preferred to a_2, then $U(a_1) > U(a_2)$, and vice versa, for all a_1 and a_2 in a.

2. The utility of a risky prospect (a_j) is equal to the expected utility of its outcome, that is, $U(a_j) = E[U(a_j)]$.

3. The utility value of each risky prospect is assigned an arbitrary origin and unit of scale.

The EUH establishes a basis for comparing risky prospects in a manner that is consistent with the decision maker's preferences. The theorem implies that agents select risky prospects so as to maximize expected utility (von Neumann & Morgenstern, 1944).

Although human behavior is goal oriented and no single attribute, such as income, alone can completely describe an agent's true utility function, it is common in decision problems

---

[4]An English translation of Bernoulli's 1738 paper has been subsequently published (Bernoulli, 1954).

[5]The term "risky prospect," in this context, means an act or choice that has a probability distribution of outcomes associated with it.

to express monetary outcomes (m) as the sole attribute of the utility function. This is used primarily because of the difficulty analysts have had with using multiple objective utility functions. Monetary outcomes have been quantified in several ways, which include such variables as wealth, income, gains and losses, or rate of return.

Given a range of risky prospects defined over monetary outcomes, values of expected utility can be determined by integrating a utility function over this range, or can be approximated with a Taylor series expansion using derivatives and moments of the utility function and probability distribution of outcomes (Anderson et al., 1977). To represent certain classes of economic behavior and risk preferences, it is necessary to look at the shape and curvature of the utility of money function. It is usually assumed that all "rational" decision makers prefer more money to less, which implies that the utility function increases monotonically, that is, $U'(m) > 0$. However, to represent different classes of risk preferences, look at the curvature of the utility function. The three classes of risk preferences are risk aversion, risk neutrality, and risk preferring or loving.

A utility function that is strictly concave has the property of decreasing marginal utility of money [$U''(m) < 0$]. Agents displaying this type of utility function are **risk averse** because they prefer a certain outcome to an uncertain outcome with the same expected value. For a linear utility function, utility increases in direct proportion to increases in money, which implies that the marginal utility of money is constant [$U''(m) = 0$]. Agents with this kind of utility function are **risk neutral** because they are indifferent between a certain outcome to an uncertain outcome with the same expected value. A utility function that is strictly convex has the property of increasing marginal utility of money [$U''(m) > 0$]. Agents displaying this type of utility function are **risk lovers** because they prefer an uncertain outcome to a certain outcome when the expected value of the uncertain outcome equals the value of the certain outcome. Figure 10.1 illustrates graphically the shapes of these three classes of utility functions.

To measure the degree of risk aversion, the concept of the **risk premium** (RP) has been used. To illustrate the RP, consider panel (a) of Figure 10.1. Suppose there is a lottery with two risky outcomes, m_1 with probability p, and m_2 with probability $(1-p)$. The utility function in panel (a) is strictly concave and therefore the agent is risk averse. The **expected monetary value** (EMV) of this lottery is EMV = $pm_1 + (1-p)m_2$. The RP is defined as the difference between the EMV and the certainty equivalent (CE), which is the amount of money exchanged with certainty that makes the agent indifferent between this exchange and the lottery (Anderson et al., 1977). In panel (a), the EMV is given by point c, the CE is given by point b, and the RP is equal to the difference. The RP will be positive, zero, or negative for agents who are risk averse, risk neutral, or risk lovers, respectively. Panel (b) and (c) of Figure 10.1 illustrate the above concepts for risk neutrality and risk preferring.

There have been many studies that have tried to determine the risk preferences of farmers. The empirical results are inconclusive, particularly with respect to farmers in the United States. Moreover, only a small sample of producers was used to elicit risk postures. Nevertheless, the majority of risk analyses in agricultural economics have assumed farmers are risk averse.

Since von Neumann and Morgenstern's development of the EUH in the 1940s, much attention has been given to operationalizing the hypothesis for empirical applications. The empirical research that has surfaced has tended to follow two directions. One direction has centered on devising elicitation procedures (**direct elicitation approach**) to directly estimate decision makers' utility functions. The other approach has been to focus on the probability distribution and moments of the risky prospect (**moment method**) and to approximate the utility of a prospect as a function of its mean and higher moments of the distribution (Anderson et al., 1977).

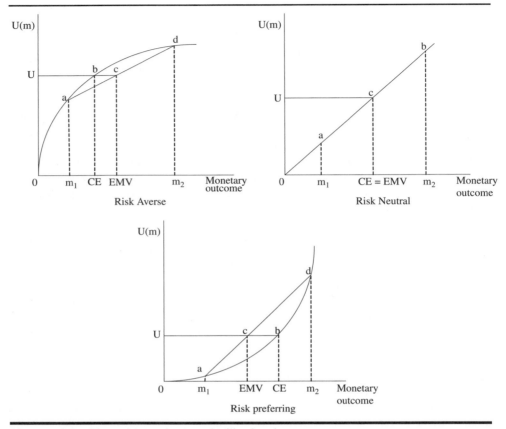

Figure 10.1   Three classes of expected utility functions.

The direct elicitation approach has not been very successful in predicting actual behavior of agents. This approach has been criticized because the results are often biased due to agents lacking familiarity dealing with probabilities, agents' preferences for specific probabilities, differences in interview procedures, and attitudes towards gambling (Young, 1980). Consequently, some researchers have turned to approximating utility as a function of the moments of the probability distribution. Under this approach, the primary concern is the probability distribution rather than the estimation of the utility function. Due to the limitations of the former approach, this chapter focuses on the moment approach.

When it is assumed or specified that utility depends solely on one argument, expected utility can be restated in terms of the moments of the probability distribution of the sole attribute. The moment method is equivalent to direct estimation of utility functions under certain circumstances and is a fair approximation for others. In addition, the required computations to derive utility functions using this approach are usually less than the direct method (Anderson et al., 1977). To develop this reasoning, consider the case where utility depends solely upon profit, Z. Taking the Taylor series expansion on the expected value of profit yields:

$$U(z) = U[E(z)] + U_1[E(z)] [z - E(z)] + U_2[E(z)] [z - E(z)]2/2! + \cdots$$
$$+ U_n[E(z)] [z - E(z)]n/n! \qquad (10.1)$$

where U_i means the ith derivative of U.

By the EUH, equation (10.1) may be restated, carrying through the expectations operator (E), as:

$$U(z) = U[E(z)] + U_1[E(z)] \ E[z - E(z)] + U_2[E(z)] \ E[z - E(z)] \ 2/2! + \cdots$$
$$+ U_n[E(z)] \ E[z - E(z)]n/n! \tag{10.2}$$

Since $E[z - E(z)] = 0$ and $E[z - E(z)]^2$ is the variance of Z $[V(z)]$, equation (10.2) can be further simplified to:

$$U(z) = U[E(z)] + U_2[E(z)] \ V(z)/2! + U_3[E(z)] \ M_3(z)/3! + \cdots \tag{10.3}$$

where $M_3(z)$ is the skewness of the distribution of z. Hence, when profit is the only argument in the utility function, the function can be expressed in terms of the moments of the probability distribution of z. Moreover, if profit is normally distributed, then the probability distribution of z is completely described by the mean and variance. As a result, under normality, the utility function of z can be written in general form as:

$$U(z) = U[E(z), V(z)] \tag{10.4}$$

Many empirical studies have assumed a normal distribution of z and have estimated utility as a function of the type expressed in equation (10.4) such as the case of quadratic risk programming. The EUH represents a useful and popular tool used to study decision making under risk and uncertainty.

10.3 QUADRATIC RISK PROGRAMMING

Markowitz (1959) used **quadratic programming (QP)** in an empirical application of a stock portfolio problem. Under the QP formulation of the stock portfolio problem, risk is considered solely in terms of revenue activities in the objective function, while resource endowments and technical parameters in the opportunity set are assumed to be known with certainty by the decision maker. The use of QP assumes that an agent has preferences among alternative strategies based entirely on their expected income (E) and associated variance (V), that is, E-V criterion. According to this criterion, an agent prefers a risky plan, p_1, to another risky plan, p_2, or is indifferent between them, if the expected income of p_1 is $\geq$ to the expected income of p_2 and the variance of p_1 is not larger than the variance of p_2. If one of the above weak inequalities holds as a strict inequality, then p_1 is strictly preferred to p_2. Use of QP in risk applications also assumes that the iso-utility curves of the agent are convex, which implies that the farmer is risk averse. In other words, the decision maker prefers a farm plan with a higher V only if E is also larger, and E must rise at an increasing rate relative to increases in V. In general, these conditions will hold if the agent's utility of income function is quadratic and strictly concave (Markowitz, 1959).

Given these assumptions, a rational decision maker will want to choose a strategy from a set of farm plans that have a minimum variance for alternative levels of expected income or a maximum expected income for alternative levels of variances. Quadratic programming algorithms can be formulated to derive a set of efficient E-V pairs over a set of all feasible farm plans. From the resulting efficiency set, an optimal strategy can be determined in one of two ways. For a given utility function, iso-utility curves can be derived, and the point of tangency between the E-V frontier and the highest iso-utility curve determines the optimal farm plan. Point C in Figure 10.2 illustrates the determination of an optimal farm plan graphically. One major drawback of this approach is that it requires estimation of a utility function, which is difficult. Thus a second and more frequently used approach is to derive the efficiency set and allow the farmer to choose the most preferable farm plan.

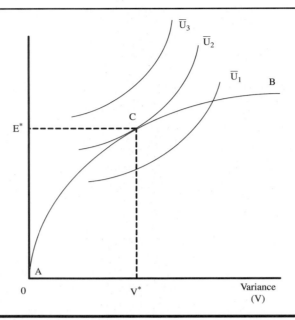

Figure 10.2   Graphical determination of the optimal farm plan in E-V analysis.

One method to derive the efficiency set is to use **parametric QP**. One formulation[6] of this technique is the following:

$$\text{Max: } U = \sum_{j=1}^{n} E(c_j)x_j - b\sum_{i=1}^{n}\sum_{j=1}^{n} V_{ij}x_i x_j \tag{0}$$

s.t.:

$$\sum_{j=1}^{n} a_{ij}x_j \leq b_i \quad i = 1, ..., n \tag{1}$$

$$x_j \geq 0 \quad j = 1, ..., m \tag{2}$$

where:

$E(c_j)$ = expected returns of the jth activity

x_j = level of jth activity

b = agent's absolute risk aversion coefficient

V_{ij} = variance of jth activity when j is equal to i, and covariance between jth and ith activity when i is not equal to j

a_{ij} = amount of resource i required per unit of the jth activity

b_i = amount of resource i available

Procedurally, a solution is derived for alternative levels of the risk aversion coefficient. The risk aversion coefficient is a link between the objective function and the agent's utility function. When b is set to zero, the solution represents the risk-free or risk-neutral case and is equivalent to maximizing expected net revenue. When b is greater than zero, the

---

[6]This formulation was first used by Freund and is based on a negative exponential utility function, which has the property of constant absolute risk aversion.

solutions are optimal for agents with various degrees of risk aversion. The resulting set of solutions may be presented to the farmer for inspection to choose a particular plan, or a utility function may be estimated and a unique optimum can be found by deriving the point of tangency between the iso-utility curve and the E-V frontier.

The application of QP and the E-V criterion is consistent with the expected utility hypothesis only under the following conditions: (1) the decision maker possesses a quadratic utility function; (2) the probability distribution of returns is normally distributed; or (3) the utility function can be truncated after the second-order moment of its Taylor series expansion. These restrictions may limit the use of E-V analysis in risk applications. Still the E-V approach has been used extensively in firm-level risk research.

An Example[7]

A farmer produces three commodities corn, wheat, and soybeans using three resources: hired labor, family labor, and machinery. Over a three-month production period, the farm is endowed with 1,000 hours of hired labor, 500 hours of family labor, 2,000 hours of machine time, and 600 acres of land. The resource requirements for each commodity are summarized below:

Resource	Soybeans	Wheat	Corn	Resource Endowment
	(hours/unit of good)			
Hired Labor	1.0	1.1	1.3	1,000 Hours
Family Labor	0.4	0.4	0.4	500 Hours
Machine Time	2.2	2.8	3.0	2,000 Hours
Land	1.0	1.0	1.0	600 Acres

There is a three-month time lag between planting and harvest for these three commodities. Because of certain random "states of nature," which are exogenous to this farm, the farmer does not know with certainty (at the time decisions must be made regarding how much of each commodity to produce) the per unit profit of each commodity. The farmer expects that the unit profit of each good will be equal to the average profit over the last 10 periods. Assume that the profit of the three commodities over the previous 10 periods is the following:

Observation	Soybeans	Wheat	Corn
		$ per acre	
1	100	93	200
2	95	99	100
3	97	97	150
4	94	110	45
5	91	111	200
6	85	120	190
7	92	100	75
8	90	121	25
9	86	127	210
10	80	129	192
Average	91	110.7	138.7

[7]This problem, solution, and corresponding sensitivity analysis are shown in the Chapter 10 supplemental materials available at www.wiley.com/college/kaiser.

If a statistical computer package is used, the variance matrix is easily found and is equal to:

	Soybeans	Wheat	Corn
Soybeans	32.6	−65.0	−101.7
Wheat	−65.0	154.6	142.2
Corn	−101.7	142.2	4,546.2

Since the farmer is risk averse, an optimal mix of commodities is needed to minimize total profit risk based on a minimum level of expected profit (E^*). The farmer measures profit risk by multiplying the variance-covariance matrix by the interaction of the individual investments in the portfolio. That is, risk is measured as:

$$R = 32.6g_1g_1 + 154.6g_2g_2 + 4{,}546.2g_3g_3 - 65.0g_1g_2 - 101.7g_1g_3 - 65.0g_2g_1 + 142.2g_2g_3 - 101.7g_3g_1 + 142.2g_3g_2$$

where: g_1 is acreage of soybeans produced, g_2 is acreage of wheat produced, and g_3 is acreage of corn produced.

Noting that the variance matrix is symmetric (i.e., $V_{ij} = V_{ji}$), the risk definition can be written more compactly as:

$$R = 32.6g_1g_1 + 154.6g_2g_2 + 4{,}546.2g_3g_3 + 2\{-65.0g_1g_2 - 101.7g_1g_3 + 142.2g_2g_3\}, \text{ or}$$

$$R = 32.6g_1g_1 + 154.6g_2g_2 + 4{,}546.2g_3g_3 - 130.0g_1g_2 - 203.4g_1g_3 + 284.4g_2g_3.$$

If it is assumed that the farmer wishes to minimize R, s.t.: $E \geq E^*$, this problem is the following:

Min: $R = 32.6g_1g_1 + 154.6g_2g_2 + 4{,}546.2g_3g_3 - 130.0g_1g_2 - 203.4g_1g_3$
$$+ 284.4g_2g_3 \tag{0}$$

s.t.:

$$91.0g_1 + 110.7g_2 + 138.7g_3 \geq E^* \tag{1}$$

$$1.0g_1 + 1.1g_2 + 1.3g_3 \leq 1{,}000 \tag{2}$$

$$0.4g_1 + 0.4g_2 + 0.4g_3 \leq 500 \tag{3}$$

$$2.2g_1 + 2.8g_2 + 3.0g_3 \leq 2{,}000 \tag{4}$$

$$1.0g_1 + 1.0g_2 + 1.0g_3 \leq 600 \tag{5}$$

$$g_1, \quad g_2, \quad g_3 \geq 0 \tag{6}$$

This problem can be easily solved using Solver. Figure 10.3 illustrates the Solver worksheet for this problem when $E^* = \$82{,}220$, which is the profit-maximizing solution. In general, it is a good practice to start by setting E^* equal to the profit-maximizing, or risk-neutral solution. This parameter is found by solving the corresponding profit maximization LP problem where risk is not included. The optimal value of this objective function gives the E^* associated with the risk-neutral solution. After substituting

	A	B	C	D	E	F	G
1		g_1	g_2	g_3			
2	**Decision Variables**	0.0	35.7	564.3			
3							
4	**Variance-Covariance**	g_1	g_2	g_3			
5	g_1	32.6	-65	-101.7			
6	g_2	-65	154.6	142.2			
7	g_3	-101.7	142.2	4546.2			
8							
9	**Objective**						
10	Min Risk	1,453,522,306					
11							
12							
13	**Constraints**	g_1	g_2	g_3	**Total**		**RHS**
14	Expected Profit	91	110.7	138.7	82,220	$\geq$	82,220
15	Hired Labor	1	1.1	1.3	773	$\leq$	1,000
16	Managerial Labor	0.4	0.4	0.4	240	$\leq$	500
17	Machine Time	2.2	2.8	3	1,793	$\leq$	2,000
18	Land	1	1	1	600	$\leq$	600

Figure 10.3   Solver formulation for quadratic risk programming example.

this E^* into the quadratic risk programming problem and solving, E^* should then be parametrically lowered in order to trace out the set of E-V efficient solutions. The range of feasibility on E^* can be used in selecting the new values of E^* because it gives a new solution to the problem in the parametric analysis.

In this example, the risk-neutral value for E^* is $82,220. If you parametrically vary E^*, an efficient set of crop mixes based on alternative levels of risk is obtained. The following table summarizes some of these efficient plans for this example.

Expected Profit ($)	Variance ($)	Standard Deviation ($)	Soybeans (acres)	Wheat (acres)	Corn (acres)
82,220	1,636,631,995	40,455	0.0	0.0	600.0
80,000	1,087,286,890	32,974	0.0	115.0	485.0
75,000	465,789,543	21,582	0.0	293.6	306.4
70,000	125,950,359	11,223	0.0	472.1	127.9
65,000	32,919,362	5,738	118.8	448.3	32.9
60,000	2,843,412	1,686	341.8	247.1	11.2

The first row in this table corresponds to the risk-neutral, or profit-maximizing, solution. This is the optimal plan for a farmer who wants to maximize profit regardless of risk. All resources should be devoted to corn, which offers the highest expected profit ($82,220), but also the highest risk (variance of $1,636,631,995). Notice that the variance gives an extremely large value of risk. Hence, the standard deviation is generally the proxy for risk which is shown to the decision maker. For the risk-neutral solution, the standard deviation is $40,455, or about one-half of its respective minimum expected return level of $82,220.

At the opposite extreme, the last row of this table provides an efficient plan for the most risk-averse farmer. If the farmer is willing to accept a lower expected profit ($60,000), then risk can be reduced to a standard deviation of $1,686. Farmers displaying this type of risk preference can achieve this minimum risk level through diversification and devoting most of their resources to the least risky commodity, soybeans.

It is important to note that all six plans given in this table are E-V efficient, or optimal. They each provide for the lowest possible risk given a certain minimum expected profit level. It is up to the farmer to choose the preferred plan from this efficient set of plans. There are several observations to glean from this table. First, there is a trade-off between E and V. To have higher expected profits, the farmer has to accept higher risk. Second, the "cost" of this trade-off is higher at the extremes. For example, for E to increase from $80,000 to $82,220, which is only a 4 percent increase in E, the farmer's risk, as measured by the standard deviation, increases from $32,974 to $40,455, which is an increase of 22.7 percent. Hence, once the farmer approaches the profit-maximizing level of E, the increase in risk to achieve the desired E becomes larger. Third, there are two types of risk-reducing strategies. The first is to switch from the riskiest crop, corn, to the less-risky wheat and least-risky soybeans. The second is to diversify the portfolio rather than rely on one crop. Diversification is a risk-reducing strategy because of the negative covariance between several crop combinations, such as wheat and soybeans, soybeans and wheat, and soybeans and corn.

Similar to LP solutions, the dual prices for each constraint have the same interpretation in QP. For example, the dual value for the minimum E constraint (for $E^* \geq 82,220$) is $188,743. This implies that if the farmer were to decrease E^* by one dollar from $82,220 to $82,219, then the minimum variance would decrease by $188,743. Put differently, the trade-off between E^* and the variance (V) for E^* in the neighborhood of $82,220 is $188,743, that is, a marginal decrease in the minimum E^* "saves" $188,743 in terms of the variance. In the context of the expected profit–standard deviation trade-off, decreasing E^* by one dollar saves $434.45 in lowering the standard deviation. In general, the magnitude of the trade-off between E and V, or E and the standard deviation will become smaller as E^* is lowered, reflecting diminishing marginal returns to risk.

The dual value on the land constraint is −20,723,200. This means that if acreage could be increased by 1 acre, the total variance could be reduced by $20,723,200. This large value is not surprising since this is the profit-maximizing level of E^*. If the farmer had one more acre of land, the high threshold $E^* = 82,220$ can be more easily achieved. Hence, the riskiest strategy of growing all 600 acres of the riskiest crop need not be followed. Indeed, if this problem is re-solved with 601 acres, the optimal solution is to grow 5 acres of wheat and the remaining 596 acres of corn. This less-risky optimal solution lowers the variance by $20,723,200.

An alternative formulation of this problem would be to maximize E subject to a maximum acceptable risk level. This formulation would yield an identical set of solutions in terms of E-V efficient plans. Under this formulation, the model becomes:

Max: $E = 91.0g_1 + 110.7g_2 + 138.7g_3$ (0)

s.t.:

$$32.6g_1g_1 + 154.6g_2g_2 + 4,546.2g_3g_3 - 130.0g_1g_2 - 203.4g_1g_3$$
$$+ 284.4g_2g_3 \leq R^* \tag{1}$$

$$1.0g_1 + 1.1g_2 + 1.3g_3 \leq 1,000 \tag{2}$$

$$0.4g_1 + 0.4g_2 + 0.4g_3 \leq 500 \tag{3}$$

$$2.2g_1 + 2.8g_2 + 3.0g_3 \leq 2,000 \tag{4}$$

$$1.0g_1 + 1.0g_2 + 1.0g_3 \leq 600 \tag{5}$$

$$g_1, g_2, g_3 \geq 0 \tag{6}$$

Note that the first constraint is equivalent to setting the objective function of the previous problem to be less than or equal to the minimum variance found in the optimal solution to that problem.

For extra practice, verify that this formulation yields the same solutions as before, and solve it using Solver.

A final and equivalent formulation of this problem would be to maximize E minus a risk term as follows:

Max: $E = 91.0g_1 + 110.7g_2 + 138.7g_3 - b\{32.6g_1g_1 + 154.6g_2g_2 + 4,546.2g_3g_3$
$\qquad -130.0g_1g_2 - 203.4g_1g_3 + 284.4g_2g_3\}$ $\qquad\qquad$ (0)

s.t.:

$$1.0g_1 + 1.1g_2 + 1.3g_3 \leq 1,000 \tag{1}$$

$$0.4g_1 + 0.4g_2 + 0.4g_3 \leq 500 \tag{2}$$

$$2.2g_1 + 2.8g_2 + 3.0g_3 \leq 2,000 \tag{3}$$

$$1.0g_1 + 1.0g_2 + 1.0g_3 \leq 600 \tag{4}$$

$$g_1, g_2, g_3 \geq 0 \tag{5}$$

where b is a risk-aversion parameter that is parametrically varied from zero (risk neutral) to some positive number (risk averse) to generate an E-V efficient set of optimal activities.

For extra practice, verify that this formulation yields the same solutions as before.

Quadratic programming (QP) models are perhaps the most commonly used method for analyzing agricultural risk. Although QP models are slightly more complicated to use and understand than LP models, most computer systems have QP algorithms, which make them readily available for use. Indeed, Solver uses a combined LP/QP algorithm as its standard engine. As in LP, the constraint set is linear, and risk is captured solely in terms of revenue activities in the objective function parameters. QP models provide specific farm planning information on optimal resource use and activity levels, making them a useful extension tool. In addition, when QP is used to derive E-V frontiers (parametric QP), the resulting farm plans are efficient for risk-averse decision makers.

There are, however, several serious limitations with using standard QP models in applied decision analysis. With regard to problem formulation, standard QP models assume a nonsequential decision environment, and therefore, all optimal decisions derived from these models are not adaptive. Agricultural production and marketing decisions are adaptive by nature and occur sequentially through time. In fact, this is one of the most important aspects of choices under uncertainty. Equally limiting is the fact that parameters in the constraint set of standard QP models are modeled nonstochastically. In reality, factors such as availability of field days, which determine when the various field operations can occur, are quite variable, and farmers do not ignore this source of production risk when planning their operations.

The determination of probability distributions in QP models has also drawn criticism. The distributions are described by means, variances, and covariances in QP, while higher moments are ignored. Thus, when the distribution is not normal, the results of QP models may not include the preferred decision strategy of some agents.

Regarding the representation of agents' risk attitudes, QP requires that either the utility function be quadratic, or that returns be normally distributed (if parametric QP is used). These assumptions are quite restrictive and rule out many types of risk prefer-ences. The most undesirable property, with respect to quadratic utility functions, is that the absolute risk aversion coefficient increases with income (Kramer & Pope, 1981). Proponents of QP, however, argue that it closely approximates a broad range of situa-tions where these assumptions do not hold (Levy & Markowitz, 1979).

10.4   LINEARIZED VERSION OF QUADRATIC RISK PROGRAMMING

One of the historical shortcomings of QP models is that due to algorithmic and other con-straints, the analyst could not formulate very large mathematical programming models.[8] Faced with these constraints, researchers developed linearized versions of QP. In this section, one such linear version, which has been widely adopted in the area of risk analysis, will be discussed. The procedure was developed by Peter Hazell (1971), who was a Ph.D. student in agricultural economics at Cornell University at the time he developed the method, and has been used extensively by agricultural economists and others.

MOTAD (Minimization of Total Absolute Deviations)

Hazell (1971) developed a linearized version of quadratic risk programming models called "minimization of total absolute deviations" (MOTAD) in 1971. The basic idea behind MOTAD is that rather than using the nonlinear variance-covariance measure of risk, the analyst can use a linear approximation of expected income variability. MOTAD models use the total absolute deviation (TAD) from expected net revenue to represent risk.

For example, consider net income as a random variable. Suppose that there are n risky prospects (risky activities) and m past observations collected for each prospect. Assuming that the agent expects the future outcome of net revenue for each prospect to be the sim-ple average of past observations, then the expected net revenue for each prospect is:

$$E(c_j) = \sum_{r=1}^{m} c_{rj}/m.$$

The absolute deviation from the mean for each observation over all possible prospects is equal to:

$$\sum_{j=1}^{n} |(c_j - E(c_j))x_j|,$$

where: $||$ is the absolute value operator. This represents the absolute deviation from the mean for one observation. Total absolute deviations from the mean are defined as the absolute deviations from the mean for all observations and all activity net rev-enues in the sample, that is:

$$TAD = \sum_{j=1}^{n} |(c_{rj} - E(c_j))x_j| \quad r = 1, \dots, m,$$

where c_{rj} is net revenue of the jth activity for the rth observation. If we define positive and negative deviations from the mean as:

$$d_j^+ = (c_{rj} - E(c_j)x_j) > 0 \text{ positive deviation}$$

$$d_j^- = (c_{rj} - E(c_j)x_j) < 0 \text{ negative deviation},$$

---

[8]This was a historical limitation that is not much of a problem with the computational capabilities of today's computers.

then total deviations from the mean can be stated as the following:

$$TAD = \sum_{j=1}^{n} (d_j^+) + (d_j^-),$$

where $d_j^+, d_j^- \geq 0$.

This implies that if $d_j > 0$, then $d_j^+ > 0$ and $d_j^- = 0$; and if $d_j < 0$, then $d_j^- > 0$ and $d_j^+ = 0$. For any random variable, it will always be true that the sum of the negative deviations will equal the sum of the positive deviations.

An Example[9]

Recall the previous farm risk problem. The QP formulation of this problem was:

Min: $R = 32.6g_1g_1 + 154.6g_2g_2 + 4{,}546.2g_3g_3 - 130.0g_1g_2 - 203.4g_1g_3$
$\qquad + 284.4g_2g_3$ (0)

s.t.:

$$91.0g_1 + 110.7g_2 + 138.7g_3 \geq E^* \tag{1}$$

$$1.0g_1 + 1.1g_2 + 1.3g_3 \leq 1{,}000 \tag{2}$$

$$0.4g_1 + 0.4g_2 + 0.4g_3 \leq 500 \tag{3}$$

$$2.2g_1 + 2.8g_2 + 3.0g_3 \leq 2{,}000 \tag{4}$$

$$1.0g_1 + 1.0g_2 + 1.0g_3 \leq 600 \tag{5}$$

$$g_1, \qquad g_2, \qquad g_3 \geq 0 \tag{6}$$

The first step in formulating this problem as a MOTAD problem is to compute the absolute deviations from the mean. The 10 observations and mean values for g_1, g_2, and g_3 are:

Observation	g_1	g_2	g_3	dev_{g1}	dev_{g2}	dev_{g3}
1	100.0	93.0	200.0	9.0	−17.7	61.3
2	95.0	99.0	100.0	4.0	−11.7	−38.7
3	97.0	97.0	150.0	6.0	−13.7	11.3
4	94.0	110.0	45.0	3.0	−0.7	−93.7
5	91.0	111.0	200.0	0.0	0.3	61.3
6	85.0	120.0	190.0	−6.0	9.3	51.3
7	92.0	100.0	75.0	1.0	−10.7	−63.7
8	90.0	121.0	25.0	−1.0	10.3	−113.7
9	86.0	127.0	210.0	−5.0	16.3	71.3
10	80.0	129.0	192.0	−11.0	18.3	53.3
Average	91.0	110.7	138.7			

The deviations from the mean for each activity net revenue (dev_{g1}, dev_{g2}, dev_{g3}) are calculated by subtracting the mean from each observation. Note that the sum of the

---

[9]This problem, solution, and corresponding sensitivity analysis are shown in the Chapter 10 supplemental materials available at www.wiley.com/college/kaiser.

positive deviations equals the sum of the negative deviation for each of the three goods. Risk (R) is now measured as the total absolute deviations.

$$\text{TAD} = d_1^+ + d_1^- + d_2^+ + d_2^- + d_3^+ + d_3^- + d_4^+ + d_4^- + d_5^+ + d_5^- \\ + d_6^+ + d_6^- + d_7^+ + d_7^- + d_8^+ + d_8^- + d_9^+ + d_9^- + d_{10}^+ + d_{10}^-$$

where:

$$d_1^+ = \max(0, 9g_1 - 17.7g_2 + 61.3g_3)$$

$$d_1^- = |\min(0, 9g_1 - 17.7g_2 + 61.3g_3)|$$

$$\vdots$$

$$d_{10}^+ = \max(0, -11g_1 + 18.3g_2 + 53.3g_3)$$

$$d_{10}^- = |\min(0, -11g_1 + 18.3g_2 + 53.3g_3)|$$

where $y = \max(a,b)$ means choose $y = a$ if $a > b$, or choose $y = b$ if $b > a$; and $y = \min(a,b)$ means choose $y = a$ if $a < b$, and choose $y = b$ if $b < a$. If the deviation for observation i is positive, for instance, $d_1 = 100$, then $d_1^+ = 100$ and $d_1^- = 0$. On the other hand, if the deviation for observation i is negative, for instance, $d_2 = -200$, then $d_2^+ = 0$ and $d_2^- = |-200| = 200$. Since non-negativity is required, all negative deviations are expressed in terms of absolute values. The MOTAD problem for this example is the following:

Min: $\text{TAD} = d_1^+ + d_1^- + d_2^+ + d_2^- + d_3^+ + d_3^- + d_4^+ + d_4^- + d_5^+ + d_5^- + d_6^+$

$\qquad + d_6^- + d_7^+ + d_7^- + d_8^+ + d_8^- + d_9^+ + d_9^- + d_{10}^+ + d_{10}^-$ (0)

s.t.:

$$-d_1^+ + d_1^- + 9g_1 - 17.7g_2 + 61.3g_3 = 0 \quad \text{(Define } d_1^+ \text{ and } d_1^-) \tag{1}$$

$$-d_2^+ + d_2^- + 4g_1 - 11.7g_2 - 38.7g_3 = 0 \quad \text{(Define } d_2^+ \text{ and } d_2^-) \tag{2}$$

$$-d_3^+ + d_3^- + 6g_1 - 13.7g_2 + 11.3g_3 = 0 \quad \text{(Define } d_3^+ \text{ and } d_3^-) \tag{3}$$

$$-d_4^+ + d_4^- + 3g_1 - 0.7g_2 - 93.7g_3 = 0 \quad \text{(Define } d_4^+ \text{ and } d_4^-) \tag{4}$$

$$-d_5^+ + d_5^- + 0g_1 + 0.3g_2 + 61.3g_3 = 0 \quad \text{(Define } d_5^+ \text{ and } d_5^-) \tag{5}$$

$$-d_6^+ + d_6^- - 6g_1 + 9.3g_2 + 51.3g_3 = 0 \quad \text{(Define } d_6^+ \text{ and } d_6^-) \tag{6}$$

$$-d_7^+ + d_7^- + 1g_1 - 10.7g_2 - 63.7g_3 = 0 \quad \text{(Define } d_7^+ \text{ and } d_7^-) \tag{7}$$

$$-d_8^+ + d_8^- - 1g_1 + 10.3g_2 - 113.7g_3 = 0 \quad \text{(Define } d_8^+ \text{ and } d_8^-) \tag{8}$$

$$-d_9^+ + d_9^- - 5g_1 + 16.3g_2 + 71.3g_3 = 0 \quad \text{(Define } d_9^+ \text{ and } d_9^-) \tag{9}$$

$$-d_{10}^+ + d_{10}^- - 11g_1 + 18.3g_2 + 53.3g_3 = 0 \quad \text{(Define } d_{10}^+ \text{ and } d_{10}^-) \tag{10}$$

$$91.0g_1 + 110.7g_2 + 138.7g_3 \geq E^* \quad \text{(Minimum E)} \tag{11}$$

$$1.0g_1 + 1.1g_2 + 1.3g_3 \leq 1,000 \quad \text{(Hired Labor)} \tag{12}$$

$$0.4g_1 + 0.4g_2 + 0.4g_3 \leq 500 \quad \text{(Manager Labor)} \tag{13}$$

$$2.2g_1 + 2.8g_2 + 3.0g_3 \leq 2,000 \quad \text{(Machine)} \tag{14}$$

$$1g_1 + \quad 1g_2 + \quad 1g_3 \le 600 \text{ (Land)} \tag{15}$$

$$d_i^+, d_i^-\, g_1, \qquad g_2, \qquad g_3 \ge 0 \text{ (Non-negativity)} \tag{16}$$

In the above model, equations (1) through (10) define the positive and negative deviations for the three activities for each year. The number of positive and negative deviation variables will always be equal to the number of observations one has collected. The interpretation of these equations becomes clearer when solving for d_i^+ or d_i^-. For instance, solving equation (1) for d_1^+ yields:

$$d_1^+ = d_1^- + 9g_1 - 17.7g_2 + 61.3g_3, \text{ or}$$

$$d_1^+ = d_1^- + dev_{g1},$$

where:

$$dev_{g1} = 9g_1 - 17.7g_2 + 61.3g_3.$$

If $dev_{g1} > 0$, then $d_1^+ = dev_{g1}$ and $d_1^- = 0$ because d_1^- is being minimized in the objective function. If $dev_{g1} < 0$, then $d_1^+ = 0$ and $d_1^- = -dev_{g1} > 0$.

An approximation of the standard deviation using TAD is given by the following formula:

$$\text{TAD SD} = (1/s)\ \text{TAD}\ [(\pi s)\ /2(s-1)]^{1/2},$$

where:

s = sample size (number of observations),

π = the mathematical constant, pi, that is, 3.14...,

TAD = total absolute deviations.

The formula that converts the value of TAD into an approximation of the variance is simply:

$$\text{TAD VAR} = (\text{TAD SD})^2.$$

In the previous section, this model was solved using QP for the following values of E^*: $82,220, $80,000, $75,000, $70,000, $65,000 and $60,000. Figure 10.4 illustrates the Excel worksheet for the Solver formulation of the problem corresponding to $E^* = \$82,220$. The QP and the corresponding MOTAD solutions for all values of E^* are summarized below.

QP Results

Expected Profit ($)	Variance ($)	Standard Deviation ($)	Soybeans (acres)	Wheat (acres)	Corn (acres)
82,220	1,636,632,000	40,455	0	0	600
80,000	1,087,286,890	32,974	0	115	485
75,000	465,789,543	21,582	0	293.6	306.4
70,000	125,950,359	11,223	0	472.1	127.9
65,000	32,919,362	5,738	118.8	448.3	32.9
60,000	2,843,412	1,686	341.8	247.1	11.2

	A	B	C	D	E	F	G	H	I	J	K	L	M	N	O	P	Q	R	S	T	U	V	W	X	Y	Z	AA
1		d_1^+	d_1^-	d_2^+	d_2^-	d_3^+	d_3^-	d_4^+	d_4^-	d_5^+	d_5^-	d_6^+	d_6^-	d_7^+	d_7^-	d_8^+	d_8^-	d_9^+	d_9^-	d_{10}^+	d_{10}^-	g_1	g_2	g_3			
2	**Decision Variables**	36,780	0	0	23,220	6,780	0	0	56,220	36,780	0	30,780	0	0	38,220	0	68,220	42,780	0	31,980	0	0	0	600			
3	**Objective**	1	1	1	1	1	1	1	1	1	1	1	1	1	1	1	1	1	1	1	1	0	0	0			
4	**Min TAD**	371,760																							**Total**		**RHS**
5	**Constraints**	-1	1																			9	-17	61.3	0	=	0
6				-1	1																	4	-11.7	-38.7	0	=	0
7						-1	1															6	-13.7	11.3	0	=	0
8								-1	1													3	-0.7	-93.7	0	=	0
9										-1	1											0	0.3	61.3	0	=	0
10												-1	1									-6	9.3	51.3	0	=	0
11														-1	1							1	-10.7	-63.7	0	=	0
12																-1	1					-1	10.3	-113.7	0	=	0
13																		-1	1			5	16.3	71.3	0	=	0
14																				-1	1	-11	18.3	53.3	0	=	0
15																						91	110.7	138.7	83,220	≤	83,220
16																						1	1.1	1.3	780	≥	1,000
17																						0.4	0.4	0.4	240	≥	500
18																						2.2	2.8	3	1,800	≥	2,000
19																						1	1	1	600	≥	600

Figure 10.4 Solver formulation for MOTAD example.

MOTAD Results

Expected Profit ($)	TAD ($)	Standard Deviation ($)	Soybeans (acres)	Wheat (acres)	Corn (acres)
82,220	371,760	49,098	0.0	0.0	600.0
80,000	303,450	40,079	0.0	115.0	485.0
75,000	198,497	26,217	0.0	293.6	306.4
70,000	102,393	13,524	0.0	472.1	127.9
65,000	51,074	6,746	126.5	435.2	38.3
60,000	13,537	1,788	349.2	234.4	16.4

In this case, comparisons between the optimal activity values from the MOTAD model and those of the QP model show that they are quite similar. However, the variance estimates from the MOTAD model are higher than the ones given in the QP model. This is due to the fact that the TAD estimate of the variance is not as efficient as the traditional nonlinear variance estimate. Hence, there is a trade-off when using a MOTAD model between the advantage of being a linear problem and disadvantage of the TAD not being as efficient of an estimate of the variance and standard deviation.

Note that the sensitivity analysis results for the case of $E^* = 82,220$ indicates the shadow price (SP) on the minimum E constraint is 21.2: that is, if the RHS value for the minimum E constraint were reduced by $1, total absolute deviations could be reduced by $21.20. The SP on the land constraint is –2,319.4: that is, an increase of one acre would lead to a $2,319.40 reduction in total absolute deviations.

Since the total negative deviations (TND) equals the total positive deviations, MOTAD models can be reduced substantially in size by minimizing total negative deviations and multiplying the resulting objective function value by 2, that is,

$$TAD = 2 \ TND.$$

Now the deviations from the mean are calculated using the negative deviations formula, that is:

$$d_j^- = |\min (0, c_{jr} - E(c_j))|,$$

for the jth activity's net revenue, rth observation. The resulting smaller model (minimizing total negative deviations) will yield similar results to the larger model (minimizing total absolute deviations, positive and negative). The negative deviations formulation to this problem is:

$$\text{Min: TND} = d_1^- + d_2^- + d_3^- + d_4^- + d_5^- + d_6^- + d_7^- + d_8^- + d_9^- + d_{10}^- \tag{0}$$

s.t.:

$$-d_1^- + 0g_1 + 17.7g_2 + \quad 0g_3 = 0 \quad (\text{Define } d_1^-) \tag{1}$$

$$-d_2^- + 0g_1 + 11.7g_2 + 38.7g_3 = 0 \quad (\text{Define } d_2^-) \tag{2}$$

$$-d_3^- + 0g_1 + 13.7g_2 + \quad 0g_3 = 0 \quad (\text{Define } d_3^-) \tag{3}$$

$$-d_4^- + 0g_1 + \ 0.7g_2 + 93.7g_3 = 0 \quad (\text{Define } d_4^-) \tag{4}$$

$$-d_5^- + 0g_1 + \quad 0g_2 + \quad 0g_3 = 0 \quad (\text{Define } d_5^-) \tag{5}$$

$$-d_6^- + 6g_1 + \quad 0g_2 + \quad 0g_3 = 0 \quad (\text{Define } d_6^-) \tag{6}$$

$$-d_7^- + 0g_1 + 10.7g_2 + 63.7g_3 = 0 \quad \text{(Define } d_7^-) \tag{7}$$

$$-d_8^- + 1g_1 + 0g_2 + 113.7g_3 = 0 \quad \text{(Define } d_8^-) \tag{8}$$

$$-d_9^- + 5g_1 + 0g_2 + 0g_3 = 0 \quad \text{(Define } d_9^-) \tag{9}$$

$$-d_{10}^- + 11g_1 + 0g_2 + 0g_3 = 0 \quad \text{(Define } d_{10}^-) \tag{10}$$

$$91.0g_1 + 110.7g_2 + 138.7g_3 \geq E^* \quad \text{(Minimum E)} \tag{11}$$

$$1.0g_1 + 1.1g_2 + 1.3g_3 \leq 1{,}000 \quad \text{(Hired Labor)} \tag{12}$$

$$0.4g_1 + 0.4g_2 + 0.4g_3 \leq 500 \quad \text{(Manager Labor)} \tag{13}$$

$$2.2g_1 + 2.8g_2 + 3.0g_3 \leq 2000 \quad \text{(Machine)} \tag{14}$$

$$1g_1 + 1g_2 + 1g_3 \leq 600 \quad \text{(Land)} \tag{15}$$

$$d_i^- \quad g_1, \quad g_2, \quad g_3 \geq 0 \quad \text{(Non-negativity)} \tag{16}$$

There are two differences between this formulation and the one that minimizes TAD. First, there are 10 fewer activities since the 10 d_i^+ activities have been eliminated. Second, the definition of the deviations from the mean (equations (1) through (10)) now includes only the negative deviations (the positive deviations are set to zero). Note that the absolute value of the negative deviations is used in equations (1) through (10).

The general form of the MOTAD model, which minimizes TAD is:

$$\text{Min: TAD} = \sum_{j=1}^{n} d_j^+ + \sum_{j=1}^{n} d_j^- \tag{0}$$

s.t.:

$$\sum_{j=1}^{n} (c_{rj} - E(c_j))x_j - d_j^+ + d_j^- = 0 \quad (r = \text{number of observations}) \tag{1}$$

$$\sum_{j=1}^{n} E(c_j)x_j \geq E^* \tag{2}$$

$$\sum_{j=1}^{n} a_{ij} x_j \leq b_i \quad \text{for all } i \tag{3}$$

$$x_j, d_j^+, d_j^- \geq 0 \tag{4}$$

As was true with the QP formulations, three equivalent formulations for MOTAD can be used. The first is to minimize TAD or TND s.t. $E \geq E^*$ and other constraints. The second is to maximize E s.t. TAD or TND $\leq$ target level and other constraints. The third is to maximize E – b TAD (or TND) s.t. constraints.

Historically, the main advantage of MOTAD over QP is that it could be used with LP solvers. Consequently, greater detail could be specified in the production and marketing strategies in the model formulation.[10] TAD succeeds as a measure of risk because

---

[10]As described in Chapter 9, with the Risk Solver Platform, the Standard LP/Quadratic Engine now finds optimal solutions for both linear and quadratic problems.

the approximation of the standard deviation is an unbiased estimate of the population standard deviation for a normal population (Hazell, 1985). Additionally, MOTAD models are generally a reasonable approximation of QP models and may even be superior to QP if distributions are skewed (Anderson et al., 1977). Finally, Boisvert and McCarl (1990) dispel concerns that MOTAD appears to ignore the covariance element of the variance-covariance matrix. However, the deviations in the MOTAD model exist across all activities, so that negative deviations from one activity can partially or completely mitigate positive deviations from another activity. This creates an incentive to lessen risk through diversification, much as the covariance term does in QP.

A disadvantage of MOTAD is that, even under normality, the approximation of variance is less efficient than with QP. MOTAD models also suffer some of the same limitations inherent in standard QP models, most notably treating the decision environment nonsequentially and setting the parameters of a constraint nonstochastically.

10.5 TARGET MINIMIZATION OF TOTAL ABSOLUTE DEVIATIONS

Tauer (1983) developed an alternative to MOTAD models called target MOTAD. Target MOTAD improves upon regular MOTAD in that its solutions are also efficient based on another efficiency criterion known as second-degree stochastic dominance. The model, which is similar to MOTAD models, adds a new constraint that sets a target level on total revenue. One formulation of the model is:

$$\text{Max: } E(NR) = \sum_{j=1}^{n} c_j \bar{x}_j \tag{0}$$

s.t.:

$$\sum_{j=1}^{n} a_{ij} x_j \leq b_i \quad \text{for all i} \tag{1}$$

$$\sum_{j=1}^{n} c_{kj} x_j + y_k \geq T \quad \text{for all k} \tag{2}$$

$$\sum_{k=1}^{K} p_k y_k \leq \lambda \tag{3}$$

$$x_j, \quad y_k \geq 0 \tag{4}$$

where y_k is the negative deviation in total net revenue in the kth state of nature below the targeted net revenue level, p_k is the probability of the kth state of nature, T is the target net revenue level, and λ is maximum amount of shortfall in net revenue permitted. Note that you may either use equally likely probabilities for each state of nature, or weight p_k differently for some states; for instance, more recent states could be weighted more heavily than more distant states. Constraint (2) measures the relation between state of nature k net revenue and the set target net revenue level. Notice that if net revenue falls below the target level in state of nature k, then y_k measures by how much the target is underachieved. Constraint (3) requires that, on average, the shortfall in net revenue not be larger than λ, which is parametrically altered. Hence, unlike the regular MOTAD model, the target MOTAD model requires parameterization on two parameters, λ and T instead of one.

An Example[11]

Consider the same example as was used for the QP and MOTAD examples. The target MOTAD formulation is:

$$\text{Max: E(NR)} = 91.0g_1 + 110.7g_2 + 138.7g_3 \tag{0}$$

s.t.:

$$y_1 + 100g_1 + 93g_2 + 200g_3 \geq T \tag{1}$$
$$y_2 + 95g_1 + 99g_2 + 100g_3 \geq T \tag{2}$$
$$y_3 + 97g_1 + 97g_2 + 150g_3 \geq T \tag{3}$$
$$y_4 + 94g_1 + 110g_2 + 45g_3 \geq T \tag{4}$$
$$y_5 + 91g_1 + 111g_2 + 200g_3 \geq T \tag{5}$$
$$y_6 + 85g_1 + 120g_2 + 190g_3 \geq T \tag{6}$$
$$y_7 + 92g_1 + 100g_2 + 75g_3 \geq T \tag{7}$$
$$y_8 + 90g_1 + 121g_2 + 25g_3 \geq T \tag{8}$$
$$y_9 + 86g_1 + 127g_2 + 210g_3 \geq T \tag{9}$$
$$y_{10} + 80g_1 + 129g_2 + 192g_3 \geq T \tag{10}$$

$$\sum_{i=1}^{10} 0.10\, y_i^- \leq \lambda \tag{11}$$

$$1.0g_1 + 1.1g_2 + 1.3g_3 \leq 1{,}000 \quad \text{(Hired Labor)} \tag{12}$$
$$0.4g_1 + 0.4g_2 + 0.4g_3 \leq 500 \quad \text{(Manager Labor)} \tag{13}$$
$$2.2g_1 + 2.8g_2 + 3.0g_3 \leq 2000 \quad \text{(Machine)} \tag{14}$$
$$1g_1 + 1g_2 + 1g_3 \leq 600 \quad \text{(Land)} \tag{15}$$
$$y_i \quad g_1, \quad g_2, \quad g_3 \geq 0 \quad \text{(Non-negativity)} \tag{16}$$

It should be noted that the weights used in this example treat all 10 observations as equally likely with a probability of 0.10. However, one could also assign a different weighting scheme with different weights for various observations as long as the weights sum to 1.0. Figure 10.5 presents this model for the case of $\lambda = \$20{,}000$ and $E^* = 82{,}220$. The following table gives the optimal target MOTAD results for selected values of T and λ.

T ($)	82,220	82,220	82,220	82,220	75,000	75,000	75,000
λ ($)	19,000	18,000	17,000	15,900	11,500	11,250	11,200
Soybeans (acres)	0.0	10.2	64.2	123.7	205.4	218.9	221.6
Wheat (acres)	0.0	0.0	0.0	0.0	0.0	0.0	0.0
Corn (acres)	600.0	589.8	535.8	476.3	394.6	381.1	378.4
Average Net Revenue ($)	83,220	82,936	81,422	79,757	77,469	77,090	77,015

In general, these results are similar to both the QP and MOTAD model results. The risk-neutral, or profit-maximizing solution occurs for values of λ larger than 19,000 when T is set at 82,220. As λ is reduced, average net revenue falls, and optimal crop mix becomes more diversified, and less dependent on corn. These solutions do not perfectly correspond to the QP and MOTAD solutions presented earlier because two parameters are being varied here and the selected solutions do not exactly coincide with the pairs of E and V (or E and TAD) presented earlier.

---

[11]This problem, solution, and corresponding sensitivity analysis are shown in the Chapter 10 supplemental materials available at www.wiley.com/college/kaiser.

	A	B	C	D	E	F	G	H	I	J	K	L	M	N	O	P	Q
1		g_1	g_2	g_3	y_1	y_2	y_3	y_4	y_5	y_6	y_7	y_8	y_9	y_{10}			
2	Decision Variables	0	0	600	0	23,220	0	56,220	0	0	38,220	68,220	0	0			
3	Objective	91	111	139	0	0	0	0	0	0	0	0	0	0			
4	Max E(NR)	83,220													Total		RHS
5	Constraints	100	93	200	1										120,000	≥	83,220
6		95	99	100		1									83,220	≥	83,220
7		97	97	150			1								90,000	≥	83,220
8		94	110	45				1							83,220	≥	83,220
9		91	111	200					1						120,000	≥	83,220
10		85	120	190						1					114,000	≥	83,220
11		92	100	75							1				83,220	≥	83,220
12		90	121	25								1			83,220	≥	83,220
13		86	127	210									1		126,000	≥	83,220
14		80	129	192										1	115,200	≥	83,220
15		0	0	0	0.1	0.1	0.1	0.1	0.1	0.1	0.1	0.1	0.1	0.1	18,588	≤	20.000
16		1.0	1.1	1.3											780	≤	1.000
17		0.4	0.4	0.4											240	≤	500
18		2.2	2.8	3.0											1,800	≤	2,000
19		1	1	1											600	≤	600

Figure 10.5 Solver formulation for target MOTAD example.

10.6   CHANCE-CONSTRAINED PROGRAMMING

While quadratic risk programming, MOTAD, and target MOTAD models represent significant contributions to applied decision analysis, each make certain assumptions that limit their use. First, risk is usually captured only in the objective function coefficients, while parameters in the constraint set are treated deterministically. In reality, however, resource availability and requirements in the constraint functions are also a source of risk to the farmer. Second, these models usually assume a static, nonsequential decision process. Farm production and marketing decision making, however, are adaptive processes involving a sequence of decisions over time. In models assuming a single decision stage, decision variables are not adapted to new information received over the planning horizon.

Chance-constrained programming offers a solution to the first problem. This technique, developed by Charnes and Cooper (1959), is the most popular approach for dealing with right-hand-side (RHS) risk. One of the biggest sources of RHS risk in agriculture is availability of field time, which depends upon the weather. If the fields are too wet for agricultural equipment, then operations such as plowing, disking, planting, harvesting, and so on cannot be completed. In reality, this source of risk is very important to farmers.

This approach is fairly simple to model. Assume that the source of risk in the RHS variable (b_i) has a probability distribution known by the decision maker. Then, a chance constraint can be added to the mathematical programming model, which puts a lower limit (α) on the probability that the constraint will be satisfied, for instance:

$$P\left(\sum_{j=1}^{n} a_{ij}x_j \leq b_i\right) \geq \alpha.$$

Denoting the average or expected value of b_i as $E(b_i)$, subtracting it from both sides of the above equation, and dividing both sides by the standard deviation of b_i (σ_i) yields:

$$P\left[\left(\sum_{j=1}^{n} a_{ij}x_j - E(b_i)\right)\right]/\sigma_j \leq [b_i - E(b_i)]/\sigma_i \geq \alpha.$$

Denoting $Z = [b_i - E(b_i)]/\sigma_i$ and denoting Z_α as the critical value on the probability distribution such that a lower value than this has a chance of occurring α percent of the time, rewrite the above equation as:

$$P\left[\left[\left(\sum_{j=1}^{n} a_{ij}x_j - E(b_i)\right)\right]/\sigma_j \leq Z_\alpha\right] \geq \alpha.$$

Finally, this expression can be rewritten and included in an LP model as:

$$\sum_{j=1}^{n} a_{ij}x_j \leq E(b_i) - Z_\alpha \alpha_i.$$

This constraint says that the resource usage, $a_{ij}\,x_j$, must be $\leq$ the mean value of the RHS parameter minus the product of its standard deviation and the critical value associated with the set probability level. In other words, chance-constrained programming deals with RHS risk by setting the availability of the resource (b_i) to a lower limit, rendering the probability of meeting this minimum level of resource availability so high that the decision maker can depend upon it.

The advantage of chance-constrained programming is that it is fairly simple to use and does not add much complexity to the linear or nonlinear programming model (Boisvert & McCarl, 1990). Subsequently, it could be combined with MOTAD or quadratic risk programming in order to incorporate RHS risk into the model. Its main drawback is its assumption that the decision process is static and nonsequential. In models assuming a single decision stage, decision variables are not adapted to new information received over the planning horizon.

An Example

Consider the following model:

Max: $Z = 10x + 5y$ (0)

s.t.:

$x + y \leq b$		(1)
$5x + y \leq 100$		(2)
$x, \quad y \geq 0$		(3)

Assume that b is distributed normally with a mean of 50 and a standard deviation of 10. To find a value for the RHS parameter b, call it b', use the following relationship:

$$b' = E(b) - Z_\alpha = 50 - 10\,Z_\alpha.$$

Parametric programming on Z_α can be used to derive solutions for this problem. The table on the next page gives some of the solutions.

Z_{α}	b	Value of Objective Function	x	y
0	50	312.5	12.5	37.5
1	40	275	15	25
2	30	237.5	17.5	12.5
3	20	200	20	0
4	10	100	10	0

It is clear from this table that the more conservative the value of b, the lower the objective function value illustrating the classic trade-off of expected revenue and risk.

Chance Constraints with the Risk Solver Platform[12]

The Risk Solver Platform offers a simple way to deal with uncertainty in the constraints. Using distribution functions, any cell can be defined as an uncertain value drawn from a defined probability distribution. Solver offers over three dozen built-in distributions and also allows for custom definition of distributions. The normal distribution with a mean of 50 and a standard deviation of 10 used in the above example would be defined with the function "=psinormal(50, 10)". Additional distributions can be found under the "Distributions" menu on the Risk Solver tab on the Ribbon.

A cell defined with a probability distribution can be identified in a constraint like a normal cell. The only problem is that it would try to generate a certain answer, or in other words it would essentially treat the distribution as the lowest possible value of that distribution. In the case of a normal distribution, that value would be negative infinity. Instead a **Chance Constraint** needs to be defined.

The process for defining Chance Constraints in Solver is the same as for normal constraints except that an additional option must be set in the Add/Change Constraint window. Use the pull-down menu to the right of the box where the RHS of the constraint is input. Usually the pull-down menu is set to Normal, but it can be changed to one of several options to make it a Chance Constraint. The options available are **VaR** (Value at Risk), **CVaR** (Conditional Value at Risk, and **USet** (Uncertainty Set), which are measures of risk commonly used in finance. By selecting VaR the chance value specified will be the probability that the value on the left-hand-side (LHS) of the constraint will be satisfied by the uncertain RHS value. The probability is defined in the box labeled "Chance:" underneath the pull-down menu. So if the constraint were defined as a VaR constraint with a Chance of 0.4, the objective function would be maximized subject to the constraint being satisfied 40% of the time. Once this constraint is added, it will appear in the Model Specification window under Chance Constraints as "$VaR_{0.4}(LHS) <= RHS$". For the above example this would lead to a solution of $x^* = 15$, $y^* = 25$, and $Z^* = 275$.

10.7   DISCRETE STOCHASTIC SEQUENTIAL PROGRAMMING

Discrete stochastic sequential programming (DSSP) is a mathematical programming technique capable of overcoming many of the limitations cited with the previous models.

---

[12]This problem, solution, and corresponding sensitivity analysis are shown in the Chapter 10 supplemental materials available at www.wiley.com/college/kaiser.

However, there have been relatively few empirical applications of DSSP to farm production and marketing problems. Despite its intuitive appeal, DSSP is often overlooked in empirical research because the programming matrix becomes quite large as the number of states of nature and decision stages increases. A related concern is the potentially formidable data requirements associated with models that capture dynamic characteristics of decision problems with many sources of risk. However, with the proliferation of flexible data management software, mathematical programming solvers for large problems, technical and economic databases, and simulation techniques for generating data, the cost of the added accuracy that techniques such as DSSP afford may be declining. Experience in the construction of DSSP models should reduce the cost further.

Discrete stochastic sequential programming was developed by Cocks (1968) and refined by Rae (1971a; 1971b) as a technique for modeling decision making as a multistage decision process characterized by a discrete specification of random problem parameters. As the name implies, DSSP models consider the stochastic and sequential nature of resource endowments, resource requirements, and objective function coefficients. This technique requires that technical coefficients, objective function parameters, and/or resource endowments be specified separately for each stochastic state of nature. Each state is then assigned a probability of occurrence, based on the subjective assessment by the decision maker. The solution to the DSSP problem is then found, which depends, in part, on the way the states and probabilities are defined and assessed.

The DSSP technique is a probabilistic decision model, based on Bayesian decision theory. Decisions in any stage are made with probabilistic knowledge of the occurrence of the states of nature in future stages of the decision process. The stages in the decision process are therefore interdependent. Decisions in later stages are restricted not only by the occurrence of particular random events in this stage, but also by random outcomes and decisions made in earlier stages. For example, the decision of when and how much corn and soybeans to market in the fall not only depends upon current market conditions and expectations on future conditions, but also on the past crop mix decision and the outcome of yield events.

In addition to defining the possible states and possible activities for each stage that the decision maker must choose, an information structure must be specified. The information structure describes the flow and extent of information regarding the occurrence of events in the various stages of the decision process. Decisions are assumed to be made at the beginning of each stage. For any stage, the decision maker may either have perfect or probabilistic knowledge of events in past, present, and future stages. For example, an information structure of complete knowledge of the past and present implies that at the beginning of stage t, the decision maker knows the outcome of random events in stages $t, t - 1, t - 2, \ldots, 1$ with certainty, but only has probabilistic knowledge of the outcome of events in stages $t + 1, t + 2, \ldots, t + n$. An information structure of complete knowledge of the past implies that the decision maker knows with certainty the outcome of events of stages $t - 1, t - 2, \ldots 1$, but only has probabilistic knowledge of outcomes in stages $t, t + 1, \ldots, t + n$. Agricultural production and marketing decisions are best described by assuming an information structure of complete knowledge of the past and present, and probabilistic knowledge of the future.

The nature of a sequential decision environment under risk characterized in DSSP can be illustrated with a decision tree, which illustrates the stages in the decision process and the states of nature in each stage. An exemplary decision tree for a two-stage, two states of nature, decision problem assuming complete knowledge of the past is shown in Figure 10.6. The interpretation of the notation is the following: e_{ijk} represents the occurrence of the ith state of nature in stage k and the jth set of activities. For example, at the beginning of

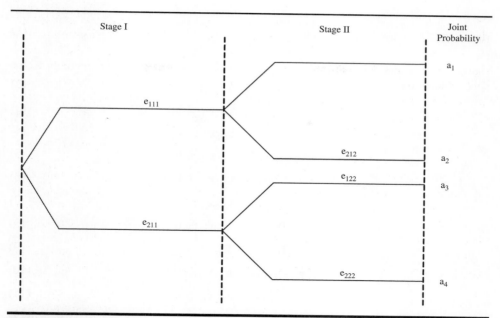

Figure 10.6   Decision tree for two-stage, two-state DSSP problem assuming complete knowledge of the past.

stage 2, assuming complete knowledge of the past and that event e_{ijl} has occurred in stage 1, the decision maker knows with certainty e_{ij1} and must decide, based on probabilistic knowledge of events in stage 2, which activities to select. A general LP formulation for this problem is constructed in Figure 10.7.

The objective function of maximization of expected net revenue will first be assumed. Stochastic components of the problem are accounted for in the constraint function coefficients (A_{ijk}), the resource endowments (b_{ijk}), and the objective function coefficients (c_{ijk}). The vector of activity levels x_{11}, x_{12}, and x_{22} form a strategy, which is derived from optimal solutions to the problem. At the beginning of the process, assume that decision vector x_{11} is selected. Vector x_{11} must be permanently feasible because the outcome of stage 1 random events is unknown when vector x_{11} is selected, which is implied by constraints (2) and (3) being satisfied regardless of which stage 1 event occurs. Stage 2 decisions must be permanently feasible as well; however, two stage 2 decision vectors (x_{12} and x_{22}) are included since the decision maker, having complete knowledge of the past, will know at the beginning of stage 2 which stage 1 state of nature has occurred. Thus stage 2 decisions are made subject to the opportunities afforded jointly by stage 2 random events, by decisions made in stage 1, and by the outcome of random events in stage 1. Hence, the decision rule that is followed at the beginning of stage 2 is to "follow x_{12} if e_{111} occurred in stage 1, or follow x_{22} if e_{211} occurred in stage 1."

Constraints (8) and (9) imply that the two stages are interdependent. Through these constraints, the continuance of stage 1 activities into stage 2 and the transfer of resources between the first and the second stage activities are insured. Matrices D_{ijk} and E_{jk} are constructed in such a way as to preserve these relationships between stages. Given the outcome of random events in stage 1, constraints (4) to (7) render decision vectors x_{12} and x_{22}, respectively, permanently feasible.

Max: $Z = a_1 y_1 + a_2 y_2 + a_3 y_3 + a_4 y_4$ $\qquad$ (1)

s.t.:

$$\mathbf{A}_{111}\mathbf{x}_{11} \leq \mathbf{b}_{111} \qquad (2)$$

$$\mathbf{A}_{211}\mathbf{x}_{11} \leq \mathbf{b}_{211} \qquad (3)$$

$$\mathbf{A}_{112}\mathbf{x}_{12} \leq \mathbf{b}_{112} \qquad (4)$$

$$\mathbf{A}_{212}\mathbf{x}_{12} \leq \mathbf{b}_{212} \qquad (5)$$

$$\mathbf{A}_{122}\mathbf{x}_{22} \leq \mathbf{b}_{122} \qquad (6)$$

$$\mathbf{A}_{222}\mathbf{x}_{22} \leq \mathbf{b}_{222} \qquad (7)$$

$$-\mathbf{D}_{111}\mathbf{x}_{11} + \mathbf{E}_{12}\mathbf{x}_{12} \leq 0 \qquad (8)$$

$$-\mathbf{D}_{211}\mathbf{x}_{11} + \mathbf{E}_{22}\mathbf{x}_{22} \leq 0 \qquad (9)$$

$$y_1 - \mathbf{c}'_{111}\mathbf{x}_{11} - \mathbf{c}'_{112}\mathbf{x}_{12} \leq 0 \qquad (10)$$

$$y_2 - \mathbf{c}'_{111}\mathbf{x}_{11} - \mathbf{c}'_{212}\mathbf{x}_{12} \leq 0 \qquad (11)$$

$$y_3 - \mathbf{c}'_{211}\mathbf{x}_{11} - \mathbf{c}'_{122}\mathbf{x}_{12} \leq 0 \qquad (12)$$

$$y_4 - \mathbf{c}'_{211}\mathbf{x}_{11} - \mathbf{c}'_{222}\mathbf{x}_{12} \leq 0 \qquad (13)$$

$$y_1, y_2, y_3, y_4, \mathbf{x}_{11}, \mathbf{x}_{12}, \mathbf{x}_{22} \geq 0 \qquad (14)$$

Figure 10.7   Linear programming model for two-stage, two-state DSSP problem assuming complete knowledge of the past.

Activities y_1 through y_4 are total net revenue associated with each possible sequence of random events in the two stages (i.e., joint events (e_{111}, e_{112}), (e_{111}, e_{212}), (e_{211}, e_{122}) and (e_{211}, e_{222}), respectively). The $\mathbf{c}_{ijk}$ vectors are objective function coefficients corresponding to the associated events. Thus, through constraints (10) to (13), net revenue levels associated with the occurrence of each combination of events are summed into y. Joint probabilities a_1, a_2, a_3, and a_4 are objective function coefficients for y, so the objective (1) is expected net revenue, which is maximized.[13] With the problem formulated in this way, the optimal stage 1 vector is then selected with consideration of the expected explicit and implicit values of stage 2 decision vectors.

Because the probability distributions of monetary outcomes are explicitly considered in DSSP, the modeling technique can be easily extended from the expected net revenue formulation presented above to a formulation for the maximization of expected utility. The extension of expected utility concepts into the DSSP model is similar to those with other risk programming models (e.g., QP, MOTAD, and expected utility functions), except for special considerations of timing in the decision-making process. The following discussion will focus on incorporating a MOTAD objective function into a DSSP

---

[13]You could also use the appropriate marginal and joint probabilities to weight the vectors $\mathbf{c}_{ijk}$ and these coefficients could be placed directly in the objective function. However, use of the vectors y_i provides useful solution information and facilitates later discussions of expected utility models.

framework (DSSP/MOTAD approach). The incorporation of other variants of the EUH in a DSSP model has been addressed elsewhere in the literature (see, for example, Rae (1971a; 1971b)).

The modifications in the DSSP problem necessary to convert the problem into a MOTAD model are straightforward. The DSSP/ED approach requires the measurement of expected net revenue and absolute negative deviations from expected net revenue. The occurrence of a particular joint event in the DSSP model is characterized by the multi-nomial distribution (Cocks, 1968). That is, one of m joint events will occur (for each trial) with probabilities a_j, $j=1, \ldots, m$. The expected value of the jth event is a_j, where $a_j \geq 0$ and $\Sigma \, a_j = 1$.[14] The absolute negative deviation, for any state i, is equal to the minimum value of either $y_i - E(y)$ or 0, where y_i is net revenue under joint event i, and $E(y)$ is expected net revenue for all joint events. The ED objective function corresponding to the problem in Figure 10.7 is:

$$\text{Max: } U = \sum_{i=1}^{4} a_i y_i - \Phi r \sum_{i=1}^{4} d_i \tag{29}$$

where Φ is the coefficient that converts total negative deviations into an approximation of the standard deviation, r is the marginal risk coefficient, and d_i is the negative deviation for state i.

Although DSSP models, in theory, conform quite well to how farm production and marketing decisions are actually made, they have not been frequently used in empirical applications to agricultural problems due to the size and complexity of DSSP models. A stochastic programming matrix will generally grow in size more than proportionally with increases in the number of sources of risk (random variables), the number of discrete values taken by the random variables, and the number of stages in the decision process. The formulation of an empirical DSSP/ED model, for example, that incorporates the risk inherent in farming is more than an ambitious task.

Therefore, the central focus of model building using DSSP must be on selecting an economical representation of the problem with the greatest level of detail specified in components critical to the analysis. Although dimensionality remains a major problem inherent in DSSP, it is becoming less of a barrier to implementing these models due to recent advances in linear and nonlinear programming software.

In addition, the impediment of very large matrix data files common with DSSP may be overcome by the use of matrix-generating computer programs. The replication in coefficient placement and parameter use inherent with these models makes the use of matrix generators a fairly straightforward process. Also, when matrix generators have been written for a deterministic (i.e., nonstochastic) version of a particular system, modification of the software to allow for stochastic parameters and a sequential decision process may be a relatively easy task.

Similarly, report generating computer programs, which may be written in conjunction with the matrix generator, are useful in overcoming the problem of analyzing the formidable set of solution values associated with a DSSP model. Report generators are basically used to search through output and find and organize key components of the output critical to the analysis.

---

[14]The variance of this distribution is $V_{ii} = a_i(1 - a_i)$, and the covariance is $V_{ij} = -a_i a_j$ (for all i not equal j).

A DSSP Example in Solver[15]

Solver offers the ability to model multiple-stage decisions with uncertainty in early-stage parameters. Consider the simple example of a 750-acre farm that can choose to plant corn and soybeans and raise cattle. The yields of the crops depend on the weather for that season. A wet season will offer higher yields, 100 bushels per acre of corn and 55 bushels per acre of soybeans, while a dry season will only offer yields of 50 bushels per acre of corn and 45 bushels per acre of soybeans. It is believed that there is a 60% chance of a wet year and 40% chance of a dry year. Soybeans can be sold for $8 per bushel. Corn may be sold for $5 per bushel, fed to cattle, and bought for $6 per bushel.

Each cattle requires 120 bushels of corn and can be sold for $500. Before the growing season begins, the farmer must first decide on the number of acres to plant and the number of cattle to raise. Once the season begins and its type is revealed, the farmer must determine the amount of corn to buy or sell. These decisions are made to maximize expected profit subject to acreage and the amount of corn necessary to feed the cattle.

The set-up of this problem in Excel is fairly straightforward, with a couple of new elements. First, the probability of the weather outcome for the year must be modeled using a distribution function. Since two discrete outcomes are offered, a discrete distribution can be specified with the function "=PsiDiscrete({1,0}, {0.6,0.4})". This will return a 1, indicating a wet year 60% of the time, and a 0, indicating a dry year 40% of the time. Second, in addition to the profit cell an expected profit cell must be defined since that is what is being maximized. This is done with the function "=psiMean(Profit)" where "Profit" refers to the cell containing the profit function.

The model definition is similar to static models. When defining the Decision Variables, the first-stage variables should be defined as usual; however, the second-stage variables should be defined as Recourse Variables. This is done in the Add/Change Variables window by changing the pull-down menu next to the "Cell Reference:" field from "Normal" to "Recourse." The expected profit cell is set as the objective function, and the objective type must be changed from Normal to Expected. This is done in the Add/Change Objective window by changing the pull down menu next to the "Set Cell:" field from "Normal" to "Expected". The model is then run as usual. In this case, the optimal solution is to plant all 750 acres with corn and use it to raise 625 head of cattle for a total expected profit of $312,500.

10.8   ISSUES IN MEASURING RISK IN RISK PROGRAMMING

As was stated earlier in this chapter, expected utility is a function of all statistical moments of the probability distribution of the random variable on which utility depends upon. However, the variance of a one-dimensional utility function (e.g., net income) has been often used in empirical studies as the measure of risk. When the variance, or a linear approximation of the variance, is used to measure risk, it is usually assumed that either the decision maker's utility function is quadratic or that net income is normally distributed, thereby reducing utility to a function of the first two moments of the probability distribution. Often, these two moments are calculated from historical time series data and the result is an "objective" measure of risk.

According to some decision theorists, objectively based measures of risk are not relevant in decision analyses since decision makers subjectively perceive risk. These theorists contend

---

[15]This problem, solution, and corresponding sensitivity analysis are shown in the Chapter 10 supplemental materials available at www.wiley.com/college/kaiser.

that risk must be measured subjectively by eliciting probabilities directly from the decision maker (Anderson et al., 1977). Additionally, supporters of this argument contend that risk should be measured subjectively for positive, as well as normative, behavioral applications.

Young (1980) argues, however, that because of difficulties and costliness of obtaining accurate subjective risk preferences, analysts should use historical indices of risk in normative (but not positive) applications. He argues that in normative applications, researchers use the most accurate time series data possible to use as objective risk measures. The results can then be presented to managers, and they can revise them in accordance with their own personal preferences.

The subjective measures of risk, at least conceptually, are the only relevant probabilities for decision making for both positive and normative purposes. But because of such difficulties associated with (1) developing elicitation procedures free of bias, (2) time-consuming and expensive process of elicitation, and (3) lack of adequate methods of multivariate elicitation schemes, historical data must be used in many applications as proxies for production and price risk. Because of these problems, elicitation procedures are likely to result in inaccurate utility functions and subjective probability distributions. Due to these limitations, most risk studies use an objective measure (variance) to estimate risk.

10.9   RESEARCH APPLICATION: QUADRATIC RISK PROGRAMMING

Falatoonzadeh et al. (1985) evaluated the optimality of various risk-management strategies available to farmers using quadratic risk programming. The authors simultaneously examined the optimality of five risk management strategies: (1) hedging in the futures markets, (2) crop diversification, (3) forward pricing to lock into certain prices, (4) call options, and (5) participation in the Federal Crop Insurance Program (FCIP). All of these options provide means for lowering net income risk. The authors included four levels of participation in the FCIP: (1) nonparticipation, (2) participation at 50%, (3) participation at 75%, and (4) full participation. The higher the level of participation in the FCIP, the greater the benefits and costs. A case study of a dry land cotton, wheat, and grain sorghum farm in Knox County, Texas, was used. Risk was incorporated into the model using an E-V approach, and it was assumed that output prices and production were the two sources of risk, while all other parameters were assumed to be known with certainty.

Time series regression and Monte Carlo simulation techniques were used to generate the probability distributions for net revenue for each of the risky price and production activities in the model. For instance, for yield risk, the authors regressed yield on a constant and a time trend term over the annual period 1965–1979, and then used the predicted equation and residual term from the regression to generate random yields for the simulation model. Similar procedures were used to generate random cash and futures prices. Expected net revenues and a variance-covariance matrix of net revenue were then calculated for each of the five risk-management strategies.

The authors used a standard E-V formulation for the objective function of the QP model. The objective for each case is:

$$\text{Max: } E(U) = s'x - \left(\frac{a}{2}\right)x'Qx,$$

where s' is a row vector of net income per unit for each production and marketing activity, x is a column vector of all marketing, production, and FCIP participation activities, x' is its transpose, a is a risk aversion coefficient, and Q is the variance-covariance matrix of net income for each activity in x. By parametrically varying the risk-aversion parameter, a, from zero (risk neutral) to a positive number (risk averse), an E-V efficient frontier can be determined.

The first set of constraints represents the usual technical and resource endowment restrictions faced by the representative farmer:

$$\mathbf{Ax} \leq \mathbf{b},$$

where $\mathbf{A}$ is a matrix of technical coefficients converting resource endowments into units of each activity, and $\mathbf{b}$ is a vector of resource endowments. Following the technique of chance-constrained programming, the authors constrained the volume of sales activities to be less than or equal to total expected production adjusted downward by $(1 - \gamma)$, where $0 \leq \gamma \leq 1$. This constraint was included to represent the risk associated with an outcome where production is lower than sales.

In the model, there are 76 activities for cotton, wheat, and grain sorghum, with only 3 representing participation levels in the FCIP. An additional 3 of the 76 are production activities, while the remaining are marketing activities. The model was solved separately for each participation scenario to generate efficient sets of activities for each scenario.

The main conclusion from the study is that full participation in the FCIP is the best option available to farmers regardless of the level of risk aversion (including the risk-neutral case). Basically, full participation offers the highest level of expected net income regardless of risk aversion level. This is evident by examining Figure 10.8, which plots the expected utility-risk aversion coefficient frontier for nonparticipation and full participation in the FCIP. In this figure, participation dominates nonparticipation for every level of risk aversion. Hence, participation in the FCIP is an optimal strategy for the representative Texas farm.

Not surprisingly, the study found that the degree of risk aversion greatly impacted the optimal marketing and production strategy. For example, diversification across crops increased with the level of risk aversion. The use of futures markets was found to be an excellent strategy for minimizing risk. For example, wheat and grain sorghum hedging via hedging in the futures markets and cotton sales via call options was found to be an excellent strategy for minimizing price risk.

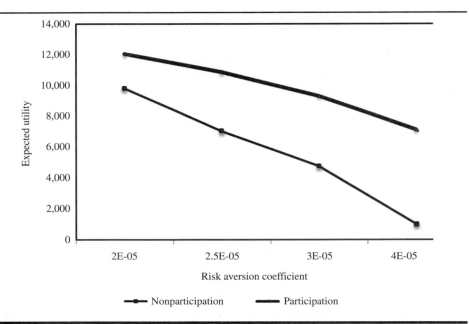

Figure 10.8   Expected utility by level of risk aversion for participation and nonparticipation in the FCIP.

The results of this study are useful for farmers in this region. At the time this study was done, the policy implications were that farmers are better off by participating in the FCIP and by using various marketing strategies to sell their crops. The authors suggested that educational programs should be conducted on these various strategies to better educate farmers on how to use these options to improve their net income and reduce its variability.

10.10   RESEARCH APPLICATION: DISCRETE STOCHASTIC SEQUENTIAL PROGRAMMING

A model developed by Kaiser and Apland (1989) is described to illustrate the formulation and use of DSSP models. One of the purposes of Kaiser and Apland's study was to identify risk-efficient production and marketing plans for a representative corn–soybean farm in Minnesota. A subset of the production and marketing solutions will be reported here. A MOTAD-type objective function was used in the model, which included random yields, prices, harvest field rates, and harvest field time.

The model's production activities and resource constraints were defined over 11 discrete time periods for tillage, planting, cultivation, and harvest operations. The inclusion of several periods is essential to capture critical timeliness characteristics of crop production. Preharvest production decisions were assumed to be made in stage 1, harvest production decisions were set in stage 2, and marketing decisions were made in stage 3. The 11 intrayear production periods were defined from April 7 through November 30. Preharvest (stage 1) operations included spring plowing, disking, herbicide application, planting, and post-planting operations, which occur in periods 1 through 6 from April 7 to June 8. Stage 2 operations included harvesting and fall plowing, which take place in periods 7 through 11 from September 15 to November 30. The constraining resources for both stages were full-time and part-time labor by production period, machine time by production period, crop acreage, and on-farm storage capacity.

Six corn and six soybean marketing alternatives were considered based on common marketing practices in the region (Gois, 1983). These included a cash market sale at harvest, a storage hedge placed at harvest and lifted in May, and four alternative postharvest cash market sales activities. Under the harvest sales activities, it was assumed that soybeans were sold in mid-October and corn was sold in mid-November. The storage hedge option consisted of two separate transactions for corn and soybeans in the cash and futures markets. First, July futures contracts were sold at harvest, and the contracted grain was placed in on-farm storage. Then, in May, the July contracts were purchased back, and the grain was sold in the cash market to lift the hedge. The postharvest sales activities involved selling the stored crops in mid-February, April, May, and June.

Four important sources of risk to corn-belt farmers were to be included in the model: crop yields, output prices, field time, and field rates. Yield variability was modeled in a conventional manner using a 10-year time series of farm-level data. Each observation was used as an equally likely yield state. Although yield time series are sometimes de-trended to remove the effects of technological change, no statistically significant trend was found in the corn or soybean yields. Production risk associated with field time variability was also included in the model.

Field time is defined here as the time during which weather and soil conditions are suitable for performing field operations. When field time is measured in days, as was the case here, the RHSs of labor and machinery constraints are calculated as the product of field days, working hours per day, and the number of units of the resource (workers or machines). Thus with field time as a random variable, the RHSs of the resource constraints are random. As with yields, 10 years of observations of field time were used to define the 10 discrete states of nature. Since at the time planting decisions (and thus crop mix decisions) are made, harvest periods are relatively distant, it was decided that field time would be modeled as a random variable in the harvest stage (stage 2). As a practical matter, stage 1 field time was modeled deterministically. Harvest rates were adjusted based upon the yield per acre and therefore varied by state of nature as well as planting and harvest period.

Output price states were defined to reflect the sequential nature of decisions and the flow of market information. The model is of a decision process that begins in April. As such, prevailing cash prices for corn and soybeans were assumed to be known at that time. To define harvest price states of nature, 10 years of harvest price data were normalized to the previous April cash price. These 10 observed price ratios were used to calculate 10 equally likely harvest price states for each crop based on the given April price, each corresponding to the yield and field time state for the same year. At the beginning of stage 3, it was assumed that the farmer knows the prevailing cash prices. However, only probabilistic knowledge regarding postharvest prices is assumed. It was also assumed that current market information would be available to farmers at harvest and that information would influence price expectations. Therefore, rather than using relative historical values as with harvest prices, simple regression equations were estimated including market indicators, which would be known at harvest. The 10 observed error terms were added to the values of the regression equations for each of the 10 observed harvest states to define 100 postharvest price states of nature.

The general structure of the model is illustrated by the decision tree in Figure 10.9. In this figure, $S_{i,t}$ represents the occurrence of the ith state of nature in stage t. The mathematical formulation of the model is as follows:

$$\text{Max: } U = E - \text{ur}\sum_{i=1}^{10}\sum_{j=1}^{10} d_{ij} \tag{1}$$

s.t.:

Accounting Constraints

$$y_{ij} + \mathbf{c}_{11}\mathbf{x}_{11} + \mathbf{c}_{12}\mathbf{x}_{12} + \mathbf{c}_{21}\mathbf{x}_{21i} + \mathbf{c}_{22}\mathbf{x}_{22i} - \mathbf{p}_{ij}\mathbf{m}_{ij} = 0$$

$$(i = 1, \ldots, 10, j = 1, \ldots, 10) \tag{2}$$

$$\sum_{i=1}^{10}\sum_{j=1}^{10} a_i e_i y_{ij} - E = 0 \tag{3}$$

$$y_{ij} - E + d_{ij} = 0 \quad (i = 1, \ldots, 10, j = 1, \ldots, 10) \tag{4}$$

Resource Constraints

$$\mathbf{A}_{11}\mathbf{x}_{11} + \mathbf{A}_{12}\mathbf{x}_{12} \leq \mathbf{b}_1 \tag{5}$$

$$\mathbf{l}_1\mathbf{x}_{12} \leq \mathbf{b}_2 \tag{6}$$

$$\mathbf{A}_{21i}\mathbf{x}_{21i} + \mathbf{A}_{22i}\mathbf{x}_{22i} \leq \mathbf{b}_{3i} \quad (i = 1, \ldots, 10) \tag{7}$$

$$-\mathbf{h}_i\mathbf{x}_{21i} + \mathbf{m}_{ij} \leq 0 \quad (i = 1, \ldots, 10) \tag{8}$$

$$\mathbf{sm}_{ij} \leq \mathbf{b}_4 \quad (i = 1, \ldots, 10, j = 1, \ldots, 10) \tag{9}$$

Sequencing Constraints

$$-\mathbf{B}_1\mathbf{x}_{11} + \mathbf{B}_2\mathbf{x}_{12} \leq 0 \tag{10}$$

$$-\mathbf{I}\mathbf{x}_{12} + \mathbf{B}_3\mathbf{x}_{21i} \leq 0 \quad (i = 1, \ldots, 10) \tag{11}$$

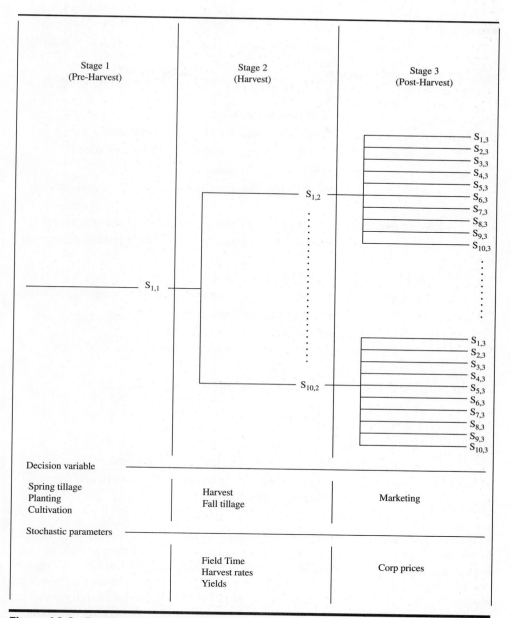

Figure 10.9   Decision tree for corn-soybean minnesota farm in research application Section 10.10.

$$-\mathbf{B}_4 x_{21i} + \mathbf{B}_5 x_{22i} \leq 0 \quad (i = 1, \dots, 10) \tag{12}$$

$$\mathbf{B}_7 x_{11} - \sum_{i=1}^{10} a_1 \mathbf{B}_6 x_{22i} \leq 0 \quad (i = 1, \dots, 10) \tag{13}$$

$$y_{ij}, d_{ij}, x_{11}, x_{12}, x_{21i}, x_{22i}, m_{ij} \geq 0 \quad (i = 1, \dots, 10, j = 1, \dots, 10) \tag{14}$$

where:

y_{ij} = total net revenue (total receipts minus total variable costs), harvest state i, post-harvest state j

d_{ij} = negative deviation from expected net revenue, harvest state i, post-harvest state j

x_{11}, x_{12} = spring tillage and planting vectors, stage 1

x_{21i}, x_{22i} = harvesting and fall tillage vectors, stage 2, harvest state i

m_{ij} = marketing decision vector, harvest state i, post-harvest state j

$c_{11}, c_{12}, c_{21}, c_{22}$ = variable cost vectors for field operations in stages 1 and 2

p_{ij} = net price vector for marketing activities, harvest state i, post-harvest state j

A_{11}, A_{12} = matrices of resource requirements for stage 1 field operations

A_{21i}, A_{22i} = matrices of resource requirements for stage 2 field operations, harvest state i

h_i = vector of crop yields, harvest state i

s = vector of zeros and ones for storage requirements of marketing activities

b_1 = vector of stage 1 resource endowments

b_2 = total crop land endowment

b_{3i} = vector of stage 2 resource endowments, harvest state i

b_4 = on-farm storage capacity

$B_1, \ldots, B_7$ = sequence preserving matrices for field operations

I = identity matrix

l_1 = vectors of ones

a_i = probability of harvest state i

e_j = probability of post-harvest state j, given harvest state i

u = (2/s)(ps/2(s−1))0.5, where s = number of joint events, and p is the mathematical constant, pi=3.141....

r = risk aversion coefficient

The objective function (1) is expected net revenue (E) adjusted for risk. By (3), E is equal to the sum of the 100 joint net revenue events (y_{ij}) each weighted by their probability ($a_i e_j$). Risk is measured as the standard deviation of net revenue as estimated by total negative deviation from the mean times the coefficient u (Hazell 1985). Constraint (2) defines net revenue activities y_{ij}, and constraint (4) defines negative deviation activities d_{ij} for each joint event. Constraint (5) restricts the use of farm labor and machinery by field operations in stage 1 to endowed levels. Constraint (6) is the land constraint. By constraint (7), the use of stage 2 resources under each of the 10 states of nature cannot exceed endowed levels. The RHSs of the flow resource constraints for both stages are equal to the number of hours of labor or machine services available for the associated production period and state of nature. Constraints (8) and (9) are output and grain storage constraints, which restrict the total amount of grain sales to the total output produced and limit the amount of grain sold from storage to no more than on-farm storage capacity. Constraints (10) through (13) preserve the proper sequence of field operations in the model. These constraints assure that spring tillage occurs prior to planting (10), planting activities are matched to harvesting activities (11), and harvesting is performed prior to fall plowing (12). Finally, constraint (13) assures that any acreage not plowed in the fall is plowed in the spring.

The case farm used in the analysis was selected to be representative of corn–soybean farms in southern Minnesota. The base year for the analysis was 1983. A farm in Jackson County with 612 tillable acres was selected from Minnesota Farm Management Association (MFMA) records for time series data on corn and soybean yields. Other production and cost data for the case farm were based on farm management studies applicable to southern Minnesota. For yield, price and field time states of nature, a 1974–1983 sample period was used. The machinery and equipment sets used in the model were typical of farms of comparable size in this region of Minnesota. It was assumed that the farm used a conventional tillage system. Field rates and other technical parameters used in the empirical model were taken from Benson and Gillard (1985). Labor requirements for all operations were assumed to be 110 percent of the machine time requirements (220% for harvest since it was assumed that two workers were required for this operation).

Variable costs were adapted from MFMA records and Benson and Gillard (1985). These costs included fuel, lubrication, and repairs for machinery and equipment as well as seed, herbicide, insecticide, fertilizer, interest on cash expenses, variable drying costs, and insurance. The operating costs (except drying) were assumed to be known by the farmer at the beginning of the decision process. Variable drying cost per acre was a function of the moisture content of the corn at harvest and the yield. Since yield and moisture content varied by planting and harvesting date, and by harvest period state of nature, the drying cost per acre was stochastic. Data on the number of field days by period were based on records from the Southwest Experiment Station in Lamberton, Minnesota, for 1974–1983 (Nelson & Straesser, 1985). Field days represent the number of days per period that farmers can perform field operations. The hours per field day were based, in part, on the number of hours from sunrise to sunset in this region. A maximum of 11 hours was assumed, and the number of daylight hours was rounded down to the next lowest hour for each time period.

For each Stage 2 state of nature, the observed corn and soybean yields were adjusted for each combination of planting and harvesting period to reflect the effects of timeliness on yields. The coefficients used to adjust yields for timeliness were based on a study by Fuller and Hasbargen (1973). These adjusted yields were incorporated in the model by specifying a separate harvesting activity for each planting-harvesting period combination. A moisture content level for corn adapted from Fuller and Hasbargen's (1973) study was set for each planting/harvest period combination to estimate per acre variable drying costs.

The 100 postharvest price states of nature (i.e., 10 for each of the 10 harvest states) consisted of net selling prices for each of the postharvest marketing activities in the model. The marketing year began at harvest and extended to mid-June of the following calendar year. To represent expected price distributions prior to planting decisions, all monthly prices (for the months corresponding to the marketing activities) were expressed as indices by dividing the price associated with each marketing activity by the preceding April cash price.

As discussed earlier, the 10 observed price ratios (harvest price/April price) were applied to the 1983 April cash price to define harvest price states of nature for both corn and soybeans. For the postharvest marketing activities, price probability distributions were desired that reflected market information available to farmers at harvest. The goal was not to develop a forecasting model or a behavioral price expectations model. Rather, what was sought was a representation of market risk faced by a farmer after harvest has been completed (yield time series data are frequently de-trended for similar reasons). To accomplish this, the postharvest price states were generated using the following model:

$$p_{mt} = e^{b0} \, p_{ht}^{b1} m_t^{b2} e^{Ut},$$

where p_{mt} is the price of postharvest marketing activity m, p_{ht} is the harvest cash price, m_t is a harvest market indicator, b0, b1, and b2 are parameters to be estimated, and Ut is a stochastic disturbance term assumed to be normally distributed with a mean of zero and constant variance. The market indicators (m_t) initially considered were: USDA forecasted carryover stocks at harvest, ratios of expected supply to expected disappearance as reported by the USDA at harvest, expected exports from USDA reports at harvest, monthly cash and futures prices before harvest, and other lagged cash prices. This price equation was estimated by ordinary least squares. All variables in the corn price expectations models were transformed into natural logarithms. The soybean price expectations models were estimated with all the same variables, except m_t, transformed into natural logarithms. m_t was expressed as its actual value in the soybean models since it ranged from negative to positive numbers. Corn and soybean monthly cash and futures prices for the 1974–1975 through 1984–1985 marketing years were obtained from the All American Cooperative, Stewartville, Minnesota, and the Chicago Board of Trade. The error terms from each of these equations were used to generate the 10 postharvest states conditional on each of the 10 harvest states of nature.

There is a fairly wide range of values on the 10 states for corn and soybean yields and prices. For example, corn yields range from a low of 50.5 bushels per acre in State 10 to a high of 148 bushels per acre in State 6. With respect to harvest prices, corn prices range from $2.01 per bushel (State 8) to $4.09 per bushel (State 1), and soybean harvest prices range from $5.98 per bushel (State 2) to $10.77 per bushel (State 7). The values of these harvest states, as well as the postharvest price states, significantly influence the crop mix and marketing activities in the optimal solution to the problem, as is demonstrated in the results.

A risk frontier was generated by adjusting the risk coefficient (r) from zero to 1.5 in increments of 0.5. Expected net revenue ranged from $109,498, in the risk-neutral case, to $76,167 for the highest risk coefficient considered. The corresponding standard deviations of net revenue were $52,987 and $21,189. Corn production tended to decline as the risk coefficient was increased, which is not surprising since corn yields had a higher coefficient of variation than soybean yields. The percentage of land planted to corn went from 63% to 45% as the r value was increased from 0.0 to 1.5. This result appears to adequately depict actual ranges of crop mix in southern Minnesota. For example, based on the 1983 MFMA annual report, the crop mix averaged 44% corn and 56% soybeans for southern Minnesota.

Optimal marketing activities varied substantially across the 10 harvest states of nature. For example, in the risk-neutral case, a significantly higher proportion of corn relative to soybeans is produced and sold under State 6 than in the rest of the states. The relatively high corn-soybean sales ratio of 8.2 in this case is due to a very high corn–soybean yield ratio and a favorable corn storage hedge price. On the other hand, the greatest proportion of soybeans relative to corn is produced and sold under State 3 than in the other states. The relatively low corn-soybean sales ratio of 3.5 in this case is due to a very low corn–soybean yield ratio and a favorable soybean May cash price. The distribution of marketing activities also varies across the 10 states of nature. For example, under harvest State 6 in the risk-neutral case, all corn marketing occurs at harvest with 29,134 bushels sold in the cash market and 32,467 bushels hedged, while all soybeans are sold from storage in the cash market in May. The reverse marketing strategy occurs under State 9. In this case, the storage capacity constraint is binding for corn with 40,000 bushels being sold from storage in May and the remaining corn (15,248 bushels) being sold at harvest. All soybeans are sold at harvest since none can be placed in storage. As with the production activities, significant adjustment occurred in the marketing activities as risk aversion increased.

On average, use of the storage hedge was greater for the risk-averse case than the risk-neutral case, 55.5% versus 30.2% of corn production and 36.2% versus 18.2% of soybean production. In the risk-neutral case, the storage hedge activity was used in 4 of the 10 harvest states for corn and 2 of the 10 harvest states for soybeans. In the risk-averse case, the storage hedge was used in 7 of the 10 harvest states for corn and 4 of the 10 harvest states for soybeans. The marketing activities tended to be more diverse in the risk-averse case. For the profit maximizer, of the 6 marketing activities for corn, only one activity was used in 4 out of the 10 harvest states—2 were used in the other 6 states. Only one soybean marketing activity was used for all 10 harvest states. On average in the risk-neutral case, 2.6 of the 12 marketing activities were optimal. While the average number of marketing instruments used in the risk-averse case was only slightly greater at 2.9, under 3 harvest states of nature, 4 or more of the 12 marketing activities were used.

The optimal levels of other activities in the model provide detail regarding the levels of various production operations by time period. In the case of harvest and fall tillage activities, the schedule of operations is provided for each of the 10 fall states of nature. Thus, a notion of variability of harvest completion dates and fall tillage levels can be gained.

A DSSP model provides detailed information to the decision maker about optimal contingency plans. For instance, the model developed in this article provides the farmer with different marketing strategies that depend upon the observed harvest production and marketing conditions. The optimal marketing strategies vary greatly over harvest conditions. Hence marketing strategies based on average harvest conditions may deviate significantly from the "true" optimal solution. In such cases, the DSSP model may be preferable for the farmer for use as a decision aid.

10.11 RESEARCH APPLICATION: AGRICULTURE AND CLIMATE CHANGE

Kaiser et al. (1993) linked a DSSP model along with climate and agronomic models to examine the potential economic and agronomic impacts of gradual climate warming at the farm level. A grain farm in southern Minnesota was used as a case study. This region is part of the northern fringe of the United States corn belt, which could be affected by climate warming. Several climate warming scenarios were analyzed, which vary in severity to simulate how sensitive crop yields, crop mix, and farm revenue are to climate change. Climate change was simulated as a gradual and dynamic process rather than the more traditionally used comparative static approach of comparing a "doubled CO_2" induced change in climate with our present climate. Given the authors' focus on farm-level adaptation issues, it was important that the climate, crop, and economic models be dynamic. There was an emphasis on simulating the effects on "tactical" farm-level decisions. For example, the model allows for adaptive management strategies such as changing plant cultivar (variety) selection and changing planting and harvesting dates in response to a gradually changing climate. Finally, changes in the variability, as well as in the averages of climatic variables, are modeled. Climate change may result in more than changes in mean values for climatic variables, with potentially important consequences.

The overall model consisted of three components: atmospheric, agronomic, and economic. The atmospheric component simulated daily values for minimum and maximum temperature, precipitation, and solar radiation over a 100-year period for several different climate change scenarios. Based on the values of the climatic variables, the agronomic component estimated crop yields, grain moisture content, and field time availability (i.e., the span of time during which weather-related soil moisture conditions allow farmers to perform field operations). In turn, crop yields, grain moisture content, and field time availability became inputs in the farm-level economic component. Crop prices, which are also a function of climate scenario, were generated by price reduced-form equations based on supply and demand variables. Finally, the output of the economic model included optimal crop mix, scheduling of field operations, and expected net farm income. In the discussion that follows, details of the economic model are discussed.

As the climate changes, farmers will be forced to re-evaluate their production decisions, and in particular, the mix of crops. This study simulated future decisions for each climate scenario using DSSP. The economic model divided the decision-making process into two stages: Stage 1 (preharvest) and stage 2 (harvest). Stage 1 decisions included spring plowing and planting operations, which can take place in four periods. Stage 2 decisions include fall plowing and harvesting, which can take place in four periods. The constraining resources for both stages included full and part-time labor by production period and crop acreage. Risk was captured by a Freund-type (1956) objective function, which maximizes expected net revenue minus a risk term adjusted by a risk-aversion coefficient.

Four important sources of risk were included in the model: field time availability, crop yields, grain drying costs, and crop prices. At the beginning of stage 1, the farmer makes spring plowing and planting decisions while facing three states of nature on field time availability. Each of these states differed by field time availability in each of the four stage 1 periods. It was assumed that the farmer expects each of the three stage 1 states to be equally likely. At the beginning of stage 2, the farmer makes harvest and fall plowing decisions. Here the farmer has perfect knowledge of which stage 1 state has occurred, but only probabilistic knowledge of which stage 2 state will occur. The stage 2 states (10 states, conditional on each of the three stage 1 states) consisted of discrete random parameters for field time availability, crop yields, grain drying costs, and crop prices. Each of these states consisted of field hours available in each of the four stage 2 periods, crop yields and drying costs associated with each of six planting and harvest dates, and output prices for the three crops. Again, it was assumed that the farmer expected each of the stage 2 states to be equally likely. There were 30 joint net revenue events possible, each corresponding to a unique sequence of a stage 1 state followed by a stage 2 state.

Values for the field time availability, yield, and crop moisture states of nature were generated by the agronomic model. Grain-drying cost states of nature were determined by grain moisture content at harvest, yield level, and costs per bushel per percentage point of moisture using the following formula:

$$dc_i = 0.024 y_i (m_i - 17) i17_i,$$

where dc_i is drying cost per acre, harvest state i, y_i is yield per acre, harvest state i, m_i is grain moisture content, harvest state i, and $i17_i$ is equal to one if the moisture content in harvest state i is greater than 17%, or equal to zero otherwise. It was assumed that it costs $0.024 per bushel to remove one percentage point of moisture, and that grain must be dried only if it has a moisture content of 17% or higher. Values for each crop's price states of nature were generated by a price reduced-form equation of supply and demand that includes crop yield as an explanatory variable (procedures are discussed in the next section).

The mathematical formulation of the economic model is as follows:

$$\text{Maximize: } E - r\Phi \sum_{i=1}^{3} \sum_{j=1}^{10} d_{ij} \qquad (1)$$

s.t.:

Accounting Constraints

$$r_{ij} + c_{11}x_{11} + c_{12}x_{12} + c_{21ij}x_{21i} + c_{22}x_{22i} - pm_i = 0 \quad (i = 1, \dots, 3, j = 1, \dots, 10) \qquad (2)$$

$$\sum_{i=1}^{3} \sum_{j=1}^{10} a_i b_j r_{ij} - E = 0 \qquad (3)$$

$$r_{ij} - E - d_{ij} \leq 0 \quad (i = 1, \dots, 3, j = 1, \dots, 10) \qquad (4)$$

Resource Constraints

$$\mathbf{A}_{11}\mathit{x}_{11} + \mathbf{A}_{12}\mathit{x}_{12} \leq \mathbf{b}_{1i} \tag{5}$$

$$\mathbf{l}_1\mathit{x}_{12} \leq \mathbf{b}_2 \tag{6}$$

$$\mathbf{A}_{21}\mathit{x}_{21i} + \mathbf{A}_{22}\mathit{x}_{22i} \leq \mathbf{b}_{3ij} \quad (i = 1, \dots, 3, j = 1, \dots, 10) \tag{7}$$

$$-\mathbf{h}_{ij}\mathit{x}_{21i} + \mathit{m}_i \leq 0 \quad (i = 1, \dots, 3, j = 1, \dots, 10) \tag{8}$$

Sequencing Constraints

$$-\mathbf{B}_1\mathit{x}_{11} + \mathbf{B}_2\mathit{x}_{12} \leq 0 \tag{9}$$

$$-\mathbf{I}\mathit{x}_{12} + \mathbf{B}_3\mathit{x}_{21i} \leq 0 \quad (i = 1, \dots, 3) \tag{10}$$

$$-\mathbf{B}_4\mathit{x}_{21i} + \mathbf{B}_5\mathit{x}_{22i} \leq 0 \quad (i = 1, \dots, 3) \tag{11}$$

$$\mathbf{B}_7\mathit{x}_{11} - \sum_{j=1}^{3} a_i\mathbf{B}_6\mathit{x}_{22i} \leq 0 \quad (i = 1, \dots, 3) \tag{12}$$

$$r_{ij}, d_{ij}, \mathit{x}_{11}, \mathit{x}_{12}, \mathit{x}_{21i}, \mathit{x}_{22i}, \mathit{m}_i \geq 0 \tag{13}$$

where:

E = expected net revenue

r = risk aversion coefficient

Φ = constant that converts total negative deviations into proxy for standard deviation $(2/s)$ $(s\Pi/2(s - 1)^{0.5}$ s is the number of joint states of nature, and Π is the mathematical constant, pi, 3.14....

d_{ij} = negative deviation from expected net revenue, pre-harvest state i, harvest state j

r_{ij} = total net revenue pre-harvest state i, harvest state j

$\mathbf{c}_{11}, \mathbf{c}_{12}, \mathbf{c}_{22}$ = variable cost vectors for spring plowing, planting, and fall plowing

$\mathit{x}_{11}, \mathit{x}_{12}$ = spring plowing and planting vectors, pre-harvest stage

$\mathbf{c}_{21ij}$ = variable cost vector for harvest, pre-harvest state i, harvest state j

$\mathit{x}_{21i}, \mathit{x}_{22i}$ = harvest and fall plowing vectors, pre-harvest state i

$\mathbf{p}$ = output price vector

m_i = marketing decision vector, pre-harvest state i

a_i = probability of pre-harvest state i occurring

b_j = probability of harvest state j occurring, given pre-harvest state i

$\mathbf{A}_{11}, \mathbf{A}_{12}, \mathbf{A}_{21}, \mathbf{A}_{22}$ = matrices of resource requirements for all field operations in stages 1 and 2

$\mathbf{b}_{1i}$ = vector of stage 1 resource endowments, pre-harvest state i

$\mathbf{l}_1$ = vector of ones

$\mathbf{b}_2$ = total crop land endowment

$\mathbf{b}_{3ij}$ = vector of stage 2 resource endowments, pre-harvest state i, harvest state j

$\mathbf{h}_{ij}$ = vector of crop yields, pre-harvest state i, harvest state j

$\mathbf{B}_1, \dots, \mathbf{B}_7$ = sequence preserving matrices for field operations

$\mathbf{I}$ = identity matrix

The objective function in (1) is to maximize expected net revenue (gross revenue minus variable costs) minus a risk adjustment term, where risk is measured as the standard deviation of net revenue calculated by the product of total absolute negative deviations from the mean times the parameter Φ (Hazell, 1985). The risk measure is discounted by the risk aversion coefficient, r. Constraints (2) through (4) are accounting constraints that define the 30 joint net revenue events (2), define expected net revenue (3), and define negative deviation from expected net revenue (4).

Constraint (5) restricts the use of farm labor (both full and part-time) by field operations in the preharvest stage to endowed levels. Note that the RHS parameter in this constraint is stochastic, corresponding to available field hours by production period for the three stage 1 states of nature. Constraint (6) is the land constraint, which limits acres planted to endowed levels. Constraint (7) restricts the use of farm labor by field operations in the harvest stage to endowed levels. The RHS parameters for this constraint are also stochastic, corresponding to the available field hours by production period for the 30 stage 2 states of nature. Crop output constraints are represented by (8), which limits the amount of crop that can be sold to the amount that is harvested for each preharvest state of nature. Finally, constraints (9) through (12) are sequencing restrictions, which preserve the proper sequence of field operations in the model. These constraints guarantee that spring plowing occurs prior to planting (9), that planting activities are matched with harvest activities (10), that harvesting occurs before fall plowing for each stage 1 state (11), and that any acreage not plowed in the fall is plowed in the spring (12).

The simulation procedures began with a Monte Carlo simulation of the stochastic weather model to generate daily weather values for climates (scenarios) changing over the 100-year period 1980–2079. These results were used by the agronomic model to generate annual values for crop yields, grain moisture, and field time availability. These values were tabulated decade by decade, yielding 10 sets of agronomic results for each scenario. Finally, the resulting crop yields, grain moisture, and field time availability parameters were used by the farm-level economic model, which generated optimal management strategies and expected net revenue. Costs, technical parameters, and resource endowments were held constant at their 1980 values. However, cultivar selection, crop yields, grain moisture content, and grain drying costs were different for each decade of each climate change scenario, according to results of the agronomic simulations.

For each crop and each possible planting–harvesting combination, three cultivars were simulated: early-, mid-, and late-maturing varieties. However, only one cultivar was used in the economic model for each planting–harvesting combination for each decade based on the following decision rule. It was assumed that farmers make cultivar decisions on the basis of yield performance in the previous decade. Specifically, the cultivar having the highest average yield for a particular plant–harvest period in the previous decade was selected by the farmer for the current decade.

To generate the 30 joint events (crop yields, grain moisture content, and field hours) for the economic component, the weather component produced 30 realizations ("years") of daily data representative of each decade for the agronomic component. Based on these 30 weather realizations, the agronomic component then produced 30 yield, field time availability, and grain moisture states of nature per decade for the economic model. Since these 30 joint states of nature were assumed to be equally representative of a decade, the solution to the economic model could be thought of as a representative year within the decade. This process was repeated 10 times for each climate change scenario to generate solutions for each decade for 1980–2079.

Because this is a micro-level model, crop prices could not be endogenously determined. Yet it is unlikely that climate change would not affect crop prices over time. To construct price trajectories by decade for each crop based on each climate scenario, price reduced-form equations were estimated using annual time series data from 1960 through 1988. In addition to several exogenous demand and supply shifters, county average crop yield was included as an explanatory variable in the price reduced-form equation. In theory, crop yield at the micro-level should not explain output price, but Minnesota yields were highly correlated with national yields, which do influence price (correlation coefficients for corn and soybeans between county and national yields were 0.84 and 0.75, respectively). The three estimated crop price reduced-form equations are:

$$\ln pc_t = -46.33 - 0.44 \ln yc_t + 9.11 \ln pop_t + 0.34 \ln pc_{t-1} + 0.59 \ln spc_t + 0.52 \, dum \, 73\text{--}75 - 0.08 \, t_t$$
$$\quad (26.5) \quad (0.21) \qquad (5.05) \qquad\quad (0.15) \qquad\quad (0.29) \qquad\quad (0.09) \qquad\qquad (0.05)$$
$$\qquad\quad R^2 = 0.87 \qquad\qquad\qquad D.W. = 1.81$$

$$\ln psb_t = 0.90 - 0.33 \ln ysb_t + 0.16 \ln pop_t + 0.53 \ln psb_{t-1} + 0.10 \ln spsb_t + 0.29 \, dum \, 73\text{--}75$$
$$\quad (7.10) \, (0.33) \qquad\quad (1.34) \qquad\quad (0.17) \qquad\qquad (0.39) \qquad\quad (0.17)$$
$$\qquad\quad R^2 = 0.68 \qquad\qquad\qquad D.W. = 2.33$$

$$\ln psg_t = 3.32 - 0.99 \ln ysg_t + 0.07 \ln inc_t - 1.20 \ln psg_{t-1} + 0.56 \ln spsg_t + 0.47 \, dum \, 73\text{--}75$$
$$\quad (3.19) \, (0.34) \qquad\quad (0.34) \qquad\quad (0.16) \qquad\qquad (0.23) \qquad\quad (0.10)$$
$$\qquad\quad R^2 = 0.88 \qquad\qquad\qquad D.W. = 1.61$$

where:

pc_t = real corn price per bushel (nominal price divided by Consumer Price Index where 1988 = 1.0), year t

yc_t = county average corn yield, year t

pop_t = U.S. civilian population, year t

spc_t = real corn support price per bushel, year t

dum 73–75 = dummy variable equal to 1 for 1973–1975, equal to zero otherwise

t_t = time trend, 1960 = 1, 1961 = 2, . . .

psb_t = real soybean price per bushel, year t

ysb_t = county average soybean yield, year t

$spsb_t$ = real soybean support price per bushel, year t

psg_t = real sorghum support price per bushel, year t

ysg_t = national average sorghum yield, year t

inc_t = U.S. per capita real income, year t

$spsg_t$ = real sorghum support price per bushel, year t

R^2 = coefficient of variation,

D.W. = Durbin–Watson statistic and

( ) = standard error

To simulate crop prices from 1990 to 2079, future values for the exogenous variables were necessary. It was assumed that U.S. civilian population increased by 1% per year throughout the simulation period. Real support prices for the three crops were assumed to decrease by 1% per year, reflecting a trend towards a market-oriented farm policy. Real income was forecasted based on a regression equation with income in the two previous years and a time trend as explanatory variables. For each decade, the price states of nature were generated by substituting the annual average yield states of nature (generated by the agronomic model) and the values for the other exogenous variables into the price reduced-form equations.

A hypothetical farm based on characteristics of southern Minnesota (Redwood County) was used to illustrate the type of analyses that can be conducted using this model. This region is the northern limit of corn production in southern Minnesota (Murray). The predominant soil series in this region is Ves, which is deep and has excellent water-holding capacity. Most of the data for the economic component for the hypothetical Minnesota farm not generated by the agronomic model are presented in Kaiser (1985), including resource requirements and variable costs. It was assumed that the farm is endowed with 600 acres of tillable land on which corn, soybeans, and/or sorghum can be grown. There were two full-time workers and one additional part-time worker that could be hired at a cost of $6.00 per hour. The farm used a conventional tillage system. Data for sorghum, which currently is not commonly grown in this region, were based on national average statistics.

The climatic, agronomic, and economic components were solved for four climate scenarios. Scenario 1 is the no-climate-change situation, scenario 2 is the mildly warmer (2.5°C) and wetter (10%) case, scenario 3 is the mildly warmer (2.5°C) and drier (10%) situation, and scenario 4 is about twice as warm and dry as scenario 3.

The mildly warmer and wetter scenario 2 had no adverse impact on crop yields at this relatively cool location. While the climate gradually warms, the accompanying increase in precipitation prevents the crops from experiencing water stress for the simulation period. In fact, sorghum and soybean yields increase over time, while corn yields remain relatively stable, increasing slightly from 2000 to 2060. For all three crops, the model predicted adoption of later maturing, higher yielding cultivars over time as field time availability increases due to climate warming. The robustness of yields to climate change in this scenario is a result of the relatively mild change in climate assumed, the relatively cool location, the introduction of later-maturing cultivars later on in the simulation period, and the absence of plant water-stress due to the excellent water-holding capacity of this soil and the wetter climate.

As was the case for the wetter climate scenario, the drier climate scenario appeared to have no adverse effects on soybean or sorghum yields. However, corn yields decreased marginally over time. In this case the drier climate had two impacts on crop yields. First, the growing season was lengthened due in part to the drier climate, allowing greater flexibility in access to the field. This had a positive effect since later-maturing cultivars can be adopted. Second, the decrease in precipitation caused some water stress, which had a negative effect on yields.

In the last and most severe scenario, the average decrease in precipitation was twice as large as in the previous scenario, and the increase in temperature was 4.2°C rather than 2.5°C in year 2060. Average soybean and sorghum yields were not adversely impacted by even this relatively severe change in climate. Average corn yields, on the other hand, trend downward. Moreover, the magnitude of decrease is larger than the previous scenario.

Two observations emerged from these results. First, the model predicted that soybean and sorghum yields at this relatively cool location were not adversely affected by these three climate change scenarios. Second, corn yields were somewhat adversely affected by the more severe warmer and drier scenarios, but not by the mild warmer and wetter scenario. At this location, corn appeared to be the most climate-sensitive of the three crops.

In all three scenarios, the real corn price increased, real sorghum price decreased, and the real soybean price was relatively constant. The average prices were quite similar between scenarios 2 and 3 because yields were comparable between the two cases. However, the real corn price path was higher in the most severe climate scenario compared to scenarios 2 and 3, because corn yields were lower in climate scenario 4.

For the risk-neutral case, net revenue was slightly lower for all three climate change scenarios compared with the no-climate change case (scenario 1) between 1990 and 2010. However, after 2010, this pattern reversed with net revenue somewhat higher for the three climate change scenarios compared to no-climate change. Net revenue actually increased over time for the three climate change scenarios. The optimal crop mix changed little over time for all three scenarios with corn acreage representing about 70% of total acreage planted, and soybean acreage comprising the rest of the acreage. Under none of these scenarios did sorghum become profitable enough to replace corn production.

The standard deviation of net farm revenue was lower for the two mild climate change scenarios (scenarios 2 and 3) relative to no-climate change for all decades. The increased stability of net farm revenue was due to a decline in yield variability for all crops compared to the baseline no-climate change case. Crop yield variability, in turn, was lower than in the no-climate change scenario because of the assumed decrease in the variance of temperature. However, the 4.2° warmer, 20% drier climate (scenario 4) resulted in higher fluctuations in revenue risk. Under this scenario, revenue risk was higher in five decades and lower in four decades, compared to the no-climate change scenario.

To represent a risk-averse solution, a risk aversion coefficient of 1.25 was used, which is in the range that Brink and McCarl (1978) found representative for Cornbelt farmers. In this case, less corn and more soybeans were grown as compared to the risk-neutral case. The difference was due to soybeans being less risky than corn in terms of net revenue variability. However, under the risk-averse case, the share of corn as a percent of total acreage increased over time under all three scenarios, perhaps due to the accompanying decrease in variability of corn yields for the climate change scenario. Sorghum, which was the most stable of all three crop yields, was still not grown under any scenario. Expected revenue was lower in the risk-averse case than in the risk-neutral case because of the positive trade-off between risk and income.

The results of this research are not intended as a basis for general conclusions about climate change and agriculture across the nation. For example, while the relatively cool location chosen for illustration generally benefits from warming climates, more southern locations could suffer considerable declines in yields and revenue. Instead, the results are intended to illustrate the importance of adaptive strategies in predicting outcomes. The results indicate that grain farmers in the southern region of Minnesota can effectively adapt to a mildly and gradually changing climate (warmer and either wetter or drier). Adaptive strategies include adopting later-maturing cultivars, changing crop mix, and altering the timing of field operations to take advantage of a longer growing season due to climate warming.

SUMMARY

Risk and uncertainty is pervasive for most agricultural decisions. Empirical evidence suggests that farmers adjust their farm plans according to their risk posture. Specifically, studies indicate that profit-maximizing models, which ignore risk preferences by farmers, fail to give accurate normative or positive economic results when applied to many farming situations. Thus, in order to properly study most farm-level decision-making problems, the decision environment must be formulated in such a way that risk and uncertainty is a critical component in the model. In this chapter, we presented several alternative mathematical programming models that relax the assumption of parameter certainty. All models are based on the EUH (expected utility hypothesis).

The first method discussed was quadratic risk programming. Quadratic programming models use the variance-covariance matrix for net revenue as the measure of risk that is faced by the decision maker. The model was initially developed by Markowitz (1959) and applied to the stock market to derive optimal portfolio selection when risk is explicitly considered by the decision maker. Quadratic programming models are characterized by linear equations for all terms except the variance-covariance equation, which is quadratic. Two methods for solving the model were illustrated. When the risk posture of the decision maker is known, a unique solution to the problem can be generated by using the specific risk-aversion parameter in the model. Alternatively, when the risk parameter is not known, which is usually the case, parametric programming can be done to derive an efficient set of E-V solutions, which can be presented to the decision maker for selection.

A popular alternative to QP is the MOTAD model. MOTAD is a linear version of QP, where absolute deviations from mean net revenue are used rather than the variance-covariance matrix as the risk measure. The advantage of MOTAD is that LP can be used to solve the problem. However, the variance estimate from the MOTAD model is larger than the variance given in the QP model. This is due to the fact that the TAD estimate of the variance is not as efficient as the traditional nonlinear variance estimate. Hence, there is a trade-off when using a MOTAD model between the advantage of being a linear problem and the disadvantage of the TAD not being as efficient as an estimate of the variance and standard deviation.

The target MOTAD approach was discussed, which improves upon regular MOTAD in that its solutions are also efficient based on another efficiency criterion known as second-degree stochastic dominance. The model is similar to MOTAD models and simply adds a new constraint that sets a target level for total revenue.

One problem with these three methods is that risk is captured only in the objective function coefficients, while parameters in the constraint set are treated deterministically. However, resource availability and requirements in the constraint functions are also a source of risk to the farmer. A very important source of RHS risk in agriculture is availability of field time, which depends upon the weather. Chance-constrained programming deals with RHS risk by artificially reducing the availability of the resource to a lower limit whereby the decision maker can be confident, in a probabilistic sense, of it being achieved. The advantage of chance-constrained programming is that it is fairly simple to use and does not add a lot of complexity to the linear or nonlinear programming model. So it could be combined, for instance, with MOTAD or quadratic risk programming so that RHS risk is incorporated into the model. The main drawback of it is that it assumes a static, nonsequential decision process. In models assuming a single decision stage, decision variables are not adapted to new information received over the planning horizon.

Discrete stochastic sequential programming is a mathematical programming technique capable of overcoming many of the limitations cited with the previous models. DSSP is a technique for modeling decision making as a multistage decision process characterized by a discrete specification of random problem parameters. As the name implies, DSSP models consider the stochastic and sequential nature of resource endowments, resource requirements, and objective function coefficients. This technique requires that technical coefficients, objective function parameters, and/or resource endowments be specified separately for each stochastic state of nature. Each state is then assigned a probability of occurrence, based on the subjective assessment by the decision maker. The solution to the DSSP problem is then found, which depends, in part, on the way the states and probabilities are defined and assessed. The DSSP technique is a probabilistic decision model, based on Bayesian decision theory. Decisions in any stage are made with probabilistic knowledge of the occurrence of the states of nature in future stages of the decision process. The stages in the decision process are therefore interdependent. Decisions in later stages are restricted not only by the occurrence of particular random events in this stage, but also by random outcomes and decisions made in earlier stages.

The chapter concludes with a summary of three empirical applications of risk programming. The first involves applying quadratic risk programming to identify optimal production and marketing plans for a representative cotton–grain farm in Texas. The second application develops a DSSP model for a representative Minnesota corn–soybean farm. The third applied a DSSP model to examine climate change and optimal farm adaptation strategies for corn–soybean farmers in the Upper Midwest.

REFERENCES

Anderson, J. R., Dillon, J. L., & Hardaker, B. (1977). *Agricultural Decision Analysis*. Ames: Iowa, IA State University Press.

Benson, F. J., & Gillard, S. (1985). Minnesota farm machinery economic cost estimates for 1985. Agricultural Economics Bulletin AG-FO-2308. University of Minnesota, Agricultural Extension Service.

Bernoulli, D. (1954, 1738). Exposition of a new theory on the measurement of risk (Louise Sommer Trans.). *Econometrica*, *22*, 22–36.

Boisvert, R. N., & McCarl, B. (1990). Agricultural risk modeling using mathematical programming. *Regional Research Bulletin*, no. 356 (Southern Cooperative Series), 1–103.

Brink, L., & McCarl, B. (1978). The trade-off between expected return and risk among Cornbelt farmers. *American Journal of Agricultural Economics*, *60*, 259–263.

Charnes, A., & Cooper, W. W. (1959). Chance-constrained programming. *Management Science*, *6*, 73–79.

Cocks, K. D. (1968). Discrete stochastic programming. *Management Science*, *15*, 72–79.

Falatoonzadeh, H., Conner, J. Richard, & Pope, Rulon D. (1985). Risk management strategies to reduce net income variability for farmers. *Southern Journal of Agricultural Economics*, *17*, 117–130.

Freund, R. J. (1956). The introduction of risk into a programming model. *Econometrica*, *24*, 253–263.

Fuller, E. I., & Hasbargen, P. R. (1973). The Minnesota corn-soybean scheduling model: Computer decision aids. 418. University of Minnesota, Department of Agricultural and Applied Economics, Agricultural Extension Service.

Gois, M. (1983). Production and price risk management in agricultural: An application to a southwestern Minnesota farm. Master's thesis, University of Minnesota, unpublished.

Hazell, P. B. R. (1971). A linear alternative to quadratic and semivariance programming for farm planning under uncertainty. *American Journal of Agricultural Economics*, *53*, 153–162.

Kaiser, H. M. (1985). An analysis of farm commodity programs as risk management strategies for Minnesota corn and soybean producers. Doctoral dissertation, University of Minnesota, unpublished.

Kaiser, H. M., & Apland, J. (1989). DSSP: A model of production and marketing decisions on a midwestern crop farm. *North Central Journal of Agricultural Economics*, *11*, 157–169.

Kaiser, H. M., Riha, S. J., Wilks, D. S., Rossiter, D. G., & Sampath, R. (1993). A farm-level analysis of the economic and agronomic impacts of gradual climate change. *American Journal of Agricultural Economics*, *75*, 387–398.

Kramer, R. A., & Pope, R. D. (1981). Participation in farm commodity programs: A Stochastic dominance analysis. *American Journal of Agricultural Economics*, *63*, 119–128.

Levy, H., & Markowitz, H. (1979). Approximating expected utility by a function of mean and variance. *American Economic Review*, *69*, 308–317.

Markowitz, H. (1959). *Portfolio Selection: Efficient Diversifications of Investments*. New York, NY: John Wiley & Sons.

Nelson, W. W., & Straesser, A. (1985). Field day data, 1974–1983. University of Minnesota, *Southwest Experiment Station*, unpublished.

Rae, A. N. (1971a). An empirical application and evaluation of discrete stochastic programming in farm management. *American Journal of Agricultural Economics*, *53*, 625–638.

Rae, A. N. (1971b). Stochastic programming, utility, and sequential decision problems in farm management. *American Journal of Agricultural Economics*, *53*, 448–460.

Tauer, L. M. (1983). Target MOTAD. *American Journal of Agricultural Economics*, *65*, 606–610.

von Neumann, J., & Morgenstern, O. (1944). *Theory of Games and Economic Behavior*. Princeton, NJ: Princeton University Press.

Young, D. L. (1980, January 16–18). Evaluating procedures for computing objective risk from historical time series. Paper presented at annual meeting of Western Regional Research Project W-149, Tucson, Arizona.

EXERCISES

1. List three reasons why risk programming is a good choice for agricultural decision makers.

2. Given the following variance-covariance matrix, write out the expression for profit risk.

	Soybeans	Wheat	Corn
Soybeans	45.4	−75.0	−115.3
Wheat	−75.0	134.6	182.2
Corn	−115.3	182.2	2,400.5

3. Suppose we have data for the unit costs of four agricultural products from 1998 to 2008.

Year	Pork ($/lb)	Beef ($/lb)	Chicken ($/lb)	Duck ($/lb)
1998	1.20	2.10	1.50	2.81
1999	1.45	2.40	1.35	3.10
2000	1.35	2.67	1.40	3.50
2001	1.23	2.78	1.56	3.67
2002	1.40	2.34	1.58	3.30
2003	1.20	2.55	1.53	3.52
2004	1.50	2.65	1.38	3.68
2005	1.55	2.87	1.62	3.70
2006	1.70	2.44	1.68	3.71
2007	1.60	2.60	1.70	3.73
2008	1.68	2.62	1.65	3.53

Use Excel or a statistical package to derive the variance-covariance matrix from the data above.

4. A farmer is deciding between planting several vegetable (or fruit) crops on a 225-acre farm. Based on the annual profits per acre returns for lettuce, tomatoes, peppers, and cucumbers listed below, formulate a MOTAD problem and graph the E-TAD efficiency horizon.

Year	Lettuce	Tomatoes	Peppers	Cucumbers
		(annual profit per acre)		
2000	12.5	−2.5	−12.5	−125
2001	125	50	10	75
2002	25	37.5	7.5	25
2003	−62.5	25	−50	−125
2004	275	87.5	137.5	125
2005	−50	50	50	37.5
2006	25	10	2.5	200
2007	−100	12.5	30	−150
2008	187.5	62.5	−62.5	250

5. A farmer produces four kinds of crops: x_1, x_2, x_3, and x_4. The land and labor requirements are summarized below. Due to uncertainty in the weather, the farmer uses the last six years' average profit as the unit profit for each crop. Formulate the following problem as a MOTAD problem where the objective function is to minimize the total negative deviations from the mean with the expected profit constraint provided.

Resource	x_1	x_2	x_3	x_4	Resource Endowment
Land	1	1	1	1	200 Acres
Labor	25	36	27	87	10,000 Hours
Year 1 profit	292	−128	420	579	
Year 2 profit	179	560	187	639	
Year 3 profit	114	648	366	379	
Year 4 profit	247	544	249	924	
Year 5 profit	426	182	322	5	
Year 6 profit	259	850	159	569	
Average profit	253	443	284	516	

Deviations from the mean:

Dev x_1	Dev x_2	Dev x_3	Dev x_4
39	−571	136	63
−74	117	−97	123
−139	205	82	−137
−6	101	−35	408
173	−261	38	−511
6	407	−125	53

6. A speculator in the futures market for corn, wheat, and sugar would like to construct a marketing portfolio. Assume that the cost of each position is $20 (corn), $10 (wheat), and $12 (sugar) per share respectively. The investor has a total of $100,000 to invest. The investor has observed the following rates of return for each commodity over the past four years:

Year	Corn	Wheat	Sugar
1	−5%	10%	25%
2	15%	0%	12%
3	−2%	1%	2%
4	15%	2%	−30%

a. Compute the variance-covariance matrix for this problem.
b. Formulate this problem as a quadratic risk programming problem, where the objective function is to minimize the total variance-covariance matrix subject to a minimum expected return constraint, which should be parametrically varied.
c. Formulate a MOTAD model to maximize return (where the expected return is the four-year simple average).

7. A farmer produces corn, wheat, and soybeans using three resources: hired labor, family labor, and machine time. Over a three-month production period, the farm is endowed with 1,200 hours of hired labor, 800 hours of family labor, 2,000 hours of

machine time, and 1,000 acres of land. The resource requirements for each commodity are summarized below:

Resource	Soybeans	Wheat	Corn	Resource Endowment
		(Hours/unit of good)		
Hired Labor	1.0	1.1	1.3	1,200 hours
Family Labor	0.7	0.6	0.8	800 hours
Machine Time	2.2	2.8	3.0	2,000 hours
Land	1.0	1.0	1.0	1,000 acres

Assume that the per acre unit profit of the three commodities over the previous 10 periods is the following:

Observation	Soybeans	Wheat	Corn
1	420	186	400
2	390	198	200
3	194	194	300
4	188	220	90
5	182	222	400
6	170	240	380
7	184	200	150
8	180	242	50
9	172	254	420
10	160	258	384
Average	224	221	277

The variance-covariance matrix of expected profit for this example is:

	Soybeans	Wheat	Corn
Soybeans	8,318	−1,507	742
Wheat	−1,507	687	569
Corn	742	569	18,185

a. Formulate and solve the following QP problem using Solver: minimize risk subject to a minimum expected profit constraint and all the other structural constraints given in this example. Trace out an E-V frontier by parametrically varying the RHS value for the minimum expected profit constraint.

b. Formulate and solve the following QP problem using Solver: maximize expected profit subject to a maximum risk constraint and all the other structural constraints given in this example. Trace out an E-V frontier by parametrically varying the RHS value for the maximum risk constraint.

c. Formulate and solve the following QP problem using Solver: maximize expected profit minus the risk term times a risk aversion coefficient subject to all the structural constraints given in this example. Trace out an E-V frontier by parametrically varying the RHS value for the minimum expected profit constraint.

8. A feed dealer can purchase corn, soybeans, sorghum, and wheat that can be stored and sold to livestock farmers later in the year. Assume that in the past 10 years, the unit profits/losses ($) per bushel on the sale of each commodity are as follows:

Year	Corn	Soybeans	Sorghum	Wheat
1	−0.05	−0.01	0.05	−0.25
2	0.50	0.30	0.04	1.00
3	0.10	0.15	0.03	0.50
4	−0.25	−0.20	0.10	−0.50
5	0.55	0.35	0.15	−0.20
6	0.50	0.20	0.20	1.10
7	0.10	0.12	0.05	0.80
8	−0.40	0.01	0.04	−0.60
9	0.75	0.60	0.15	1.00
10	0.25	0.25	0.10	−0.50
Average	0.21	0.18	0.09	0.24

Variance-Covariance Matrix of Expected Unit Profit

	Corn	Soybeans	Sorghum	Wheat
Corn	0.139	0.067	0.012	0.159
Soybeans	0.067	0.049	0.005	0.078
Sorghum	0.012	0.005	0.003	0.008
Wheat	0.159	0.078	0.008	0.531

Assume the feed dealer can buy and store 500,000 bushels of each of the grains.

a. Formulate the LP problem that maximizes expected profit, where the expected profit is the average profit from the 10 years of observations.

b. Formulate Part a as a quadratic risk programming problem, where the objective function is to minimize the total variance-covariance matrix subject to a minimum expected return constraint. Use parametric programming and start off by setting the minimum expected return RHS value to the profit-maximizing solution found in part a.

c. Trace out the E-V frontier for this problem using parametric programming.

9. Use Excel or a statistical package to derive the variance-covariance matrix given in Exercise 8. To do this in Excel, use the DVARP function for the variance terms and the COVAR function for the covariance terms.

10. Solve Exercise 8 assuming the objective is to maximize expected profit subject to a maximum constraint on the variance-covariance matrix. Use parametric programming by first setting the RHS value for the risk constraint to a very large number, such as 999,999,999,999. Then, by using sensitivity analysis, systematically lower this number to trace out an E-V frontier.

11. Solve Exercise 8 assuming the objective is to maximize expected profit minus the variance-covariance matrix times a risk coefficient (b). Note that setting b = 0 gives the profit-maximizing (risk-neutral) solution. Trace out an E-V frontier by parametrically altering (b).

12. Climate change will impact agricultural yields, prices, profits, and other factors affecting this sector. Suppose that you are looking at the impact of climate change on corn

and soybean farming in the Upper Midwest. Working with agronomists and climatologists, you estimate the following series for a representative farm:

Observation	Corn	Soybeans
1	313	250
2	250	275
3	363	288
4	188	163
5	125	163
6	388	313
7	313	250
8	250	275
9	238	250
10	63	63
Average	249	229

Net revenue per acre with climate change		
1	263	350
2	289	275
3	302	425
4	175	225
5	188	200
6	328	426
7	263	344
8	289	275
9	263	261
10	66	75
Average	242	286

The resource requirements for each commodity are summarized below:

Resource	Soybeans	Corn	Endowment Resource
Hired Labor	1.0	1.3	1,200 Hours
Family Labor	0.7	0.8	800 Hours
Machine Time	2.2	3.0	2,000 Hours
Land	1.0	1.0	1,000 acres

Formulate two LP models that maximize expected average profit with and without a climate change. Is the farmer better off with or without climate change?

13. Reformulate Exercise 12 as two quadratic risk programming problems, with and without climate change. Assume the objective function is to minimize total risk as measured by the variance-covariance matrix subject to a minimum expected profit constraint. You will need to compute the variance-covariance matrix using Excel. Use the DVARP and COVAR functions. Set the RHS value to the corresponding profit-maximizing solutions for the two scenarios found in Exercise 12. Then, use parametric programming to trace out an E-V frontier for with and without climate change. Is the farmer better off with climate change under all pairs of E and V?

14. Reformulate Exercise 12 assuming the objective is to maximize expected profit subject to a maximum constraint on the variance-covariance matrix. Use parametric programming by first setting the RHS value for the risk constraint to a very large number, such as 999,999,999,999. Then, by using sensitivity analysis, systematically lower this number to trace out an E-V frontier. Compare it to the E-V frontier found in Exercise 7.

15. Solve Exercise 12 assuming the objective is to maximize expected profit minus the variance-covariance matrix times a risk coefficient (b). Note that setting b=0 gives the profit-maximizing (risk-neutral) solution. Trace out an E-V frontier by parametrically altering (b).

16. Formulate Exercise 7 as a MOTAD problem where the objective function is to minimize total absolute deviations from the mean and where there is a minimum expected profit constraint. Use parametric programming on the minimum expected profit constraint to trace out a set of E-TAD efficient farm plans. Graph E and the estimated standard deviation for TAD.

17. Formulate Exercise 7 as a MOTAD problem where the objective function is to minimize total negative deviations from the mean and where there is a minimum expected profit constraint. Use parametric programming on the minimum expected profit constraint to trace out a set of E-TND efficient farm plans. Graph E and the estimated standard deviation for TND.

18. Formulate Exercise 8 as a MOTAD problem where the objective function is to minimize total absolute deviations from the mean and where there is a minimum expected profit constraint. Use parametric programming on the minimum expected profit constraint to trace out a set of E-TAD efficient farm plans. Graph E and the estimated standard deviation for TAD.

19. Formulate Exercise 8 as a MOTAD problem where the objective function is to minimize total negative deviations from the mean and where there is a minimum expected profit constraint. Use parametric programming on the minimum expected profit constraint to trace out a set of E-TND efficient farm plans. Graph E and the estimated standard deviation for TND.

20. Formulate Exercise 12 as a MOTAD problem where the objective function is to minimize total absolute deviations from the mean and where there is a minimum expected profit constraint. Use parametric programming on the minimum expected profit constraint to trace out a set of E-TAD efficient farm plans. Graph E and the estimated standard deviation for TAD.

21. Formulate Exercise 12 as a MOTAD problem where the objective function is to minimize total negative deviations from the mean and where there is a minimum expected profit constraint. Use parametric programming on the minimum expected profit constraint to trace out a set of E-TND efficient farm plans. Graph E and the estimated standard deviation for TND.

22. Formulate Exercise 7 as a target MOTAD problem. Set T to a level that is less than the profit-maximizing solution and experiment with differing levels of λ. Analyze the resulting plans that are derived with varying levels of T and λ.

23. Formulate Exercise 8 as a target MOTAD problem. Set T to a level that is less than the profit-maximizing solution, and experiment with differing levels of λ. Analyze the resulting plans that are derived with varying levels of T and λ.

24. Formulate Exercise 12 as a target MOTAD problem. Set T to a level that is less than the profit-maximizing solution and experiment with differing levels of λ. Analyze the resulting plans that are derived with varying levels of T and λ.

25. A farmer can grow three crops on the 1,000 acres of land he owns: corn, soybeans, and wheat. In producing these three crops, the farmer has to do the following steps: (1) plowing (*pl*), (2) plant corn (*pc*), (3) plant soybeans (*ps*), (4) plant wheat (*pw*), (5) harvest corn (*hc*), (6) harvest soybeans (*hs*), (7) harvest wheat (*hw*), (8) sell the corn after harvest (*sc*), (9) sell the soybeans after harvest (*ss*), and (10) sell the wheat after harvest (*sw*). The farmer must plow the land prior to planting and must plant the crops prior to harvesting the crops. In addition to his land endowment of 1,000 acres, the farmer expects to have a total of 1,200 hours available to perform all of the above production operations. However, due to uncertain weather conditions, those 1,200 hours are not certain. His expectations regarding the labor requirements (hours per acre) and variable costs for each operation, as well as his expected price and yield (bushels per acre) at harvest for the three crops are presented below.

Operation	Labor Requirement (hours per acre)	Variable Cost ($/acre)	Crop Yields and Prices
Plowing (*pl*)	0.40	$4.00	
Plant Corn (*pc*)	0.39	$114.00	
Plant Soybeans (*ps*)	0.30	$80.00	
Plant Wheat (*pw*)	0.30	$78.00	
Harvest Corn (*hc*)	0.60	$48.00	$120.00
Harvest Soybeans (*hs*)	0.30	$17.00	$40.00
Harvest Wheat (*hw*)	0.28	$10.00	$70.00
Corn Price ($/bushel)			$2.90
Soybean Price ($/bushel)			$5.75
Wheat Price ($/bushel)			$3.00

Assume that the farmer's objective is to maximize net revenue from corn, soybean, and wheat production. Assume that the RHS value for the available field time (b = 1,200) is distributed normally with a mean of 1,200 and a standard deviation of 50. To find a value for the RHS parameter b, call it b', you can use the following relationship:

$$b' = E(b) - Z_\alpha \ \alpha = 1{,}200 - 50 \ Z_\alpha.$$

Formulate this problem using a chance-constrained programming model and use parametric programming on Z_α to derive solutions for this problem. Summarize the efficient plans for various levels of Z_α.

26. Solve Exercise 12 as a chance-constrained programming problem. Assume that the farmer's objective is to maximize net revenue from corn and soybean production under both climate change scenarios. Assume that the RHS value for the hired labor (b = 1,200) is distributed normally with a mean of 1,200 and a standard deviation of 200. To find a value for the RHS parameter b, call it b', you can use the following relationship:

$$b' = E(b) - Z_\alpha \ \alpha = 1{,}200 - 200 \ Z_\alpha.$$

Formulate this problem using a chance-constrained programming model and use parametric programming on Z_α to derive solutions for this problem. Summarize the efficient plans for various levels of Z_α.

11

Price Endogenous Mathematical Programming Models

All of the models examined thus far have been firm-level models, where it is assumed that firms are "price-takers." As a result both output and input prices have been treated as constants. This approach is consistent with the assumption of **perfect competition**, where there are so many sellers in the market that no one seller can influence output or input price levels by altering output levels. When moving from the individual firm level to the market level, which is composed of all sellers and buyers within some defined location, the assumption of constant price is no longer valid. At the market level, price is determined by the interaction of market supply (the collection of all individual firms' supply curves in the market) and market demand (the collection of all individual consumers' demand curves in the market). Consequently, if the goal is to model a market or sector rather than an individual firm, then a "price endogenous" or "sector programming" model is necessary. Price endogenous models are also necessary at the firm level if the firm has some degree of market power, because in such cases, the firm can influence price by altering its output.

The purpose of this chapter is to examine price endogenous mathematical programming models at both the market and firm levels. The chapter begins with an overview of the market under perfect competition. A simple price endogenous model is presented that features a quadratic objective function composed of consumer and producer surplus (social welfare), which when maximized yields a quantity and price solution that is equivalent to the market equilibrium values found by equating the market supply and demand functions. Next, the assumption of perfect competition is relaxed for both the output and input side of the market. Five price endogenous models are presented including (1) **monopoly** on the output side and **monopsony** on the input side, (2) monopoly on the output side and **perfect competition** on the input side, (3) perfect competition on the output side and monopsony on the input side, (4) perfect competition on the output side and perfect competition on the input side, and (5) a general formula that approximates any degree of market competition from perfect competition to monopoly in mathematical programming models. Recall from microeconomics that a monopolistic market features one seller and many buyers, a monopsonistic market features one buyer and many sellers, and a perfectly competitive market features many buyers and many sellers.

One of the most prominent uses of price endogenous mathematical programming models is **spatial equilibrium analysis**. Takayama and Judge's (1964a; 1964b) perfectly competitive spatial equilibrium model is presented, followed by a numerical example of their model. The model is similar to transportation models, except that demand and supply in each region are not fixed, but determined endogenously, as are optimal trade flows and prices.

Next, sector-level mathematical programming is extended to industry models that encompass multiple markets from the farm to retail levels. Industry models can accommodate multiple inputs and outputs as well. Finally, the chapter concludes with two research examples of price endogenous mathematical programming models. The first is applied to the U.S. dairy sector, which can be used for any degree of market competition, while the second examines the potential impacts of climate change on the entire U.S. agricultural sector.

11.1 THE MARKET UNDER PERFECT COMPETITION

Under perfect competition, all firms are price-takers, and the equilibrium price is determined at the market-level by the interaction of supply and demand. At the market level, demand is downward sloping, and supply is upward sloping, and equilibrium occurs at their intersection. To illustrate how to model this as a mathematical programming problem, consider the following simple example of the U.S. orange market.

Suppose that the market for oranges in the United States can be characterized by the following demand and supply functions:

$$q_d = 100 - 5p,$$

$$q_s = 25 + 10p,$$

where q_d is market quantity demand, p is market price, and q_s is market quantity supply. In applied analyses, q_d and q_s are often estimated using econometric techniques and market-level data. Assuming that the market is competitive, the market equilibrium can be solved by imposing the following condition for p:

$$q_d = q_s = q^*.$$

Substituting q^* for q_s and q_d, and solving for p^* yields:

$$p^* = 5.$$

Substituting $p^* = 5$ into either the q_d or q_s equation yields:

$$q^* = 75.$$

Hence, in equilibrium, the U.S. orange market would sell 75 units of oranges at a market price of $5 per unit.

This problem could also be solved by transforming the supply and demand equations into price inverse form. It is often more convenient in programming problems to express market supply and demand in inverse form. To do this, simply solve each equation for P:

$$p_d = 20 - 0.2q_d,$$

$$p_s = -2.5 + 0.1q_s.$$

To solve for the equilibrium, impose the equilibrium condition:

$$p_s = p_d = p^* \text{ and solve for } q^*,$$

$$q^* = 75,$$

$$p^* = 5,$$

which is the same solution as in the quantity-dependent original form of the problem.

This simple problem can also be solved as an endogenous mathematical programming problem by maximizing the sum of consumer and producer surplus,[1] which is sometimes called "social surplus" because it measures the welfare of both consumers and producers in the market. Samuelson (1952) and others have shown graphically that the sum of consumer and producer surplus is equivalent to the area between the demand and supply curves and to the left of their respective intersection.[2] The optimal solution to maximize social welfare is equivalent to the market equilibrium solution determined above. To provide a more general representation of this problem, assume that the inverse market demand and supply curves are linear:

$$p_d = a - bq_d, \text{ where a and b} > 0, \text{ and}$$

$$P_s = c + dq_s, \text{ where c and d} > 0.$$

Solving the following maximization problem yields the equilibrium q^*:

$$\text{Max: } Z = aq_d - 0.5bq_d^2 - cq_s - 0.5dq_s^2 \tag{0}$$

s.t.:

$$q_d \qquad\quad - q_s \le 0 \tag{1}$$

$$q_d, \qquad\quad q_s \ge 0 \tag{2}$$

In the objective function, the first term, $aq_d - 0.5bq_d^2$, is the area under the demand curve from 0 to q^* in Figure 11.1 (area A + B). The second term in the objective function, $cq_s + 0.5dq_s^2$, is the area under the supply curve from zero to q^* (area B). Therefore, the

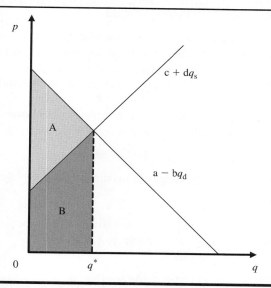

Figure 11.1 Graphical depiction of objective function areas in the maximization of social surplus (consumer plus producer surplus) problem.

---

[1] Consumer surplus, which is a measure of consumer welfare, is the area underneath the demand curve and above the equilibrium price. Producer surplus, which is a measure of producer welfare, is the area above the supply curve and below the equilibrium price.
[2] Samuelson (1952) also stated this as being equivalent to the area under the excess demand curve (demand minus supply), or the negative of the area under the excess supply curve (supply minus demand).

difference between the first and second terms in the objective function is the area between the demand and supply curves from 0 to q^*, which is the sum of consumer and producer surplus (area A).

The equilibrium p^* can be determined in two ways. First, simply substitute q^* into either the supply or demand equation to get p^*. Alternatively, p^* is equal to the shadow price (SP) of the demand and supply balancing constraint (1).

In the previous example where a = 20, b = 0.2, c = -2.5, and d = 0.1, the optimal solution to this problem yields $q_d = q_s = q^* = 75$, and the SP of the first constraint is 5, which is the same as p^*.

Why is the SP on the first constraint equal to the market equilibrium price? To answer that, suppose that the right-hand-side (RHS) value was increased from 0 to 1. This would mean that demand is permitted to be larger than supply by one unit. The value on that extra unit of demand to the consumer would be approximately equal to the market equilibrium price times 1 (the increase in quantity), or the market equilibrium price. Graphically, this is equivalent to extending a vertical line from the quantity axis one unit above the equilibrium quantity to the market demand curve. Its intersection with the demand curve would be very close to the market equilibrium price.

This simple example illustrates how endogenous mathematical programming can be applied to an economic sector to determine market equilibrium output price and quantity. This constrained optimization is a quadratic programming (QP) model, and was first developed by several economists including Enke (1951), Samuelson (1952), and Takayama and Judge (1964a; 1964b).

11.2 THE MARKET UNDER MONOPOLY/MONOPSONY AND IMPERFECT COMPETITION

The extreme opposites of perfect competition are monopoly and monopsony, which are markets characterized by a single seller or a single buyer. Under this form of market structure, the firm has complete control over its output price (monopoly) or input price (monopsony). To illustrate the market equilibrium in this case, consider the following example of a monopoly.

Assume that an agricultural cooperative has complete control over its members' output and price in the market. It faces the following inverse demand curve:

$$p_d = 100 - 2q_d \tag{11.1}$$

Assume the cooperative's sole objective is to maximize profit, which is equal to:

$$p_d q_d - w q_s \tag{11.2}$$

where p_d is price, q_d is quantity demanded, w is variable cost to the firm, and q_s is quantity supplied. Assume in this example that variable cost w = 10.

Substituting (11.1) into (11.2) yields:

$$(100 - 2q_d)q_d - 10q_s, \text{ or}$$

$$100q_d - 2q_d^2 - 10q_s, \text{ or assuming } q_d = q_s$$

$$90q_d - 2q_d^2 \tag{11.3}$$

To maximize profit, take the first derivative of (11.3), set it equal to 0, and solve for q_d^*:

$$90 - 4q_d = 0, \text{ or}$$

$$q_d^* = 22.5 \tag{11.4}$$

The equilibrium output price for the monopolist is found by substituting $q_d^* = 22.5$ into (11.1):

$$p_d^* = 100 - 2q_d^* = 55 \tag{11.5}$$

Maximum profit in this example is:

$$\Pi^* = (22.5)(55) - (10)(22.5) = 1{,}012.50 \tag{11.6}$$

This problem can be solved using price endogenous mathematical programming, as follows:

Max: $Z = 100q_d - 2q_d^2 - 10q_s$ (0)

s.t.:

$$q_d \quad\quad - q_s \le 0 \tag{1}$$

$$q_d, \quad\quad q_s \ge 0 \tag{2}$$

The solution yields $q_d^* = q_s^* = 22.5$, which can be substituted back into the demand equation to get the equilibrium price of $55.

More generally, consider the following mathematical programming problem:

Max: $Z = aq_d - bq_d^2 - cq_s - dq_s^2$ (0)

s.t.:

$$q_d \quad\quad - q_s \le 0 \tag{1}$$

$$q_d, \quad\quad q_s \ge 0 \tag{2}$$

where:

$$p_d = a - bq_d \text{ (output inverse demand equation), and}$$

$$p_s = c + dq_s \text{ (input inverse supply equation).}$$

The Kuhn–Tucker conditions to this problem can be found by forming the LaGrange function and differentiating it with respect to q_d, q_s, and λ:

$$L = aq_d - bq_d^2 - cq_s - dq_s^2 + \lambda(q_s - q_d) \tag{11.7}$$

$$L_{qd} = a - 2bq_d - \lambda \le 0,\ L_{qd}q_d = 0,\ q_d \ge 0 \tag{11.8}$$

$$L_{qs} = -c - 2dq_s + \lambda \le 0,\ L_{qs}q_s = 0,\ q_s \ge 0 \tag{11.9}$$

$$L_\lambda = q_s - q_d \ge 0,\ L_\lambda \lambda = 0,\ \lambda \ge 0 \tag{11.10}$$

Assuming an interior solution ($q_d > 0$ and $q_s > 0$), conditions (11.8) and (11.9) can be rewritten as equalities:

$$a - 2bq_d = \lambda, \text{ and} \tag{11.11}$$

$$c + 2dq_s = \lambda, \text{ or combining the two conditions:} \tag{11.12}$$

$$a - 2bq_d = c + 2dq_s \tag{11.13}$$

Condition (11.13) is the optimality condition from economic theory that marginal revenue equals marginal cost. Nelson and McCarl (1984) summarize four possible market solutions involving monopoly, monopsony, and perfectly competitive behavior on the output and input side:

Case 1. Monopoly on the output side and monopsony on the input side. In this case, the mathematical programming problem is:

$$\text{Max: } Z = (a - bq_d)q_d - (c + dq_s)q_s \tag{0}$$

s.t.:

$$q_d \qquad\qquad - q_s \leq 0 \tag{1}$$

$$q_d, \qquad\qquad q_s \geq 0 \tag{2}$$

Case 2. Monopoly on the output side and perfect competition on the input side. In this case, the mathematical programming problem is:

$$\text{Max: } Z = (a - bq_d)q_d - (c + 0.5dq_s)q_s \tag{0}$$

s.t.:

$$q_d \qquad\qquad - q_s \leq 0 \tag{1}$$

$$q_d, \qquad\qquad q_s \geq 0 \tag{2}$$

Note that the difference here is the 0.5 added to the slope term in the supply equation.

Case 3. Perfect competition on the output side and monopsony on the input side. In this case, the mathematical programming problem is:

$$\text{Max: } Z = (a - 0.5bq_d)q_d - (c + dq_s)q_s \tag{0}$$

s.t.:

$$q_d \qquad\qquad - q_s \leq 0 \tag{1}$$

$$q_d, \qquad\qquad q_s \geq 0 \tag{2}$$

Case 4. Perfect competition on the output and the input side. In this case, the mathematical programming problem is:

$$\text{Max: } Z = (a - 0.5bq_d)q_d - (c + 0.5dq_s)q_s \tag{0}$$

s.t.:

$$q_d \qquad\qquad - q_s \leq 0 \tag{1}$$

$$q_d, \qquad\qquad q_s \geq 0 \tag{2}$$

Finally, Nelson and McCarl (1984) demonstrate the most general model that encompasses any degree of market competition including monopoly, perfect competition, and everything in between. This is done by substituting the following expression for the 0.5 coefficient in the supply and demand terms of the objective function:

$$(n + 1)/2n,$$

$$(m + 1)/2m,$$

where n is the number of equally sized firms in the input market, and m is the number of equally sized firms in the output market. When n or m equal 1, this term equals 1, which is the monopoly/monopsony model. When n and m approach infinity, this term approaches 0.5, which is the perfect competition model. Finally, any number in between represents an imperfectly competitive market on a scale of 0.5 being perfect competition and 1.0 being monopoly. In this general case, the mathematical programming model is:

$$\text{Max: } Z = (a - [(m + 1)/2m]bq_d)q_d - (c + [(n + 1)/2n]dq_s)q_s \tag{0}$$

s.t.:

$$q_d \qquad\qquad - q_s \le 0 \tag{1}$$

$$q_d, \qquad\qquad q_s \ge 0 \tag{2}$$

11.3 SPATIAL EQUILIBRIUM MODELS

One of the principal applications of price endogenous sector programming models is spatial equilibrium analysis, which is concerned with the geographic alignment of market prices. Spatial price analysis is particularly important in the agricultural and food industry, where transportation costs make up a large component of food costs. The spatial equilibrium model is a partial equilibrium model of an economic sector that finds optimal trade flows among regions. Low-cost producing regions become exporters to higher-cost regions.

Takayama and Judge (1964a, 1964b) developed the following spatial equilibrium model that has become a popular model for analyzing spatial price analysis problems. Suppose that there are n regions and denote the inverse demand and supply functions for region i as:

$$p_{di} = f(q_{di}), \text{ and}$$

$$p_{si} = f(q_{si}).$$

In this model, the equilibrium solution is equivalent to the optimal solution of the following maximization problem, which Samuelson (1952) termed the "net social payoff" (NSP) from trade, which is the geometric area under each region's excess demand function (or excess supply function[3]) minus the sum of transportation costs for trade flows between all regions. This is equivalent to:

$$\text{Max: NSP } = \sum_{i=1}^{n}\left\{ \int_{0}^{q_{di}} p_{di} dq_{di} - \int_{0}^{q_{si}} p_{si}\, dq_{si} \right\} - \sum_{i=1}^{n}\sum_{j=1}^{n} c_{ij}t_{ij} \tag{0}$$

s.t.:

$$q_{di} - \qquad\qquad \sum_{j=1}^{n} t_{ij} \le 0 \ \text{ for all } i \tag{1}$$

$$-q_{si} + \qquad \sum_{j=1}^{n} t_{ij} \le 0 \ \text{ for all } i \tag{2}$$

$$q_{di}, \qquad q_{si}, \qquad\qquad t_{ij} \ge 0 \text{ for all } i \text{ and } j$$
$$(i = 1, \ldots, n, j = 1, \ldots, n) \tag{3}$$

where c_{ij} is unit transportation cost of shipping from region i to j, and t_{ij} is the level of shipments from region i to j.

---

[3]A region's excess demand function is defined as the region's demand minus the region's supply function. A region's excess supply function is defined as the region's supply minus the region's demand function.

The objective function is to maximize the NSP from trade among regions. Constraint (1) is identical to the transportation model that requires the sum of shipments from all regions to region i to be at least as large as region i's demand level. Constraint (2) limits total shipments from region i to not exceed its supply, and constraint (3) is the non-negativity restriction on the variables. The solution to this problem gives equilibrium trade between regions (t_{ij}^*), equilibrium consumption (q_{di}^*), and equilibrium supply (q_{si}^*).

Some interesting insights are gleaned from examining some of the Kuhn–Tucker first-order necessary conditions to this problem. The LaGrange function for this problem is:

$$L = \sum_{i=1}^{n}\left\{ \int p_{di}dq_{di} - \int p_{si}\,dq_{si} \right\} - \sum_{i=1}^{n}\sum_{j=1}^{n}c_{ij}t_{ij} + \sum_{i=1}^{n}\lambda_i\left(\sum_{j=1}^{n}t_{ij} - q_{di}\right) + \sum_{i=1}^{n}\theta_i\left(q_{si} - \sum_{j=1}^{n}t_{ij}\right) \quad (0)$$

The Kuhn–Tucker conditions are:

$$L_{qdi} = p_{di} - \lambda_i \le 0,\ (L_{qdi})q_{di} = 0,\ q_{di} \ge 0,\ i=1,\ldots,n \quad (1)$$

$$L_{qsi} = -p_{si} + \theta_i \le 0,\ (L_{qsi})q_{si} = 0,\ q_{si} \ge 0,\ i=1,\ldots,n \quad (2)$$

$$L_{tij} = -c_{ij} + \lambda_i - \theta_i \le 0,\ (L_{tij})t_{ij} = 0,\ t_{ij} \ge 0,\ i=1,\ldots,n,\ j=1,\ldots,n \quad (3)$$

$$L_{\lambda i} = \sum_{j=1}^{n}t_{ji} - q_{di} \ge 0,\ (L_{\lambda i})\lambda_i = 0,\ \lambda_i \ge 0,\ i=1,\ldots,n \quad (4)$$

$$L_{\theta i} = q_{si} - \sum_{j=1}^{n}t_{ij} \ge 0,\ (L_{\theta i})\theta_i = 0,\ \theta_i \ge 0,\ i=1,\ldots,n \quad (5)$$

The first condition implies that for any positive demand ($q_{di} > 0$) the SP (λ_i) of constraint (1) is equal to the demand price. Likewise, the second condition implies that for any positive supply ($q_{si} > 0$) the SP (θ_i) of constraint (2) is equal to the supply price. Hence, the solution to the dual problem gives the equilibrium supply prices (θ_i^*) and demand prices (λ_i^*). The third condition implies that the demand price in region j cannot exceed the supply price in region i plus the transportation cost, c_{ij}.

In the case of a linear demand ($f(q_{di}) = a_i - b_iq_{di}$) and supply function ($f(q_{si}) = c_i + d_iq_{si}$), the objective function for this problem is:

$$\text{Max: NSP} = \sum_{i=1}^{n}a_iq_{di} - 0.5b_iq_{di}^2 - c_iq_{si} - 0.5d_iq_{si}^2 - \sum_{i=1}^{n}\sum_{j=1}^{n}c_{ij}t_{ij} \quad (0)$$

An Example[4]

Consider the following numerical example involving four regions of the United States: North (n), South (s), East (e) and West (w). The inverse market demand functions for each region are:

$$p_{dn} = 300 - 0.5q_{dn}$$

$$p_{ds} = 275 - 1.0q_{ds}$$

---

[4]This problem, solution, and corresponding sensitivity analysis are shown in the Chapter 11 supplemental materials available at www.wiley.com/college/kaiser.

$$p_{de} = 200 - 0.3q_{de}$$

$$p_{dw} = 155 - 0.75q_{dw}$$

Suppose the four regions have the following inverse supply functions:

$$p_{sn} = 30 + 1.0q_{sn}$$

$$p_{ss} = 75 + 0.9q_{ss}$$

$$p_{se} = 20 + 0.8q_{se}$$

$$p_{sw} = 15 + 1.10q_{sw}$$

Further assume the following unit transportation costs across all regions:

	North	South	East	West
North	0	4	2	3
South	4	0	2	3
East	2	2	0	8
West	3	3	8	0

The objective function value, NSP, is 62,681. Demand levels in each of the regions are 278.4 in the North, 114.2 in the South, 137.1 in the East, and 0 in the West. Supply levels are 130.8 in the North, 95.4 in the South, 173.6 in the East, and 129.8 in the West. Hence, the North and South are net importers, while the East and West are net exporters. The East exports 36.6 units in total, exporting 36.5 to the North and 0.1 to the South. The West exports its entire supply of 129.8, shipping 111.1 to the North and 18.7 to the South.

The equilibrium demand prices are $160.81 in the North, $160.78 in the South, $158.87 in the East, and $157.83 in the West. Notice that the equilibrium demand prices are lowest in the exporting regions and highest in the importing regions. The equilibrium supply price in the East is approximately $2.00 lower than the demand prices in the North and South, which is equal to the transportation costs to these two regions from the East. The equilibrium supply price in the West is $3.00 lower than the demand price in the North and South, which is also approximately the transportation cost to those regions from the West.

It may not be realistic for the West to export all their supply and consume nothing. If that is the case, a minimum consumption constraint can be added for the West. For instance, by adding a minimum demand constraint of 75 units for the West, the objective function is reduced from 62,681 to 60,093. The new results indicate a different level of shipments, but are not reported here.

The above spatial equilibrium models assume perfect competition. However, some researchers have modified the model above for imperfectly competitive markets (e.g., Kawaguchi, et al., 1997; Nelson & McCarl, 1984). The first research application in this chapter summarizes the Kawaguchi et al. (1997) model for imperfectly competitive milk markets in the United States.

11.4   INDUSTRY MODELS

Often in agricultural-food problems, researchers are interested in modeling an entire industry from the farm-to-processing-to-retail markets. Such "industry" models are very useful for tracing out impacts from one market to the other. For instance, what happens to retail and processor prices if the farm price increases or decreases? Sector-level mathematical programming can be extended to industry models in a similar fashion as the previous applications.

Consider the following industry model. A farm commodity (F) can be used to make two food products: (1) Product 1 (P_1) and (2) Product 2 (P_2). Suppose the industry is composed of two markets: (1) farm market, where F is produced, and (2) a processing-retail market, where F is processed and sold as P_1 and P_2 directly to consumers.

Assume the inverse supply function for F is:

$$p_F = 0.5 + 0.004q_F.$$

Assume the inverse demand functions for P_1 and P_2 are:

$$p_{P1} = 0.75 - 0.009q_{P1},$$

$$p_{P2} = 0.90 - 0.004q_{P2}.$$

One unit of the farm input produces 8 units of P_1 or 10 units of P_2 ignoring processing costs. Hence, the production functions for converting F into P_1 and P_2 are:

$$q_{P1} = 8q_F,$$

$$q_{P2} = 10q_F.$$

Assume the producers in this manufacturing market are price-takers in the input and output markets. The objective function for this problem is:

$$\text{Max: } Z = (0.75 - (0.5)(0.009)q_{P1})q_{P1} + (0.90 - (0.5)(0.004)q_{P2})q_{P2}$$
$$- (0.5 + (0.5)(0.004)q_F)q_F.$$

The farm input needs to be converted into Product 1 and Product 2 for this problem, and this is accomplished by the following constraint:

$$1/8q_{P1} + 1/10q_{P2} - q_F \leq 0.$$

11.5   RESEARCH APPLICATION: A SPATIAL EQUILIBRIUM MODEL FOR IMPERFECTLY COMPETITIVE MILK MARKETS

Spatial equilibrium models have been frequently used to analyze interregional competition problems in agriculture, including regional competition issues associated with the dairy industries in the United States (e.g., Chavas, et al., 1994; McDowell, 1982; Yavuz et al., 1996), as well as other countries such as Japan (e.g., Sasaki, 1969; Kobayashi, 1983; Hayashi, 1984). These models are a class of nonlinear programming (NLP) known as QP, which feature a quadratic objective function and a linear constraint set. Originally developed by Enke (1951) and Samuelson (1952) and then refined by Takayama and Judge (1964a; 1964b), spatial price equilibrium models have assumed that markets are either perfectly competitive or monopolistic. However, the structure of dairy markets in most countries is often neither. Therefore, a more plausible model for analyzing interregional milk movements would be a spatial imperfect competition equilibrium model. Accordingly, the purpose of the research of Kawaguchi et al. (1997) was to develop a generalization of Takayama and Judge's (1964a; 1964b) spatial equilibrium model that allows for the incorporation of any degree of market structure from perfect competition to monopoly. The usefulness of the model was demonstrated by applying it to interregional milk movements in the Japanese dairy industry and comparing the solutions for alternative scenarios regarding the degree of market competition.

Dairy policy in Japan features a quota system in the manufacturing milk market to prevent excess milk production from occurring because of higher-than-competitive market prices. As a result, the Japanese dairy industry can be divided into three distinct markets: fluid milk market, manufacturing market within-payment quotas, and manufacturing market over-payment quotas. Prices in the manufacturing markets are set by the government based on a deficiency payment program. For manufacturing milk sold within-payment quotas, prefectural milk marketing boards (the consignment milk sellers for farmers) receive deficiency payments equal to the difference between the guaranteed price and the standard transaction price for manufacturing milk. Both prices are determined by the national government: the guaranteed price is based on milk production costs, while the standard transaction price is based on dairy product market conditions, and all buyers of manufacturing milk are required to pay this price. To discourage excess production, over-payment quota manufacturing milk receives the lower standard transaction price. Payment quotas for the guaranteed price are not given to individual producers, but to each prefectural milk marketing board. Individual producers are paid the prefecture-wide uniform pooled price (weighted average prices for milk sold in the fluid and manufacturing milk markets).

Given manufacturing milk prices determined by the government, discriminated price formation for fluid milk occurs through negotiations between each prefectural milk-marketing board and the processors it supplies. Since the fluid milk market is more price inelastic than the manufacturing milk market, the fluid market has higher prices. The structure of the Japanese milk market includes an oligopolistic group of consignment milk sellers (prefectural milk marketing boards) who allocate milk to maximize sales revenue, and a large number of perfectly competitive producers who receive pooled returns (blend prices). We refer to this situation as a "dual structure" because dairy farmers are perfectly competitive in producing milk, but they are oligopolistic in selling it through their milk marketing boards. Previous spatial price equilibrium models have not accounted for this "dual structure" in the Japanese milk market.

Nonlinear Price Endogenous Programming Model

Consider n milk producing and consuming regions with the geographical scope of producing region i the same as consuming region i. In each consuming region, there are three administratively different markets: fluid milk market (fmm_i), manufacturing milk market within-payment quota (wpq_i), and manufacturing milk market over-payment quota (opq_i). Unit transportation cost for shipping raw milk from producing region i to consuming region j (t_{ij}) is assumed to be the same for both fluid and manufacturing milk.[5]

Buyers of fluid milk in each consuming region are assumed to behave as price-takers, which is reasonable considering the many fluid processors in Japan. Within-payment quota milk is traded at the fixed guaranteed price, fp_1, and the quantity of milk is limited to the fixed-payment quota. Over-payment quota milk is traded at the lower fixed standard transaction price, fp_2, and it is assumed that the demand for this milk is perfectly elastic. It is also assumed that each region has a linear marginal raw milk cost function and a linear fluid demand function, with all functions known by all agents (or consignment sellers).

---

[5]Unlike cooperatives in the United States, Japanese cooperatives only have a small share of the milk manufacturing market and function primarily as raw milk shippers who negotiate a price for their farmer members. Therefore, the focus of our model is on transactions between raw milk shippers (cooperatives) and manufacturing companies. Consequently, it is realistic to assume the same transportation costs for raw milk being shipped for fluid and manufacturing product processing.

Milk producers in region i consign their annual milk supply, fs_i, to agent i. Agent i's role is to allocate farmers' milk among the 3n markets to maximize sales revenues net of transportation costs. The following notation is used based on the variables described above:

d_j = quantity of milk demanded in fluid market j (j = 1, 2, ..., n)

fs_i = quantity of raw milk supplied and consigned in region i (i = 1, 2, ..., n)

ps_i = marginal revenue net of transportation costs for each market for region i (i = 1, 2, ..., n)

x_{ij} = quantity of raw milk shipped from region i to market j (i = 1, 2, ..., n; j = 1, 2, ..., 3n)

$x_{i(n+j)}$ = quantity of raw milk shipped from region i to the manufacturing milk market within-payment quotas (wpq$_j$) (i = 1, 2, ..., n; j = 1, 2, ..., n)

$x_{i(2n+j)}$ = quantity of raw milk shipped from region i to the manufacturing milk market over-payment quotas (opq$_j$) (i = 1, 2, ..., n; j = 1, 2, ..., n)

pd_j = demand price in the fluid market j (j = 1, 2, ..., n)

ppp_i = producer's pooled (blend) price in region i (i = 1, 2, ..., n)

$d_j = \alpha_j - \beta_j pd_j$ = demand function in fluid market j (j = 1, 2, ..., n)

$fs_i = -v_i + \eta_i ppp_i$ = marginal cost function for raw milk in region i (i = 1, 2, ..., n), where ppp_i means marginal cost

t_{ij} = unit transportation cost of shipping raw milk from producing region i to consuming region j (i = 1, 2, ..., n; j = 1, 2, ..., 3n)

q_i = limited quantity (payment-quota) paid the differences between the guaranteed price (fp$_1$) and the standard transaction price (fp$_2$) (i = 1, 2, ..., n)

sp_j = SP of the right to sell a unit of milk in the manufacturing milk market within-payment quotas (wpq$_j$) (i = 1, 2, ..., n)

R_i = total milk sales revenue net of transportation costs in region i (i = 1, 2, ..., n)

Using the above notation, agent i's milk sales revenue maximization problem net of transportation costs can be expressed as:

$$\text{Max: } R_i = \sum_{j=1}^{n} pd_j x_{ij} + \sum_{j=1}^{n} fp_1 x_{i(n+j)} + \sum_{j=1}^{n} fp_2 x_{i(2n+j)} - \sum_{j=1}^{3n} t_{ij} x_{ij} \qquad (1)$$

Total revenue maximization problem for all n agents is expressed as:

$$\text{Max: } \sum_{i=1}^{n} R_i \qquad (2)$$

Agent i's fluid sales revenue in market j ($pd_j x_{ij}$) can be written as:

$$pd_j x_{ij} = [\alpha_j/\beta_j - (1/\beta_j) d_j] x_{ij}$$

$$= \left[\alpha_j/\beta_j - (1/\beta_j)(\sum_{i=1}^{n} x_{ij})\right] x_{ij}$$

$$= \left[\alpha_j/\beta_j - (1/\beta_j)(\sum_{m \neq i}\sum x_{mj} + x_{ij})\right] x_{ij} \qquad (3)$$

where m (m $\neq$ i) indicates all agents other than i. When agent i believes that a change in his fluid supply to market j will cause changes in all other agents' fluid supply to market j, agent i's "perceived" marginal fluid revenue in market j is:

$$\partial(pd_j x_{ij})/\partial x_{ij} = [\alpha_j/\beta_j - (1/\beta_j)d_j] - (1/\beta_j)(\partial \sum_{m \neq i} x_{mj}/\partial x_{ij} + 1)x_{ij}$$
$$= pd_j - (1/\beta_j)(r_{ij} + 1)x_{ij} \tag{4}$$

where r_{ij} is agent i's conjectural variation regarding changes in all other agents' fluid supply to market j caused by a change in agent i's supply.

Using the relationship (4), the total revenue maximization problem for all n agents can be respecified as the following NSP maximization problem adjusted for imperfectly competitive markets (ANSP):

$$\text{Max: ANSP} = \sum_{j=1}^{n} \int [\alpha_j/\beta_j - (1/\beta_j)d_j] dd_j + \sum_{j=1}^{n}\sum_{i=1}^{n} fp_1 x_{i(n+j)} + \sum_{j=1}^{n}\sum_{i=1}^{n} fp_2 x_{i(2n+j)}$$
$$- \sum_{j=1}^{n}\sum_{i=1}^{n}(1/\beta_j)(r_{ij}+1)\int x_{ij} dx_{ij} - \sum_{j=1}^{3n}\sum_{i=1}^{n} t_{ij} x_{ij} \tag{5}$$

s.t.:
$$d_j \leq \sum_{i=1}^{n} x_{ij}, \quad \text{for all j} \tag{6}$$

$$\sum_{i=1}^{n} x_{i(n+j)} \leq q_i, \quad \text{for all j} \tag{7}$$

$$\sum_{j=1}^{3n} x_{ij} \leq fs_i, \quad \text{for all i} \tag{8}$$

$$d_j \geq 0, x_{ij} \geq 0, \quad \text{for all i and j} \tag{9}$$

The difference between ANSP in (5) and the NSP in the conventional spatial competitive equilibrium model by Takayama and Judge (1964a; 1964b) is the term:

$$-\sum_{j=1}^{n}\sum_{i=1}^{n}(1/\beta_j)(r_{ij}+1)\int x_{ij} dx_{ij}.$$

When the market is perfectly competitive ($r_{ij} = -1$), the term is zero and (5) is equal to the original Takayama and Judge (1964a; 1964b) model. When Cournot–Nash behavior is assumed ($r_{ij} = 0$), the term is equivalent to:

$$-\sum_{j=1}^{n}\sum_{i=1}^{n}(1/\beta_j)\int x_{ij} dx_{ij},$$

which is shown in Hashimoto's (1985) spatial Nash equilibrium model. Cournot–Nash behavior means that agent i believes that the other agents will not change their supply in response to the agent's action.

Using the LaGrange function (L) with the multipliers, λ, ω, and θ for the constraints (6), (7), and (8), respectively, the Kuhn–Tucker optimality conditions for the maximization problem can be expressed as follows:

$$\partial L/\partial d_j = \alpha_j/\beta_j - (1/\beta_j)d_j - \lambda_j \leq 0, d_j(\partial L/\partial d_j) = 0, \quad \text{for all j} \tag{10}$$

$$\partial L/\partial x_{ij} = -(1/\beta_j)(r_{ij}+1)x_{ij} - t_{ij} + \lambda_j - \theta_i \leq 0, x_{ij}(\partial L/\partial x_{ij}) = 0, \quad \text{for all i and j} \tag{11}$$

$$\partial L/\partial x_{i(n+j)} = fp_1 - t_{ij} - \omega_j - \theta_j \le 0, \ x_{i(n+j)}(\partial L/\partial x_{i(n+j)}) = 0, \quad \text{for all i and j} \tag{12}$$

$$\partial L/\partial x_{i(2n+j)} = fp_2 - t_{ij} - \theta_i \le 0, \ x_{i(2n+j)}(\partial L/\partial x_{i(2n+j)}) = 0, \quad \text{for all i and j} \tag{13}$$

$$-\partial L/\partial \lambda_j = d_j - \sum_{i=1}^{n} x_{ij} \le 0, \lambda_j(\partial L/\partial \lambda_j) = 0, \quad \text{for all j} \tag{14}$$

$$-\partial L/\partial \omega_j = \sum_{i=1}^{n} x_{i(n+j)} - q_j \le 0, \ \omega_j(\partial L/\partial \omega_j) = 0, \quad \text{for all j} \tag{15}$$

$$-\partial L/\partial \theta_i = \sum_{j=1}^{3n} x_{ij} - fs_i \le 0, \ \theta_i(\partial L/\partial \theta_i) = 0, \quad \text{for all i} \tag{16}$$

The LaGrange multipliers (or dual variables), λ, ω, and θ, measure the fluid demand price (pd_j), the SP for the right to sell milk in the within-payment quota manufacturing market (sp_j), and marginal revenue net of transportation costs for each market (ps_i), respectively. The Kuhn–Tucker conditions, represented by (11), (12), and (13), indicate that each agent must equalize marginal revenue net of transportation costs across all markets where it sells milk. The equilibrium values can be calculated by the QP model solution.

The term $(1/\beta_j)(r_{ij} + 1)x_{ij}$ in (11) indicates the difference between the fluid demand price and agent i's marginal revenue in market j. The greater the degree of market power by agents, the larger this difference. For example, in the case of perfect competition, the term becomes zero because $r_{ij} = -1$. On the other hand, the term becomes $(1/\beta_j)x_{ij}$ when Cournot–Nash behavior ($r_{ij} = 0$) is assumed. In this research, Cournot–Nash behavior is assumed to illustrate the imperfect competition solution, and coalition among agents is treated as follows. To illustrate, consider Cournot–Nash agent 1 whose "perceived" marginal revenue in fluid market j is $pd_j - (1/\beta_j)x_{1j}$. If agent 1 forms a coalition with agent 2, then marginal revenue for agent 1's and agent 2's coalition is $pd_j - (1/\beta_j)(x_{1j} + x_{2j})$. In the case of monopoly where agent 1 forms a coalition with all other agents, marginal revenue for agent 1 is $pd_j - (1/\beta_j)(\sum_{i=1}^{n} x_{ij})$. Because any agent can sell the consigned milk individually or in coalition with other agents, as a price-taker or according to Cournot–Nash behavior, many combinations of agents' marketing behavior can be simulated.

To complete the model, individual farmers' milk supply needs to be incorporated. Unlike the oligopolistic marketing behavior of agents, individual farmers' milk production is competitively determined. Producers in region i, as price-takers, determine their supply given the producer pooled price. That is, their production level is determined by equating marginal cost to the producer pooled price. Thus,

$$ppp_i = R_i/fs_i \quad \text{for all i,} \tag{17}$$

$$fs_i = -v_i + \eta_i ppp_i \quad \text{for all i.} \tag{18}$$

In the comparative-static equilibrium, fs_i in (18) must be equal to fs_i given in the above milk sales maximization problem. To solve the model, the following iterative solution process is used to find equilibrium values for fs_i.

First, the QP model is used to generate equilibrium fluid milk prices and equilibrium quantities of milk shipments in the sales maximization problem expressed by (5) to (16), based on initial values for fs_i and given patterns of behavior of agents in the oligopolistic milk market. Second, producer pooled prices are calculated in (17). Third, new values of fs_i for the next iteration are computed based on the calculated producer pooled prices and marginal cost functions of producing regions, and the assumption that producers behave as price-takers in (18). Finally, the QP problem is solved again with new parameter values for fs_i to obtain new equilibrium fluid milk prices and quantities of milk shipments. This iteration process is continued until values for fs_i become stationary.

This model is applied to the Kyushu area of Japan as a case study. Region 1 includes Fukuoka, Saga, and Nagasaki prefectures, region 2 is the Kumamoto prefecture, region 3 is the Oita prefecture, and region 4 includes Miyazaki and Kagoshima prefectures.

Based on the long-run price elasticity of Kyushu milk supply by Ito (1989) (0.429), the Kyushu fluid demand price elasticity by Suzuki and Kobayashi (1993) (-0.77), and the regional price and quantity observations in Table 11.1, linear marginal cost and fluid milk demand functions for each region are specified as follows:

$$fs_1 = 135.162 + 0.967ppp_1, \qquad d_1 = 361.434 - 1.438pd_1,$$

$$fs_2 = 118.078 + 0.832ppp_2, \qquad d_2 = 181.071 - 0.666pd_2,$$

$$fs_3 = 43.490 + 0.293ppp_3, \qquad d_3 = 88.146 - 0.324pd_3,$$

$$fs_4 = 119.121 + 0.874ppp_4, \qquad d_4 = 163.371 - 0.639pd_4,$$

where: fs_i and d_j are measured by tons (in thousands), and ppp_i and pd_j are measured by yen per kilogram. Unit transportation costs, t_{ij}, are:

$$t_{12} = t_{21} = ¥4.58/kg,$$

$$t_{13} = t_{31} = ¥3.95/kg,$$

$$t_{14} = t_{41} = ¥7.80/kg,$$

$$t_{23} = t_{32} = ¥4.71/kg,$$

$$t_{24} = t_{42} = ¥6.11/kg,$$

$$t_{34} = t_{43} = ¥6.00/kg.$$

Table 11.1   Observations in 1989 (unit: 1,000 tons and ¥/kg)

| From/To | Fluid Milk Market | | | | Manufacturing Milk Market | | | | | | | | Total |
| | | | | | Within Quota | | | | Over Quota | | | | |
	1	2	3	4	1	2	3	4	1	2	3	4	
1	128.4	19.1	12.0	1.7	34.0	0	0	0	10.9	0	0	0	206.1
2	33.1	74.7	1.5	1.6	0	32.9	0	0	0	0	0	0	143.8
3	31.4	0	34.3	0	0	0	8.1	0	0	0	1.2	0	75.0
4	11.3	8.5	2.0	89.0	0	0	0	39.1	0	0	0	8.3	158.2
Total	204.2	102.3	49.8	92.3	34.0	32.9	8.1	39.1	10.9	0	1.2	8.3	583.1

Region (i or j)	Fluid Milk Price (pd_j)	Producer's Pooled Price (ppp_i)
1	109.35	101.62
2	118.22	103.21
3	118.22	107.75
4	111.20	99.07
Average	112.75	102.91

Because little milk is traded between Kyushu and other regions of Japan, this milk is treated as exogenous to simplify the model. Payment quotas q_i for the four regions are $q_1 = 34.0$, $q_2 = 32.9$, $q_3 = 8.1$, and $q_4 = 39.1$ thousand tons. The fixed guaranteed price for within-payment quota is $fp_1 = ¥79.83/kg$, and the fixed standard transaction price for over-payment quota is $fp_2 = ¥67.25/kg$.

Results

To demonstrate how solutions vary based on the assumption of market structure, the model is solved for perfect competition, monopoly, and imperfect competition scenarios. To represent the perfectly competitive solution, the model is solved assuming that the four agents are all price-takers. For the monopoly solution, the model is solved with the assumption that there is a coalition of four agents. To represent imperfect competition, 15 separate combinations of price-takers and Cournot–Nash players are solved. In the first case, the four agents are all individual Cournot–Nash players (Cournot–Nash equilibrium). In the next four cases, one agent is a price-taker, and the other three are individual Cournot–Nash players, thereby creating four combinations of market structure. For cases 6 to 11, two agents are price-takers, and the other two are individual Cournot–Nash players, thereby creating six new combinations of market structure. Finally, in the last four cases, three agents are price-takers, and the other agent is a Cournot–Nash player, thereby creating four combinations. Although there are other combinations with coalitions, they are not analyzed since the purpose here is to simply demonstrate examples of imperfectly competitive solutions.

The "dual structure" spatial perfect competition solution is shown in Table 11.2. In this case, virtually all raw milk is allocated to the fluid market, except for a trivial

Table 11.2   "Dual-Structure" Spatial Perfect Competition Equilibrium (unit: 1,000 tons and ¥/kg)

| | | | | | Manufacturing Milk Market | | | | | | | | |
| | | Fluid Milk Market | | | Within Quota | | | | Over Quota | | | | |
From/To	1	2	3	4	1	2	3	4	1	2	3	4	Total
1	193.2	0	0	0	0	0	0	0	0	0	0	0	193.2
2	4.8	125.8	0	0	0	0	0	0	0	0	0	0	130.6
3	7.0	0	61.0	0	0	0	0	0	0	0	0	0	68.0
4	30.5	0	0	112.4	0	0	0	1.4	0	0	0	0	144.3
Total	235.5	125.8	61.0	112.4	0	0	0	1.4	0	0	0	0	536.1

Region (i or j)	Fluid Milk Price (pd_j)	Agent's Marginal Revenue[a] (ps_i)	Producer's Pooled Price[b] (ppp_i)
1	87.63	87.63	88.23
2	83.05	83.05	87.19
3	83.68	83.68	83.68
4	79.83	79.83	83.32
Average	83.55	83.55	85.61

[a]ps_i is agent i's "perceived" marginal revenue (net of transportation costs) equalized in each market (marginal revenue = market price in perfect competition).
[b]Exogenously given milk shipments from each region to the outside of Kyushu are taken into account in calculating ppp_i.

amount shipped to the within-payment quota manufacturing milk market in region 4. Also, there is only a small amount of interregional shipments of fluid milk, mostly to region 1. The amount of milk allocated to the fluid market in the perfect competition solution is substantially higher than the actual amount allocated (see Table 11.1). This is due to the assumption that agents act as price-takers, which results in equality of price across markets net of transportation costs instead of equality of perceived marginal revenue across markets net of transportation costs. Consequently, fluid milk prices and producer pooled prices in the perfect competition case are much lower than actual levels.

The "dual structure" spatial monopoly solution is shown in Table 11.3. In this case, the allocation of raw milk to the fluid market is about one-half of the amount allocated under perfect competition and is also less than actual levels (Table 11.1). Instead, the monopoly solution allocates significant amounts of raw milk to the within-payment and over-payment quota manufacturing milk markets. The model predicts no interregional shipment of milk in all three markets. Because the demand for fluid milk is inelastic, restricting allocations to the fluid milk market results in higher pooled returns to farmers. In fact, producer pooled prices under monopoly are 30% higher than in the perfect competition case, as well as 10 percent higher than actual prices. It should be noted that the monopoly distribution of pooled returns to farmers is based on the assumption that the differences in producer pooled prices among regions are the same as the differentials generated in the perfect competition solution. Alternatively, one national producer pooled price for all regions could have been allocated. It should also be noted that total milk supply is largest in monopoly equilibrium under the "dual structure." Unless agents have power to control supply, individual producers increase milk supply as higher blend prices are given. Consequently, real monopoly rents cannot be realized under the "dual structure."

Table 11.3 "Dual-Structure" Spatial Monopoly Equilibrium (unit: 1,000 tons and ¥/kg)

| | Fluid Milk Market | | | | Manufacturing Milk Market | | | | | | | | |
| | | | | | Within Quota | | | | Over Quota | | | | |
From/To	1	2	3	4	1	2	3	4	1	2	3	4	Total
1	132.4	0	0	0	34.0	0	0	0	51.8	0	0	0	218.2
2	0	68.1	0	0	0	32.9	0	0	0	50.9	0	0	151.9
3	0	0	33.2	0	0	0	8.1	0	0	0	34.3	0	75.6
4	0	0	0	60.2	0	0	0	39.1	0	0	0	67.5	166.8
Total	132.4	68.1	33.2	60.2	34.0	32.9	8.1	39.1	51.8	50.9	34.3	67.5	612.5

Region (i or j)	Fluid Milk Price (pd_j)	Agent's Marginal Revenue (ps_i)	Producer's Pooled Price[a] (ppp_i)
1	159.30	67.25	114.03
2	169.56	67.25	112.99
3	169.65	67.25	109.48
4	161.46	67.25	109.13
Average	164.99	67.25	111.41

[a]Estimated ppp differentials in the perfect competition equilibrium are used to allocate monopoly pooled returns and to calculate ppp_i of each region.

Table 11.4   "Dual-Structure" Spatial Cournot–Nash Equilibrium (unit: 1,000 tons and ¥/kg)

| From/To | Fluid Milk Market | | | | Manufacturing Milk Market | | | | | | | | Total |
| | | | | | Within Quota | | | | Over Quota | | | | |
	1	2	3	4	1	2	3	4	1	2	3	4	
1	64.6	29.5	14.5	24.7	34.0	6.6	8.1	0	20.5	0	0	0	202.5
2	51.4	29.5	12.8	22.9	0	26.3	0	0	0	0	0	0	142.9
3	33.5	17.6	10.1	14.6	0	0	0	0	0	0	0	0	75.8
4	50.6	27.2	13.2	28.5	0	0	0	39.1	0	0	0	0	158.6
Total	200.1	103.8	50.6	90.7	34.0	32.9	8.1	39.1	20.5	0	0	0	579.8

Region (i or j)	Fluid Milk Price (pd_j)	Agent's Marginal Revenue (ps_i)	Producer's Pooled Price (ppp_i)
1	112.18	67.25	97.94
2	116.10	71.83	102.09
3	115.99	84.96	109.91
4	113.76	69.20	99.70
Average	114.51	73.31	102.41

The Cournot–Nash equilibrium is shown in Table 11.4. The regional fluid milk and producer pooled prices in this solution are the closest to actual prices for the four regions (Table 11.1). Not surprisingly, the allocation of raw milk among the three markets in this case is somewhere between the perfect competition and monopoly cases. Unlike the two previous cases, however, the Cournot–Nash equilibrium solution results in the same two regions shipping milk to each other: for instance, region 2 ships 51,400 tons of fluid milk to region 1, and region 1 ships 29,500 tons of fluid milk to region 2. While these shipping patterns are unintuitive, they do occur in reality as shown in Table 11.1. The other two spatial competition models did not predict these interregional milk shipment patterns. This suggests that the current complicated interregional milk movements may be caused by imperfectly competitive behavior.

To explain why the actual situation in Japan conforms more closely to the Cournot–Nash solution than the monopoly or perfectly competitive solutions, the market power of the prefectural milk marketing boards should be examined. Each prefectural milk marketing board controls total milk supplied in the prefecture, and therefore has some market power, particularly within the prefecture. However, at the national level, the prefectural milk marketing boards compete with one another, which lessens the market power of each marketing board. This suggests that the prefectural marketing boards are neither price-takers nor pure monopolists, but rather behave at some intermediate level between the two market power extremes. Therefore, it is reasonable that the actual situation conforms more with the Cournot–Nash solution than other solutions.

Compared with the other imperfect competition cases where at least one region is assumed to be a price-taker (an example is given in Table 11.5), fluid and producer pooled prices in the Cournot–Nash equilibrium solution in Table 11.4 are closer to actual prices. Price-takers' returns tend to be greater than Cournot–Nash players' when both price-takers and Cournot–Nash players exist as shown in Table 11.5. This is because Cournot–Nash agents try to keep fluid milk prices higher based on their "perceived" marginal revenues,

Table 11.5 "Dual-Structure" Spatial Equilibrium in the Case Where Agent 1 Is a Price-Taker and the Others Are Individual Cournot–Nash Players (unit: 1,000 tons and ¥/kg)

| From/To | Fluid Milk Market | | | | Manufacturing Milk Market | | | | | | | | Total |
| | | | | | Within Quota | | | | Over Quota | | | | |
	1	2	3	4	1	2	3	4	1	2	3	4	
1	107.9	47.8	24.1	28.1	0	0	0	0	0	0	0	0	207.9
2	36.0	22.8	9.4	20.0	16.3	32.9	0	0	0	0	0	0	137.4
3	30.6	16.7	9.5	17.3	0	0	0	0	0	0	0	0	74.1
4	36.0	20.9	10.0	26.0	13.0	0	8.1	39.1	0	0	0	0	153.1
Total	210.5	108.2	53.0	91.4	29.3	32.9	8.1	39.1	0	0	0	0	572.5

Region i or j	Fluid Milk Price pd_j	Agent's Marginal Revenue ps_i	Producer's Pooled Price ppp_i
1	104.89	104.89	103.45
2	109.47	75.25	95.45
3	108.84	79.68	104.16
4	112.69	72.03	93.37
Average	108.97	82.96	99.11

and price-takers obtain benefits by moving their milk to the fluid milk markets. In this case, acting as a price-taker is like "cheating" in a cartel agreement.

11.6 RESEARCH APPLICATION: CLIMATE CHANGE AND U.S. AGRICULTURE

A study by Adams et al. (1990) was one of the first comprehensive analyses of potential agronomic and economic impacts of climate change on the U.S. agricultural sector. The authors combined the results of two climate scenarios generated from global circulation models with agronomic crop yield models. The yield results were then used in a mathematical programming model of the U.S. agricultural sector.

The two climate models simulated the impact of a doubling of atmospheric concentrations of carbon dioxide from industrial revolution levels on daily temperatures and precipitation levels for various representative locations in the United States. The Goddard Institute for Space Studies (GISS) model predicted that a doubling of carbon dioxide in the atmosphere would increase mean winter and summer temperatures by 5.46°C and 3.50°C and mean winter and summer precipitation by 0.13 mm and 0.24 mm per day. The Princeton Geophysical Fluid Dynamics Laboratory (GFDL) model predicted that a doubling of carbon dioxide in the atmosphere would be more severe with an increase in mean winter and summer temperatures by 5.25°C and 4.95°C and an increase in mean winter precipitation of 0.19 mm and a decrease in mean summer precipitation of 0.08 mm per day.

The temperature and precipitation changes due to climate change were used as model inputs in three crop yield models: SOYGRO for soybeans, CERES-Maize for corn, and CERES-Wheat for wheat. These models simulated crop yields based on soil characteristics, water availability (including irrigation), temperature, precipitation, and solar radiation. The yields of other crops such as cotton, barley, sorghum rice, and alfalfa were adjusted by the average yield changes of the modeled crops. Because a doubling of carbon dioxide would also have some beneficial effects on crop yields (i.e., the CO_2 "fertilizer effect"), the authors assumed a 35%, 25%, and 10% increase in photosynthesis rates for soybeans, wheat, and corn, respectively. The average crop yields in most regions were predicted to increase in the milder GISS scenario, with the CO_2 fertilizer effect more than offsetting the warmer climate effects. However, the hotter and drier GFDL scenario generally resulted in lower crop yields for most regions, even with the CO_2 fertilizer effect. Not surprisingly, rain-fed crop yields were both lower and more variable than irrigated crop yields.

The authors also simulated the impact of the two climate scenarios on irrigation water use in the United States. The authors found considerable increases in irrigated crop water use, especially in the Southeast and Delta States particularly under the GFDL scenario. Small declines were predicted for the Southern Plains and for other regions in the wetter GISS scenario.

The authors used a price endogenous spatial equilibrium model of the U.S. agricultural sector similar to the ones presented in this chapter in order to simulate the economic impacts of the two climate change scenarios. The objective function of the model maximized NSP, and it was assumed that agricultural markets are perfectly competitive. The outputs of the model included consumer and producer welfare, equilibrium prices, quantities supplied and demanded, agricultural exports and imports, and food processing. Most crops (irrigated and non-irrigated) and livestock produced in the United States were included in the model. A total of 1,683 primary (farm) and secondary (processing) activities were represented in the model.

The model divided the United States into 64 geographic regions based on resource endowments. The model was then aggregated into 10 larger regions on the basis of land, labor, and water supplies. Water supply was of interest in this study since climate change will have a major impact on available water for agricultural uses. Both irrigated and non-irrigated crops were included in the model.

The model offered a comparative static depiction of the equilibrium of the U.S. agricultural sector. In other words, the model compared two or more market equilibrium scenarios, which in this case involved a baseline scenario and the two climate scenarios. The baseline scenario was based on no climate change and market conditions in 1982. The two climate change scenarios used the crop yield and irrigated water availability and requirements results from the agronomic models along with the GISS and GFDL climate change forecasts. Hence, the comparison among the three scenarios could be thought of as long-term market equilibriums for these three different climate situations.

Table 11.6 provides a broad summary of U.S. economic results in terms of changes in prices and quantities. The economic impacts were vastly different for each climate change scenario. In the mildly warmer and wetter GISS scenario, consumer surplus, producer surplus, and social welfare increased by $9.30 billion, $1.59 billion, and $10.89 billion, respectively. Under this climate scenario, real crop and livestock prices decreased by almost 20%, and crop and livestock outputs increased by 9% and 6%, respectively. The

Table 11.6   Agricultural Commodity Price and Quantity Bases for Climate Change Scenarios (base = 1.00)

Climate Model	Field Crop Price	Field Crop Quantity	Livestock Price	Livestock Quantity
GISS	0.83	1.09	0.84	1.06
GFDL	1.34	0.80	1.08	0.98

output gains were mainly increased crop yields caused by the warmer, but wetter climate forecasted by GISS.

On the other hand, under the more severe (hotter and drier) GFDL scenario, consumer surplus and social welfare decreased by $13.89 billion and $10.33 billion, respectively, while producer surplus increased by $3.55 billion. In this scenario, crop yields decreased substantially, even with the CO_2 fertilizer effect, and costs increased due to increased irrigation of crops. Overall, field crop output fell by 20% and livestock output declined by 2%. Consumers were the group most severely affected under this scenario because of the substantial price increase for food. Producers actually gained under the severe climate scenario because the production decreases were more than offset by price increases for both crops and livestock. The average increase in field crop prices was 34%, while livestock price increased 8% in this climate scenario.

The authors found that climate change results in lower crop acreage in the United States. In the GISS scenario, yields are higher, and therefore total acreage declines. In the GFDL scenario, acreage is reduced because of shifts in cropping patterns, that is, some land and/or resources are no longer productive for agricultural output, and hence are removed. This is particularly true of non-irrigated cropland. A general pattern for crops is a shift north and northwest. These regional changes in agricultural production will have major environmental implications for regions in terms of ground water and soil quality, as well as quality of wildlife habitats. Since crop production under climate change will favor irrigation, this change will also have a major impact on water availability and uses in most regions.

The authors made several overall conclusions based on the model results for the two climate scenarios. First, especially with respect to the more adverse climate scenario, climate change may imply a major reduction in the role of the United States as a major agricultural exporter. Second, climate change may result in major shifts in regional agricultural production in the United States, which would be due to major irrigation requirements. Third, there are important environmental concerns for any major changes in agricultural land use as a result of climate change impacts on water availability for irrigated crop agriculture, especially in the GFDL scenario. Fourth, climate change does not appear to present a major food insecurity problem for the United States. In the GISS scenario, agricultural output actually increases, while in the GFDL scenario output is reduced, and prices rise, but much of the loss is borne upon foreign consumers since the United States is a major agricultural exporter.

One major shortcoming of the analysis is that the authors did not include any technological innovations or major adaptations to climate change. In reality, climate change will be more gradual in nature than the comparative static model presented in this study. Even under the most rapid climate change scenario, the agricultural sector will have some chance at adapting, especially in developed countries like the United States. New plant cultivars will be developed to take advantage of warmer climates, such as longer growing conditions.

While these estimates of gains and losses are relatively large from a sector perspective, for instance, the $10 billion figure was about 8% of the value of the 1982 crop and livestock sector total, they are relatively small compared with actual Gross Domestic Product in the United States.

SUMMARY

In this chapter, we examined price endogenous mathematical programming models at both the market and firm levels. The chapter began with a price endogenous model for a perfectly competitive market with linear supply and demand functions. The model included a quadratic objective function composed of consumer and producer surplus (social welfare), which when maximized yields a quantity and price solution that is equivalent to the market equilibrium values found by equating the market supply and demand functions.

Five price endogenous models for imperfectly competitive markets were then presented. The first assumed monopoly on the output side and monopsony on the input side. The second assumed monopoly on the output side and perfect competition on the input side. The third assumed perfect competition on the output side and monopsony on the input side. The fourth assumed perfect competition on the output side and perfect competition on the input side. Finally, a general formula that approximates any degree of market competition from perfect competition to monopoly in mathematical programming models was presented.

Takayama and Judge's (1964a; 1964b) perfectly competitive spatial equilibrium model was presented. The model is similar to transportation models, except that demand and supply in each region is not fixed, but determined endogenously, as are optimal trade flows and prices.

Next, sector-level mathematical programming was extended to industry models that encompass multiple markets from the farm to retail levels. Industry models can accommodate multiple inputs and outputs as well.

Finally, the chapter concluded with two research examples of price endogenous mathematical programming models. The first was applied to the Japanese dairy sector, while the second examined the potential impacts of climate change on the entire U.S. agricultural sector.

REFERENCES

Adams, R. M., Rosenzweig, C., Peart, R. M., Ritchie, J. T., McCarl, B. A., Glyer, J. D., Curry, R. B., Jones, J. W., Boote, K. J., & Allen, L. H., Jr. (1990). Global climate change and U.S. agriculture. *Nature, 345*, 219–224.

Chavas, J. P., Cox, T. L., & Jesse, E. V. (1994, May). *Regional impacts of reducing dairy price supports and removing milk marketing orders in the U.S. dairy sector*, staff paper *367*, University of Wisconsin, Department of Agricultural Economics.

Enke, S. (1951, January). Equilibrium among spatially separated markets: Solution by electric analogue. *Econometrica, 19,* 40–47.

Hashimoto, H. (1985). A spatial Nash equilibrium model. In *Spatial Price Equilibrium: Advances in Theory, Computation and Application*, ed. P. T. Harker, 20–40. New York, NY: Springer-Verlag.

Hayashi, M. (1984). Market adjustment of milk in Japan. In *Excessive Supply and Market Adjustment of Agricultural Products*, ed. K. Tsuchiya, 145–165. Tokyo: Norin-Tokei-Kyokai.

Ito, F. (1989). *A spatial equilibrium analysis on regional milk trade*, Snow Brand Research Institute Report No. *54*. Sapporo, Japan.

Judge, G. G., & Takayama, T. (1973). *Studies in economic planning over space and time.* Amsterdam: North-Holland.

Kawaguchi, T., Suzuki, N., & Kaiser, H. M. (1997). A spatial equilibrium model for imperfectly competitive milk markets. *American Journal of Agricultural Economics, 79*, 851–159.

Kobayashi, K. (1983). *Prices and Market Adjustment of Milk.* Tokyo: Taimei-do.

McDowell, F. H. (1982). Domestic dairy marketing policy: An interregional trade approach. Doctoral dissertation, University of Minnesota, unpublished.

Nelson, C. H., & McCarl, B. (1984). Including imperfect competition in spatial equilibrium models. *Canadian Journal of Agricultural Economics, 32*, 55–70.

Samuelson, P. A. (1952, June). Spatial price equilibrium and LP. *American Economic Review, 42*, 283–303.

Sasaki, K. (1969, December). Spatial equilibrium in eastern Japan milk market. *Journal of Rural Economics, 41*, 106–116.

Suzuki, N., & Kobayashi, K. (1993, March). Equilibrium in interregional milk transportation. *Journal of Rural Economics, 64*, 221–232.

Takayama, T., & Judge, G .G. (1964a, October). Equilibrium among spatially separated markets: A reformulation. *Econometrica, 32*, 510–524.

Takayama, T., & Judge, G. G. (1964b, February). Spatial equilibrium and quadratic programming. *Journal of Farm Economics, 46,* 67–93.

Yavuz, F., Zulauf, C., Schnitkey, G., & Miranda, M. (1996, October 18). A spatial equilibrium analysis of regional structural change in the U.S. dairy industry. *Review of Agricultural Economics,* 693–703.

EXERCISES

1. Suppose that the U.S. market for apples is characterized by the following supply and demand functions:

$$q_s = 6p_s,$$
$$q_d = 100 - 10p_d.$$

   a. Solve for the equilibrium price and quantity.
   b. Put the supply and demand equations in the price-inverse form. Re-solve the equilibrium price and quantity and verify that the values are the same as in part a.
   c. Solve for the equilibrium values using Solver to maximize the difference between consumer and producer surplus.
   d. Verify that the SP from the Solver sensitivity analysis is the same as the equilibrium price found in part a.

2. The market for avocados is depicted by the following market supply and demand functions:

$$p_s = 10 + 5q_s,$$
$$p_d = 100 - 1q_d.$$

   a. Solve this problem by hand for the equilibrium price and quantity.
   b. Solve for the equilibrium values using Solver to maximize the difference between consumer and producer surplus.

3. Consider the following inverse demand and supply functions for a market:

$$q_d = 50 - 5p_d,$$
$$q_s = 20 + 10p_s.$$

   Find the equilibrium price and quantity by hand and then by using Solver to maximize the difference between consumer and producer surplus.

4. Consider the following inverse demand and supply functions for a market:

$$p_d = 50 - 2q_d,$$
$$p_s = 100 + q_s.$$

   Use Solver to solve the case when the market is a monopoly on the output-side and a monopsony on the input-side.

5. Consider the following inverse demand and supply functions for a market:

$$q_d = 500 - 5p_d,$$
$$q_s = 50 + 10p_s.$$

   Solve this problem as a monopolist on the output-side and monopsonist on the input-side using Solver.

6. Solve the following problem by hand and with Solver:

$$q_s = 2p_s,$$

$$q_d = 100 - 25p_d.$$

7. Solve the following problem using Solver and verify by hand:

$$p_d = 100 - 15q_d,$$

$$p_s = 10q_s.$$

8. Consider the following supply and demand functions:

$$q_s = -360 + 40p_s,$$

$$q_d = 240 - 16p_d.$$

Solve for the equilibrium price and quantity using Solver.

9. Why are spatial equilibrium models particularly applicable for agricultural and food industry analysis?

10. Find the equilibrium quantity and price by solving the following supply and demand system algebraically and by using Solver to maximize the difference between consumer and producer surplus.

$$p_d = 1,000 - 100q_d,$$

$$p_s = 25q_s.$$

11. Solve the following problem by hand and with Solver:

$$p_d = 100 - 15q_d,$$

$$p_s = 25 + 20q_s.$$

12. Consider the following market supply and demand functions for onions:

$$q_s = 2.5p_s,$$

$$q_d = 100 - 0.75p_d.$$

a. Solve this problem for the equilibrium price and quantity.
b. Put the supply and demand equations in the price-inverse form. Re-solve the equilibrium price and quantity and verify that the values are the same as in part a.
c. Solve this problem using Solver to maximize the difference between consumer and producer surplus.
d. Verify that the SP from the Solver sensitivity analysis is the same as the equilibrium price found in part a.

13. Compute the equilibrium values for the following problem using Solver:

$$p_d = 1,000 - 10q_d,$$

$$p_s = 200 + 40q_s.$$

14. A regional feed dealer has an effective monopoly on the feed he sells to local farmers. Assume he faces the following demand curve and average variable cost for his product:

$$p_d = 100 - 15q_d,$$

$$AC = 10q_s.$$

Solve for the profit maximizing output ($q_s = q_d = q^*$) and price (p^*) by hand.

15. Solve the monopoly problem in Exercise 14 using Solver and the follow functional form:

$$\text{Max: } Z = (a - bq_d)q_d - (c + dq_s)q_s \tag{0}$$

s.t.:

$$q_d \qquad - q_s \leq 0 \tag{1}$$

$$q_d, \qquad q_s \geq 0 \tag{2}$$

16. A monopsonist milk buyer buys milk from a lot of dairy farmers, which makes it a perfect competition-monopsonist market situation. Suppose that the demand and supply curves are:

$$p_d = 250 - 2.5q_d,$$

$$p_s = 10 + 1q_s.$$

Solve for the equilibrium p and q values using Solver and the mathematical programming formula given in this chapter.

17. Consider the following supply and demand functions:

$$q_s = 25 + 1p_s,$$

$$q_d = 100 - 0.5p_d.$$

Solve this problem as a monopolist on the output-side and monopsonist on the input-side using Solver.

18. Solve Exercise 17 as a monopolist on the output-side and as perfect competition on the input-side using Solver.

19. Solve Exercise 17 as a perfect competition on the output-side and a monopsonist on the input-side using Solver.

20. Solve Exercise 17 as a perfect competition on the output-side and a perfect competition on the input-side using Solver. Which of the previous four solutions result in the highest and lowest output price and highest and lowest input price?

21. Consider the following supply and demand functions:

$$q_s = 100 + 0.5p_s,$$

$$q_d = 1,500 - 0.25p_d.$$

Suppose that there are four equally sized firms in the market and 100,000 consumers. Solve this problem using Solver with the following formula for market competition:

$$(n + 1)/2n, \text{ and}$$

$$(m + 1)/2m.$$

22. Suppose that the market for cheese has the following supply and demand functions:

$$p_s = 10q_s,$$

$$p_d = 2,000 - 5q_d,$$

There are four equally sized firms in the market, and two large buyers of the firms' cheese. Solve this price endogenous, imperfectly competitive problem using Solver.

23. Consider the following spatial example involving trade and the following five countries: U.S. (u), Japan (j), Canada (c), England (e), and Russia (r). The inverse market demand functions for each country are:

$$p_{du} = 300 - 1q_{du}$$
$$p_{dj} = 275 - 1q_{dj}$$
$$p_{dc} = 200 - 1q_{dc}$$
$$p_{de} = 155 - 1q_{de}$$
$$p_{dr} = 220 - 1q_{dr}$$

Suppose the five countries have the following inverse supply functions:

$$p_{su} = 30 + 1q_{su}$$
$$p_{sj} = 75 + 1q_{sj}$$
$$p_{sc} = 20 + 1q_{sc}$$
$$p_{se} = 15 + 1q_{se}$$
$$p_{sr} = 45 + 1q_{sr}$$

Further assume the following unit transportation costs across all countries:

	u	j	c	e	r
u	0	8	1	4	5
j	8	0	9	10	6
c	1	9	0	3	4
e	4	10	3	0	2
r	5	6	4	2	0

Write the spatial equilibrium mathematical programming model corresponding to this problem assuming perfect competition.

24. Solve Exercise 23 using Solver.

25. Summarize the optimal solution to Exercise 23 using a map with the equilibrium trade flows, production, demand, and prices.

26. Consider Exercise 23 with the following modification. The United States decides to implement an import quota that restricts total imports into the United States to no more than 50 units. Modify the problem to account for this policy and solve the new problem using Solver. By how much is social welfare reduced due to this trade restriction by the United States?

27. Reconsider Exercise 23 with the following modification. England decides to implement an export quota that restricts the total exports from England to not exceed 50 units. Modify the exercise to account for this policy and solve the new problem using Solver. By how much is social welfare reduced due to this trade restriction by England?

28. Reconsider Exercise 23 with the following modification. England decides to implement a fixed-rate export tariff of $10 per unit of exports. Modify the exercise to account for this policy and solve the new problem using Solver. By how much is social welfare reduced due to this trade restriction by England? By how much are exports reduced from England?

29. Reconsider Exercise 23 with the following modification. England decides to implement a fixed-rate export subsidy of $3 per unit of exports. Modify the exercise to account for this policy and solve the new problem using Solver. By how much is social welfare changed due to this trade policy by England? By how much are exports increased from England?

12

Goal Programming

One important, and sometimes limiting, assumption of mathematical programming models is that the objective or goal of the decision maker is optimizing a single objective, such as maximize profits or minimize costs. In reality, individuals and institutions usually have multiple objectives. For example, an agricultural producer is not only concerned with maximizing profits, but is also interested in maximizing the probability of staying in business, maintaining worker morale, increasing the size of the business, and promoting good environmental stewardship. Some of these objectives may in fact be in conflict with maximizing profits. Likewise, in environmental and natural resource problems, such as pollution abatement, managing fisheries, and managing forest, there are often multiple competing goals such as environmental quality goals and economic growth goals.

To address the single objective limitations of mathematical programming models, Charnes et al. (1955) developed **goal programming (GP)**, which is a technique that relaxes the sole objective assumption. Under this approach, the analyst can specify multiple goals or targets for the decision maker and minimize the deviations from not achieving each goal. Both linear and nonlinear programming models can incorporate multiple goals using this approach, but the majority of GP problems have been linear.

Goal programming has been used extensively in environmental, natural resource, and agricultural economics as a planning tool for forestry management, land use planning, pollution mitigation, and farm planning. In this chapter, the usefulness of GP is illustrated by an example of a parasite control program and an example of forest protection.

The objectives of this chapter are to:

1. Provide an overview of the concepts behind GP, including how to set up and solve such problems.

2. Present several illustrations in agricultural and resource economics of how a linear programming (LP) model can be extended to have multiple objectives rather than a single objective.

3. Provide two research examples of GP to illustrate its usefulness in decision analysis.

12.1 GOAL PROGRAMMING

Goal programming was developed in order to model situations where the decision maker has multiple objectives. Goal programming is similar to other programming models in that it usually has constraints on resources controlled by the decision maker. However, GP also specifies certain goals or **targets** of the decision maker such that all goals cannot be satisfied simultaneously. Goal programming solutions then provide the optimal solution that comes closest to achieving all goals.

The basic idea of GP is to specify a numeric target value for each goal and then formulate a model that minimizes the weighted sum of the unwanted deviations from each goal. There are two types of GP: **nonpreemptive** and **preemptive** (also referred to as **lexicographic**). Nonpreemptive GP should be used when the decision maker has multiple goals, and the goals can be weighted "cardinally" by preferences: for instance, goal 1 is twice as important as goal 2. Cardinally ordered preferences imply the decision maker knows not only which goals are preferred, but also by how much each goal is preferred. For example, as described in Chapter 7, land conservation officials in Maryland gave a parcel's measured ecological and habit value three times more value than its proximity to other protected lands, while giving a parcel's size a value twice that of the proximity to other protected lands (Messer, 2006).

Preemptive GP is used when the decision maker has a clear hierarchy of priority levels as goals, where the first tier goal is substantially more important than the second tier goal, and so on. This is sometimes referred to as lexicographic preferences; for instance, a farmer's ultimate goal is to have more profit, but if there is a strategy that yields the same level of profit, then the farmer's next goal is to choose the strategy that is least harmful to the environment. The following examples illustrate nonpreemptive and preemptive GP using LP.

12.2 NONPREEMPTIVE GOAL PROBLEM

Suppose a semi-retired farmer has four goals: (1) spend 43 hours per week with his family, (2) work enough to earn $2,000 per week, (3) spend 15 hours per week volunteering at the local food pantry, which supplies free food to low-income families, and (4) have 70 hours of sleep per week. Let:

x_1 = hours spent with family per week

x_2 = hours worked per week (wage = $40/hour)

x_3 = hours spent at the food pantry per week

x_4 = hours spent sleeping

Goal 1: Spend 43 hours per week with family

Goal 2: Make $2,000 per week from working

Goal 3: Spend 15 hours per week at the food pantry

Goal 4: Spend 70 hours per week on sleep

The four goals in this problem are expressed in terms of **target values** (43 hours of leisure, $2,000 of income, 15 hours volunteering, and 70 hours sleeping). Target values are included in all GP problems. Rather than maximizing or minimizing a single goal, the objectives are expressed in terms of reaching or coming as close as possible to the desired level for each goal.

To formulate a GP problem, every similar goal must be expressed in the same unit. In this example, if goal 2 is converted from income into hours, then all four goals will be stated in terms of hours. Assume that the farmer makes $40 per hour, which means goal 2 can be restated as spending 50 hours ($2,000/$40=50) per week on work. Since there is a total of 168 hours each week, it is clear that the farmer cannot achieve all four goals in this problem because they sum to 178 hours. Hence, it is clear that all goals will not be met.

The first goal could be written as the following constraint:

$$x_1 = 43 \tag{12.1}$$

However, if goal 1 is expressed in this manner, then goal 1 would have to be met, which may not be possible given the other three goals. To remedy this, deviation variables are used. Let:

d_1^+ = number of hours of family time above the desired 43 hours,

d_1^- = number of hours of family time under the desired 43 hours.

Both d_1^+ and d_1^- must be non-negative. Does this sound familiar? This is similar to the positive and negative deviation variables introduced in Chapter 10 when the MOTAD model was presented.

Constraint (12.1) can be restated as:

$$x_1 = 43 + d_1^+ - d_1^-, \text{ or}$$

$$x_1 - d_1^+ + d_1^- = 43 \tag{12.2}$$

For example, suppose $x_1 = 53$. Then $d_1^+ = 10$ and $d_1^- = 0$.

Likewise, the second goal could be written as:

$$40x_2 = 2,000 \tag{12.3}$$

Recall that in order to get the goals or target levels in the same units, the right-hand side (RHS) of the second constraint should be transformed into hours required to earn $2,000. Dividing both sides of (12.3) by $40.00 per hour yields:

$$x_2 = 50 \tag{12.4}$$

Again, if goal 2 is expressed in this manner, then goal 2 would have to be met, which may not be possible given the other goals. Let:

d_2^+ = number of hours of work above the targeted 50 hours

d_2^- = number of hours of work under the targeted 50 hours

Then, the second constraint becomes:

$$x_2 = 50 + d_2^+ - d_2^-, \text{ or}$$

$$x_2 - d_2^+ + d_2^- = 50 \tag{12.5}$$

For the third goal, let:

d_3^+ = number of hours spent at the food pantry above the desired 15 hours

d_3^- = number of hours spent at the food pantry under the desired 15 hours

In the model, this goal can be accounted for by the following constraint:

$$x_3 = 15 + d_3^+ - d_3^-, \text{ or}$$

$$x_3 - d_3^+ + d_3^- = 15 \tag{12.6}$$

For the fourth goal, let:

$$d_4^+ = \text{number of hours spent sleeping above the desired 70 hours}$$

$$d_4^- = \text{number of hours spent sleeping under the desired 70 hours}$$

In the model, this goal can be accounted for by the following constraint:

$$x_4 = 70 + d_4^+ - d_4^-, \text{ or}$$

$$x_4 - d_4^+ + d_4^- = 70 \tag{12.7}$$

Finally, a last constraint is necessary to assure that total hours spent on the various activities for the week do not exceed 168 hours:

$$x_1 + x_2 + x_3 + x_4 \leq 168 \tag{12.8}$$

Thus, there are eight structural constraints and 12 decision variables for this problem.

It is clear that given 168 hours available per week, all four goals of the farmer cannot be met. In order to determine the objective function, create a weighting scheme for the deviation variables. A simple weighting scheme would be to assign a 0 value to the positive deviations and a value of 1 to the negative deviations and then minimize the total deviations, that is:

$$\text{Min: } Z = \sum_{i=1}^{4} \left(0\,d_i^+ + 1d_i^-\right).$$

The reason for assigning a 0 coefficient for the positive deviations is that there is no penalty associated with over-achieving a goal.[1] The reason for assigning a coefficient of 1 on the negative deviations is that there is a penalty associated with under-achieving a goal. By assigning a 1 to the negative deviations for the four goals, it is assumed that the farmer gives equal importance to achieving all goals. On the other hand, if one goal is more important than the other goals, then a different weighting scheme could be used.

The nonpreemptive GP problem is:

$$\text{Min: } Z = \sum_{i=1}^{4} \left(0\,d_i^+ + 1d_i^-\right) \tag{0}$$

s.t.:

$$1x_1 - 1d_1^+ + 1d_1^- = 43 \tag{1}$$

$$1x_2 - 1d_2^+ + 1d_2^- = 50 \tag{2}$$

$$1x_3 - 1d_3^+ + 1d_3^- = 15 \tag{3}$$

$$1x_4 - 1d_4^+ + 1d_4^- = 70 \tag{4}$$

$$\sum_{i=1}^{4} x_i \leq 168 \tag{5}$$

$$x_i, d_i^+, d_i^- \geq 0 \quad i=1,\ldots,4 \tag{6}$$

---

[1]For goals where over-achievement is bad and under-achievement is good, such as lower risk or lower pollution, one should assign a 0 coefficient to the negative deviation and a positive coefficient such as 1 to the positive deviation.

The optimal solution to this problem is:

$$x_1^* = 43, \ x_2^* = 50, \ x_3^* = 15, \ x_4^* = 60, \ d_1^{+*} = 0, \ d_1^{-*} = 0, \ d_2^{+*} = 0, \ d_2^{-*} = 0,$$
$$d_3^{+*} = 0, \ d_3^{-*} = 0, \ d_4^{+*} = 0, \ d_4^{-*} = 10.$$

According to the solution, the farmer can meet the target levels of the first three goals of family time, income, and volunteer service, but will get 10 hours less sleep per week than desired. This result is based on equal weights. Suppose that the farmer values the sleeping goal twice as much as the other goals, and values the volunteer goal one-half as much as goals 1 and 2. Then the objective function of the problem becomes:

$$\text{Min: } Z = \sum_{i=1}^{4} 0 d_i^+ + 1 d_1^- + 1 d_2^- + 0.5 d_3^- + 2 d_4^-.$$

In nonpreemptive GP, the objective function, Z, is sometimes referred to as the **achievement function**, since it gives a numerical value to the goals that are unmet. The new solution to this problem achieves goals 1, 2, and 4, but under-achieves goal 3, the volunteer service, by 10 hours.

There are three characteristics of this problem:

1. Each goal appears as a separate constraint with the right-hand-side (RHS) value reflecting the target level of the goal.

2. Positive and negative deviation variables are included for each goal to reflect over-achievement and under-achievement of the goal.

3. The objective function requires minimization of the weighted sum of deviation variables, where the weights represent the relative preferences for achieving each goal. The non-negative weights are applied to the undesirable deviations, that is, weighting positive deviations for goals where under-achievement is preferred and weighting negative deviations where over-achievement is preferred.

This problem is fairly trivial since nonpreemptive GP is not necessary to determine the optimal solution to this problem. That is, if the farmer prefers the nonwork activities to income, then the nonwork goals would be achieved at the expense of the income goal, and vice versa. However, this is a very simplistic GP problem, and in more complicated problems, the optimal solutions may not be as clear without the use of non-preemptive GP.

12.3   PREEMPTIVE GOAL PROGRAMMING

In many decision problems, goals are lexicographic in nature. In these cases, preemptive GP should be used, as it is the same as nonpreemptive GP except that substantially higher weights are placed on under-achieving the most important goals.

Consider the example of a centrally planned developing country that is currently formulating its agricultural plan for the next several years.[2] The Minister of Agriculture has under her control 200,000 acres of agricultural land to be put into production, and the government has several goals. First, the country is in dire need of foreign currency, and therefore, the Minister would like to export food to other countries in exchange for currency. Second, in order to develop the industrial sector of the economy, the Minister knows that a significant amount of food that is grown on this land is needed to feed industrial workers

---

[2] This problem, solution, and corresponding sensitivity analysis are shown in the Chapter 12 supplemental materials available at www.wiley.com/college/kaiser.

in the urban areas of the country. Finally, the Minister also has environmental concerns about water pollution caused by agricultural production and wants to limit this because it has negative consequences for the nation's drinking water supply. Assume there are four basic food commodities to be grown on this acreage: rice, beans, wheat, and maize. Assume the following parameters apply to this problem:

Item	Rice	Beans	Wheat	Maize	Endowment
Net Revenue—Domestic Sales	400	600	300	500	
Net Revenue—Foreign Sales	500	650	350	450	
Yield (bushels/acre)	50	25	75	100	
Land (per acre)	1	1	1	1	200,000
Labor (hours/acre)	15	20	8	25	4,000,000
Water pollution (contaminants/acre)	10	1	2	5	

Suppose that the Minister has a hierarchy of preferences for attaining the goals: goal 1 (gaining foreign currency) is the most important goal, goal 2 (domestic production to feed workers) is the next most important goal, and goal 3 (environmental quality) is the least important. So there are three levels of priorities in this case. Assume the Minister has the following target levels for the three goals:

Goal 1: at least $40 million in foreign sales

Goal 2: at least 13 million bushels of rice, beans, wheat, and maize for domestic sales

Goal 3: no more than 1 million contaminants of pollution

The first goal is expressed as the following constraint:

$$500 frice + 650 fbeans + 350 fwheat + 450 fmaize + d_1^- - d_1^+ = 40,000,000,$$

where:

$$d_1^+ = \text{over-achievement of goal 1}$$

$$d_1^- = \text{under-achievement of goal 1}$$

$frice, fbeans, fwheat, fmaize$ = acres devoted to foreign sales of the commodities

Since the top priority is to achieve goal 1, the form of the model should place top priority on forcing d_1^- to zero.[3] The second goal can be expressed as:

$$50 drice + 25 dbeans + 75 dwheat + 100 dmaize + d_2^- - d_2^+ = 13,000,000,$$

where:

$$d_2^+ = \text{over-achievement of goal 2}$$

$$d_2^- = \text{under-achievement of goal 2}$$

$drice, dbeans, dwheat, dmaize$ = acres devoted to domestic sales of the commodities[4]

---

[3]Deviation variables are always complements in the sense that if one variable is positive the other must be zero.

[4]The coefficients on the commodities are the per-acre yields given in the problem.

The third goal is expressed as:

$$10\,drice + 1\,dbeans + 2\,dwheat + 5\,dmaize + 10\,frice + 1\,fbeans$$
$$+ 2\,fwheat + 5\,fmaize + d_3^- - d_3^+ = 1{,}000{,}000,$$

where:

$$d_3^+ = \text{over-achievement of goal 3}$$
$$d_3^- = \text{under-achievement of goal 3}$$

Finally, the model requires two additional structural constraints for land and labor. The land constraint is:

$$drice + dbeans + dwheat + dmaize + frice + fbeans + fwheat + fmaize \leq 200{,}000.$$

The labor constraint is:

$$15\,drice + 20\,dbeans + 8\,dwheat + 25\,dmaize + 15\,frice$$
$$+ 20\,fbeans + 8\,fwheat + 25\,fmaize \leq 4{,}000{,}000.$$

The full model is:

Min: $Z = G_1 d_1^- + G_2 d_2^- + G_3 d_3^+$ (0)

s.t.:

$$500\,frice + 650\,fbeans + 350\,fwheat + 450\,fmaize + d_1^- - d_1^+ = 40{,}000{,}000 \quad (1)$$

$$50\,drice + 25\,dbeans + 75\,dwheat + 100\,dmaize + d_2^- - d_2^+ = 13{,}000{,}000 \quad (2)$$

$$10\,drice + 1\,dbeans + 2\,dwheat + 5\,dmaize$$
$$+ 10\,frice + 1\,fbeans + 2\,fwheat + 5\,fmaize + d_3^- - d_3^+ = 1{,}000{,}000 \quad (3)$$

$$drice + dbeans + dwheat + dmaize$$
$$+ frice + fbeans + fwheat + fmaize \leq 200{,}000 \quad (4)$$

$$15\,drice + 20\,dbeans + 8\,dwheat + 25\,dmaize$$
$$+ 15\,frice + 20\,fbeans + 8\,fwheat + 25\,fmaize \leq 4{,}000{,}000 \quad (5)$$

$$\text{non-negativity} \quad (6)$$

Notice that the objective function coefficients penalize the negative deviations for the first two goals since under-achievement is undesirable, and penalize the positive deviation for the third goal since over-achievement is undesirable. Let $G_1 = 1{,}000{,}000$, $G_2 = 500{,}000$, and $G_3 = 1$.

In this case, it is optimal for the government to devote 40,723.98 acres to domestic wheat production, 97,737.56 acres to domestic maize production, and 61,538.46 acres to foreign bean production, thus utilizing all 200,000 acres of land.

Has goal 1 been satisfied? Yes, substitute $fbean^* = 61{,}538.46$ into (1):

$$d_1^{+*} = 0.$$

In this case, goal 1 is exactly achieved.

Has goal 2 been satisfied? No, plug $dwheat^* = 40{,}723.98$, $dmaize^* = 97{,}737.56$, and $fbean^* = 61{,}538.46$ into (2):

$$d_2^{-*} = 171{,}946.$$

Hence, goal 2 is under-achieved by 171,946 bushels.

Has goal 3 been satisfied? Yes, plug $dwheat^* = 40{,}723.98$, $dmaize^* = 97{,}737.56$, and $fbean^* = 61{,}538.46$ into (3):

$$d_3^{-*} = 368{,}326 \text{ and}$$
$$d_3^{+*} = 0.$$

Thus, goal 3 is over-achieved. Notice that even though the decision maker prefers goal 2 to goal 3, goal 3 is achieved while goal 2 is not achieved. This is due to the fact that both goals 1 and 2 cannot be achieved simultaneously in this example, but goals 1 and 3 can be achieved simultaneously.

A Risk Example of Preemptive Goal Programming

Goal programming can also be used to model decision making under risk. Consider the following example of a food manufacturer who faces price and income uncertainty.

A food manufacturer sells five products to supermarkets: A, B, C, D, and E. Due to input seasonality, the firm is forced to purchase all five commodities from farmers in the summer. The firm may then process and sell the products immediately in October after harvest or store them in a warehouse for later sale. Suppose that the commodities are harvested in October and either immediately processed and sold in October or processed and stored for later sale in February or June. Sales in either February or June require storage, and the firm has a storage capacity of 1,000 units of either A, B, C, D, or E. The firm knows with certainty all technical parameters (e.g., technical coefficients and resource endowments) but only has probabilistic knowledge of the net profitability of each commodity due to price uncertainty. The firm has the following expectations regarding profit and risk (standard deviation):

Month	Product	Profit	Risk
October	A	100	0
October	B	200	20
October	C	150	10
October	D	350	80
October	E	250	40
February	A	120	10
February	B	275	35
February	C	200	20
February	D	400	100
February	E	100	10
June	A	125	5
June	B	235	25
June	C	200	15
June	D	360	90
June	E	250	60

The firm perceives the total profit risk it faces as the sum of the number of products it sells each month multiplied by its standard deviation in profit per unit. Let the marketing activities be denoted as m_{ij}, where i refers to product (A, B, C, D, or E) and j refers to month (October, February, or June). Then risk (R) is equal to:

$$R = \sum_{i=1}^{5} \sum_{j=1}^{3} m_{ij} SD_{ij},$$

where SD_{ij} is the per unit profit standard deviation for product i in month j.

The firm's production technology is summarized in the table below. Each food product requires certain amounts of machinery, labor, management, and nonmachinery capital.

Resource (unit)	Unit Resource Requirement					Resource Endowment
	A	B	C	D	E	
Labor (L) (hours)	0.90	0.80	1.20	1.50	0.85	1,350
Machinery (M) (hours)	1.35	1.00	0.30	1.30	0.25	2,500
Management (MG) (hours)	0.10	0.05	0.00	0.20	0.15	190
Capital (C) (hours)	1.00	0.90	0.95	1.10	0.75	3,000

This problem could be formulated in at least three different ways. The first would be to maximize profit ignoring risk, by using the expected profit for each marketing activity as the relevant objective function coefficient. The second would be to maximize profit subject to a constraint on risk, which could be parametrically varied in order to generate a risk-efficient set of solutions. Finally, GP can be used where the two goals are to maximize income and to minimize risk.

The profit maximizing formulation is:

Max: $Z =$ $100oa + 200ob + 150oc + 350od + 250oe$
   $+ 120fa + 275fb + 200fc + 400fd + 100fe$
   $+ 125ja + 235jb + 200jc + 360jd + 250je$ (0)

s.t.:

   $0.90oa + 0.80ob + 1.20oc + 1.50od + 0.85oe$
   $+0.90fa + 0.80fb + 1.20fc + 1.50fd + 0.85fe$
   $+0.90ja + 0.80jb + 1.20jc + 1.50jd + 0.85je \leq 1,350$ (1)

   $1.35oa + 1.00ob + 0.30oc + 1.30od + 0.25oe$
   $+ 1.35fa + 1.00fb + 0.30fc + 1.30fd + 0.25fe$
   $+ 1.35ja + 1.00jb + 0.30jc + 1.30jd + 0.25je \leq 2,500$ (2)

   $0.10oa + 0.05ob + 0.00oc + 0.20od + 0.15oe$
   $+ 0.10fa + 0.05fb + 0.00fc + 0.20fd + 0.15fe$
   $+ 0.10ja + 0.05jb + 0.00jc + 0.20jd + 0.15je \leq 190$ (3)

   $1.00oa + 0.90ob + 0.95oc + 1.10od + 0.75oe$
   $+ 1.00fa + 0.90fb + 0.95fc + 1.10fd + 0.75fe$
   $+ 1.00ja + 0.90jb + 0.95jc + 1.10jd + 0.75je \leq 3,000$ (4)

   $fa + fb + fc + fd + fe + ja + jb + jc + jd + je \leq 1,000$ (5)

   $oa, ob, oc, od, oe, fa, fb, fc, fd, fe, ja, jb, jc, jd, je \geq 0$ (6)

where $o =$ October, $f =$ February, $j =$ June, and $a,b,c,d,$ and e are the five products.

The profit-maximizing solution is to process 1,000 units of product B, which is stored and sold later in February, and to process 647.06 units of product E, which is sold immediately in October. The total expected profit from this solution is \$436,764.71. The binding constraints include labor, which has a shadow price (SP) of \$249.12, and storage capacity, which has an SP of \$39.71.

The model with the risk constraint is:

$$\text{Max: } Z = 100oa + 200ob + 150oc + 350od + 250oe$$
$$+ 120fa + 275fb + 200fc + 400fd + 100fe$$
$$+ 125ja + 235jb + 200jc + 360jd + 250je \tag{0}$$

s.t.:

$$0oa + 20ob + 10oc + 80od + 40oe$$
$$+ 10fa + 35fb + 20fc + 100fd + 10fe$$
$$+ 5ja + 25jb + 15jc + 90jd + 60je \le 40{,}000 \tag{1}$$

$$0.90oa + 0.80ob + 1.20oc + 1.50od + 0.85oe$$
$$+ 0.90fa + 0.80fb + 1.20fc + 1.50fd + 0.85fe$$
$$+ 0.90ja + 0.80jb + 1.20jc + 1.50jd + 0.85je \le 1{,}350 \tag{2}$$

$$1.35oa + 1.00ob + 0.30oc + 1.30od + 0.25oe$$
$$+ 1.35fa + 1.00fb + 0.30fc + 1.30fd + 0.25fe$$
$$+ 1.35ja + 1.00jb + 0.30jc + 1.30jd + 0.25je \le 2{,}500 \tag{3}$$

$$0.10oa + 0.05ob + 0.00oc + 0.20od + 0.15oe$$
$$+ 0.10fa + 0.05fb + 0.00fc + 0.20fd + 0.15fe$$
$$+ 0.10ja + 0.05jb + 0.00jc + 0.20jd + 0.15je \le 190 \tag{4}$$

$$1.00oa + 0.90ob + 0.95oc + 1.10od + 0.75oe$$
$$+ 1.00fa + 0.90fb + 0.95fc + 1.10fd + 0.75fe$$
$$+ 1.00ja + 0.90jb + 0.95jc + 1.10jd + 0.75je \le 3{,}000 \tag{5}$$

$$fa + fb + fc + fd + fe + ja + jb + jc + jd + je \le 1{,}000 \tag{6}$$

$$oa, ob, oc, od, oe, fa, fb, fc, fd, fe, ja, jb, jc, jd, je \ge 0 \tag{7}$$

The risk constraint here is set at \$40,000, that is, the sum of the product of the standard deviation times marketing activity for all activities cannot exceed \$40,000. The right-hand side of this constraint can be parametrically altered to derive a set of plans that are efficient for expected profit and standard deviation. This formulation leads to a more diversified production-marketing plan for the firm. In this case, 687 units of product B are processed and sold immediately in October, and an additional 1,000 units of B are processed and stored for sales of 125 units in February and 875 units in June. Diversification of marketing or production activities is a classic way to reduce risk in portfolio problems. Total expected profitability in this case is \$222,500, which is lower than the profit maximizing solution of \$436,764.71; however, the risk is also lower (\$40,000 in this case versus \$60,882.40 in the profit-maximizing case). The binding constraints include labor, which has an SP of \$294.12, and storage capacity, which has a SP of \$39.71.

The third way to formulate this problem is to use GP, which can be done in two ways. The first formulation weights risk (*risk*) as being the top priority followed by income (*income*), as follows:

Min: $Z = din + 10{,}000drp$ $\quad\quad$ (0)

s.t.:

$$100oa + 200ob + 150oc + 350od + 250oe$$
$$+\ 120fa + 275fb + 200fc + 400fd + 100fe$$
$$+\ 125ja + 235jb + 200jc + 360jd + 250je - income = 0 \quad\quad (1)$$

$$0oa + 20ob + 10oc + 80od + 40oe$$
$$+\ 10fa + 35fb + 20fc + 100fd + 10fe$$
$$+\ 5ja + 25jb + 15jc + 90jd + 60je - risk = 0 \quad\quad (2)$$

$$din + income - dip = 600{,}000 \quad\quad (3)$$

$$-\ drp + risk + drn = 40{,}000 \quad\quad (4)$$

$$0.90oa + 0.80ob + 1.20oc + 1.50od + 0.85oe$$
$$+\ 0.90fa + 0.80fb + 1.20fc + 1.50fd + 0.85fe$$
$$+\ 0.90ja + 0.80jb + 1.20jc + 1.50jd + 0.85je \le 1{,}350 \quad\quad (5)$$

$$1.35oa + 1.00ob + 0.30oc + 1.30od + 0.25oe$$
$$+\ 1.35fa + 1.00fb + 0.30fc + 1.30fd + 0.25fe$$
$$+\ 1.35ja + 1.00jb + 0.30jc + 1.30jd + 0.25je \le 2{,}500 \quad\quad (6)$$

$$0.10oa + 0.05ob + 0.00oc + 0.20od + 0.15oe$$
$$+\ 0.10fa + 0.05fb + 0.00fc + 0.20fd + 0.15fe$$
$$+\ 0.10ja + 0.05jb + 0.00jc + 0.20jd + 0.15je \le 190 \quad\quad (7)$$

$$oa + 0.90ob + 0.95oc + 1.10od + 0.75oe$$
$$+\ 1.00fa + 0.90fb + 0.95fc + 1.10fd + 0.75fe$$
$$+\ 1.00ja + 0.90jb + 0.95jc + 1.10jd + 0.75je \le 3{,}000 \quad\quad (8)$$

$$fa + fb + fc + fd + fe + ja + jb + jc + jd + je \le 1{,}000 \quad\quad (9)$$

$$oa,\ ob,\ oc,\ od,\ oe,\ fa,\ fb,\ fc,\ fd,\ fe,\ ja,\ jb,\ jc,\ jd,\ je,\ din,\ dip,\ rpp,\ drn \ge 0 \quad\quad (10)$$

This formulation would be useful to a decision maker who is risk averse. The two goals are modeled by putting a positive deviation variable for risk (drp) and a negative deviation variable for income (din) into the objective function. Since risk is the top priority, drp has a higher valued objective function coefficient than din in this formulation. Constraints (1) and (2) define *income* and *risk* for the problem. Constraint (3) models the income goal with the target-level set at a higher figure than is obtainable ($600,000). Constraint (4) models the risk goal with the target-level set at $40,000. Note that both negative and positive deviation variables are included in constraints (3) and (4) so as to allow for the possibility of over- or under-achieving each goal (din, dip, drp, and drn). The structural constraints (5) through (9) are all the same as in the profit-maximizing case. The Solver solution to this problem gives an identical solution to the model with the risk constraint.

The second formulation is identical to the first, except it weights income as being the top priority followed by risk, that is:

$$\text{Min: } Z = 10{,}000din + drp \quad\quad (0)$$

s.t.: the same constraint set as before.

This formulation gives an identical solution to the profit-maximizing case. Hence, by using GP the results of two different models can be derived by simply altering the weight of the two objective function coefficients.

Many different ways to extend this current model to include other goals are simple in conception. For example, perhaps the food manufacturer would have an additional goal of using all of one resource, or a goal of using all resources in certain proportions. While not illustrated here, this extension would be relatively simple, that is, just add the appropriate deviation variables, constraints, and penalties in the objective function.

12.4   DERIVING WEIGHTS FOR GOAL PROGRAMMING

When using GP, the analyst needs to devise a way to elicit weights that accurately and consistently depict the decision maker's relative preferences for achieving each of the goals. For preemptive GP, the priority levels must be determined and then large weights must be assigned to the top priorities. However, for nonpreemptive GP, a more precise cardinal weighting of the goals is necessary. There are a number of techniques that various researchers have used to elicit these weights, including multiple regression and "hybrid rating" regression analysis (Bell 1976). Other common approaches of systematically measuring and ranking preferences include the Analytic Hierarchy Process (Saaty, 1990; Duke & Hyde, 2002) and the Logic Scoring of Preferences approach (Dujmovic, 2007; Dujmovic et al., 2010). One of the more popular techniques—Churchman-Ackoff—is discussed here.

To illustrate this technique, consider the following example. Assume a decision maker has four goals: G_1, G_2, G_3, and G_4, and needs to develop a relative preference ranking to achieve each goal. For this example, assume G_1 is the most preferred goal followed by G_2, G_3, and finally G_4. To elicit relative numeric weights for each goal, this technique assigns a value of 1.0 for the most preferred goal, G_1. G_2 is then given a numeric value less than 1.0 to reflect how much the decision maker values achieving it relative to G_1, for instance, if G_2 is valued one-half as much as G_1, then G_2 receives a weight of 0.5. Suppose in this example, the decision maker expresses the following values for Goals 1 to 4: $W_1 = 1.0$, $W_2 = 0.6$, $W_3 = 0.5$, and $W_4 = 0.1$.

The next step in the process is to conduct a test to determine if these elicited weights are consistent with the decision maker's true preferences. The decision maker is asked whether achieving G_1 is preferred over achieving all three other goals. Assuming the answer is yes, then W_1 should be greater than $W_2 + W_3 + W_4$ combined. In this example, W_2, W_3, and W_4 sum to greater than W_1, and therefore would need to be adjusted downward so that the sum is less than 1.0, for instance, $W_2 = 0.4$, $W_3 = 0.3$, and $W_4 = 0.05$. So those become the new adjusted weights given $W_1 > W_2 + W_3 + W_4$.

Next, compare the weight given to the second goal to that of the third and fourth goal combined. Assume that the decision maker indicates that G_2 is more important than achieving G_3 and G_4. This means that $W_2 > W_3 + W_4$. Currently, this condition holds, so no adjustment needs to be made. If this condition did not hold, one would need to decrease the relative values of W_3 and W_4 so that $W_2 > W_3 + W_4$. Finally, the fact that $W_3 > W_4$ means that no further adjustments in the weights need to be made.

Based on this technique, a relative set of weights has been constructed that are consistent with the decision maker's true preferences for achieving each of the four goals. These weights can be used as objective function coefficients in the GP model on the underachievement deviation variables.

There are a couple of caveats with this approach that need mentioning. First, this approach elicits relative rather than absolute weights for achieving each goal. Second, it assumes that preferences are additive. Finally, the approach is not precise since the method of adjustment only requires the weights to sum to the appropriate strict inequality.

12.5   RESEARCH APPLICATION: OPTIMAL PARASITE CONTROL PROGRAMS

Johnson et al. (1991) used nonpreemptive GP to examine parasite control programs for sheep in developing countries. Increasing livestock production is one way to reduce global hunger, but a severe limitation on livestock production systems in many developing countries is poor animal nutrition, which is aggravated by parasites. Parasite control programs are essential for improving livestock production efficiency. The authors present a hypothetical example of a GP model to illustrate its usefulness for finding efficient parasite control plans.

GP has been used in many natural resource applications to address forestry decision problems.[5] In contrast, there have been relatively few applications of GP to livestock production systems. Several exceptions include: Wit et al. (1988) who developed a GP model to evaluate sheep husbandry possibilities in a semiarid area of the Mediterranean basin; Bong-Soon (1983) who demonstrated a farm-planning GP model for subsistence farms in South Korea; El-Shishiny (1988) who developed a single-time-period GP model for planning the development of reclaimed lands in Egypt; Rehman and Romero (1973) who proposed treating livestock diet formulation problems as a GP model; and Neal et al. (1986) who took a similar approach to formulate least-cost rations for pregnant ewes.

The GP model of Johnson et al. (1991) assumes three goals, to (1) maximize the number of geographic zones involved in a parasite control program to increase farmer participation, (2) maximize total wool production from all zones, and (3) minimize personnel required for program administration. All three goals cannot be simultaneously satisfied.

Success of parasite control is partially dependent on the number of annual strategic anthelmintic (i.e., drugs that eliminate parasitic worms) treatments. Production losses decrease as the frequency of appropriate treatments increases up to a saturation point of control. With unlimited resources, wool production and participation would be maximized by applying the saturation number of anthelmintic treatments to every zone. However, the model constructed here contains more realistic scenarios of limited resources: given finite resources and conflicting goals, how many zones should receive 0, 1, 2, ... , x anthelmintic treatments each year, where x is the treatment saturation point? It is assumed each zone will use only one treatment strategy to maximize treatment effects.

The model requires that the impact of anthelmintic treatment frequency on wool production be determined for each control program area. Investigations and observations from parasitologists can be used to determine these impacts. In addition to production effects, treatment cost and field personnel requirements also need to be quantified.

The model is illustrated using a hypothetical case based on the senior author's work with a parasite control program in Peru. The region can be subdivided into 200 zones, each having approximately 3,500 untreated native sheep. Wool production averages two pounds of wool per year per sheep when no anthelmintic treatment is administered. Albendazole, a broad-spectrum anthelmintic, was chosen for the program at a cost of $0.20 per sheep for one treatment dose.

Field personnel administer the program. The first time a zone is treated, an extension agent will need three days to complete the work. The second treatment requires two days since efficiency is improved. If a zone receives a third or fourth treatment, no extension agents will be needed because the villagers can administer treatments.

The impact of anthelmintic treatment on wool production is based on aggregating 12 investigations reported in Barger (1982) and the senior author's personal observations. Table 12.1 summarizes the numerical information for the hypothetical case providing program cost, expected wool production, and personnel required for the different strategies.

---

[5] See, for example, Arp and Lavigne (1982); Bare and Anholt (1976); Bell (1976); Buongiorno et al. (1981); Chang and Buongiorno (1981); Dane et al. (1977); Field et al. (1980); Hansen (1977); Hotvedt et al. (1982); Kao and Brodie (1979); Mitchell and Bare (1981); Porterfield (1973); Rustagi (1973); Rustagi (1976); Schuler et al. (1973).

Table 12.1   Cost, Wool Production, and Personnel by Anthelmintic Treatments for a Hypothetical Parasite Control Program (one zone = 3,500 sheep)

	Number of Anthelmintic Treatments Per Year				
	0	1	2	3	4
Cost/Sheep ($)	0	0.20	0.40	0.60	0.80
Cost/Zone ($)	0	700	1,400	2,100	2,800
Increased Wool Production (%)	0	5	14	28	35
Wool production/zone (lb)	7,000	7,350	7,980	8,960	9,450
Extension agents/zone (person days)	0	3	5	5	5

Table 12.2   Linear Programming Solutions for a Number of Zones by Number of Treatments and Different Objective Functions

Number of Treatments/Year	Objective Function		
	Maximum Wool	Maximum Participation	Minimum Personal
0	105	0	129
1	0	190	0
2	0	0	0
3	93	8	0
4	2	2	71
Value of Objective Function	1,587,180 lb	200 zones	355 people days

The model includes the following constraints:

1. The total program budget is set at $200,000.
2. No more than 200 zones may participate since this is the maximum number in the area.
3. A certain number of areas should act as showcases to publicize benefits. The model incorporates the condition that at least 2 zones receive 4 treatments per year, and at least 10 zones receive either 3 or 4 treatments per year.

Table 12.2 indicates the available strategies if only one goal were to be considered where $200,000 must be spent. Based on these results, the following target levels were set:

Goal 1: 200 zones

Goal 2: 1,587,180 pounds of wool

Goal 3: 355 people days

The GP model minimizes deviations between the achievement of goals and their target levels. By attaching numerical weights to each goal, their relative priority levels can be quantified. For example, if maximum participation is desired, the negative deviation corresponding to the number of zones receiving treatment would be weighted higher than those for goals 2 or 3. When wool production is considered a top priority, the negative deviation for goal 2 will be weighted higher and so forth. Equal weighting of goals may be used to strive to equitably balance all three goals.

The mathematical formulation for this nonpreemptive GP problem is as follows:

Min: $Z = 0.5W_1 d_1^- + 0.63W_{22}d_2^- + 0.28W_{31}d_3^+$ (0)

s.t.:

$$x_0 \quad + x_1 \quad + x_2 \quad + x_3 \quad + x_4 \qquad\qquad = 200 \quad (1)$$

$$700x_1 + 1{,}400x_2 + 2{,}100x_3 + 2{,}800x_4 \qquad\qquad \leq 200{,}000 \quad (2)$$

$$x_1 \quad + x_2 \quad + x_3 \quad + x_4 + d_1^- \qquad = 200 \quad (3)$$

$$0.7x_0 + 0.735x_1 + 0.789x_2 + 0.896x_3 + 0.945x_4 - d_2^+ + d_2^- = 158.718 \quad (4)$$

$$3x_1 \quad + 5x_2 \quad + 5x_3 \quad + 5x_4 - d_3^+ + d_3^- = 355 \quad (5)$$

$$x_4 \qquad\qquad \geq 2 \quad (6)$$

$$x_3 \quad + x_4 \qquad\qquad \geq 10 \quad (7)$$

$$x_0, \quad x_1, \quad x_2, \quad x_3, \quad x_4, d_1^-, d_2^+, d_2^-, d_3^+, d_3^- \geq 0 \quad (8)$$

where:

x_0 = number of zones receiving zero anthelmintic treatments per year

x_1 = number of zones receiving one anthelmintic treatment per year

x_2 = number of zones receiving two anthelmintic treatments per year

x_3 = number of zones receiving three anthelmintic treatments per year

x_4 = number of zones receiving four anthelmintic treatments per year

d_1^- = number of zones not included in the project

d_2^+ = pounds of wool above the goal

d_2^- = pounds of wool below the goal

d_3^+ = people days above the goal

d_3^- = people days below the goal

W_1 = weighted priority for goal 1

W_{22} = weighted priority for goal 2

W_{31} = weighted priority for goal 3

The objective function (0) minimizes the weighted deviations from the three goals' target levels. Penalties (the W coefficients) are positive for each deviation variable that is not preferred by the decision maker, that is, production of wool that is less than the target level (d_2), or number of people days above the target level (d_3), and zero for over-achievement deviation variables. To make the different units of goals comparable (zones, pounds of wool, people days), the deviations from the three goals are expressed as percentages (Romero & Rehman, 1984). This corresponds to 100/200=0.5, 100/158=0.63, 100/355 = 0.28, for goals 1, 2 and 3, respectively.

Constraint (1) forces the number of zones receiving the various treatment strategies to be exactly 200. Constraint (2) is the budget constraint that limits the entire plan to not exceed $200,000. The next three constraints quantify the definition of the three goals. The first goal, maximizing the number of zones receiving at least one treatment, is reflected by constraint (3). Similarly, constraints (4) and (5) define the desired wool production (pounds/10,000) and field personnel for the problem. Finally, constraints (6) and (7) apply to the showcase requirement specified by the decision maker. The non-negativity constraint (8) restricts all problem activities from taking on negative values.

The model was solved using different priority levels for the three goals. Several methods have been proposed for selecting goal weights. Weights can be arbitrarily chosen and adjusted until output coincides with the decision maker's actual behavior (Candler & Boehlje, 1971). Goal weights can be inferred from past activities and adjusted until a satisfactory solution is produced. Preferences for goals may be directly elicited from decision makers with surveys (Barnett et al., 1982). Willis and Perlack (1980) suggest avoiding objective weights and presenting an efficient set of solutions to the decision maker. This is the approach selected here.

The authors first ran the model by assigning a proportional goal weight of 1:1:1 to the three goals. Based on sensitivity analysis of the objective function coefficients, deviation from Goal 1 had maximum and minimum limits on its weights of 0.817 and 0.176, respectively; d_2^- had a maximum limit on its weight of 1.785; and d_3^+ had a minimum limit on its weight of 0.1740. Therefore, four goal priority changes could change the optimal solution. The model was run assigning numerical goal priorities of 0.01:0.63:0.28, for which another optimal solution was generated. For the new solution, four goal priority changes were indicated to search for other optimal solutions. Recall that this type of sensitivity analysis is called the **range of optimality** for objective function coefficients. This range gives the values that the objective function coefficients can take without changing the optimal solution.

The process of reviewing objective row ranges for the three variables whose deviations were being minimized, and then running the model with new priorities suggested by the maximum limits, continued until all optimal solutions were identified. This coincides with the process reported by Romero and Rehman (1984).

For a GP model with n goals, the number of different optimal solutions will be less than or equal to n!, depending on model constraints. For this three-goal model (participation = N, wool = W, personnel = P), a maximum number of six goal permutations exist (N > W > P; N > P > W; W > N > P; W > P > N; P > N > W; P > W > N). Fourteen model runs were needed to determine the results. Six different optimal solutions were identified (Table 12.3). One solution corresponds to the case of assigning equal priorities to the three goals (N = W = P). The same optimal solution was generated for priority levels W > N > P and W > P > N.

By comparing Tables 12.2 and 12.3, it is evident that the solution for a goal priority of P > W > N is the same as the solution obtained by simple minimization of the personnel objective function. Simple maximization of wool production coincides with the solution for a priority of W > N > P or W > P > N. The simple maximization of participation

Table 12.3   Linear Goal Programming Solutions for Number of Zones and Deviation from Goals for Different Goal Preferences

	Goal priorities: participation (N)/wool(W)/personal(P)					
	N > P > W	N > W > P	N = W = P	W > P > N W > N > P	P > N > W	P > W > N
x_0	0	0	0	105	88	129
x_1	190	158	171	0	102	0
x_2	0	0	0	0	0	0
x_3	0	40	0	93	0	0
x_4	10	2	29	2	10	71
d_1^-	0	0	0	105	88	129
d_2^-	95,340	47,970	56,340	0	126,300	12,390
d_3^+	265	329	302	118	0	0

is not duplicated when goals are simultaneously considered, although a priority level of $N > P > W$ closely approximates it.

Maximum participation can be completely achieved with three choice solutions ($N > P > W; N > W > P; N = W = P$). Wool production and personnel requirements vary for the three solutions: wool production can fall 95,340, 47,970, or 56,340 pounds short of its goal and personnel can exceed its goal by 265,329, or 302 people days.

When wool maximization is given priority over participation and personnel, preference assigned to the other two goals is irrelevant. The wool target level will be reached with a participation level of 47.5% and 118 extra people days needed to administer the program.

If personnel use is given priority over the other two goals, two different solutions are obtained depending on whether the second-highest priority is put on wool or participation. The personnel goal is achieved under both scenarios, but wool production can be 12,390 or 126,300 pounds below its goal, and the level of participation can be 36% or 56% of its goal.

While the model is set up as a nonpreemptive GP model, the solutions partially mimic the results expected from a preemptive or lexicographic model. That is, the model satisfies the highest priority goal first, and then considers the lesser-ranked goals.

The model generated six "choice" solutions that provide different degrees of multiple goal realization. Decision makers can be presented with Tables 12.2 and 12.3 and choose an optimal solution that coincides with their goal realization preferences. Trade-offs for different options can be easily understood. The model was illustrated for sheep, but the same principles are applicable to other farm animals. For example, instead of using wool production, weight gain or milk production could be incorporated.

Sensitivity analysis for model constraints could provide additional information for decision makers. For example, the current model assumes that $200,000 would be spent on parasite control. One unanswered question is whether $200,000 is the most efficient amount. The expected pay-offs for control programs using fewer dollars could be addressed through sensitivity analysis of the budget constraint. The constraint representing the number of showcase areas desired could be handled in a similar manner. When doing sensitivity analysis, guard against generating dominated solutions by not allowing target levels to fall to pessimistic levels.

Three conflicting goals are included in the model and seven constraints. Additional goals could be incorporated in the same manner. For most situations, however, four goals or less should be adequate. The number of allowable constraints is very flexible. Yazdanian and Peralta (1986) found that linear GP worked well with up to 90 constraints.

12.6   RESEARCH APPLICATION: FOREST LAND PROTECTION

The Forest Legacy Program (FLP) is a program administered by the USDA Forest Service to support the acquisition of conservation easements and other voluntary protection mechanisms on privately held forest land.[6] FLP is the largest federal forest protection program in the U.S. with an annual acquisition budget in the range of $50 to $55 million per year. From 2006 to 2009, FLP has helped to protect nearly 1.9 million acres across 41 states and Puerto Rico.

One aspect of the FLP, which is also true for many other conservation programs, is that applicants must offer some in-kind cost sharing. FLP requires that states provide at least 25% of the total project costs, but does not place a cap on the percentage of in-kind cost sharing

---

[6] This research example comes from Fooks and Messer (2010).

that can be offered by the applicants. In the past, some applicants to the FLP have submitted in-kind cost shares of nearly 90%. For instance, in 2009, the State of Utah submitted a project to FLP to protect 4,868 acres of forest in Northern Utah, including a segment of a river which is host to a large population of wild Bonneville Cutthroat Trout, a popular sport fish and the only salmonid native to the region. The total cost for the project was $6,935,000, the conservation partners in Utah offered an in-kind cost share of $1,727,000, or 25% of total costs of funding out of its own budget. Thus, to protect these 4,968 acres, FLP would only have to pay $5,208,000 (USDA, 2009).

Several justifications are given for factoring in the amount of the in-kind costs share into the selection of which forest projects should received FLP funding. A primary justification provided is that asking applicants to commit to pay for some of the costs themselves ensures that the project has local benefits and the local partners are committed to the project and consider it worthy of being protected. In other words, the applicant is willing to "put their money where their mouth is" for the project. Another justification is that cost sharing helps extend the reach of FLP as it can be considered a type of discount on the total project price. From the point of view of the funding agency, this could mean that there would be more money available to fund additional projects and also help increase the number of forest acres protected with FLP funds.

FLP program managers have been interested in demonstrating how they have used their funds to leverage funds from other agencies and organizations. Thus, these managers seek to maximize *both* environmental benefits and in-kind cost sharing. This research example sets up a GP model that seeks to maximize these two objectives. This GP problem can best be understood by looking at the two extreme cases. The first case is where only environmental benefits are maximized. This model is similar to the binary linear program introduced in Chapter 7, which selected projects to maximize benefits subject to the constraint that the total costs are less than the available budget. With this type of model, in-kind cost sharing only affects the bottom-line costs of the project being considered by the FLP. In this case, binary linear programming will still select high-quality projects that can be protected for a relatively good price; however, this is unlikely to yield the highest total in-kind cost share possible. The second case is where only the total in-kind cost share is maximized.

This research example uses data from the 83 proposed forest projects considered for funding by FLP in 2009. For that year, the acquisition budget was $53 million. To determine the environmental benefit score for each project, the FLP assembled a group of 10 experts and asked each of them to score the projects from 0 to 30 based on importance, threat, and strategic value. FLP administrators then added these three scores such that each project had a score between 0 and 90 from each reviewer. The FLP administrator then discarded the lowest and highest of the ten scores and averaged the remaining eight to give the project a final score.

When the goal is to maximize the total environmental benefit score, the maximum aggregate environmental benefit obtained is 3,024. In this case, the total project cost for the 45 selected projects is $99.4 million, with $45.6 million (or 46.8%) being in-kind cost shares. In the other case, where the goal is to maximize the total in-kind cost share, the model selects 22 projects, such that the total in-kind cost share is $124.3 million (70.1% of the $177.3 million total cost). In this case, the total environmental benefits are just 1,534. Thus, when in-kind cost share is the sole goal to be optimized, only 51% of the possible environmental benefit score is achieved, and when environmental benefits are the sole goal to be optimized, then only 37% of possible cost share is achieved.

Using these two extremes as targets, a GP problem can be set up, and then the weight between the two can be varied parametrically to examine the tradeoffs. This problem is stated as follows:

$$\text{Min: } Z = \lambda(d_C^-/124{,}272{,}474) + (1 - \lambda)(d_B^-/3{,}024) + \Sigma 0 x_i \tag{0}$$

s.t.:

$$\sum_{i=1}^{83} c_i x_i - d_C^+ + 1d_C^- = 124{,}272{,}474 \tag{1}$$

$$\sum_{i=1}^{83} b_i x_i - d_B^+ + d_B^- = 3{,}024 \tag{2}$$

$$\sum_{i=1}^{83} p_i x_i \le 53{,}000{,}000 \tag{3}$$

$$x_i \in \{0,1\}; d_i^+, d_i^-, \ge 0 \tag{4}$$

where x_i is a binary variable indicating whether project i is chosen. c_i is the pledged in-kind cost sharing amount, b_i is the environmental benefit score, and p_i is the funding request amount. In the objective function, the negative deviations are divided by the target value so that each represents a percentage deviation and, thus, are of comparable magnitudes. The weight factor, λ, can be varied parametrically from 0 to 1 and represents the percentage of the optimization given to in-kind costs, such that when λ is 0, 100% of the priority will be given to maximizing environmental benefits, and when λ is 1, 100% of the priority will be given to maximizing in-kind cost share. For example, when λ is 0.3, this means that 30% of the weight will be on in-kind cost share, and 70% will be on environmental benefits. The results of varying λ from 0 to 1 in increments of 0.1 are shown in the table below.

λ	Benefits	Total In-Kind Cost Share ($)	Total Project Cost ($)	Cost in In-Kind (%)	Number of Projects	Total Acres
1.0	1,534	124,272,474	177,268,827	70	22	196,554
0.9	1,919	123,103,470	176,071,810	70	28	197,455
0.8	2,169	121,081,918	173,985,818	70	32	193,528
0.7	2,301	119,440,251	172,414,151	69	34	192,933
0.6	2,604	112,792,251	165,751,151	68	39	195,287
0.5	2,689	109,685,751	162,564,651	67	40	194,405
0.4	2,753	105,830,251	158,829,151	67	41	193,120
0.3	2,753	105,830,251	158,829,151	67	41	193,120
0.2	2,937	84,600,001	137,473,651	62	44	87,628
0.1	2,997	70,600,001	123,573,651	57	45	88,743
0.0	3,024	46,515,001	99,413,651	47	45	100,975

Figure 12.1 displays the results as an Efficiency Frontier, or Pareto Efficient Set[7] for this problem such that at each point along the curve one of the criteria cannot be improved without making the other worse. This type of figure can provide decision makers with an intuitive visual representation of the trade-off between cost sharing and environmental benefits.

For instance, it can be readily seen that as the priority of cost sharing in the decision process is increased with increases in λ there are large improvements at relatively little

---

[7]Note that solutions to goal programs are not necessarily Pareto efficient, particularly when the target values are very easily achieved (see Tamiz et al., 1999). In this research example, this is not likely to be a problem as the target values represent two jointly exclusive extremes.

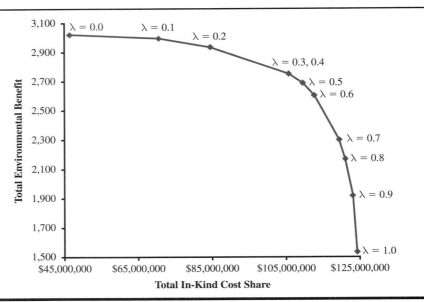

Figure 12.1   Goal programming results for forest legacy example by changes in the weight of goals.

cost in terms of total environmental benefits. For example, when $\lambda = 0.2$, the program could increase its cost share by 82% (from $46.5 million to $84.6 million) while decreasing benefits by less than 3% (from 3,023 to 2,936). Interestingly, if the program manager also cared about the number of acres protected, then a good result might be $\lambda = 0.3$. In this case, there would be a 127% increase in cost share (from $46.5 million to $105.8 million) *and* a 91% increase in total acres protected (from 100,975 acres to 193,120 acres), with only a 9% decrease in total environmental benefits (from 3,023 to 2,936). These examples illustrate how GP can be used by conservation managers to balance priorities and to make better-informed decisions for their limited conservation dollars.

SUMMARY

This chapter has focused on GP models, which relax the sometimes limiting assumption of mathematical programming models that the decision maker wants to optimize only a single goal. In many situations, decision makers have multiple goals that cannot all be achieved simultaneously.

Using GP, multiple goals or targets can be specified for the decision maker and the deviations from not achieving each goal are minimized. The basic idea of GP is to specify a numeric value for each goal and then formulate a model, which minimizes the weighted sum of the unwanted deviations of each goal. This chapter provided an overview of the concepts behind GP, including how to set up and solve such problems. Both nonpreemptive and preemptive GP were examined. Nonpreemptive GP is used when the decision maker has a concrete ranking scheme for the multiple goals. Preemptive GP is used when the decision maker has lexicographic preferences among the multiple goals.

This chapter presented several illustrations of how an LP model can be extended to have multiple objectives rather than a single objective. Also, two research applications were examined. The first application of GP was for developing countries interested in implementing parasite control for sheep. The second application was for forest protection where the conservation agency sought to both maximize the conservation value and the in-kind cost share for the projects being considered.

REFERENCES

Arp, P. A., & Lavigne, D. (1982). Planning with goal programming: A case study for multiple use of forested land. *Forestry Chronicle*, *58*, 225–232.

Bare, B. B., & Anholt, B. (1976). Selecting forest residue treatment alternatives using goal programming. USDA Forest Service General Technical Report PNW-43, Portland, Oregon.

Barger. I. A. (1982). Helminth parasites and animal production. In *Biology and Control of Endoparasites*, ed. L. Symons et al., 133–155. New York, NY: Academic Press.

Barnett, D., Blake, B., & McCarl, B. (1982). Goal programming via multidimensional scaling applied to Senegalese subsistence farms. *American Journal of Agricultural Economics*, *64*, 720–727.

Bell, E. F. (1976). Goal programming for land use planning. USDA Forest Service, General Technical Report PNW-53, Portland, Oregon.

Bong-Soon, K. (1983). A linear goal programming model for model planning of semi-subsistence farms. *Journal of Rural Development*, *6*, 87–105.

Buongiorno, J., Svanquist, N., & Wiroatmodjo, P. (1981). Forestry sector development planning: A model for Indonesia. *Agricultural Systems*, *7*, 113–135.

Candler, W., & Boehlje, M. (1971). Use of LP in capital budgeting with multiple goals. *American Journal of Agricultural Economics*, *5*, 325–330.

Chang, S. J., & Buongiorno, J. (1981). A programming model for multiple use forestry. *Journal of Environmental Management*, *13*, 41–54.

Charnes, A., Cooper, W. W., & Ferguson, R. (1955). Optimal estimation of executive compensation by LP. *Management Science*, *1*, 138–151.

Dane, C. W., Meador, N., & White, J. (1977). Goal programming in land use planning. *Journal of Forestry*, *75*, 325–329.

de Wit, C. T., van Keulen, H., Seligman, N. G., & Spharim, I. (1988). Application of interactive multiple goal programming techniques for analysis and planning of regional agricultural development. *Agricultural Systems*, *3*, 211–230.

Dujmovic, J. J., Tre, G. D., & Weghe, N. V. (2010). LSP Suitability Maps. *Soft Computing*, *14*, 421–434.

Dujmovic, J. J. (2007). Continuous Preference Logic for System Evaluation. *IEEE Transactions on Fuzzy Systems*, *15*, 1082–1099.

Duke, J. M., & Hyde, R. A. (2002). Identifying public preferences for land preservation using the analytic hierarchy process. *Ecological Economics*, *42*, 131–145.

El-Shishiny, H. (1988). A goal programming model for planning the development of newly reclaimed lands. *Agricultural Systems*, *26*, 245–261.

Field, R. C., Dress, P., & Fortson, J. (1980). Complementary linear and goal programming procedures for timber harvest scheduling. *Forest Science*, *26*, 121–133.

Fooks, J., & Messer, K. D. (2010). Maximizing conservation and in-kind cost share: Applying goal programming to forest protection. *Alfred Lerner College of Business and Economics Working Paper*, University of Delaware, Newark, Delaware, 2010-08.

Hansen, B. G. (1977). Goal programming: A new tool for the Christmas tree industry. USDA Forest Service. Research Paper NE-378, Upper Darby, Pennsylvania.

Hotvedt, J. E., Leuschner, W., & Buyhoff, U. (1982). A heuristic weight determination procedure for goal programs used for harvest scheduling models. *Canadian Journal of Forest Research*, *12*, 292–298.

Johnson, P. J., Oltenacu, P. A., Kaiser, H. M., & Blake, R. W. (1991). Modeling parasite control programs for developing nations using goal programming. *Journal of Agricultural Production*, *4*, 33–38.

Kao, C., & Brodie, J. D. (1979). Goal programming for reconciling economic, even-flow, and regulation objectives in forest harvest scheduling. *Canadian Journal of Forest Research*, *9*, 525–531.

Messer, K. D. (2006). The conservation benefits of cost-effective land acquisition: A case study in Maryland. *Journal of Environmental Management*, *79*, 305–315.

Mitchell, B. R., & Bare, B. (1981). A separable goal programming approach to optimizing multivariate sampling designs for forest inventory. *Forest Science*, *27*, 147–162.

Neal, H. D. St. C., France, J., & Treacher, T. (1986). Using goal programming in formulating rations for pregnant ewes. *Animal Production*, *42*, 97–104.

Porterfield, R. L. (1973). Predicted and potential gains from tree improvement programs: A goal programming analysis of program efficiency. Doctoral dissertation, Yale University, New Haven, Connecticut (unpublished).

Rehman, T., & Romero, C. (1973). Goal programming with penalty functions and livestock ration formulation. *Agricultural Systems*, *23*, 117–132.

Romero, C., & Rehman, T. (1984). Goal programming and multiple criteria decision-making in farm planning: An expository analysis. *Journal of Agricultural Economics*, *35*, 177–190.

Rustagi, K. P. (1973). Forest management planning for timber production: A goal programming approach. Doctoral dissertation, Yale University, New Haven, Connecticut (unpublished).

Rustagi, K. P. (1976). Forest management planning for timber production: A goal programming approach. Yale University School of Forestry and Environmental Studies Bulletin No. 89.

Saaty, T. L. (1990). How to make a decision: The analytic hierarchy process. *European Journal of Operational Research*, *48:1*, 9–26.

Schuler, A. T., Webster, H., & Meadows, J. (1973). Goal programming in forest management. *Journal of Forestry*, *75*, 320–324.

Tamiz, M., Mirrazavi, S. K., & Jones, D. F. (1999). Extensions of Pareto efficiency analysis to integer goal programming. *Omega*, *27*, 179–188.

U.S. Department of Agriculture (USDA). (2009). Fiscal year 2009—forest legacy program funded projects. Retrieved from www.fs.fed.us/spf/coop/library/fy09_funded_projects.pdf.

Willis, C. E., & Perlack, R. (1980). Generating techniques and goal programming. *American Journal of Agricultural Economics*, *62*, 66–74.

Yazdanian, A., & Peralta, R. (1986). Maintaining target groundwater levels using goal programming: Linear and quadratic methods. *Transactions of the American Society of Agricultural Engineers*, *29*, 995–1004.

EXERCISES

1. Suppose a farmer has four equally ranked goals: (1) spend 50 hours per week with his family, (2) work enough to earn $3,000 per week, (3) spend 15 hours per week volunteering at the local food pantry, which supplies free food to low-income families, and (4) sleep for 70 hours per week. Assume that the farmer earns $50 per hour of work. Assuming that each goal is equally ranked, set up this exercise as a GP problem and solve it using Solver.

2. Resolve Exercise 1 as a preemptive GP problem with the following weights:

   Spend 50 hours per week with family = 0.5

   Earn $3,000 per week = 1

   Spend 15 hours per week volunteering at food pantry = 0.1

   Spend 70 hours per week sleeping = 0.7

3. Decide how many units of product 1 and product 2 should be produced by using GP given the following information:

Item	Product 1	Product 2	Endowment
Profit/unit	16	12	
Labor	3	6	72
Material/unit	2	1	30

   The company has the following goals:

   a. Total profit should be at least 260.

   b. At least 5 units of product 2 should be produced.

4. Explain the difference between preemptive and nonpreemptive GP. Give an example where preemptive is more appropriate than nonpreemptive GP.

5. A corn–soybean farmer has three goals for his farm: maximize profits, minimize risk, and minimize the carbon footprint. He currently uses a conventional tillage system but is considering using a minimum tillage system that is less costly, less risky, and has a substantially lower carbon footprint, but is also less profitable. Assume he expects the following profits, risk, and index of carbon footprints:

Crop/Tillage System	Profit/acre	Risk/acre	Carbon emission/acre
Conventional Corn	415	200	500
Conventional Soybeans	350	100	350
Minimum Tillage Corn	325	75	75
Minimum Tillage Soybeans	250	25	45

The farmer owns 1,000 acres of land and can rent up to an additional 2,000 acres from his neighbors for $275 per acre. He also has 10,000 hours of family labor available and can hire an additional 30,000 hours of hired labor at $14 per hour. Assume it takes 10 hours to grow an acre of conventional corn, 6 hours to grow an acre of conventional soybeans, 7 hours to grow an acre of minimum tillage corn, and 3 hours to grow an acre of minimum tillage soybeans.

Formulate this scenario as a profit-maximizing LP model where risk and carbon footprint are not considered. Note there should be separate activities for conventional corn, conventional soybeans, minimum tillage corn, and minimum tillage soybeans. Solve this exercise using Solver.

6. Modify Exercise 5 by adding a maximum risk constraint setting an upper limit of $350,000 on total risk. Solve the exercise using Solver, and parametrically alter the RHS value of the risk constraint downward using sensitivity analysis.

7. In addition to the maximum risk constraint of $350,000 in Exercise 6, add a maximum carbon emission constraint of 500,000. Solve the exercise using Solver, and parametrically alter the RHS value of the maximum emission constraint downward using sensitivity analysis.

8. Set up and solve Exercise 7 using nonpreemptive GP, where the three goals are:

Profit = at least $690,000

Total risk = no more than $350,000

Total emission = no more than 500,000

9. Solve Exercise 8 using preemptive GP and the following weights: the total profit goal is the most important, followed by risk, followed by carbon emission.

10. Smalltown is a city of 20,000 inhabitants. The city council is in the process of developing an equitable city rate tax table. Taxes come from a combination of four sources:

  a. Property taxes ($550M base)

  b. Food and Drugs ($35M base)

  c. Other Sales ($55M base)

  d. Gasoline (Consumption: 7.5 million gallons per year)

Smalltown would like to come up with a "fair" city tax with the following conditions:

a. Tax revenues must be at least $16M

b. The property tax rate should be $\leq 1\%$

c. Food and drug taxes must be $\leq 10\%$ of all taxes collected

d. Sales taxes must be $\leq 20\%$ of all taxes collected

e. The gasoline tax must be $\leq \$0.02/$gallon

Use nonpreemptive GP to minimize the sum of possible tax-types (property, food and drug, etc).

11. John, a farmer in Maryland, is preparing a farming business plan for next spring. He owns 5,000 acres and plants three crops on the farm: corn, beans, and rice. This coming spring, he decides to plant one more crop on his farm: wheat. Assume the following parameters apply to this exercise:

	Corn	Beans	Rice	Wheat	Endowment
Profit ($/bushel)	$4.50	$3.50	$2.50	$3.00	
Risk ($/acre)	$200.00	$100.00	$50.00	$80.00	
Yield (bushels/acre)	100	50	60	85	
Labor (hour/acre)	25	20	10	15	80,000
Land (per acre)	1	1	1	1	5,000

Although John wants to diversify his farm products, he is reluctant to change the current planting pattern of 1,500 acres of corn, 500 acres of beans, and 2,000 acres of rice, because changes in planting will cause an unknown risk and cost. Use GP to determine what John should do for next spring. He would like to make at least $1.07 million in profit and not bear more than $0.45 million in risk.

12. A centrally planned developing country is currently drafting its agricultural plan for the next several years. The Minister of Agriculture controls 300,000 acres of agricultural land to be put into production and has several goals. First, the country is in dire need of foreign currency, and therefore the Minister would like to export food to other countries in exchange for currency. Second, in order to develop the industrial sector of the economy, the Minister knows that a significant amount of food that is grown on this land is needed to feed industrial workers in the urban areas of the country. Third, there are also environmental concerns about water pollution caused by agricultural production, which the Minister wants to limit because it has negative consequences for the nation's water supply. Fourth, the Minister wants to limit the total risk associated with growing the commodities. Assume there are four basic food commodities to be grown on this acreage: rice, beans, wheat, and maize. Assume the following parameters apply to this exercise:

	Rice	Beans	Wheat	Maize	Endowment
Item		thousands of bushels			
Net Revenue Domestic Sales	400	600	300	500	
Net Revenue Risk Domestic	100	150	75	125	
Net Revenue Foreign Sales	500	650	350	450	
Net Revenue Risk Foreign	115	165	100	150	
Yield (bushels/acre)	50	25	75	100	
Land (per acre)	1	1	1	1	300,000
Labor (hours/acre)	15	20	8	25	2,400,000
Water Pollution (contaminants/acre)	10	1	2	5	

Assume that the Minister's objective is to maximize net revenue from domestic and foreign sales of the four commodities. Formulate the LP model for this exercise, and solve it using Solver.

13. Modify Exercise 12 as follows: add a maximum water pollution constraint of 1,200,000, and solve it using Solver. Use parametric programming, and alter the RHS value for this constraint to see how the solution changes.

14. Modify Exercise 13 to include a minimum constraint for domestic production. Specifically, assume that the country needs at least 10 million bushels of rice, beans, wheat, and maize in total.

15. Modify Exercise 14 to also include a maximum amount of risk for domestic and foreign production of $45 million.

16. Solve Exercise 15 as a nonpreemptive GP problem with the following goals:

    Goal 1: at least $45 million in foreign sales

    Goal 2: at least 13 million bushels of rice, beans, wheat, and maize for domestic sales

    Goal 3: no more than 1 million contaminants of pollution

    Goal 4: no more than $25 million in total risk

17. Solve Exercise 16 as a preemptive GP problem with the following weights:

    Goal 1 = 1, Goal 2 = 100, Goal 3 = 1, and Goal 4 = 1

18. Solve Exercise 16 as a preemptive GP problem with the following weights:

    Goal 1 = 1, Goal 2 = 100, Goal 3 = 1, and Goal 4 = 50

19. A firm manufactures two types of wood beams. Each top-quality super beam (x_1) nets the firm $200 profit and each lower-grade beam (x_2) nets $100 profit. Each super beam requires two Grade-1 logs and one Grade-2 log. Each lower grade beam requires one Grade-1 log and three Grade-2 logs (Grade-1 logs are better quality than Grade-2 logs). The firm can acquire a total of 15 Grade-1 logs and 20 Grade-2 logs per week. Formulate this as an LP problem assuming the firm's goal is to maximize profits.

20. Now suppose that rather than maximizing profits in Exercise 19, the firm has the following preemptive goals:

    Goal 1: Achieve a $2,000 weekly profit

    Goal 2: Completely utilize all Grade-1 logs

    Goal 3: Completely utilize all Grade-2 logs

    Formulate this as a nonpreemptive GP problem. Solve it using Solver.

21. Formulate Exercise 20 as a preemptive GP problem where the priorities are Goal 1 > Goal 2 > Goal 3.

22. Formulate Exercise 20 as a preemptive GP problem where the priorities are Goal 3 > Goal 2 > Goal 1.

23. A new energy business is considering selling and installing three types of alternative energy systems for residential homes in addition to its conventional system (*conv*). The first is a thermal solar (*solar₁*) hot water system that provides both hot water and heat to the home. The second is a solar system (*solar₂*) that provides electricity to the home. The

third is a geothermal (*geo*) heating and cooling system. The business has 10 installers and faces the following parameters for the year:

	conv	*solar*$_1$	*solar*$_2$	*geo*	Endowment
Profit/unit	$2,000	$1,500	$2,500	$1,000	
Labor (weeks)	0.5	1.0	1.5	0.75	520

The firm has four goals:

Goal 1: Make at least $1,500,000 per year from the sale of these four systems

Goal 2: Sell at least 160*solar*$_1$ systems each year

Goal 3: Sell at least 160*solar*$_2$ systems each year

Goal 4: Sell at least 160*geo* systems each year

Formulate and solve the profit maximizing solution to this exercise ignoring the four goals.

24. Formulate and solve Exercise 23 as a nonpreemptive GP problem.

25. Formulate and solve Exercise 23 as a preemptive GP problem with the following weights:

$$\text{Goal } 1 = 1, \text{ Goal } 2 = 5, \text{ Goal } 3 = 5, \text{ and Goal } 4 = 1$$

26. Reconsider the first preemptive GP example in this chapter, for the food manufacturer with profit and risk goals. Modify this exercise by incorporating the following additional goals besides profit and risk:

Goal 3: Provide at least 100 units of A, B, C, D, or E in October

Goal 4: Provide at least 100 units of A, B, C, D, or E in February

Goal 5: Provide at least 100 units of A, B, C, D, or E in June

Formulate and solve this new GP model.

27. A farmer is considering planting some combinations of corn, soybeans, and rice. She has 10,000 acres of land available and 200,000 hours of labor. The following table details the profit, risk, pollution, and labor requirements (all per-acre) associated with each activity.

	Corn	Soybeans	Rice
Revenue	350	525	400
Risk	75	90	65
Pollution	5	7	12
Land	1	1	1
Labor	20	17	15

Given that she wants to maximize profit and minimize risk and pollution:

a. Find the optimal feasible level for each goal independently, assuming that all of the land is used.

b. Find the optimal solution for a GP using the results from part a as target values, assuming that she places about two-thirds more emphasis on minimizing risk than on the other objectives.

c. Find the optimal solution for a preemptive GP assuming her first priority is risk, followed by revenue, then pollution.

13

Dynamic Programming

With the exception of the dynamic farm models presented earlier in this book, the previously presented mathematical programming models have all been static models. In general, the majority of linear, integer, binary, goal, sector, and nonlinear programming models are static, one-period models. These static problems, in essence, assume all activities occur at once. Although these models have been extremely useful and used widely in real-world problem solving, some problems require treating the decision process dynamically. The technique of **dynamic programming (DP)**, which is the focus of this chapter, is particularly well suited for solving such problems.

Dynamic programming is a method used to solve large and complicated problems by splitting them into smaller subproblems that are both easier to solve and yield the same optimal solution as the original large problem. These smaller subproblems are referred to as **stages** in DP nomenclature. Dynamic programming problems are solved most commonly by working backwards. It is not a solution algorithm like the simplex method, but rather a solution approach that varies with each problem.

As the term "dynamic" implies, DP is often used in applications where each stage has a time dimension, and there is a certain timing sequence for each stage. For example, crop agricultural decision problems might be divided into planting, preharvest, harvest, and postharvest stages. Within each stage, certain decisions need to be made, and DP models provide optimal solutions for these decisions. In addition, DP can also be used in static problems where stages correspond to something else besides time, such as cities in a network. Because DP is often used for problems without a time dimension, the term "dynamic programming" is a bit misleading. Perhaps a better term than DP would be "recursive programming," as the essence of DP is the relationship between multiple stages in the problem, and a stage may or may not involve time.

Dynamic programming is used both in solving mathematical programming optimization problems and in solving huge problems involving a lot of computational power in computer science. Of course, the focus here is on mathematical programming problems. There are two general categories of DP: deterministic and probabilistic. Deterministic, which is the focus in this book, means that there is certainty regarding all parameters of the model, whereas probabilistic or stochastic DP relaxes this assumption.

This chapter illustrates several problems that can be solved using DP. As will be seen, the solution approach varies with each problem presented in this chapter. Three examples are presented, including a network problem, an inventory-purchase problem, and a capital budgeting problem. The chapter also includes a discussion of the general elements of DP, and the advantages and disadvantages of DP. The chapter concludes with two research applications of DP. The first deals with animal health control policies in Malawi, while the second illustrates the use of DP in converting conventional farm land to organic farming.

13.1   A NETWORK PROBLEM

Consider the following simple network example displayed in Figure 13.1, where a person wants to go from node 1 (e.g., Washington, D.C.) to node 9 (e.g., San Francisco). Assume the person wants to travel the shortest distance to get from 1 to 9. In this example, the number inside each circle represents a city, and the number on each arc represents the distance between each city.

A critical underpinning of any DP problem is the **principle of optimality**, which is attributed to the creator of DP, Richard Bellman. This principle, which is also called the **Bellman equation**, is the basis for decomposing a complex problem into much simpler subproblems that can be solved "recursively." **Recursive** means each stage is interconnected, and hence can be broken up and solved in subproblems. In the context of this example, the principle of optimality implies that if a node in the network is part of the optimal arc, then the shortest distance from that node to the final node is also part of the optimal arc. This is important because it allows us to split the larger problem into smaller ones that, when solved, give the same solution as solving the larger problem by itself.

The DP method solves each stage by working backwards. For instance, in the problem at hand, the first stage of the problem consists of the nodes that are just prior to the final destination: nodes 7 and 8 in Figure 13.1. The second stage includes the nodes just prior to stage 1 nodes (e.g., 5 and 6). Note that 5 and 6 are the **input nodes**, and 7 and 8 are the **output nodes** for stage 2. Stage 3 includes input nodes that are just prior to stage 2 (2, 3, and 4) and output nodes that are the input nodes to stage 3 (5 and 6). Finally, stage 4 includes one input node (1) and three output nodes (2, 3, and 4).

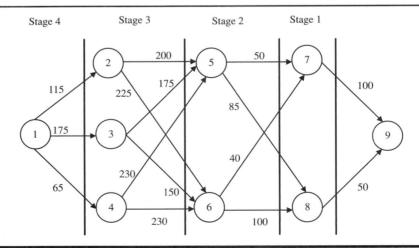

Figure 13.1   Shortest route problem by stage in the network.

Stage 1. There are two input nodes in stage 1: 7 and 8. For each of these nodes, there is only one way to get to the final destination, and so each represents the shortest route. If travelling from 7 to 9 is 100 miles and if travelling from 8 to 9 is 50 miles, the stage 1 solutions can be summarized as:

Input Nodes	Route	Minimum Distance to 9
7	7–9	100 miles
8	8–9	50 miles

Stage 2. For stage 2, the traveler can start at either input nodes 5 or 6. Consider node 5 first. There are two possible routes leaving node 5, from 5 to 7 with a distance of 50 miles, and from 5 to 8 at a distance of 85 miles. If arc 5–7 is selected, the total distance to the destination is $50 + 100 = 150$ miles. If arc 5–8 is selected, the total distance to the destination is $85 + 50 = 135$ miles. Hence, if we are at node 5, the minimum distance to the destination is through arc 5–8.

Next consider node 6. There are two possible routes leaving node 6, from 6 to 7 with a distance of 40 miles, and from 6 to 8 at a distance of 100 miles. If arc 6–7 is selected, the total distance to the destination is $40 + 100 = 140$ miles. If arc 6–8 is selected, the total distance to the destination is $100 + 50 = 150$ miles. Hence, if we are at node 6, the minimum distance to the destination is through arc 6–7. The stage 2 solutions can be summarized as:

Input Nodes	Route	Output Nodes	Minimum Distance to 9
5	5–8	8	135 miles
6	6–7	7	140 miles

Stage 3. There are three input nodes to consider in stage 3: 2, 3, and 4. Considering node 2 first, there are two possible routes: from 2 to 5 at a distance of 200 miles, or from 2 to 6 at a distance of 225 miles. When going from 2 to 5, the minimum distance is 200 miles to get to 5 plus the minimum distance from 5 to the final destination 9, which is given in the table above as 135 miles, that is, $200 + 135 = 335$ miles to get from 2 to 9. Going from 2 to 6, the minimum distance is 225 miles to get to 6 plus the minimum distance from 6 to the final destination 9, which is given in the table above as 140 miles, that is, $225 + 140 = 365$ miles to get from 2 to 9. In this case, because 335 miles is less than 365 miles, route 2–5 is the optimal route.

Next consider node 3, which has two possibilities: from 3 to 5 at a distance of 175 miles, or from 3 to 6 at a distance of 150 miles. Going from 3 to 5, the minimum distance is 175 miles to get to 5 plus the minimum distance from 5 to the final destination 9, which is given in the table above as 135 miles, that is, $175 + 135 = 310$ miles to get from 3 to 9. Going from 3 to 6, the minimum distance is 150 miles to get to 6 plus the minimum distance from 6 to the final destination 9, which is given in the table above as 140 miles, that is, $150 + 140 = 290$ miles to get from 3 to 9. In this case, because 290 miles is less than 310 miles, route 3–6 is the optimal route.

Finally, node 4 also has two possibilities: from 4 to 5 at a distance of 230 miles, or from 4 to 6 at a distance of 230 miles. Going from 4 to 5, the minimum distance is 230 miles to get to 5 plus the minimum distance from 5 to the final destination 9, which is given in the table on page 455 as 135 miles, that is, $230 + 135 = 365$ miles to get from 4 to 9. Going from 4 to 6, the minimum distance is 230 miles to get to 6 plus the minimum distance from 6 to the final destination 9, which is given in the table above as 140 miles, that is, $230 + 140 = 370$ miles to get from 4 to 9. In this case, because 365 miles is less than 370 miles, route 4–5 is the optimal route. The stage 3 solutions are summarized as:

Input Nodes	Route	Output Nodes	Minimum Distance to 9
2	2–5	5	335 miles
3	3–6	6	290 miles
4	4–5	5	365 miles

Stage 4. There are three possibilities: from 1 to 2 at a distance of 115 miles, from 1 to 3 at a distance of 175 miles, and from 1 to 4 at a distance of 65 miles. Now it is easy to see that the optimal solution to this problem is given as:

$$\min(115 + 335 = 450, 175 + 290 = 465, 65 + 365 = 430) = 430 \text{ miles.}$$

The optimal route is from 1 to 4 to 5 to 8 to 9 at a total distance of 430 miles.

While it would have been possible to evaluate all possible routes by hand to come up with the optimal solution, this example serves to illustrate the efficiency of DP. That is, with DP, evaluation of all possible routes is not necessary to find the optimal solution. In this network, there are $(3)(2)(2) = 12$ possible paths to get from node 1 to node 9. Using DP, fewer than 12 computations were made because by moving backwards, not all possible routes needed to be considered. Moreover, in substantially larger networks, the number of computations to evaluate all routes explodes exponentially, and the use of DP becomes significantly more economical.

Components of a Dynamic Programming Problem

Consider the following notation in reference to the problem above. Let:

x_i = input for stage i, and output for stage i+1

d_i = decision at stage i

For example, x_3 in the problem above is the input for stage 3, which represents the location in the network at stage 3, and the output for stage 4, which represents the node reached due to the decision in the previous stage; d_3 is the decision variable at stage 3, which is the route selected. In general, the input/output variables (x_i) in a DP are called the **state variables**. State variables connect the subproblems or stages together in the DP problem. These variables define the condition or state in the current stage of the system. For example, if currently in Cincinnati on a trip from Washington, D.C., to San Francisco, the passenger's next stage involves deciding to go to Chicago, Kansas City, or Ames. In this case, Cincinnati represents the current state of the system.

The **decision variables** (d_i) in a DP problem, which are sometimes called **control variables**, are more analogous to the decision variables in mathematical programming models. The term "control" implies that the decision maker has control over these variables. In the simple network problem above, the control variables are the routes chosen, which are conditional on the state variable. For example, the decision to go to either Chicago, Kansas City, or Ames is made conditional upon being in Cincinnati.

More generally, in DP, a state variable can be written as:

$$x_{i-1} = f_i(x_i, d_i),$$

where f_i is a function at stage i that transforms the input into stage i output. This function is sometimes called the **stage transformation function**. The functional form depends upon the particular problem. For example, the following would be the transformation for stage 2 of the previous problem ($x_1 = f_2(x_2, d_2)$), where the routes are the decision variables in the middle cells of the table:

	Output State x_1	
Input State x_2	Node 7	Node 8
Node 5	5–7	5–8
Node 6	6–7	6–8

To evaluate the various outcomes, a value must be placed on each possibility, such as miles of the various routes. In DP, this is called a **return function**, and can be expressed as:

$$r_i = r_i(x_i, d_i).$$

Hence, there are two inputs and two outputs for each stage; the two inputs are the state (x_i) and the decision variable (d_i), and the two outputs are the new values for the state variable (x_{i-1}) and the return for that stage (r_i). Figure 13.2 illustrates this process, where each box represents a stage of the problem, and the value of the state variable is determined by the transformation function, $x_{i-1} = f_i(x_i, d_i)$ and the value of the return for each stage is determined by the return function, $r_i = r_i(x_i, d_i)$. The total return function, which links all stages together, given the total return from stage T-1, (where T is the total number of stages) given input state T-1 and the decision made in stage T-1 is:

$$r_T = r_T(x_T, d_T).$$

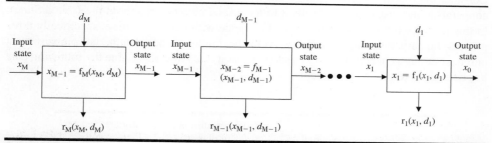

Figure 13.2   Decomposition of dynamic programming problem by decision and state variables and return and stage transformation functions.

13.2 CHARACTERISTICS OF DYNAMIC PROGRAMMING PROBLEMS

There are several characteristics of all DP problems. First, the problem can be divided into multiple **stages** in which decisions are made in each stage, and each stage can be solved separately. In the example above, each stage represented nodes in a network, and the decision for each stage was the route to choose to get to the next set of nodes. In a crop farm problem, the planting stage would include a host of decisions related to how much acreage should be planted with the various crops and the timing of planting for each crop.

A second attribute of DP is that certain states are associated with each stage of the problem. A state in DP provides a means of going from one stage to another. For instance, in the network problem, each state represents a node within the network.

Third, decisions made in any individual stage transform the state in that stage into the state in the next stage. That is, stage i decisions transform the state in stage i into the state in the next stage. If a decision has been made to travel from Washington, D.C., to Cincinnati, then that transforms the state in the next stage such that Cincinnati becomes the state to travel from in subsequent decisions involving routes.

Fourth, conditioned on the current state, the optimal solution for remaining stages does not depend on previous states attained or previous decisions made. This is the so-called principle of optimality. Consider the network example. Suppose that it is known that path M is the shortest route from node 1 to 9, and it is known to pass through node 4. Then the path of M going from 4 to 10 must be the shortest path from 4 to 10. If this were not the case, then it would be possible to find an alternative path from 1 to 10 that was shorter than M by adding the shortest path from 4 to 10 to the portion of M from 1 to 4. This would result in a shorter route than M, which contradicts the proposition that M is the shortest path.

Finally, the problem involves a **recursive** relationship determining the optimal decision for stage i, given that stage i+1 has already been solved, and the final stage must be solvable by itself. For example, suppose the initial state for Stage 1 is S_1. In DP, we work backwards beginning with the last stage, and use recursion to determine $f_i(x_i, d_i)$ and $r_i(x_i, d_i)$ starting with stage T-1. In other words, start by determining $f_{T-1}(x_{T-1}, d_{T-1})$ and $r_{T-1}(x_{T-1}, d_{T-1})$ for each stage T-1 state, then use recursion to determine $f_{T-2}(x_{T-2}, d_{T-2})$ and $r_{T-2}(x_{T-2}, d_{T-2})$ for each stage T-2 state, and so on until f_1 is solved.

13.3 A PRODUCTION INVENTORY PROBLEM

Consider the following problem for a large food distributor who buys cheese from a dairy cooperative and distributes and sells it to local food outlets throughout the United States. In this problem the stages in the DP will coincide with periods of time, specifically three months: January, February, and March. At the beginning of each month, the company must purchase enough cheese to satisfy the demand of its customers. Assume the following data has been gathered for this problem:

Month	Demand (millions of pounds)	Cost per million lbs. of cheese ($ million)	Storage cost per million lbs. ($ million)
January	1	1.00	0.10
February	3	1.25	0.10
March	2	1.50	0.10

Define the following:

d_n = decision variable at stage n, which is the amount of cheese purchased for month n = 1, 2, 3

x_n = state variable at stage n, which is the amount of cheese inventory at the beginning of month n

D_n = demand for cheese in month n

$r_n(x_n, d_n)$ = sum of costs of acquiring and storing cheese in month n

$r_T(x_T, d_T)$ = overall return function, which is the total sum of costs of acquiring and storing cheese for all three months

To solve this DP, work backwards letting March be stage 1, February be stage 2, and January be stage 3. Denoting the cost of acquiring the cheese in stage n as p_n, and the cost of storing it as c_n, the total cost of acquiring and storing the cheese in stage n is:

$$r_n(x_n, d_n) = p_n d_n + c_n(x_n + d_n - D_n).$$

Note that the second term represents total storage costs for that month. Also, note that the term in parentheses, $x_n + d_n - D_n$, represents the amount by which beginning inventories plus cheese purchases exceed the current month's demand for cheese. Therefore, this amount is also equal to the beginning inventory level in the subsequent month, that is:

$$x_{i-1} = x_i + d_i - D_i.$$

This is the stage transformation function for this problem, which simply states that beginning cheese inventories for month i-1 are equal to cheese inventories in the previous month, plus cheese purchases in the previous month, minus demand in the previous month. Figure 13.3 illustrates the stage transformations for this problem across the three stages.

The last element of the DP is an expression for the overall return function from this recursive problem, $r_T(x_T, d_T)$:

$$r_T(x_T, d_T) = r_n(x_n, d_n) + r_{T-1}(x_{T-1}, d_{T-1}).$$

This recursion says that the overall return of this problem is the sum of the current decision plus cumulative earlier decisions made.

Stage 1 (March). Assume no inventories are desired for the final month in this problem. Demand for March is given as 2 million pounds of cheese. Hence, at most, we cannot have more than 2 million pounds of cheese in inventories at the beginning of March. The least amount of cheese that can be in inventories at the beginning of March is 0. Assume that the food company can only purchase cheese in integer amounts from the dairy cooperative

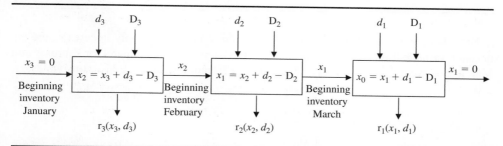

Figure 13.3   Stage transformation process for three-stage cheese inventory DP problem.

(e.g., 1, 2, 3, 4, ... , million pounds). In this case, there are three possibilities for the state variable in this stage, x_1 can be 0, 1, or 2. For each possible state, the total purchase and storage costs can be computed from the following function:

$$r_1(x_1, d_1) = 1.50d_1 + 0.10(x_1 + d_1 - D_1).$$

The following table calculates values of r_1 given each possible cheese inventory state:

Cheese Inventory States x_1	$d_1 = 0$	$d_1 = 1$	$d_1 = 2$	$r_1^*(x_1, d_1)$	d_1^*
0			3.0	3.0	2
1		1.5		1.5	1
2	0			0	0

Note that not all the cells in this table have values; the cells are empty in cases where March demand (2) is not exactly met. In this table, if March begins with zero cheese inventories, the optimal solution is to purchase 2 million pounds of cheese, which costs $3 million. If March begins with 1 million pounds of cheese inventories, the optimal solution is to purchase 1 million pounds of cheese, which costs $1.5 million. If March begins with 2 million pounds of cheese inventories, the optimal solution is to purchase no cheese since demand is exactly met by the 2 million pounds of cheese in inventories. In the final state, costs are equal to zero.

Stage 2 (February). Again, we start out by defining the possible states. Now there are six possibilities for the state variable in order to meet demand in February and March (i.e., $3 + 2 = 5$), that is, x_2 can be 0, 1, 2, 3, 4, or 5. For each possible state, the total purchase and storage costs can be computed from the following function:

$$r_2^*(x_2, d_2) = 1.25d_2 + 0.10(x_2 + d_2 - D_2) + r_1^*(x_2 + d_2 - D_2).$$

The following table calculates values of r_2 given each possible cheese inventory state:

Cheese Inventory States x_2	$d_2 = 0$	$d_2 = 1$	$d_2 = 2$	$d_2 = 3$	$d_2 = 4$	$d_2 = 5$	$r_2^*(x_2, d_2)$	d_2^*
0				6.75	6.60	6.45	6.45	5
1			5.50	5.35	5.20		5.20	4
2		4.25	4.10	3.95			3.95	3
3	3.00	2.85	2.70				2.70	2
4	1.60	1.45					1.45	1
5	0.20						0.20	0

Note that not all the cells in this table have values; the cell is empty in cases where the decision variable levels are either too small or too large to satisfy demand for February and March. For example, consider $x_2 = 3$ and $d_2 = 2$. Denoting that as $r_2(3, 2)$, this yields:

$$r_2(3, 2) = 1.25(2) + 0.10(3 + 2 - 3) + r_1^*(3 + 2 - 3),$$

$$= 2.50 + 0.20 + r_1^*(2),$$

$$= 2.70 + 0,$$

$$= 2.70.$$

Notice that the second term, $r_1^*(2)$, was actually derived already in stage 1, and here is equal to 0.

As another example, consider $x_2 = 1$ and $d_2 = 2$. Denoting that as $r_2(1, 2)$, this yields:

$$r_2(1, 2) = 1.25(2) + 0.10(1 + 2 - 3) + r_1^*(1 + 2 - 3),$$

$$= 2.50 + r_1^*(0),$$

$$= 2.50 + 3 = 5.5.$$

Here, the second term, $r_1^*(0)$, derived already in stage 1, is equal to 3.

Stage 3 (January). In stage 3, there is one state variable assuming beginning inventories in January are zero. Hence, $x_1 = 0$. The decision variable, cheese purchases, however, can vary between 1 (January demand) to 6 (January + February + March demand) units. For each possible state, the total purchase and storage costs can be computed from the following function:

$$r_3(x_3, d_3) = 1.00d_3 + 0.10(x_3 + d_3 - D_3) + r_2^*(x_3 + d_3 - D_3), \quad \text{or since}$$
$$x_3 = 0 \text{ and } D_3 = 1,$$
$$r_3(x_3, d_3) = 1.10d_3 - 0.10 + r_2^*(d_3 - 1),$$

where the $*$ indicates the optimal solution in this stage.

The following table calculates values of r_3 given each possible cheese inventory state:

Cheese Inventory States x_3	$d_3 = 1$	$d_3 = 2$	$d_3 = 3$	$d_3 = 4$	$d_3 = 5$	$d_3 = 6$	$r_3^*(x_3, d_3)$	d_3^*
0	7.45	7.30	7.15	7.00	6.85	6.70	6.70	6

Consider the $d_3 = 1$ entry for zero beginning inventories:

$$r_3^*(0, 1) = 1.10(1) - 0.10 + r_2^*(1 - 1),$$

$$= 1 + r_2^*(0),$$

$$= 1 + 6.45,$$

$$= 7.45.$$

Consider the $d_3 = 2$ entry for zero beginning inventories:

$$r_3^*(1, 2) = 1.10(2) - 0.10 + r_2^*(2 - 1),$$

$$= 2.10 + r_2^*(1),$$

$$= 2.10 + 5.20,$$

$$= 7.30.$$

Consider the $d_3 = 3$ entry for zero beginning inventories:

$$r_3^*(1, 3) = 1.10(3) - 0.10 + r_2^*(3 - 1),$$

$$= 3.20 + r_2^*(2),$$

$$= 3.20 + 3.95,$$

$$= 7.15.$$

Consider the $d_3 = 4$ entry for zero beginning inventories:

$$r_3^*(1, 4) = 1.10(4) - 0.10 + r_2^*(4 - 1),$$
$$= 4.30 + r_2^*(3),$$
$$= 4.30 + 2.70,$$
$$= 7.00.$$

Consider the $d_3 = 5$ entry for zero beginning inventories:

$$r_3^*(1, 5) = 1.10(5) - 0.10 + r_2^*(5 - 1),$$
$$= 5.40 + r_2^*(4),$$
$$= 5.40 + 1.45,$$
$$= 6.85.$$

Consider the $d_3 = 6$ entry for zero beginning inventories:

$$r_3^*(1, 6) = 1.10(6) - 0.10 + r_2^*(6 - 1),$$
$$= 6.50 + r_2^*(5),$$
$$= 6.50 + 0.20,$$
$$= 6.70.$$

The optimal decision in stage 3 is $d_3^* = 6$ for zero beginning inventories ($x_3^* = 0$).
   The optimal decision in the previous two stages can be found by the stage transformation function:

$$x_2^* = x_3 + d_3 - D_3 = 0 + 6 - 1 = 5$$

d_2^* is equal to:

$$d_2^* = \min(D_2 - x_2^*, 0),$$
$$= \min(3 - 5, 0),$$
$$= 0.$$

The optimal solution for stage 1 is:

$$x_1^* = x_2^* + d_2^* - D_2 = 5 + 0 - 3 = 2.$$

d_1^* is equal to:

$$d_1^* = \min(D_1 - x_1^*, 0),$$
$$= \min(2 - 5, 0),$$
$$= 0.$$

So the optimal solution to this problem is to purchase 6 million pounds of cheese in January, sell 1 million pounds, and store the remaining 5 million pounds for sale in February and March. In February and March, the company will not purchase any cheese, and will meet demand exactly from inventory sales. This least-cost strategy costs a total of $6.70 million.

13.4   A CAPITAL BUDGETING PROBLEM

A large agribusiness firm is considering funding four possible projects for the future and has a total budget of $20 million. Suppose that the four projects have the following expected costs and returns:

Project	Expected cost ($ million)	Expected returns ($ million)
1	11	12
2	13	19
3	7	14
4	6	6

This problem could be solved as a binary programming problem, as funding fractional amounts of projects is not a possibility. Rather, the decision is whether to fund or not to fund each project. This problem could also be solved using DP, which is formulated as follows.

Here there is no time element involved, but the stages correspond to the projects. Hence, there are four stages.

The state variables are the amounts of money available at each stage. Define the state variable, x_i, as the amount of capital available for stage i.

The decision variable for each stage is whether or not to fund the project. Hence, $d_i = 0$ or 1 for each stage.

The stage transformation function for this problem relates the amount of capital available from one stage to the next. This can be written mathematically as:

$$x_{i-1} = x_i - c_i d_i,$$

where c_i is the cost of project i. This equation states that the amount of capital in stage $i-1$ depends on how much was spent in the previous stages.

The final component of the DP problem is the return function. Define the return function as:

$$r_i(x_i, d_i) = p_i d_i,$$

where p_i is the expected return for project i. The total return function is:

$$r_i(x_i, d_i) = p_i d_i + r_{i-1}^*(x_i - c_i d_i), \text{ and}$$

$$d_i = 0 \text{ or } 1.$$

Stage 1: The return function is:

$$\text{Max: } r_1(x_1, d_1) = 12 d_1$$

$$d_1 = 0 \text{ or } 1.$$

As we are working backwards, we do not know how much capital is available for stage 1; we do know, however, that $20 million is available for Stage 4. We also know that the available capital for stage 1 will range between $0 and $20 million. Assume that the managers must fund at least one project prior to stage 1. This helps us define the states for stage 1. Because the manager must fund at least one project prior to stage 1, there are five possibilities for available capital in stage 1:

$$20 - 6 = 14,$$
$$20 - 6 - 7 = 7,$$
$$20 - 7 = 13,$$
$$20 - 6 - 13 = 1,$$
$$20 - 7 - 13 = 0.$$

The return function for stage 1 is $r_1 = 12d_1$. The following table calculates the values of r_1 given each possible available capital state:

Available Capital States x_1	$d_1 = 0$	$d_1 = 1$	$r_1^*(x_1, d_1)$	d_1^*
0	0		0	0
1	0		0	0
7	0		0	0
13	0	12	12	1
14	0	12	12	1

Stage 2: Available capital in stage 2 will depend upon whether projects 3 or 4 were funded. There are four possibilities:

$$20 - 6 - 0 = 14,$$
$$20 - 6 - 7 = 7,$$
$$20 - 7 = 13,$$
$$20 - 0 = 20.$$

The stage 2 return function is:

$$r_2(x_2, d_2) = 19d_2 + r_1^*(x_2 - 13d_2), \text{ and}$$
$$d_2 = 0 \text{ or } 1.$$

The following table calculates values of r_2, given each possible available capital state:

Available Capital States x_2	$d_2 = 0$	$d_2 = 1$	$r_2^*(x_2, d_2)$	d_2^*
7	0		0	0
13	12	19	19	1
14	12	19	19	1
20	12	19	19	1

Notice that even if $d_2 = 0$, there are entries of $12 million under three of the four states in that column. This is due to the fact that there is enough capital to fund d_1 for three of these four states, and that project returns $12 million.

Stage 3: Available capital in stage 3 will depend upon whether project 4 was funded. There are two possibilities:

$$20 - 6 = 14,$$
$$20 - 0 = 20.$$

The stage 3 return function is:

$$r_3(x_3, d_3) = 14d_3 + r_2^*(x_3 - 7d_3), \text{ and}$$
$$d_3 = 0 \text{ or } 1.$$

The following table calculates values of r_3 given each possible available capital state:

Available Capital States x_3	$d_3 = 0$	$d_3 = 1$	$r_3^*(x_3, d_3)$	d_3^*
14	19	20	14	0
20	19	33	33	1

Notice again that even when $d_3 = 0$, there are entries of $19 million under all two states since there is enough budget to fund project 2, which returns $19 million.

Stage 4: Available capital in stage 4 is $20 million.
The stage 4 return function is:

$$r_4(x_4, d_4) = 6d_4 + r_3^*(x_4 - 6d_4), \text{ and}$$
$$d_4 = 0 \text{ or } 1.$$

The following table calculates values of r_4, given each possible available capital state:

Available Capital States x_4	$d_4 = 0$	$d_4 = 1$	$r_4^*(x_4, d_4)$	d_4^*
20	33	20	33	0

So the optimal solution is as follows: $d_4^* = 0$, which means that $x_3^* = 20$ because none of the budget is spent on project 4.
From the stage 3 table, when $x_3^* = 20$, $d_3^* = 1$, and the remaining budget is therefore $20 - 7 = 13 = x_2^*$.
From the stage 2 table, when $x_2^* = 13$, $d_2^* = 1$, and the remaining budget is therefore $20 - 7 - 13 = 0 = x_1^*$.
From the stage 1 table, for $x_1^* = 0$, $d_1^* = 0$.
Hence, the optimal solution is to fund projects 2 and 3 and to not fund projects 1 and 4. The expected total return is $33 million.

13.5   COMMENTS ON DYNAMIC PROGRAMMING

All three examples of DP that were presented here are very simple applications of DP, which were chosen by design in order to illustrate the logic behind recursion in the simplest manner possible. Even these simple applications can be fairly tricky to comprehend. Nevertheless, the logic behind larger and more realistic DP is essentially the same. The reader can imagine network, inventory, and capital budgeting problems with hundreds or even thousands of stages that would require solving by computer. Indeed, there are DP software packages available for the types of problems illustrated in this chapter.

When setting up a DP problem, it should be clear from these three examples that there are five basic elements of the problem. First, determine the stages. Second, define the possible states in the system. Third, indentify the decision variables. Fourth, specify the stage transformation function mathematically. Finally, determine the total return function and the recursive relationship. These five elements comprise any DP.

The primary disadvantage of DP is that, unlike most mathematical programming models, DP is not very generalizable like the simplex method, or some of the NLP algorithms discussed in previous chapters. As a result, each problem requires a unique specification. DP also suffers from the **curse of dimensionality**. This means the number of computations explodes exponentially with the number of stages and state variables. Also, solving a DP is not as efficient as other mathematical programming solution techniques such as the simplex method. However, this is not as much of a problem with the large computational power of modern personal computers.

There are several advantages of DP. First, for some problems, DP is the only solution approach. For instance, DP has been widely used in inventory problems for this reason. Second, for decision-tree type problems, DP is particularly well suited. Third, DP is applicable to a wide host of problems, so it is quite flexible.

13.6   RESEARCH APPLICATION: ANIMAL HEALTH IN DEVELOPING COUNTRIES

Hall et al. (1998) developed a DP model to determine efficient animal health control for developing countries. Many economically important animal diseases remain a problem in developing countries despite expensive control attempts. In such cases, there remains a need for modeling more cost-effective control method alternatives. A good example is the East Coast Fever (ECF) situation in eastern and central Africa. ECF is a protozoan disease caused by *Theileria parva parva*, transmitted to cattle by the *Rhipicephalus appendiculatus* tick. Annual calf mortality from ECF can be greater than 75% and closely parallels the tick burden on pasture, which is greatest in the rainy season. Most parts of East Africa control ECF by dipping cattle at acaricide dip tanks, but tick resistance to acaricides and concerns regarding the cost-effectiveness and human health risk of frequent dipping have prompted consideration of alternative disease control methods (Pegram & Chizyuka 1990).

Several methods have been used to evaluate the economics of animal health control programs, including cost-benefit analysis and LP. Hall et al. (1998) developed the first DP model to look at this issue. Their DP model estimated the economic benefits of controlling ECF in Malawi Zebu cattle herds in the Lilongwe plateau of Malawi. For alternative treatment scenarios, the model determined optimal net benefits of treatment and optimal treatment frequency.

Dynamic Programming Model

The investigative unit of their model is the herd of all cattle within the same ecozone, at which the disease control program is targeted. The objective function of the model maximizes the net present value (NPV) of the difference between savings from reductions in mortality and the costs of the control program for a specific planning horizon. Thus, the objective function considers both government costs of disease control and the producer benefits in terms of the value of reduced mortality. The value of the objective function therefore acts as a proxy for the net social benefits of the control program.

The mathematical formulation of the DP model is:

Max: $Z = \sum_i \sum_j d_i \{(PDTBDZ_{ij})(VALUE_j)(XN_{ij}) - (RES_{ij})(XN_{ij})\}$ (13.1)

s.t.:

Herd inventory and structural constraints:

$$N_{ij}/\sum_j N_{ij} \leq a_{ij} \tag{13.2}$$

$$\sum_j N_{ij} \leq b_i \tag{13.3}$$

Nutritional constraints:

$$\sum_i \sum_j (ME_{ij})(N_{ij}) \leq (FORGHAS_j)(MEFORG_i)(DMI_{ij}) \tag{13.4}$$

$$\sum_i \sum_j (MP_{ij})(N_{ij}) \leq (FORGHAS_i)(MPFORG_i)(DMI_{ij}) \tag{13.5}$$

$$\sum_i \sum_j (NDF_{ij})(N_{ij}) \leq (FORGHAS_i)(NDFFORG_i)(DMI_{ij}) \tag{13.6}$$

Budgetary constraints:

$$\sum_i \sum_j d_i (RES_{ij})(XN_{ij}) \leq BUDG \tag{13.7}$$

where:

Z = net benefits of mortality savings – programs costs (the objective function)

$d_i = 1/(1 + \text{interest})^i$ in time period i, where interest = 10%

$PDTBDZ_{ij}$ = the difference in the probability of death from the target disease between control and treatment groups, calculated for cohort j in time period i

$VALUE_j$ = total sales (milk, meat, and livestock) from one animal in cohort j – salvage value (i.e., potential sale value – salvage value if animal had died)

XN_{ij} = variable number of animals to be treated, recommended by the model; decision variable

RES_{ij} = vector of drug and labor costs to carry out control measure in time period i

a_{ij} = chosen percent of animals of age cohort j in the herd at time period i

b_{ij} = chosen total number of animals in the herd at the time of period i

ME_{ij} = metabolizable energy requirements of an animal of age cohort j at time i

MP_{ij} = metabolizable protein requirements of an animal of age cohort j at time i

NDF_{ij} = neutral detergent fiber intake limit of an animal of age cohort j at time i

$MEFORG_i$ = metabolizable energy available from forage (kg DM per ha) at time i

$MPFORG_i$ = metabolizable protein available from forage (kg DM per ha) at time i

$NDFFORG_i$ = neutral detergent fiber available from forage (kg DM per ha) at time i

$FORGHAS_i$ = hectare of forage available at time i

DMI_{ij} = dry matter intake of an animal in cohort j at time i

N_{ij} = number of animals of cohort j in the herd at time period i, state variable

$BUDG$ = total control program budget

The stages in the model correspond to time periods for the animal treatment program. The state variables in the model are the number of cattle of each age cohort in the herd (N_{ij}), and the decision variables are the number of cattle of each age cohort to treat (XN_{ij}). Thus, in order to maximize overall net social benefits, the model solves for the number of animals to treat at each stage. The model also uses rates of culling and offtake (sales) to maintain herd structure and to meet nutritional constraints.

The mortality savings portion of the objective function considers the difference in mortality between the control group and the treatment group, using the variable number of animals to be treated generated by the model. The difference in the probability of death from the target disease between control and treatment group (PDTBDZ) is thus defined to value only the mortality losses from the target disease that are prevented by treatment and not the losses that would occur despite treatment. The model is solved separately for each alternative treatment scenario, and the net benefits between treatments are compared.

The herd is divided into four cohorts based on age and immune status (susceptible calves, immune calves, susceptible adults, and immune adults), each of which is a state variable. The dynamic nature of the herd size, composition, and immune status is described by a series of interdependent equations, which differ slightly depending on treatment. As calves mature, they are transferred into the appropriate adult cohorts. Using susceptible calves (SCALF) as an example, the basic form of the equations describing changes in the size of a cohort at time $i + 1$ is:

$$N(i + 1, \text{SCALF}) = N(i, \text{SCALF}) - \{(1 - \text{PDT}(i, \text{SCALF}))(1 - \text{OFFT}(i, \text{SCALF}))\} \\ + (\text{CR})(N(i, \text{SADULT}))$$

where:

      PDT = total probability of death (from the target disease plus other causes, where mortality varies depending on treatment group)

   OFFT = rate of offtake

      CR = calving rate of dams (proportion of cows calving per time period)

SADULT = susceptible adult cohort

An aggregate value was derived for each of the four cohorts, which reflected the market prices and offtake potential of each cohort, using weighted mean market prices (e.g., sale price of steers) and proportions of animals sold in each cohort (e.g., percentage of adults sold as steers). These aggregate values are used in estimating the value of the offtake loss prevented by the treatment of one animal, in each of the four cohorts. It is perhaps clearer to think of this as the replacement value of an animal, or as the shadow price (SP) of treatment, since it is indicative of the potential increased value of offtake from treating one animal.

The structure of the herd is maintained subject to constraints consistent with the assumed management capabilities of the herder and the assumed reproductive potential of the herd. Following the start of the control program, it is expected that the rate of growth of the herd will increase due to improved health. To prevent oversized herds, a constraint is set on the maximum size of each cohort, consistent with historical patterns of feed and other resources, current and future objectives of herders, and herd management. Excess cattle are sold as meat or replacement stock. The ratio of calves to adults is also monitored to prevent undesirable stratification of the herd. The proportions of steers, bulls, heifers, and cows are assumed to remain constant in the adult cohorts. The right-hand-side (RHS) values (the assumed desired levels of herd composition, a_{ij} of equation (13.2) described below, are set prior to running the model. Culling and offtake rates are adjusted by the model to meet these constraints.

In order to prevent herd growth to a size inconsistent with available feed resources, constraints are set based on estimated available nutrients of feeds and nutrient requirements of cattle. The nutrient requirements of the herd are evaluated based on the nutrient requirements of each age and sex cohort. If the available nutrients are less than the minimum nutrient requirements, growth of the herd to expected levels predicted by disease control is prevented. In this case, the level of offtake is increased to adjust herd size to a level consistent with available nutrients. In the event that the level of available nutrients is greater than or equal to the level required, the herd expands in a way consistent with other constraints and reproductive parameters.

The cost of the control program over the entire length of the planning horizon is constrained so as to not exceed the budget. There is no constraint on costs per time period, other than not exceeding the total budget, which allows the costs of control to be higher in earlier time periods of the program.

Calvings per month are calculated by the multiplication of a known monthly calving rate by the number of dams of calving age, and by a vector of dummy variables to indicate in which month calvings occur. This avoids having to code a separate model for seasonal and nonseasonal calvings. The calving rate is not adjusted as dams are treated. This is not realistic since reproductive rates are known to decrease in diseased dams, although without a quantitative relationship expressing this decrease, this factor could not be taken into account.

The model was developed using secondary data from Malawi tick dipping field trials (Soldan & Norman, 1994) in which the cost-effectiveness of preventing ECF by dipping cattle in ticks acaricide was investigated. Data were collected from approximately 1,800 Malawi Zebu cattle belonging to 143 farmers, monitored at six dip tanks in the Lilongwe plateau using a cohort study type of investigative approach. The *Rhipicephalus appendiculatus* tick occurred in all trial areas, and ECF was endemic to the region. The data set included details of herd size, age, and status of animals, sale prices, and labor and treatment costs. The nutritional parameters of the model were based on the Cornell Net Carbohydrate and Protein System (CNCPS) microcomputer software program and data described in Hall et al. (1998).

Model Results

Hall et al. (1998) demonstrated the model with an initial hypothetical herd size of 500 adults and 100 calves. A representative time series of the results of the decision variables and herd structure for a chlorfenvinphos treatment group over a five-year time horizon is presented in Table 13.1. The costs, mortality savings, NPV benefits (objective function values), and offtake values of the model run for the control and treatment groups over five time horizons are presented in Table 13.2.

Calves and adults in Table 13.1 are considered susceptible until dipped and immune when treated until the next treatment period. The decision variables generated by the model recommended that adults never be dipped, which is reasonable since the probability of mortality is low in adults facing continuous field challenge. Calves are dipped in periods with a higher probability of mortality. Interestingly, either all or no calves were dipped in each period, reflecting a high marginal value of treatment and adequate availability of resources. Thus, the calves were either all in the susceptible cohort or all in the immune cohort. Similarly, adults were always in the susceptible cohort, since treatment of adults was never recommended. The percentage of animals treated and the percentage in each cohort changed when nutrition was restricted and when costs of treatment were changed, reflecting a change in the marginal value of treatment. As calves mature to adulthood, in month 5 for example, they are transferred to the adult cohorts.

Treatment resulted in higher NPV benefits than did no treatment. In all scenarios vaccination treatment resulted in the highest net benefits, exceeding NPV benefits of dipping treatment by between 10 and 41%. This finding runs counter to the current government policy of dipping cattle for ticks. The net benefits of treating total cattle at risk as a percentage of the

Table 13.1   Number of Animals per Cohort and Percent Treated Over 5 Years Using Chlorfenvinphos on the Lilongwe Plateau, Malawi

Month	Calves[a]	Treated[b]	Adults	Treated	Month	Calves	Treated	Adults	Treated
1	100	0	500	0	31	83	100	533	0
2	99	100	495	0	32	80	0	523	0
3	96	100	489	0	33	78	0	515	0
4	93	100	483	0	34	77	0	509	0
5	26	100	563	0	35	76	100	503	0
6	56	0	556	0	36	74	0	498	0
7	86	100	548	0	37	13	0	491	0
8	84	0	537	0	38	72	100	485	0
9	81	0	530	0	39	70	100	479	0
10	80	0	523	0	40	68	100	474	0
11	79	100	517	0	41	24	100	531	0
12	77	0	511	0	42	53	0	524	0
13	76	0	504	0	43	81	100	516	0
14	75	100	499	0	44	78	0	506	0
15	73	100	493	0	45	76	0	499	0
16	71	100	487	0	46	75	0	493	0
17	32	100	546	0	47	74	100	487	0
18	70	0	539	0	48	73	0	482	0
19	107	100	531	0	49	71	0	475	0
20	104	0	521	0	50	70	100	470	0
21	101	0	514	0	51	68	100	464	0
22	99	0	507	0	52	67	100	458	0
23	98	100	501	0	53	24	100	514	0
24	96	0	496	0	54	54	0	508	0
25	95	0	489	0	55	83	100	500	0
26	93	100	484	0	56	80	0	490	0
27	91	100	478	0	57	78	0	483	0
28	88	100	472	0	58	77	0	477	0
29	23	100	548	0	59	16	100	472	0
30	53	0	541	0	60	74	0	467	0

[a] Calves and adults reported as number of animals.
[b] Treated reported as percent of animals treated.

short-term projected agriculture sector GDP were 0.80% and 0.89% for dipping and vaccinating, respectively. Long-run benefits (25 years) as a percentage of the projected agricultural GDP were 0.74% and 1.10%, respectively.

The costs of vaccination in the short run (5 to 10 years) were six to seven times the costs of dipping with chlorfenvinphos. In contrast, mortality savings from vaccination were more than twice the mortality savings when using chlorfenvinphos. If the cost of treatment is the sole criterion for treatment choice due to, for example, severe budget constraints, and the benefits of mortality savings are ignored, then vaccination appears to be a worse option than dipping. Mortality savings (deaths prevented) directly affect total offtake per time horizon and herd population structure. If mortality savings are high and the herd is expanding, higher offtake rates are possible. However, observation of offtake rates alone can also lead to false conclusions regarding the economic efficiency of treatment. The chlorfenvinphos treatment option in the shortest term (five years) resulted in the highest offtake at nearly 50% greater than the level of the control group, although the same treatment

Table 13.2 Costs, Mortality Savings, Maximum NPV and Offtake From Tick Control Using Chlorfenvinphos Dip or Vaccination on the Lilongwe Plateau, Malawi

Treatment	Costs	Mortality Savings	NPV	Offtake
5-year planning horizon				
Control	0	0	0	151,536
Chlorfenvinphos	1,997	8,377	6,380	231,026
Vaccine	11,238	18,307	7,069	190,202
10-year planning horizon				
Control	0	0	0	206,650
Chlorfenvinphos	3,223	13,533	10,331	363,918
Vaccine	23,575	35,939	12,364	308,485
15-year planning horizon				
Control	0	0	0	229,371
Chlorfenvinphos	3,977	16,737	12,760	444,827
Vaccine	26,742	43,720	16,979	394,571
20-year planning horizon				
Control	0	0	0	239,131
Chlorfenvinphos	4,445	18,711	14,267	494,494
Vaccine	31,017	51,113	20,096	461,649
25-year planning horizon:				
Control	0	0	0	241,621
Chlorfenvinphos	4,736	19,938	15,203	525,522
Vaccine	35,019	56,088	21,069	511,100

resulted in lower NPV benefits than from vaccinating. Clearly neither maximum mortality savings nor minimum costs should be considered as a sole criterion for treatment choice.

In the long term (15 to 25 years), the costs of vaccination were higher than the costs of chlorfenvinphos treatment. Despite the high cost, vaccination treatment provided very high levels of protection from ECF, resulting in the highest calf survival rates and greatest mortality savings. This strongly influenced the high net benefits of vaccination. Vaccination was more attractive as a treatment option where production costs were low and cattle market prices were stable. Both treatments improved offtake over the control group across all planning horizons. In all cases offtake rates paralleled mortality savings.

Herd structure was considerably more stable for both treatments than for the control, although minor variations resulted between treatments. Dipping frequency was reduced in longer planning horizons. Optimal dipping frequencies suggested by the model were less frequent than is currently the practice. Vaccination was only indicated by the model for calves, which was not surprising since vaccination costs were high and previously exposed adults were relatively immune to ECF.

13.7 RESEARCH APPLICATION: CONVERSION TO ORGANIC ARABLE FARMING

Due to its increasing popularity and potential market growth, organic farming attracts increasing attention in modern agriculture operations. Consumer demand has stimulated the conversion of many conventionally farmed lands into organic farming systems.

However, due to a series of factors and restrictions set by the requirements of organic farming, farmers have to take the risk of undergoing an economically difficult conversion period in order to arrive at the profitable phase of organic farming. Though in reality this conversion is a very complicated biological and technical process, which requires professional expertise and strict operations, the example examined here is a simplified version from a previous problem discussed by Acs, Berentsen, and Huirne (2007). The purpose here is to demonstrate the capabilities of DP to assist in decision making in farmland conversion.

Consider a farmer, who is contemplating a conversion to organic farming on a 48-hectare typical arable farm in the Netherlands central clay region, which is currently cultivated in conventional ways. The conversion is not a simple one-step task, but requires careful planning and is a continuous endeavor, requiring things such as training labor, improving soil condition, accumulating experience, and developing markets. As a result, the farmer may have to go through an economically difficult "conversion period," which involves farming organically with relatively high input costs and low revenues before being eligible to receive the higher organic prices. In this example, the entire planning horizon is limited to three stages, which are one year of conventional farming, one year of conversion, and one year of organic farming. A DP model can be developed to help determine how much land should go from conventional cultivation to organic farming and how many hectares of different crops should be grown at each stage.

Decision Variables

In this dynamic model, the key decision is to determine which crops and how much of them should be produced at each stage in order to maximize total revenues over the three stages. As shown in Tables 13.3a to 13.3c, suppose that the farmer has two crop options in the first conventional period: seed potatoes and carrots. During the following conversion and organic periods, more diverse crop selections are available.

Table 13.3a   Crop Selection Options at Conventional Stage

	Seed Potatoes	Carrots
Total Land	Yes	Yes

Table 13.3b   Crop Selection Options at Conversion Stage

	Seed Potatoes	Carrots	Winter Wheat	Alfalfa
Seed Potatoes	No	Yes	Yes	Yes
Carrots	Yes	No	Yes	Yes

Table 13.3c   Crop Selection Options at Organic Stage

	Seed Potatoes	Carrots	Winter Wheat	Alfalfa	Spring Barley	Kidney Beans
Seed Potatoes	No	Yes	Yes	Yes	Yes	Yes
Carrots	Yes	No	Yes	Yes	Yes	Yes
Winter Wheat	Yes	Yes	No	Yes	No	Yes
Alfalfa	Yes	Yes	Yes	No	Yes	Yes

A basic concern of organic farming is crop rotation, the practice of growing a series of dissimilar types of crops in the same area in sequential seasons. In this example, assume:

1. One crop planted in one stage cannot be planted again on the same land in the following stage. For example, if 12 hectares of land are planted with winter wheat at the conversion stage, no winter wheat can be planted on this area at the organic stage.

2. Crops of the same type cannot be planted on the same land in two sequential stages. For example, if 12 hectares of land are planted with winter wheat at the conversion stage, spring barley cannot be planted on this area at the organic stage either. This 12-hectare land should be shifted to other types of crops at the organic stage.

Based on the crop rotation requirements, more specific crop selection options at each stage are provided below.

Table 13.3a shows that at the first stage, the farmer has two possible crops to grow on his land. Table 13.3b shows that at the conversion stage, the area where seed potatoes were planted during the first stage can be cultivated with carrots, winter wheat, or alfalfa. Similarly, the area where carrots were planted at the first stage can be cultivated with seed potatoes, winter wheat, or alfalfa. In the same way, the "Yes" in any cell of Table 13.3c indicates that the crop in its column can replace the crop in its row at the organic stage. Therefore, the decision variables should measure the hectares of land cultivated for each possible crop at each stage. These variables are defined as follows:

$$x_{1i} = \text{hectares for crop i at conventional period} \qquad i = 1, 2 \qquad (13.8)$$

$$x_{2ij} = \text{hectares for crop j from crop i at conversion period} \qquad i = 1, 2; \ j = 1, \dots, 4 \quad (13.9)$$

$$x_{3jk} = \text{hectares for crop k from crop j at organic period} \quad j = 1, \dots, 4; \ k = 1, \dots, 6 \quad (13.10)$$

where i, j, k = 1 for seed potatoes; i, j, k = 2 for carrots; j, k = 3 for winter wheat; j, k = 4 for alfalfa; k = 5 for spring barley and k = 6 for kidney beans.

Crop Rotation

Constraints on crop rotation ensure the flow of land from stage to stage. For example, at the conversion stage, the total area of land planted with carrots, winter wheat, or alfalfa should not exceed the area of land allocated to seed potatoes in the previous conventional stage. Therefore, a series of constraints are formulated as follows:

$$\sum_{i=1}^{2} x_{1i} \leq 48 \qquad (13.11)$$

$$\sum_{j=1}^{4} x_{2ij} - x_{1i} \leq 0 \qquad i = 1, 2 \qquad (13.12)$$

$$\sum_{k=1}^{6} x_{3jk} - \sum_{i=1}^{2} x_{2ij} \leq 0 \qquad j = 1, \dots, 4 \qquad (13.13)$$

$$x_{211} + x_{222} + x_{311} + x_{322} + x_{333} + x_{335} + x_{344} = 0 \qquad (13.14)$$

Notice that constraint (13.14) actually restricts the value of each of the variables in this equation to 0 when combined with non-negative constraints. It reflects the requirement of organic farming of growing different crops on the same land in sequential stages.

Soil Improvement

Organic farming requires that a minimum of one-sixth of the total land farmed be planted with legume crops in the organic period. Legume crops contribute to soil organic matter, nutrient supply (nitrogen fixation), and yield improvement. In our example, a minimum of 8 hectares needs to be planted with kidney beans in the organic period. The constraint is written as follows:

$$\sum_{j=1}^{4} x_{3j6} \geq 8 \tag{13.15}$$

Environmental Regulations

Environmental regulations are an import factor in the farming process. One such regulation in the Netherlands is MINAS (Dutch Mineral Accounting System). MINAS focuses on the restriction of nutrient surpluses within the farm, specifically nitrogen and phosphate, and states an acceptable level of surplus at the hectare level (100kg N and 25kg P_2O_5). If the farm is above this acceptable level, the farmer must pay a levy of €2.3/kg for nitrogen and €9/kg for phosphate. Given the nutrient requirement for each of these crops at different stages in our example, the following tables of nutrient surplus for each possible crop selection can be developed at each stage.

Levies are an essential cost during organic farming. By multiplying the per-hectare levy for each crop at each stage with the actual area of each crop at each stage, we can get the total cost of the levy for the entire conversion process:

$$c_{Levy} = \sum_{i=1}^{2}(t_{1i}x_{1i}) + \sum_{j=1}^{4}(t_{2j}\sum_{i=1}^{2}x_{2ij}) + \sum_{k=1}^{6}(t_{3k}\sum_{j=1}^{4}x_{3jk}) \quad i=1,\,2;\ j=1,\ldots,4;\ k=1,\ldots,6 \tag{13.16}$$

where t_{1i} are values of the last column of Table 13.4 and t_{2j} are values of the last column of Table 13.5, and t_{3k} are values of the last column of Table 13.6.

Table 13.4   Nutrient Surplus and Levy for Each Crop at the Conventional Stage

Conventional	N (kg/ha)	P (kg/ha)	Levy (€/ha)
Seed Potatoes	25	95	912.5
Carrots	0	95	855

Table 13.5   Nutrient Surplus and Levy for Each Crop at the Conversion Stage

Conventional	N (kg/ha)	P (kg/ha)	Levy (€/ha)
Seed Potatoes	0	22	198
Carrots	0	32	288
Winter Wheat	25	37	390.5
Alfalfa	0	108	972

Table 13.6   Nutrient Surplus and Levy for Each Crop at the Organic Stage

Organic	N (kg/ha)	P (kg/ha)	Levy (€/ha)
Seed Potatoes	0	22	198
Carrots	0	32	288
Winter Wheat	25	37	390.5
Alfalfa	0	108	972
Spring Barley	0	35	315
Kidney Beans	0	0	0

Farm Labor Supply

In general, growing organic crops requires farm laborers with some special knowledge and training. Therefore, the farmer has to hire skilled labor at the conversion and organic stages, unlike the unskilled labor used in the conventional stage. Suppose skilled labor costs €18/hr, and unskilled labor costs €9/hr. Currently, the farmer has 2,255 hours of family labor available for each stage. It is assumed that the family labor hours can be regarded as skilled labor and add no extra costs. Given the needed labor requirement for each of these crops at different stages, the total cost of hired labor can be formulated as follows:

$$c_{Labor} = (9)\max(\sum_{i=1}^{2}(l_{ai}x_{1i}) - 2,255, 0) + (18)\max(\sum_{j=1}^{4}(l_{bj}\sum_{i=1}^{2}x_{2ij}) - 2,255, 0)$$

$$+ (18)\max(\sum_{k=1}^{6}(l_{bk}\sum_{j=1}^{4}x_{3jk}) - 2,255, 0) \quad i = 1, 2; \; j = 1, \ldots, 4; \; k = 1, \ldots, 6 \quad (13.17)$$

where l_{ai} are values of the column titled as "Labor" in Table 13.7, l_{bj} and l_{bk} are values of the column titled as "Labor" in Table 13.8. Notice that there is a nonlinear component of "max" function appearing in this equation.

Table 13.7 Revenue, Production Cost, Labor and Nutrient Requirements of Conventional Crops (per hectare per year)

Crops	Conventional				
	Revenue (€)	P-Cost (€)	Labor (hour)	Nutrient Requirement	
				N(kg)	P(kg)
Seed Potatoes	7,740	3,245	95.3	125	120
Carrots	12,320	9,450	29.3	80	120

Table 13.8 Revenue, Production Cost, Labor and Nutrient Requirements of Conversion and Organic Crops (per hectare per year)[1]

Crops	Conversion and Organic					
	Conversion Revenue (€)	Organic Revenue (€)	P-Cost (€)	Labor (hour)	Nutrient Requirement	
					N(kg)	P(kg)
Seed Potatoes	5,200	9,620	2,226	77.1	50	47
Carrots	8,800	18,700	12,450	185.7	40	57
Winter Wheat	1,246	1,926	439	13	125	62
Spring Barley	—	1,691	393	12.1	25	60
Kidney Beans	—	2,817	624	25.6	50	20
Alfalfa	840	960	169	2.2	0	133

[1]Spring wheat and kidney beans are only available at the organic stage.

Production Costs

Production costs include the cost of pesticide (conventional stage), nutrients, energy use, field operations, and other relevant costs such as insurance. Production costs in this case do not include the cost of labor. Given the production costs of each crop at each stage, the total production costs of the three stages can be formulated as follows:

$$c_{\text{Production}} = \sum_{i=1}^{2}(p_{ai}x_{1i}) + \sum_{j=1}^{4}\left(p_{bj}\sum_{i=1}^{2}x_{2ij}\right) + \sum_{k=1}^{6}\left(p_{bk}\sum_{j=1}^{4}x_{3jk}\right) \qquad (13.18)$$

$$i = 1,\ 2;\ j = 1,\ldots,4;\ k = 1,\ldots,6$$

where p_{ai} are values of the column entitled "P-Cost" in Table 13.7 and p_{bj} are values of the column entitled "P-Cost" in Table 13.8.

Revenue and Net Revenue

In our example, the farmer's revenue is the total revenue over three stages. In each stage, the yearly revenue is calculated by summing up all crop revenues, which are the product of revenue per hectare multiplied by hectares. This can be written as:

$$r_{\text{Gross}} = \sum_{i=1}^{2}(r_{ai}x_{1i}) + \sum_{j=1}^{4}\left(r_{bj}\sum_{i=1}^{2}x_{2ij}\right) + \sum_{k=1}^{6}\left(r_{bk}\sum_{j=1}^{4}x_{3jk}\right) \qquad (13.19)$$

$$i = 1,\ 2;\ j = 1,\ldots,4;\ k = 1,\ldots,6$$

where r_{ai} are values of the column titled as "Revenue" in Table 13.7 and r_{bj} and r_{bk} are values of the column titled as "Conversion Revenue" and "Organic Revenue" in Table 13.8. The final objective function of our model can be expressed as:

$$\text{Max: } Z = r_{\text{Net}} = r_{\text{Gross}} - c_{\text{Levy}} - c_{\text{Labor}} - c_{\text{Production}} \qquad (13.20)$$

In conclusion, Tables 13.7 and 13.8 show the model input data on revenues, production costs, needed labor, and nutrient requirements. Notice that even though the conversion farming stage shares the same production cost, labor, and nutrient requirements as the organic stage, the revenue at the conversion stage is assumed to be less than that at the organic stage. This is mainly because the crop yields at this stage are relatively low compared to the organic stage, and the crop price is lower than in the organic stage.

Non-negativity Constraints

All the decision variables represent allocated hectares, so they should be non-negative:

$$x_{1i},\ x_{2ij},\ x_{3jk} \geq 0 \quad i = 1,\ 2;\ j = 1,\ldots,4;\ k = 1,\ldots,6 \qquad (13.21)$$

Solution

As shown in Figure 13.4, in order to maximize the total net revenue over the three stages, the optimal plan is to plant 35.69 hectares of seed potatoes and 12.31 hectares of carrots in the conventional farming period. At the conversion stage, the area where seed potatoes was cultivated in the previous conventional stage is now planted with 1.66 hectares of

	A	B	C	D	E	F	G	H
19	Objective (Max Net Revenue)							
20		Revenue	Levy Cost	Labor Cost	P-Cost	Net Revenue		
21		785,924.65	87,124.18	75,209.21	373,474.78	250,116.47		
22								
23	Decision Variables							
24	Conventional	Seed Potato	Carrot	Total				
25		35.69	12.31	48.00				
26	Conversion	Seed Potato	Carrot	Winter Wheat	Alfalfa	Total		
27	Seed Potato	0.00	1.66	0.00	34.03	35.69		
28	Carrot	4.05	0.00	7.43	0.83	12.31		
29	Total	4.05	1.66	7.43	34.86			
30	Organic	Seed Potato	Carrot	Winter Wheat	Alfalfa	Spring Barley	Kidney Bean	Total
31	Seed Potato	0.00	0.00	0.00	0.00	0.00	4.05	4.05
32	Carrot	1.66	0.00	0.00	0.00	0.00	0.00	1.66
33	Winter Wheat	3.21	0.00	0.00	0.00	0.00	4.22	7.43
34	Alfalfa	7.13	4.62	3.23	0.00	3.16	16.72	34.86
35	Total	12.01	4.62	3.23	0.00	3.16	24.98	

Figure 13.4 Optimal solution of farmland conversion example.

	A	B	C	D	E
38	Levy Cost	Levy			
39		N(Euro/kg)	P(Euro/kg)		
40		2.3	9		
41	Conventional	N(kg/ha.)	P(kg/ha.)	Total	Levy Cost
42	Seed Potato	25	95	912.5	43,092.28
43	Carrot	0	95	855	
44	Conversion	N(kg/ha.)	P(kg/ha.)	Total	Levy Cost
45	Seed Potato	0	22	198	
46	Carrot	0	32	288	
47	Winter Wheat	25	37	390.5	38,066.79
48	Alfalfa	0	108	972	
49	Organic	N(kg/ha.)	P(kg/ha.)	Total	Levy Cost
50	Seed Potato	0	22	198	
51	Carrot	0	32	288	
52	Winter Wheat	25	37	390.5	
53	Alfalfa	0	108	972	5,965.11
54	Spring Barley	0	35	315	
55	Kidney Bean	0	0	0	

Figure 13.5 Levy cost at each stage under the optimal land conversion plan.

carrots, 4.33 hectares of winter wheat, and 29.70 hectares of alfalfa. Similar information is provided for the other lands in Figure 13.4.

According to the optimal land conversion plan, the farmer's final net revenue is €252,411. From the structure of gross revenue at the top of Figure 13.4, it can be seen that production costs are the largest of all possible costs, followed by the levy cost then labor cost.

This model also provides other useful information to the farmer. Figure 13.5 gives the farmer an idea of the amount of levy he is going to pay for each crop at each stage. Notice that the levy decreases from the conventional stage to the organic stage. This trend results from less manure use and agricultural soil improvement during the land conversion process. Similarly, the results shown in Figure 13.6 shows the farmer's labor cost, production cost, and revenues at every stage. Notice that the labor cost increases from the conventional stage

	A	B	C	D	E	F	G	H	I
57	Labor Cost								
58		Wage Rate							
59		Skilled (Euro/h)	Unskilled (Euro/h)	Family Labor					
60		18	9	2000					
61		Seed Potato	Carrot	Winter Wheat	Alfalfa	Spring Barley	Kidney Bean	Total	Cost
62	Conventional	3,401.43	360.63					1,762.06	15,858.53
63	Conversion	312.00	308.78	96.57	76.70			794.05	14,292.91
64	Organic	925.69	857.71	42.05	-	38.21	639.56	2,503.21	45,057.77
65									
66	Production Cost								
67		Seed Potato	Carrot	Winter Wheat	Alfalfa	Spring Barley	Kidney Bean	Total	
68	Conventional	115,819.91	116,312.44					232,132.35	
69	Conversion	9,008.07	20,701.65	3,261.11	5,891.67			38,862.51	
70	Organic	26,726.03	57,503.71	1,419.86	-	1,240.95	15,589.37	102,479.93	
71									
72	Revenue								
73		Seed Potato	Carrot	Winter Wheat	Alfalfa	Spring Barley	Kidney Bean	Total	
74	Conventional	276,254.58	151,636.95					427,891.53	
75	Conversion	21,043.12	14,632.49	9,255.92	29,284.04			74,215.57	
76	Organic	115,500.61	86,371.04	6,229.29	-	5,339.57	70,377.03	283,817.54	

Figure 13.6   Labor cost, production cost, and revenue at each stage under the optimal land conversion plan.

to the organic stage since organic crops generally require more labor than conventional crops. More labor input results in more work on the field, so the production costs increase over the stages. However, because of the price advantage of organic crops, revenues obtained in the organic stage are higher than those in the previous stages.

SUMMARY

This chapter provided an overview of DP. Dynamic programming is a method used to solve large and complicated problems by splitting them into smaller subproblems, called "stages," that are both easier to solve and yield the same optimal solution as the original large problem. Dynamic programming is often used in applications where each stage has a time dimension, and there is a certain timing sequence of each stage. In addition, DP can also be used in static problems, where stages correspond to something else besides time such as cities in a network. Dynamic programming is not a solution algorithm like the simplex method, but rather a solution approach that varies with each problem.

Two of the most important elements of DP are recursion and the principle of optimality or Bellman's equation. Simply put, recursion means that each stage is interconnected. The principle of optimality means that if a decision in a specific stage is part of the optimal decision, then the overall optimal decision will include that stage's decision regardless of which initial state or previous decisions occurred. This is important because it allows us to split the larger problem into smaller ones that, when solved, give the same solution as solving the larger problem itself. The DP method discussed in this chapter solves each stage by working backwards.

This chapter illustrated several problems that can be solved using DP. Three examples were presented: (1) a network problem, (2) an inventory and purchases problem, and (3) a capital budgeting problem. The chapter also discussed the general elements of DP, and the advantages and disadvantages of DP. The chapter concluded with two research applications of DP. The first applied DP to animal health control policies in Malawi. The second applied DP to conventional farm conversion to organic farming.

REFERENCES

Acs S., Berentsen, P. B. M., & Huirne, R. B. M. (2007). Conversion to organic arable farming in The Netherlands: A dynamic linear programming analysis. *Agricultural Systems, 94*, 405–415.

Hall, D., Kaiser, H. M., & Blake, R. (1998). Modeling of economics of animal health control programs using dynamic programming. *Agricultural Systems, 56*, 125–144.

Pegram, R. G., & Chizyuka, H. C. B. (1990). The impact of natural infestations of ticks in Zambia on the productivity of cattle and implications for tick control strategies in Central Africa. *Parasitologia, 32*, 165–175.

Soldan, A. W., & Norman, T. L. (1994). Livestock disease evaluation project dipping trial. Central Veterinary Laboratory, Lilongwe, Malawi.

EXERCISES

1. Use DP to solve the network problem given in figure below.

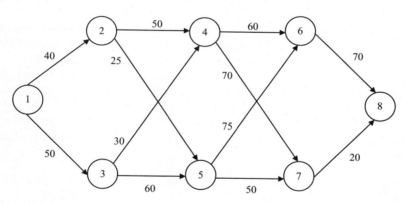

2. List all possible routes from 1 to 8 in the figure for Exercise 1 and calculate by hand each possible total distance. Explain why DP is more efficient than this approach.

3. Use DP to solve the network problem given in figure below.

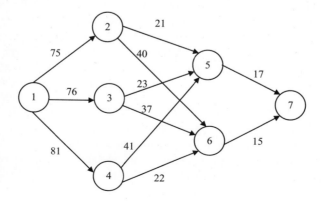

4. List all possible routes from 1 to 7 in the figure for Exercise 3, and calculate by hand each possible total distance. Explain why DP is more efficient than this approach.

5. Reconsider the network in Figure 13.1. Find the shortest route from node 1 to node 9 using DP assuming that node 4 is not a possibility.

6. A manager of the New York State Department of Conservation has four environmental projects needed to be completed. His sole objective is to minimize the total time to complete all four projects. The following data pertain to the problem:

Project	Number of Workers		
	1	2	3
1	10	6	4
2	14	10	6
3	18	16	14
4	22	16	16

Assume the manager has 12 workers to assign to each project. Formulate this as a DP problem that minimizes the completion time of all projects.

7. A food company buys butter from a dairy cooperative and distributes and sells it to restaurants. The next three months have estimated demands of 3,000, 7,000, and 5,000 pounds. The estimated purchase costs for months 1, 2, and 3 are $1,000 per thousand pounds (month 1), $1,200 per thousand pounds (month 2), and $900 per thousand pounds (month 3). The storage costs are $100 per thousand pounds (month 1), $110 per thousand pounds (month 2), and $115 per thousand pounds (month 3). Assume no inventories are desired at the end of month 3, and there are no beginning inventories in month 1. Solve this problem using DP to minimize total purchase plus storage costs.

8. The C&Y Company has five auditors available to allocate to three overseas projects. The auditors are to be allocated so that TOTAL decreased risk is maximized.

Number of auditors	Decreased Risk (in thousands of dollars)		
	Project 1	Project 2	Project 3
0	0	0	0
1	45	20	50
2	70	50	70
3	90	75	80
4	100	110	120
5	120	145	135

Formulate this as a DP problem to determine how many auditors, if any, should be allocated to each project. Define stages, state variables, decision variables, return functions and recursive functions. (Do not solve it.)

9. Given the following network with noted distances, use DP to find the shortest path from A to Z.

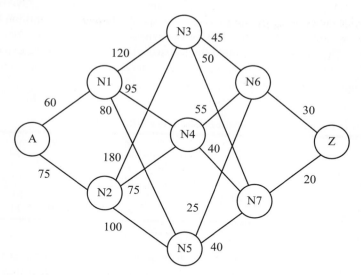

10. A farmer in Georgia has 24 hectares of land available this year. A seed salesman introduced the farmer to new products. The salesman has three different types of seeds: corn, wheat, and cotton. One bag of corn can be used to plant 4 hectares, and one bag of wheat and cotton can each be used to plant 6 and 8 hectares, respectively. The expected profits for these three seeds are shown in the following table. Formulate and solve it as a DP problem to help the farmer to maximize his expected profit.

	Corn	Wheat	Cotton
Profit ($/bag)	8,000	15,000	22,000

11. An agribusiness firm is considering funding four possible projects for the future and has a total budget of $18 million. Suppose that the four projects have the following expected costs and returns:

Project	Expected cost ($ million)	Expected returns ($ million)
1	10	15
2	14	18
3	8	13
4	7	7

Solve this as an integer programming (IP) problem, where the objective is to maximize total expected return from these projects. Assume that each project is either funded (1) or not funded (0).

12. Solve Exercise 11 using DP. Show all your calculations.

13. The State Department of Environmental Safety is considering an advertising campaign aimed at reducing littering in the state. It has a budget of $1 million that can be spent in increments of $100,000 across four types of media: print, outdoor, radio, and television. The Ad Agency has told the Department that the following exposure index (0 to 100) can be expected from various amounts spent (in $100,000 increments) on advertising by type of media:

Number of People Exposed to Advertisement Per $100,000 Spent

Media	1	2	3	4	5	6	7	8	9	10
Print	10	15	25	30	50	75	80	85	90	95
Outdoor	15	30	35	40	55	60	65	70	75	75
Radio	8	16	25	31	35	39	41	45	51	55
Television	9	25	45	50	65	80	85	90	90	90

Assume the Department's objective is to maximize exposure to the advertisement, given the $1 million budget. Formulate and solve the DP problem that maximizes exposure by choosing each media type.

14. A feed dealer buys corn from grain farmers and distributes and sells it to livestock farmers. In this exercise, the stages in the DP problem coincide with periods of times, specifically three months: January, February, and March. At the beginning of each month, the dealer must purchase enough corn to satisfy the feed demand of its customers. Assume the following data has been gathered for this exercise:

Month	Demand (million lbs)	Cost per million bushels of corn ($ million)	Storage cost per million bushels ($ million)
January	1	4.00	0.15
February	3	4.15	0.15
March	2	4.25	0.15

Assume no inventories are desired at the end of March, and there are no beginning inventories in January. Solve this exercise using DP to minimize total purchase plus storage costs.

15. A local government has five parcels of land and would like to rent these parcels to three different farmers. The profits for the three farmers when a different number of parcels are allocated are shown in the following table. The local government can get 20% revenue from each farmer's profit. Formulate and solve the DP problem that maximizes the profit for the local government.

| | Profit ($'000s) | | |
Number of Parcels	Farmer A	Farmer B	Farmer C
0	0	0	0
1	32	65	45
2	70	102	63
3	90	110	110
4	120	110	120
5	140	120	122

16. As in Exercise 15, if the local government only has four parcels to rent to these same three farmers, what is the optimal allocation solution?

17. A Maryland cannery buys fish from the local fishery and sell cans of fish to the regional supermarket. In January, one of his major fish suppliers has three tons of salmon, four tons of tuna, two tons of sardines, and two tons of mackerel. According to the experience in the market, the highest demand for the canned fish will come in April. The cannery owner is thinking that it is better to finish canning in 100 days so that he could have some time to focus on other parts of his business, such as transportation and advertising. The profits and days required for canning each kind of fish is shown in the following table. Suppose there is no budget constraint for the owner. Formulate and solve this exercise using DP to determine the optimal canning strategy. Confirm this answer by using IP.

	Supply (tons)	Canning Time (days/ton)	Profit ($/ton)
Salmon	3	30	8,000
Tuna	4	10	2,000
Sardine	2	40	11,000
Mackerel	2	70	20,000

18. Based on Exercise 17, suppose the cannery owner finds that he will only have 70 days for canning the fish and he needs to can some sardines. What will his optimal canning strategy be? Use DP to solve this exercise.

19. A farmer has a budget of $60,000, and wants to invest the money to buy crop seeds for her farm. She is considering four different crops: corn, wheat, soybeans, and cotton. The profit coming from these four crops depends on the amount the farmer invests in her land while growing these crops. The relationships between on-farm investment and her profit for these four crops are shown in the following table.

Profit	Investment ($'000s)						
	0	10	20	30	40	50	60
Corn	0	20	50	65	80	85	85
Wheat	0	20	40	50	55	60	65
Soybean	0	25	60	85	100	110	115
Cotton	0	25	40	50	60	65	70

Use DP to assist the farmer in developing the best crop and investment strategy. Confirm this answer using IP.

20. Regarding Exercise 19, suppose the farmer discovers that the demand for cotton in the market this year may be lower than initially expected. Thus, she decides to only grow corn, wheat and soybeans on her farm. Furthermore, she wants to reduce her budget to $50,000. Formulate and solve this question as a DP problem.

Index

A

Absolute stop, 337
Accuracy, 337
Achievement function, 431
Activities, linear programming (LP)
    problems formulation, 8
Additivity assumption of LP model, 7
Adjusted for imperfectly competitive
    markets (ANSP), 413
Analytical Hierarchy Process (AHP), 269
Aggregate model, 116
Agricultural Adjustment Act of 1933, 159
Agricultural decision analysis under risk
    and uncertainty, 349–353
Agricultural production decisions, 432
Agriculture, LP application, 3–5
Algebraic solution, 41–42
Algebraic way of calculating SPs, 39–40
Algorithm efficiency, 324
All-integer programming, 249
Allocation problem, 4
Allocation problem, LP application, 4
Alternative optimal solutions, 22
Analysis, 33
Answer Report, 112
Arcs, 174
Artificial variables, 67–68
Assignment model for food and agricultural
    markets, 189–191
Assignment problem, 189
Assumptions of LP models, *See* Standard
    assumptions of LP models

B

Barrier algorithm, 324
Basis, 37
Basis column, 57
Basic feasible solution (BFS), 56
Basic infeasible solution, 56
Basic solution, 56
Basic variables, 56
Bayesian decision theory, 349, 372, 392
Bellman equation, 454
Benefit targeting algorithm, 269
Binary integer programming, 248–282
    capital budgeting, 257
        transfer activities, 260–262
    research application, 268–274, *See also*
        Delaware Agricultural Lands
        Preservation Foundation program
    in Solver, 256–257
Binary linear programming, 4
Binding constraint, 146
Branch-and-bound solution procedure,
    250–255, *See also under* Integer
    programming (IP)

C

Capital budgeting problem, LP application,
    4, 257, 463–465
    transfer activities, 260–262
Carbon abatement problem, LP
    application, 3
Certainty assumption of LP model, 7

Chance-constrained programming, 369–371
    Chance constraint, 371
    CVaR (Conditional Value at Risk), 371
    USet (Uncertainty Set), 371
    VaR (Value at Risk), 371
Change in basis, 37
Choice criterion, 349
Classic interval method (in Solver), 337
Climate change, 419–421
Cobb-Douglas functional form, 312
Column vector, 120
Combinatorial optimization problem, 339
Complementary optimal solution property,
        99–100
Complementary slackness conditions, 295
    both constraints are binding (Case 1),
        296–297
    constraint 1 is binding and constraint 2 is
        not binding (Case 3), 298
    constraint 1 is not binding and constraint
        2 is binding (Case 2), 297–298
Computer software, LP, 13
Constrained nonlinear functions
        optimization
    with equality constraints, 290–293
    with inequality constraints, 293–299
Constrained quadratic maximization, 300
Constraint line, 14
Constraint set, LP model, 6
    equal-to restrictions, 6
    greater-than-or-equal-to, 6
    less-than-or-equal-to, 6
    non-negativity constraint, 6
    structural constraints, 6
Consumer surplus, 420, 421
Control variables, 456
Convergence, 336
Convex optimization, 323
Convex problems, 324
Convex set, 16, 29
Convexity, 323
Cournot–Nash behavior/equilibrium,
        414, 416–418
Covariance, 348–349
Critical point, 288
Crop diversification, 358, 377
Crop farm, static models of, 136–145
    disaggregated model, 138–141
    input demand functions, 142–144
    output supply functions, 141–142
    research application, 158–163

Crop farming in northeast Australia, 341
Crop–livestock enterprises, 148–151
Crop rotation, 473
Cross-over method, 328
Curse of dimensionality, 466

D

Decision theory, 348
Decision tree, 466
Decision variables, 456
Degeneracy, 79
Delaware Agricultural Lands Preservation
        Foundation program, 268–271
    Benefit Targeting Algorithm, 269
Demand nodes, 174
Derivatives for nonlinear function, 286–287
    first derivative, 286
    second derivative, 286
Deriving weights for GP, 438
Deterministic algorithm, 323
Development feasibility constraints, 219
Development rate, 217
Diet problem, LP application in, 3
Differential calculus, 283
Direct elicitation approach, 351
Direct rate of increase, simple maximization
        problem, 58
Discount rate, 306
Discrete stochastic sequential programming
        (DSSP), 371–376, 379–385
    resource constraints, 380
    sequencing constraints, 380
    sequential decision environment under, 372
Discrete time periods, one-year model,
        152–155
Distribution system design, 263–266
Divisibility assumption of LP model, 7
Dual feasibility experiment, 156
Dual problem, 97, 328
Dual variables, 98, 414
Duality, 96–107
    additional properties of, 98–100
        complementary optimal solution
            property, 99–100
        strong duality property, 99
        symmetry property, 99–100
        weak duality property, 99
    economic intuition behind, 105–107
    primal and dual problems, relationship
        between, 97–98, 104

primal and dual solutions, relationship
between, 100–103
Dynamic farm-level models, 151–155
one-year model with discrete time
periods, 152–155
labor constraints, 153
land constraint, 152
output constraints, 153
Dynamic models, 151
Dynamic objective function, 309
Dynamic programming (DP), 453–479
Bellman equation, 454
capital budgeting problem, 463–465
decision variables, 456
state variables, 456
conversion to organic arable farming
application, 471–478
input nodes, 454
network problem, 454–457
output nodes, 454
principle of optimality, 454
production inventory problem, 458–462
recursive, 454
research application, 466–471
animal health in developing countries,
466–471

E

Economic feasibility, 164
Economic interpretation, maximization
problem solving, 20–21
Effluent treatment plant (ETP), 314
Endogenous variables, 302
Environment, LP application, 3–5
Environmental economics applications of
LP, 211–246
Environmental index, 218
Environmentally sensitive region, 158
Equality constraints, 32, 290
Equal-to constraints, 72, 96
Equilibrium constraints, 342
Evolutionary Engine (in Solver), 339
Evolutionary Solver, using, 328–336
convergence, 336
cross-over method, 328
fitness criterion, 328
iterations, 335
limits options, 336
local search, 336
max time without improvement, 333

mutations, 328
offspring solutions, 328
parent solutions, 328
population, 333
precision, 335
random seed, 334
require bounds, 334
tolerance, 332
Exogenous variables, 289
Expected monetary value, 351
Expected profit, 435
Expected utility hypothesis, 350–353
continuity, 350
independence, 350
ordering, 350
transitivity, 350
Expected value, 348–349
Extreme points, 16

F

Farmland conservation with simultaneous
multiple-knapsack model, 271–274
Farm-level LP models, 135–172, *See also*
Crop farm, static models; Multiple-
year model, crop farm
crop–livestock enterprises, 148–151
dynamic models, 151–155, *See also*
Dynamic farm-level models
model calibration, 158
model validation, 155–158
change experiment, 156
dual feasibility experiment, 156
feasibility, 156
prediction experiment, 156
price experiments, 156
quantity, 156
tracking experiment, 156
validation by construct, 155
validation by results, 155
Feasibility, 56, 92
Feasibility experiment, 156
Feasibility tolerance, 328
Feasible region, 14, 16
convex set, 16
extreme points, 16
Feasible solution, 14
Federal Milk Marketing Order Program, 191
Field time, 379
First derivative, 286
First-order conditions (FOCs), 287–288

Fishery management using NLP, 305–309
    discount rate, 306
Fitness criterion, 328
Fixed rate tariff, 181
Forest management, LP applications, 211–214
    model development, 213–214
Formulating LP problems, 7–13
    objective function coefficients, 8
    objective function, 8
    RHS parameters, 8
    structural constraints, 8
    technical coefficients, 8

G
Gap tolerance, 327
General form of LP model, components, 5–6
Generalizable search algorithms, 328
Generalized Reduced Gradient (GRG)
        Algorithm, 299, 321
Genes (of parent solution), 330
Global optimum, 287
Goal programming (GP), 427–447
    achievement function, 431
    deriving weights, 438
    nonpreemptive GP, 428
    preemptive/lexicographic GP, 428, 431–438
    research application, 438–443
        forest land protection, 443–446
        optimal parasite control programs,
            438–443
Graphical approach to LP, 2–54, See also
        Sensitivity analysis in IP
    maximization, 2, 13–26, See also
        Maximization problem solving
    minimization, 2, 26–33
    right-hand-side-value, 42–43
Greater-than-or-equal-to constraints,
        67–72, 96
Greedy agent algorithm, 269
Grizzly bear corridor study, 233–237

H
Hybrid Rating Regression Analysis, 438

I
Imperfectly competitive milk markets,
        410–419
    spatial equilibrium model, 410–419
        "dual structure" spatial monopoly
            solution, 417

"dual-structure" spatial Cournot–Nash
    equilibrium, 418
"dual structure" spatial perfect
    competition solution, 416
nonlinear price endogenous
    programming model, 411–419
Import quota, 181
Independence axiom, EUH, 350
Independent service operator (ISO), 342
Indirect rate of decrease, simple
        maximization problem, 58
Inequality constraints, 32
Infeasible solution, 14
Information structure, 372
Initialization, 251
Input demand functions, crop farm static
        model, 142–144
Input nodes, 454
Integer activity values, 248
Integer programming, 7, 248–282, See also
        Binary integer programming; Mixed
        integer programming
branch-and-bound solution procedure,
        250–255
    branching phase, 252
    initialization phase, 251
    lower bound (LB), 252
    upper bound (UB), 252
multiple choice constraints, 262–263
mutually exclusive constraints, 262–263
sensitivity analysis, 266–268
    Major axis points, 266
    MaxVal, 266
    MinVal, 266
    Multiple optimizations report, 266
    Optimization parameter tool, 266
    Parameter Analysis process, 266
    PSI optimization parameter formula,
        266
    RangeRef, 266
Solver using, 256–257
Integer tolerance, 257
Interior point, 299, 324
Interior solution, 294
Intermediate transshipment nodes, 182
Interval branch and bound, 336
Interval function, 336
Interval Global Solver, using, 321, 336–338
    absolute versus relative stop, 337
    accuracy, 337
    LP phase II, 338
    LP test, 338

max time without improvement, 337
  resolution, 337
  second-order, 337
Inventory problem, 458, 466
Irrigation decisions, 211
Irrigation water constraint, 230
Iso-contribution line, 17, 19
Iso-cost line, 27, 29–30
Iso-profit line, 17
Iso-revenue lines, 21
Iso-utility curves, 353

K

Kuhn–Tucker conditions, 293–299, 414
  complementary slackness conditions, 295
  interior solution, 294

L

LaGrange function, 290
LaGrange multipliers, 414
Land use planning, LP applications,
    215–222
Less-than-or-equal-to constraints, range of
    feasibility, 92–93
Lexicographic, 428
Limits options (Solver), 336
Limits report (Solver), 115–116
Linear enclosure method (Solver), 337
Linear function, 6
Linear programming (LP)/LP applications,
    See also Graphical approach to LP;
    Land use planning, LP applications
    in; Natural resource applications of
    LP; Solver, solving LP problems
    using; Standard assumptions of LP
    models
  activities, LP problems formulation, 8
  additivity assumption of LP model, 7
  agriculture, 3–5
  allocation problem, LP application in, 4
  capital budgeting problem, 4, 257,
      463–465
  carbon abatement problem, 3
  certainty assumption of LP model, 7
  computer software, LP, 13
  constraint set, LP model, 6, *See also*
      individual entry
  diet problem, 3
  divisibility assumption of LP model, 7
  environment, 3–5

farm-level LP models, 135–172, *See also*
    indvidual entry
forest management, 211–214
formulating LP problems, 7–13, *See also*
    individual entry
normal form of LP model, 103–105
objective function, LP problems
    formulation, 8, 27
portfolio problem, 3
product mix problem, 3
proportionality assumption of LP model,
    7, 248
resources economics, 3–5
standard assumptions of LP models, 7
structural constraints in LP application,
    5–6, 8
technical coefficients, LP problems
    formulation, 8
transportation problem, 4
Linearized version of QP, 360–367
Local optimum, 287
Local search, 336
Lower bound, 252

M

Major axis points, 266
Management of wild game farms, 237
Marginal rate of product transformation, 21
Market equilibrium, 401–420
Market under perfect competition,
    402–404
Marketing decision, 135, 359
Mathematical program with equilibrium
    constraints (MPEC), 342
Matrix algebra, 61
Matrix notation, 117
  basic operations and notation, 119
    addition, 120
    multiplication, 121
    subtraction, 121
    transpose, 122
  diagonal elements, 120
  dimension, 120
  element, 120
  vector, 120
Max time without improvement, 333, 337
Maximization problem, LP, 56–64
  basic solution, 56
  columns, 57
  rows, 58
    direct rate of increase, 58

Maximization problem (*continued*)
   indirect rate of decrease, 58
   z_j and $c_j - z_j$ rows, 58
   simplex tableau, 57–60
   solution improvement by changing basis,
     60–61
   substitution coefficients, 59
Maximization problem solving, graphical
     approach, 2, 13–26
   constraints, 15
   economic interpretation, 20–21
   no feasible solution, 21
   multiple optimal solutions, 22
   optimal solution, finding, 16–18
   simplex method, 13, 56–64, *See also*
     Simple maximization problem
   simultaneous equations approach, 18–19
   two-activity maximization problems,
     solving, 14–16
   unbounded solution, 21
Maximize profit, 243
Maximizing revenue, 179
Maximum percentage error, 252–253
Maximum time without improvement,
     313, 337
Mean-variance analysis, 347
Metaheuristic algorithm, 321
Minimization of Total Absolute Deviations
     (MOTAD), 347, 360–367
   target MOTAD, 367–369
Minimization problem solving, graphical
     approach, 2, 26–33
   iso-cost line, 27, 29–30
   sensitivity analysis, 40–41
   simultaneous equations approach, 30–31
   standard form, 31–33
Mixed integer programming, 103, 249,
     255–256
Mixed structural constraints, 24–26
Model calibration, 158
Model validation, 155–158
Moment method, 351
Monopoly, 404–407
Monopsony, 404–407
Monte Carlo simulation, 377
Moore–Skelboe algorithm, 337
Multiple choice constraints, 262–263
Multiple goals, 427
Multiple-knapsack model, 271–274
Multiple optimal solutions, 22, 79
Multiple optimizations report, 266
Multiple-year model, crop farm, 145–148

LP tableau, 147
terminal value, 148
Multivariate nonlinear functions,
     289–290
   partial derivative, 289
Mutations, 328, 334
Mutually exclusive constraints,
     262–263

N
Natural resource applications of LP,
     211–246, *See also* Land use
     planning; Optimal stocking problem
     for game ranch
   efficient cropping patterns, 228–233
   efficient irrigation, 228–233
   forest management, 211–214
   research application, 233–237, *See also*
     Grizzly bear corridor study
Necessary condition, 288
Negative RHS values, handling, 72
Net social payoff, 407
Network optimization problem, 173
Networks sensitivity analysis, 179
No feasible solution, 21, 78
Nonbasic variables, 56, 60
Non-convex problems, 323
Nondeterministic algorithms, 323
Nonlinear functions optimization,
     283–320
   constrained optimization
     with equality constraints, 290–293
     with inequality constraints, 293–299
   constrained optimization problems (in
     Solver), 299–304
   fishery management, 305–309
   general algorithms, 299
   Kuhn–Tucker conditions, 293–299
   multivariate functions, 289–290
   research applications, 309–311
     optimal advertising, 309–311
     water pollution abatement policies,
       312–314
   shortcut formulas for derivatives,
     286–287
   slopes of functions, 283–286
   special-purpose algorithms for quadratic
     programming, 299
   unconstrained optimization, 287–288
     critical or stationary point, 288
     first-order conditions (FOCs), 287

global optimum, 287
local optimum, 287
second-order sufficient conditions
    (SOCs), 288
Nonlinear optimization, global approaches,
    321–343, *See also* Evolutionary
    solver
  forestry example, 338–340
  research applications, 341, 342
    crop farming in northeast Australia,
      341
    energy market deregulation
      analysis, 342
  SOCP barrier solver, 324–328
Nonlinear price endogenous programming
    model, 411–419
Nonlinear problems, 322–324
  convex versus nonconvex problems, 323
  convexity, 323
  deterministic versus nondeterministic
    algorithms, 323
  problem formulation, 322–323
Non-negativity constraint, 6, 11, 476
Nonpreemptive GP, 428
Nonslack variables, 64
Normal form of LP model, 103–105

O

Objective function coefficients (c_i)
    sensitivity analysis, 5, 8, 34–37, 89
  range of optimality, calculation, 35–37
Objective function, LP problems
    formulation, 8, 27
Offspring solutions, 328
Oligopoly, 411
Optimal advertising, 309–311
Optimal forest rotation, 211
Optimal integer solution, 250
Optimal parasite control program, 438–443
Optimal solution, 6, 16–18
Optimal stocking problem for game ranch,
    222–228
Optimal value, 3
Optimization of nonlinear functions, 283–315
Optimization parameter tool, 266
Ordering axiom, 350
Output nodes, 454
Output supply curve, deriving, 37–39
  parametric programming, 37
  resource endowment (b_i) sensitivity
    analysis, 37–39

shadow price (SP), 38
Output supply functions, crop farm static
    model, 141–142

P

Parameter analysis process, 266
Parametric programming, 37
Parametric quadratic programming, 354
Parent solutions, 328
Pareto-optimal solutions, 341
Partial derivative, 289
Penalty approach, 68
Perfect competition, 401
Pivot column, 60, 64
Pivot element, 61
Pivot row, 61, 64
Plant capacity, 188
Polluter's problem, 8–11
Population (Solver), 333
Population constraint, 224
Portfolio problem, LP application, 3
Power index (Solver), 328
Precision (Solver), 335
Prediction experiment, 156
Preemptive/lexicographic GP, 428, 431–438
Present value, 306
Price endogenous mathematical
    programming models, 401–422
  industry models, 409–410
  monopoly, 401
  monopsony, 401
  perfect competition, 401
  research application, 410–419, *See also*
    Imperfectly competitive milk markets
    climate change and U.S. agriculture,
      419–421
Price experiment, 156
Primal and dual problems, relationship
    between, 97–98, 100–104, *See also*
    under Duality
Primal problem, 97
Principle of optimality, 454
Probability distribution, 349
Product conversion, 186
Product mix problem, LP application, 3
Production decision, 146
Production inventory problem, 458–462
Production possibility frontier (PPF), 20
Production possibility set, 20
Proportionality assumption of LP model,
    7, 248
PSI optimization parameter, 266

Q

Quadratic risk programming, 253, 353–360, 377–379
  linearized version, 360–367
  parametric quadratic programming, 354
Quantity experiment, 156

R

Random seed (Solver), 334
Range names, 108
Range of feasibility, 92–93
  for greater-than-or-equal-to and equal-to constraints, 96
  for less-than-or-equal-to constraints, 92–93
Range of optimality, 35, 41, 442
  for basic variable, finding, 89–90
  calculation, 35–37
  for a nonbasic variable, 90–91
Recursive, 454
Relative stop, 337
Require bounds (Solver), 334
Residential land, 215
Resolution (Solver), 337
Resource endowment (b_i) sensitivity analysis, 37–39, 91–92
Resources economics, LP application, 3–5
Return function, 457
Right-hand-side (RHS) value, 6, 8, 42–43, 72, 95–96
Risk programming models, 347–392, *See also* Discrete stochastic sequential programming; Quadratic risk programming
  agricultural decision analysis, 349–353
  chance-constrained programming, 369–371
  covariance, 348–349
  direct elicitation approach, 351
  discrete stochastic sequential programming, 379–385
  expected monetary value, 351
  expected utility hypothesis, 350–353
  expected value, 348–349
  issues in measuring risk, 376–377
  moment method, 351
  research application, 377–385, *See also* Discrete stochastic sequential programming
  agriculture, 385–391

climate change, 385–391
  risk averse, 351
  risk lovers, 351
  risk neutral, 351
  risk premium, 351
Risk Solver, 107, 371
  risky prospect, 351
  variance, 348–349
Route constraints, 180

S

Saddle point, 299
Scalar multiplication, 121
Search direction (Solver), 328
Second derivative, 286
Second order (Solver), 288
Second-order conic problem barrier solver, 321, 324–328
  feasibility tolerance, 328
  gap tolerance, 327
  power index, 328
  search direction, 328
  step size factor, 328
Second-order sufficient conditions, 288
Sector programming model, 401, 407
Sensitivity analysis in IP, 2, 179, 266–268, *See also under* Integer programming (IP)
  with graphical approach, 33–43
    objective function coefficients sensitivity analysis, 34–37
Sensitivity report (Solver), 113
Separable linear programming, 299
Sequencing constraints, 139
Shadow price (SP), 38
  algebraic way of calculating, 39–40
Simplex-based sensitivity analysis, 88
  for maximization problems, 89–93
    objective function coefficients (c_i), 89
    range of feasibility for less-than-or-equal-to constraints, 92–93
    range of optimality for basic variable, finding, 89–90
    range of optimality for c_{S2}, a nonbasic variable, 90–91
    resource endowments sensitivity analysis, 91–92
  for minimization problems, 93–96
    range of feasibility for greater-than-or-equal-to and equal-to constraints, 96
    right-hand-side sensitivity analysis, 95–96

Simplex method for maximization
  problems, 64–72
  equal-to constraints, 72
  flow chart, 65
  greater-than-or-equal-to constraints,
    67–72
  negative RHS values, handling, 72
  penalty approach, 68
Simplex method for minimization problems,
  72–79
  converting minimization problem to an
    equivalent maximization problem,
    73–74
  degeneracy, 79
  no feasible solution, 78
  graphical solution, 74–75
  multiple optimal solutions, 79
  selection criterion for a new, nonbasic
    variable, 73
  special cases, 78–79
  unbounded solution, 78
Simultaneous equations approach, 18–19,
    30–31
Slack variables, 31, 33, 59, 64
Slope(s), 19
  of functions, 283–286
  of nonlinear functions, 283
  slope-intercept form, 17
Social surplus, 403
Social welfare, 403
Solver, solving LP problems using,
    107–116
  answer report, 112, 114
  binary IP, 256–257
  constrained optimization problems,
    299–304
  constraints, 111
    less-than-or-equal-to, 111
    greater-than-or-equal-to, 111
    equal-to, 111
  evolutionary engine, 339
  integer and binary programming
    options, 256
  limits options, 336
  limits report, 115–116
  population, 333
  power index, 328
  precision, 335
  random seed, 334
  range names, 108
  require bounds, 334
  resolution, 337

Risk Solver Platform for
  education, 107
  engine tab, 110
  model tab, 110
  output tab, 110
  platform tab, 110
  search direction, 328
  second order, 288
  sensitivity report, 113, 115
  step size factor, 328
  troubleshooting process, 109
  visual basic with applications (VBA), 108
Spatial equilibrium analysis, 402, 407–409
Species protection, 233
Stackleberg competition, 342
Stage transformation function, 457
Stages, 372, 453
Standard assumptions of LP models, 7
  additivity, 7
  certainty, 7
  divisibility, 7
  proportionality, 7
Standard deviation, 435
Standard form for maximization and
    minimization problems, 31–33
State variables, 456
Static models of crop farm, 136–145, *See
    also* Crop farm, static models of
Stationary, 337
Stationary point, 288
Step size factor (Solver), 328
Stopping rule, 63
Strong duality property, 99
Structural constraints, in LP application,
    5–6, 8
Substitution coefficients, 59
SUMIF function, 339
Summation notation, 117–119
SUMPRODUCT function, 108
Surplus variables, 33
Sustainable practices, 158
Symmetry property, 99–100

T
Tableau form, 32, 57
Tangent, 285
Target MOTAD, 367–369
Target values, 428
Taylor polynomial approximation, 322
Technical coefficients, LP problems
    formulation, 8

Terminal value, 148
Tolerance (Solver), 332
Total absolute deviation (TAD), 360
Tracking experiment, 156
Transfer activities, 260–262
Transhipment model, 182–189
Transitivity axiom, 350
Transportation model for food and
        agricultural markets, 173–210
    extensions of the model, 179–182
    general model, 174–179
        arcs, 174
        demand nodes, 174
        network, 174
        supply nodes, 174
    incorporating route constraints, 180
    incorporating unacceptable routes into
        network, 180–182
    maximizing revenue rather than minimizing
        transportation costs, 179–180
    research application, 191–196
    sensitivity analysis, 179
Transportation problem, LP application, 4
Transpose, 122
Transshipment model, 173, 182–189
    with product conversion, 186–189
    transshipment node, 182
    Warehouse model, 183–186
        demand constraints, 184
        supply constraints, 184
        transshipment constraints, 184
Transshipment warehouse, 183–186
Troubleshooting process, 109–110
Two-activity maximization problems,
        solving, 14–16

U
U.S. Dairy Sector Simulator (USDSS),
        191–196
    LP transportation model application in,
        191–196
U.S. Department of Agriculture (USDA),
        158–159, 192, 443
Unacceptable routes, 180
Unbounded feasible region, 44
Unbounded solution, 21, 78
Unconstrained optimization, nonlinear
        functions, 287–288
Unit column, 57
Unit vector, 57
Upper bound, 252
Utility maximization, 303–304

V
Validation by construct, 155
Validation by results, 155
Variance, 348–349
Visual basic with applications
        (VBA), 108
VLOOKUP function, 339

W
Water pollution abatement policies,
        312–314
Weak duality property, 99
Wildlife corridor, 233
Working backwards, 453, 458,
        463, 479